Lecture Notes in Mathematics 1998

FONDAZIONE
CIME
ROBERTO CONTI
CENTRO INTERNAZIONALE MATEMATICO ESTIVO
INTERNATIONAL MATHEMATICAL SUMMER CENTER

C.I.M.E. means Centro Internazionale Matematico Estivo, that is, International Mathematical Summer Center. Conceived in the early fifties, it was born in 1954 and made welcome by the world mathematical community where it remains in good health and spirit. Many mathematicians from all over the world have been involved in a way or another in C.I.M.E.'s activities during the past years.

So they already know what the C.I.M.E. is all about. For the benefit of future potential users and co-operators the main purposes and the functioning of the Centre may be summarized as follows: every year, during the summer, Sessions (three or four as a rule) on different themes from pure and applied mathematics are offered by application to mathematicians from all countries. Each session is generally based on three or four main courses (24–30 hours over a period of 6–8 working days) held from specialists of international renown, plus a certain number of seminars.

A C.I.M.E. Session, therefore, is neither a Symposium, nor just a School, but maybe a blend of both. The aim is that of bringing to the attention of younger researchers the origins, later developments, and perspectives of some branch of live mathematics.

The topics of the courses are generally of international resonance and the participation of the courses cover the expertise of different countries and continents. Such combination, gave an excellent opportunity to young participants to be acquainted with the most advanced research in the topics of the courses and the possibility of an interchange with the world famous specialists. The full immersion atmosphere of the courses and the daily exchange among participants are a first building brick in the edifice of international collaboration in mathematical research.

C.I.M.E. Director
Pietro ZECCA
Dipartimento di Energetica "S. Stecco"
Università di Firenze
Via S. Marta, 3
50139 Florence
Italy
e-mail: zecca@unifi.it

C.I.M.E. Secretary
Elvira MASCOLO
Dipartimento di Matematica "U. Dini"
Università di Firenze
viale G.B. Morgagni 67/A
50134 Florence
Italy
e-mail: mascolo@math.unifi.it

For more information see CIME's homepage: http://www.cime.unifi.it

CIME activity is carried out with the collaboration and financial support of:
- INdAM (Istituto Nationale di Alta Mathematica)
- MIUR (Ministero dell'Universit e della Ricerca)

This course is partially supported by LAMI (Laboratorio di Applicazioni della Matematica all'Ingegneria, Università della Calabria)

Marco Abate · Eric Bedford · Marco Brunella
Tien-Cuong Dinh · Dierk Schleicher
Nessim Sibony

Holomorphic Dynamical Systems

Lectures given at the
C.I.M.E. Summer School
held in Cetraro, Italy,
July 7–12, 2008

Editors:
Graziano Gentili
Jacques Guenot
Giorgio Patrizio

Springer

Fondazione CIME Roberto Conti

Editors
Prof. Graziano Gentili
Università di Firenze
Dipartimento di Matematica "U. Dini"
viale Morgagni 67/A
50134 Firenze
Italy
gentili@math.unifi.it

Prof. Giorgio Patrizio
Università di Firenze
Dipartimento di Matematica "U. Dini"
viale Morgagni 67/A
50134 Firenze
Italy
patrizio@math.unifi.it

Prof. Jacques Guenot
Università della Calabria
Dipartimento di Strutture
Ponte Pietro Bucci 39
87036 Arcavacata di Rende (CS)
Edificio 37B - 38B CDE - 39B - 39C
Italy
guenot@libero.it

ISBN: 978-3-642-13170-7 e-ISBN: 978-3-642-13171-4
DOI: 10.1007/978-3-642-13171-4
Springer Heidelberg Dordrecht London New York

Lecture Notes in Mathematics ISSN print edition: 0075-8434
ISSN electronic edition: 1617-9692

Library of Congress Control Number: 2010930280

Mathematics Subject Classification (2000): 37FXX, 32AXX, 32QXX, 32H50, 30DXX, 31BXX

Cover design: WMXDesign GmbH

Printed on acid-free paper

springer.com

Preface

The theory of holomorphic dynamical systems is a subject of increasing interest in mathematics, both for its challenging problems and for its connections with other branches of pure and applied mathematics.

A holomorphic dynamical system is the datum of a complex variety and a holomorphic object (such as a self-map or a vector field) acting on it. The study of a holomorphic dynamical system consists in describing the asymptotic behavior of the system, associating it with some invariant objects (easy to compute) which describe the dynamics and classify the possible holomorphic dynamical systems supported by a given manifold. The behavior of a holomorphic dynamical system is pretty much related to the geometry of the ambient manifold (for instance, hyperbolic manifolds do no admit chaotic behavior, while projective manifolds have a variety of different chaotic pictures). The techniques used to tackle such problems are of various kinds: complex analysis, methods of real analysis, pluripotential theory, algebraic geometry, differential geometry, topology.

To cover all the possible points of view of the subject in a unique occasion has become almost impossible, and the CIME session in Cetraro on Holomorphic Dynamical Systems was not an exception. On the other hand the selection of the topics and of the speakers made it possible to focus on a number of important topics in the discrete and in the continuous setting, both for the local and for the global aspects, providing a fascinating introduction to many key problems of the current research. The CIME Course aimed to give an ample description of the phenomena occurring in central themes of holomorphic dynamics such as automorphisms and meromorphic self-maps of projective spaces, of entire maps on complex spaces and holomorphic foliations in surfaces and higher dimensional manifolds, enlightening the different techniques used and bringing the audience to the borderline of current research topics. This program, with its interdisciplinary characterization, drew the attention and the participation of young researchers and experienced mathematicians coming from different backgrounds: complex analysis and geometry, topology, ordinary differential equations and number theory. We are sure that the present volume will serve the same purpose. We briefly describe here the papers that stemmed from the courses and constitute the Chapters of this volume.

In his lectures, Marco Abate outlines the local theory of iteration in one and several variables. He studies the structure of the stable set K_f of a selfmap f of a neighborhood U of a fixed point, describing both the topological structure of K_f and the dynamical nature of the (global) dynamical system $(K_f, f_{|K_f})$. One important way to study a local holomorphic dynamical system consists in replacing it by an equivalent but simpler system. Following a traditional approach, Abate considers three equivalence relations - topological, holomorphic and formal conjugacy - and discusses normal forms and invariants in all these cases. He starts surveying the one-dimensional theory, which is fairly complete, even though there are still some open problems, and then he presents what is known in the multidimensional case, that is an exciting mixture of deep results and still unanswered very natural questions.

The lectures of Eric Bedford provide an introduction to the dynamics of the automorphisms of rational surfaces. The first part is devoted to polynomial automorphisms of $\mathbf{C}^2$ and in particular to the complex Hénon maps, the most heavily studied family of invertible holomorphic maps. The investigations of the Hénon maps can be guided by the study of the dynamics of polynomial maps of one variable, a very rich and classical topic. Although the Hénon family is only partially understood, its methods and results provide motivation and guidance for the understanding of other types of automorphisms. In the second part of the notes, Bedford considers the geometry of compact rational surfaces with the illustration of some examples of their automorphisms. In contrast with the case of the polynomial automorphisms of $\mathbf{C}^2$, not much is known about neither the set of all rational surface automorphisms, nor about a dynamical classification of them.

The theory of foliations by Riemann surfaces is central in the study of continuous aspects of holomorphic dynamics. In his lecture notes, Marco Brunella describes the state of the art of the topic and reports on his results on the uniformisation theory of foliations by curves on compact Kähler manifolds. Each leaf is uniformized either by the unit disk or by $\mathbf{C}$ or by the projective line. Brunella explains how the universal covers may be patched together to form a complex manifold with good properties and he studies the analytic properties of this manifold, in particular regarding holomorphic convexity. In turn this leads to results on the distribution of parabolic leaves inside the foliation and to positivity statements concerning the canonical bundle of the foliation, generalizing results of Arakelov on fibrations by algebraic curves.

Sibony's course in the CIME session was based on the lecture notes by Tien-Cuong Dinh and Nessim Sibony that are included in this volume. This contribution, which could be a stand alone monograph for depth and extension, gives a broad presentation to the most recent developments of pluripotential methods, and to the theory of positive closed currents, in dynamics in Several Complex Variables. The notes concentrate on the dynamics of endomorphisms of projective spaces and the polynomial-like mappings. Green currents and equilibrium measure are constructed to study quantitative properties and speed of convergence for endomorphisms of projective spaces; equidistribution problems and ergodic properties are also treated. For polynomial-like mappings, the equilibrium measure of maximal entropy is constructed and equidistribution properties of points are proved, under suitable dynamical degree assumptions. The tools introduced here are of independent interest and can be applied in other dynamical problems. The presentation includes all the necessary prerequisites about plurisubharmonic functions and currents, making the text self-contained and quite accessible.

In his lectures, Schleicher studies iteration theory of entire holomorphic functions in one complex variable, a field of research that has been quite active in recent years. A review of dynamics of entire maps, which includes the classical and well developed theory of polynomial dynamics, serves to introduce the main topic: the consideration of transcendental maps. The notes study key dynamical properties of large classes of transcendental functions and of special prototypical families of entire maps such as the exponential family $z \mapsto \lambda e^z$ or the cosine family $z \mapsto ae^z + be^{-z}$. It turns out that some aspects of the dynamics of transcendental entire maps are inspired by the polynomial theory, others are very different and exploit all the power of deep results from complex analysis. Transcendental dynamics turns out to be a largely yet unexplored and fascinating area of research where surprising mathematical results - that sometimes had been constructed artificially in other branches of mathematics - arise in a natural way.

It is a real pleasure to thank the speakers for their great lectures and all the authors for the beautiful contributions to this volume. We also like to thank all the participants for their interest and enthusiasm that created a very warm and stimulating scientific atmosphere. We like to express our gratitude to CIME for sponsoring the summer school, and to the LAMI (Laboratorio di Applicazioni della Matematica all'Ingegneria, Università della Calabria) for its financial contribution. Our particular gratitude goes to Pietro Zecca and Elvira Mascolo for their continuous support, and to their collaborators Carla Dionisi and Maria Giulia Bartaloni for their invaluable help.

Graziano Gentili, Jacques Guenot and Giorgio Patrizio

Contents

List of Contributors

Marco Abate Dipartimento di Matematica, Università di Pisa, Largo Pontecorvo 5, 56127 Pisa, Italy, abate@dm.unipi.it

Eric Bedford Department of Mathematics, Indiana University, 47405 Bloomington, IN, USA, bedford@indiana.edu

M. Brunella IMB - CNRS UMR 5584, 9 Avenue Savary, 21078 Dijon, France, Marco.Brunella@u-bourgogne.fr

T.-C. Dinh UPMC Univ Paris 06, UMR 7586, Institut de Mathématiques de Jussieu, F-75005 Paris, France, dinh@math.jussieu.fr, http://www.math.jussieu.fr/~dinh

N. Sibony Université Paris-Sud 11, Mathématique - Bâtiment 425, 91405 Orsay, France, nessim.sibony@math.u-psud.fr

Dierk Schleicher School of Engineering and Science, Research I, Jacobs University, Postfach 750 561, D-28725 Bremen, Germany, dierk@jacobs-university.de

Discrete Holomorphic Local Dynamical Systems

Marco Abate

Abstract This chapter is a survey on local dynamics of holomorphic maps in one and several complex variables, discussing in particular normal forms and the structure of local stable sets in the non-hyperbolic case, and including several proofs and a large bibliography.

1 Introduction

Let us begin by defining the main object of study in this survey.

Definition 1.1. Let M be a complex manifold, and $p \in M$. A *(discrete) holomorphic local dynamical system* at p is a holomorphic map $f\colon U \to M$ such that $f(p) = p$, where $U \subseteq M$ is an open neighbourhood of p; we shall also assume that $f \not\equiv \mathrm{id}_U$. We shall denote by $\mathrm{End}(M, p)$ the set of holomorphic local dynamical systems at p.

Remark 1.2. Since we are mainly concerned with the behavior of f nearby p, we shall sometimes replace f by its restriction to some suitable open neighbourhood of p. It is possible to formalize this fact by using germs of maps and germs of sets at p, but for our purposes it will be enough to use a somewhat less formal approach.

Remark 1.3. In this survey we shall never have the occasion of discussing continuous holomorphic dynamical systems (i.e., holomorphic foliations). So from now on all dynamical systems in this paper will be discrete, except where explicitly noted otherwise.

To talk about the dynamics of an $f \in \mathrm{End}(M, p)$ we need to define the iterates of f. If f is defined on the set U, then the second iterate $f^2 = f \circ f$ is defined

M. Abate
Dipartimento di Matematica, Università di Pisa, Largo Pontecorvo 5, 56127 Pisa, Italy
e-mail: abate@dm.unipi.it

G. Gentili et al. (eds.), *Holomorphic Dynamical Systems*,
Lecture Notes in Mathematics 1998, DOI 10.1007/978-3-642-13171-4_1,

on $U \cap f^{-1}(U)$ only, which still is an open neighbourhood of p. More generally, the k-th iterate $f^k = f \circ f^{k-1}$ is defined on $U \cap f^{-1}(U) \cap \cdots \cap f^{-(k-1)}(U)$. This suggests the next definition:

Definition 1.4. Let $f \in \mathrm{End}(M,p)$ be a holomorphic local dynamical system defined on an open set $U \subseteq M$. Then the *stable set* K_f of f is

$$K_f = \bigcap_{k=0}^{\infty} f^{-k}(U).$$

In other words, the stable set of f is the set of all points $z \in U$ such that the *orbit* $\{f^k(z) \mid k \in \mathbb{N}\}$ is well-defined. If $z \in U \setminus K_f$, we shall say that z (or its orbit) *escapes* from U.

Clearly, $p \in K_f$, and so the stable set is never empty (but it can happen that $K_f = \{p\}$; see the next section for an example). Thus the first natural question in local holomorphic dynamics is:

(Q1) *What is the topological structure of K_f?*

For instance, when does K_f have non-empty interior? As we shall see in Proposition 4.1, holomorphic local dynamical systems such that p belongs to the interior of the stable set enjoy special properties.

Remark 1.5. Both the definition of stable set and Question 1 (as well as several other definitions and questions we shall see later on) are topological in character; we might state them for local dynamical systems which are continuous only. As we shall see, however, the *answers* will strongly depend on the holomorphicity of the dynamical system.

Definition 1.6. Given $f \in \mathrm{End}(M,p)$, a set $K \subseteq M$ is *completely f-invariant* if $f^{-1}(K) = K$ (this implies, in particular, that K is *f-invariant*, that is $f(K) \subseteq K$).

Clearly, the stable set K_f is completely f-invariant. Therefore the pair (K_f, f) is a discrete dynamical system in the usual sense, and so the second natural question in local holomorphic dynamics is

(Q2) *What is the dynamical structure of (K_f, f)?*

For instance, what is the asymptotic behavior of the orbits? Do they converge to p, or have they a chaotic behavior? Is there a dense orbit? Do there exist proper f-invariant subsets, that is sets $L \subset K_f$ such that $f(L) \subseteq L$? If they do exist, what is the dynamics on them?

To answer all these questions, the most efficient way is to replace f by a "dynamically equivalent" but simpler (e.g., linear) map g. In our context, "dynamically equivalent" means "locally conjugated"; and we have at least three kinds of conjugacy to consider.

Definition 1.7. Let $f_1\colon U_1 \to M_1$ and $f_2\colon U_2 \to M_2$ be two holomorphic local dynamical systems at $p_1 \in M_1$ and $p_2 \in M_2$ respectively. We shall say that f_1 and f_2 are *holomorphically* (respectively, *topologically*) *locally conjugated* if there are open neighbourhoods $W_1 \subseteq U_1$ of p_1, $W_2 \subseteq U_2$ of p_2, and a biholomorphism (respectively, a homeomorphism) $\varphi\colon W_1 \to W_2$ with $\varphi(p_1) = p_2$ such that

$$f_1 = \varphi^{-1} \circ f_2 \circ \varphi \qquad \text{on} \qquad \varphi^{-1}\big(W_2 \cap f_2^{-1}(W_2)\big) = W_1 \cap f_1^{-1}(W_1).$$

If $f_1 : U_1 \to M_1$ and $f_2 : U_2 \to M_2$ are locally conjugated, in particular we have

$$f_1^k = \varphi^{-1} \circ f_2^k \circ \varphi \quad \text{on} \quad \varphi^{-1}\big(W_2 \cap \cdots \cap f_2^{-(k-1)}(W_2)\big) = W_1 \cap \cdots \cap f_1^{-(k-1)}(W_1)$$

for all $k \in \mathbb{N}$ and thus

$$K_{f_2|W_2} = \varphi(K_{f_1|W_1}).$$

So the local dynamics of f_1 about p_1 is to all purposes equivalent to the local dynamics of f_2 about p_2.

Remark 1.8. Using local coordinates centered at $p \in M$ it is easy to show that any holomorphic local dynamical system at p is holomorphically locally conjugated to a holomorphic local dynamical system at $O \in \mathbb{C}^n$, where $n = \dim M$.

Whenever we have an equivalence relation in a class of objects, there are classification problems. So the third natural question in local holomorphic dynamics is

(Q3) *Find a (possibly small) class $\mathscr{F}$ of holomorphic local dynamical systems at $O \in \mathbb{C}^n$ such that every holomorphic local dynamical system f at a point in an n-dimensional complex manifold is holomorphically (respectively, topologically) locally conjugated to a (possibly) unique element of $\mathscr{F}$, called the holomorphic (respectively, topological)* normal form *of f.*

Unfortunately, the holomorphic classification is often too complicated to be practical; the family $\mathscr{F}$ of normal forms might be uncountable. A possible replacement is looking for invariants instead of normal forms:

(Q4) *Find a way to associate a (possibly small) class of (possibly computable) objects, called* invariants, *to any holomorphic local dynamical system f at $O \in \mathbb{C}^n$ so that two holomorphic local dynamical systems at O can be holomorphically conjugated only if they have the same invariants. The class of invariants is furthermore said* complete *if two holomorphic local dynamical systems at O are holomorphically conjugated if and only if they have the same invariants.*

As remarked before, up to now all the questions we asked made sense for topological local dynamical systems; the next one instead makes sense only for holomorphic local dynamical systems.

A holomorphic local dynamical system at $O \in \mathbb{C}^n$ is clearly given by an element of $\mathbb{C}_0\{z_1,\dots,z_n\}^n$, the space of n-uples of converging power series in $z_1,\dots,z_n$

without constant terms. The space $\mathbb{C}_0\{z_1,\dots,z_n\}^n$ is a subspace of the space $\mathbb{C}_0[[z_1,\dots,z_n]]^n$ of n-uples of formal power series without constant terms. An element $\Phi \in \mathbb{C}_0[[z_1,\dots,z_n]]^n$ has an inverse (with respect to composition) still belonging to $\mathbb{C}_0[[z_1,\dots,z_n]]^n$ if and only if its linear part is a linear automorphism of $\mathbb{C}^n$.

Definition 1.9. We say that two holomorphic local dynamical systems f_1, $f_2 \in \mathbb{C}_0\{z_1,\dots,z_n\}^n$ are *formally conjugated* if there is an invertible $\Phi \in \mathbb{C}_0[[z_1,\dots,z_n]]^n$ such that $f_1 = \Phi^{-1} \circ f_2 \circ \Phi$ in $\mathbb{C}_0[[z_1,\dots,z_n]]^n$.

It is clear that two holomorphically locally conjugated holomorphic local dynamical systems are both formally and topologically locally conjugated too. On the other hand, we shall see examples of holomorphic local dynamical systems that are topologically locally conjugated without being neither formally nor holomorphically locally conjugated, and examples of holomorphic local dynamical systems that are formally conjugated without being neither holomorphically nor topologically locally conjugated. So the last natural question in local holomorphic dynamics we shall deal with is

(Q5) *Find normal forms and invariants with respect to the relation of formal conjugacy for holomorphic local dynamical systems at $O \in \mathbb{C}^n$.*

In this survey we shall present some of the main results known on these questions, starting from the one-dimensional situation. But before entering the main core of the paper I would like to heartily thank François Berteloot, Kingshook Biswas, Filippo Bracci, Santiago Diaz-Madrigal, Graziano Gentili, Giorgio Patrizio, Mohamad Pouryayevali, Jasmin Raissy and Francesca Tovena, without whom none of this would have been written.

2 One Complex Variable: The Hyperbolic Case

Let us then start by discussing holomorphic local dynamical systems at $0 \in \mathbb{C}$. As remarked in the previous section, such a system is given by a converging power series f without constant term:

$$f(z) = a_1 z + a_2 z^2 + a_3 z^3 + \cdots \in \mathbb{C}_0\{z\}.$$

Definition 2.1. The number $a_1 = f'(0)$ is the *multiplier* of f.

Since $a_1 z$ is the best linear approximation of f, it is sensible to expect that the local dynamics of f will be strongly influenced by the value of a_1. For this reason we introduce the following definitions:

Definition 2.2. Let $a_1 \in \mathbb{C}$ be the multiplier of $f \in \mathrm{End}(\mathbb{C},0)$. Then

- if $|a_1| < 1$ we say that the fixed point 0 is *attracting;*
- if $a_1 = 0$ we say that the fixed point 0 is *superattracting;*

- if $|a_1| > 1$ we say that the fixed point 0 is *repelling;*
- if $|a_1| \neq 0, 1$ we say that the fixed point 0 is *hyperbolic;*
- if $a_1 \in S^1$ is a root of unity, we say that the fixed point 0 is *parabolic* (or *rationally indifferent*);
- if $a_1 \in S^1$ is not a root of unity, we say that the fixed point 0 is *elliptic* (or *irrationally indifferent*).

As we shall see in a minute, the dynamics of one-dimensional holomorphic local dynamical systems with a hyperbolic fixed point is pretty elementary; so we start with this case.

Remark 2.3. Notice that if 0 is an attracting fixed point for $f \in \mathrm{End}(\mathbb{C},0)$ with non-zero multiplier, then it is a repelling fixed point for the inverse map $f^{-1} \in \mathrm{End}(\mathbb{C},0)$.

Assume first that 0 is attracting for the holomorphic local dynamical system $f \in \mathrm{End}(\mathbb{C},0)$. Then we can write $f(z) = a_1 z + O(z^2)$, with $0 < |a_1| < 1$; hence we can find a large constant $M > 0$, a small constant $\varepsilon > 0$ and $0 < \delta < 1$ such that if $|z| < \varepsilon$ then

$$|f(z)| \leq (|a_1| + M\varepsilon)|z| \leq \delta |z|. \tag{1}$$

In particular, if Δ_ε denotes the disk of center 0 and radius ε, we have $f(\Delta_\varepsilon) \subset \Delta_\varepsilon$ for $\varepsilon > 0$ small enough, and the stable set of $f|_{\Delta_\varepsilon}$ is Δ_ε itself (in particular, a one-dimensional attracting fixed point is always stable). Furthermore,

$$|f^k(z)| \leq \delta^k |z| \to 0$$

as $k \to +\infty$, and thus every orbit starting in Δ_ε is attracted by the origin, which is the reason of the name "attracting" for such a fixed point.

If instead 0 is a repelling fixed point, a similar argument (or the observation that 0 is attracting for f^{-1}) shows that for $\varepsilon > 0$ small enough the stable set of $f|_{\Delta_\varepsilon}$ reduces to the origin only: all (non-trivial) orbits escape.

It is also not difficult to find holomorphic and topological normal forms for one-dimensional holomorphic local dynamical systems with a hyperbolic fixed point, as shown in the following result, which can be considered as the beginning of the theory of holomorphic dynamical systems:

Theorem 2.4 (Kœnigs, 1884 [Kœ]). *Let $f \in \mathrm{End}(\mathbb{C},0)$ be a one-dimensional holomorphic local dynamical system with a hyperbolic fixed point at the origin, and let $a_1 \in \mathbb{C}^* \setminus S^1$ be its multiplier. Then:*

(i) *f is holomorphically (and hence formally) locally conjugated to its linear part $g(z) = a_1 z$. The conjugation φ is uniquely determined by the condition $\varphi'(0) = 1$.*
(ii) *Two such holomorphic local dynamical systems are holomorphically conjugated if and only if they have the same multiplier.*
(iii) *f is topologically locally conjugated to the map $g_<(z) = z/2$ if $|a_1| < 1$, and to the map $g_>(z) = 2z$ if $|a_1| > 1$.*

Proof. Let us assume $0 < |a_1| < 1$; if $|a_1| > 1$ it will suffice to apply the same argument to f^{-1}.

(i) Choose $0 < \delta < 1$ such that $\delta^2 < |a_1| < \delta$. Writing $f(z) = a_1 z + z^2 r(z)$ for a suitable holomorphic germ r, we can clearly find $\varepsilon > 0$ such that $|a_1| + M\varepsilon < \delta$, where $M = \max_{z \in \overline{\Delta_\varepsilon}} |r(z)|$. So we have

$$|f(z) - a_1 z| \le M|z|^2 \qquad \text{and} \qquad |f^k(z)| \le \delta^k |z|$$

for all $z \in \overline{\Delta_\varepsilon}$ and $k \in \mathbb{N}$.
Put $\varphi_k = f^k / a_1^k$; we claim that the sequence $\{\varphi_k\}$ converges to a holomorphic map $\varphi \colon \Delta_\varepsilon \to \mathbb{C}$. Indeed we have

$$\begin{aligned} |\varphi_{k+1}(z) - \varphi_k(z)| &= \frac{1}{|a_1|^{k+1}} \left| f(f^k(z)) - a_1 f^k(z) \right| \\ &\le \frac{M}{|a_1|^{k+1}} |f^k(z)|^2 \le \frac{M}{|a_1|} \left(\frac{\delta^2}{|a_1|} \right)^k |z|^2 \end{aligned}$$

for all $z \in \overline{\Delta_\varepsilon}$, and so the telescopic series $\sum_k (\varphi_{k+1} - \varphi_k)$ is uniformly convergent in Δ_ε to $\varphi - \varphi_0$.
Since $\varphi_k'(0) = 1$ for all $k \in \mathbb{N}$, we have $\varphi'(0) = 1$ and so, up to possibly shrink ε, we can assume that φ is a biholomorphism with its image. Moreover, we have

$$\varphi\big(f(z)\big) = \lim_{k \to +\infty} \frac{f^k\big(f(z)\big)}{a_1^k} = a_1 \lim_{k \to +\infty} \frac{f^{k+1}(z)}{a_1^{k+1}} = a_1 \varphi(z),$$

that is $f = \varphi^{-1} \circ g \circ \varphi$, as claimed.
If ψ is another local holomorphic function such that $\psi'(0) = 1$ and $\psi^{-1} \circ g \circ \psi = f$, it follows that $\psi \circ \varphi^{-1}(\lambda z) = \lambda \psi \circ \varphi^{-1}(z)$; comparing the expansion in power series of both sides we find $\psi \circ \varphi^{-1} \equiv \mathrm{id}$, that is $\psi \equiv \varphi$, as claimed.

(ii) Since $f_1 = \varphi^{-1} \circ f_2 \circ \varphi$ implies $f_1'(0) = f_2'(0)$, the multiplier is invariant under holomorphic local conjugation, and so two one-dimensional holomorphic local dynamical systems with a hyperbolic fixed point are holomorphically locally conjugated if and only if they have the same multiplier.

(iii) Since $|a_1| < 1$ it is easy to build a topological conjugacy between g and $g_<$ on Δ_ε. First choose a homeomorphism χ between the annulus $\{|a_1|\varepsilon \le |z| \le \varepsilon\}$ and the annulus $\{\varepsilon/2 \le |z| \le \varepsilon\}$ which is the identity on the outer circle and given by $\chi(z) = z/(2a_1)$ on the inner circle. Now extend χ by induction to a homeomorphism between the annuli $\{|a_1|^k \varepsilon \le |z| \le |a_1|^{k-1}\varepsilon\}$ and $\{\varepsilon/2^k \le |z| \le \varepsilon/2^{k-1}\}$ by prescribing

$$\chi(a_1 z) = \frac{1}{2}\chi(z).$$

Putting finally $\chi(0) = 0$ we then get a homeomorphism χ of Δ_ε with itself such that $g = \chi^{-1} \circ g_< \circ \chi$, as required. □

Remark 2.5. Notice that $g_<(z) = \frac{1}{2}z$ and $g_>(z) = 2z$ cannot be topologically conjugated, because (for instance) $K_{g_<}$ is open whereas $K_{g_>} = \{0\}$ is not.

Remark 2.6. The proof of this theorem is based on two techniques often used in dynamics to build conjugations. The first one is used in part (i). Suppose that we would like to prove that two invertible local dynamical systems $f,\, g \in \mathrm{End}(M,p)$ are conjugated. Set $\varphi_k = g^{-k} \circ f^k$, so that

$$\varphi_k \circ f = g^{-k} \circ f^{k+1} = g \circ \varphi_{k+1}.$$

Therefore if we can prove that $\{\varphi_k\}$ converges to an invertible map φ as $k \to +\infty$ we get $\varphi \circ f = g \circ \varphi$, and thus f and g are conjugated, as desired. This is exactly the way we proved Theorem 2.4.(i); and we shall see variations of this technique later on.

To describe the second technique we need a definition.

Definition 2.7. Let $f\colon X \to X$ be an open continuous self-map of a topological space X. A *fundamental domain* for f is an open subset $D \subset X$ such that

(i) $f^h(D) \cap f^k(D) = \emptyset$ for every $h \neq k \in \mathbb{N}$;
(ii) $\bigcup\limits_{k \in \mathbb{N}} f^k(\overline{D}) = X$;
(iii) if $z_1, z_2 \in \overline{D}$ are so that $f^h(z_1) = f^k(z_2)$ for some $h > k \in \mathbb{N}$ then $h = k+1$ and $z_2 = f(z_1) \in \partial D$.

There are other possible definitions of a fundamental domain, but this will work for our aims.

Suppose that we would like to prove that two open continuous maps $f_1\colon X_1 \to X_1$ and $f_2\colon X_2 \to X_2$ are topologically conjugated. Assume we have fundamental domains $D_j \subset X_j$ for f_j (with $j = 1, 2$) and a homeomorphism $\chi\colon \overline{D_1} \to \overline{D_2}$ such that

$$\chi \circ f_1 = f_2 \circ \chi \tag{2}$$

on $\overline{D_1} \cap f_1^{-1}(\overline{D_1})$. Then we can extend χ to a homeomorphism $\tilde{\chi}\colon X_1 \to X_2$ conjugating f_1 and f_2 by setting

$$\tilde{\chi}(z) = f_2^k\bigl(\chi(w)\bigr), \tag{3}$$

for all $z \in X_1$, where $k = k(z) \in \mathbb{N}$ and $w = w(z) \in \overline{D}$ are chosen so that $f_1^k(w) = z$. The definition of fundamental domain and (2) imply that $\tilde{\chi}$ is well-defined. Clearly $\tilde{\chi} \circ f_1 = f_2 \circ \tilde{\chi}$; and using the openness of f_1 and f_2 it is easy to check that $\tilde{\chi}$ is a homeomorphism. This is the technique we used in the proof of Theorem 2.4.(iii); and we shall use it again later.

Thus the dynamics in the one-dimensional hyperbolic case is completely clear. The superattracting case can be treated similarly. If 0 is a superattracting point for an $f \in \mathrm{End}(\mathbb{C},0)$, we can write

$$f(z) = a_r z^r + a_{r+1} z^{r+1} + \cdots$$

with $a_r \neq 0$.

Definition 2.8. The number $r \geq 2$ is the *order* (or *local degree*) of the superattracting point.

An argument similar to the one described before shows that for $\varepsilon > 0$ small enough the stable set of $f|_{\Delta_\varepsilon}$ still is all of Δ_ε, and the orbits converge (faster than in the attracting case) to the origin. Furthermore, we can prove the following

Theorem 2.9 (Böttcher, 1904 [Bö]). *Let $f \in \mathrm{End}(\mathbb{C},0)$ be a one-dimensional holomorphic local dynamical system with a superattracting fixed point at the origin, and let $r \geq 2$ be its order. Then:*

(i) *f is holomorphically (and hence formally) locally conjugated to the map $g(z) = z^r$, and the conjugation is unique up to multiplication by an $(r-1)$-root of unity;*
(ii) *two such holomorphic local dynamical systems are holomorphically (or topologically) conjugated if and only if they have the same order.*

Proof. First of all, up to a linear conjugation $z \mapsto \mu z$ with $\mu^{r-1} = a_r$ we can assume $a_r = 1$.

Now write $f(z) = z^r h_1(z)$ for a suitable holomorphic germ h_1 with $h_1(0) = 1$. By induction, it is easy to see that we can write $f^k(z) = z^{r^k} h_k(z)$ for a suitable holomorphic germ h_k with $h_k(0) = 1$. Furthermore, the equalities $f \circ f^{k-1} = f^k = f^{k-1} \circ f$ yield

$$h_{k-1}(z)^r h_1\big(f^{k-1}(z)\big) = h_k(z) = h_1(z)^{r^{k-1}} h_{k-1}\big(f(z)\big). \tag{4}$$

Choose $0 < \delta < 1$. Then we can clearly find $1 > \varepsilon > 0$ such that $M\varepsilon < \delta$, where $M = \max_{z \in \overline{\Delta_\varepsilon}} |h_1(z)|$; we can also assume that $h_1(z) \neq 0$ for all $z \in \overline{\Delta_\varepsilon}$. Since

$$|f(z)| \leq M|z|^r < \delta |z|^{r-1}$$

for all $z \in \overline{\Delta_\varepsilon}$, we have $f(\Delta_\varepsilon) \subset \Delta_\varepsilon$, as anticipated before.

We also remark that (4) implies that each h_k is well-defined and never vanishing on $\overline{\Delta_\varepsilon}$. So for every $k \geq 1$ we can choose a unique ψ_k holomorphic in Δ_ε such that $\psi_k(z)^{r^k} = h_k(z)$ on Δ_ε and with $\psi_k(0) = 1$.

Set $\varphi_k(z) = z\psi_k(z)$, so that $\varphi_k'(0) = 1$ and $\varphi_k(z)^{r^k} = f^k(z)$ on Δ_ε; in particular, formally we have $\varphi_k = g^{-k} \circ f^k$. We claim that the sequence $\{\varphi_k\}$ converges to a holomorphic function φ on Δ_ε. Indeed, we have

$$\begin{aligned}\left|\frac{\varphi_{k+1}(z)}{\varphi_k(z)}\right| &= \left|\frac{\psi_{k+1}(z)^{r^{k+1}}}{\psi_k(z)^{r^{k+1}}}\right|^{1/r^{k+1}} = \left|\frac{h_{k+1}(z)}{h_k(z)^r}\right|^{1/r^{k+1}} = \left|h_1\big(f^k(z)\big)\right|^{1/r^{k+1}} \\ &= \left|1 + O(|f^k(z)|)\right|^{1/r^{k+1}} = 1 + \frac{1}{r^{k+1}} O(|f^k(z)|) = 1 + O\left(\frac{1}{r^{k+1}}\right),\end{aligned}$$

and so the telescopic product $\prod_k (\varphi_{k+1}/\varphi_k)$ converges to φ/φ_1 uniformly in Δ_ε.

Since $\varphi_k'(0) = 1$ for all $k \in \mathbb{N}$, we have $\varphi'(0) = 1$ and so, up to possibly shrink ε, we can assume that φ is a biholomorphism with its image. Moreover, we have

$$\varphi_k\big(f(z)\big)^{r^k} = f(z)^{r^k}\,\psi_k\big(f(z)\big)^{r^k} = z^{r^k}h_1(z)^{r^k}h_k\big(f(z)\big) = z^{r^{k+1}}h_{k+1}(z) = \big[\varphi_{k+1}(z)^r\big]^{r^k},$$

and thus $\varphi_k \circ f = [\varphi_{k+1}]^r$. Passing to the limit we get $f = \varphi^{-1} \circ g \circ \varphi$, as claimed.

If ψ is another local biholomorphism conjugating f with g, we must have $\psi \circ \varphi^{-1}(z^r) = \psi \circ \varphi^{-1}(z)^r$ for all z in a neighbourhood of the origin; comparing the series expansions at the origin we get $\psi \circ \varphi^{-1}(z) = az$ with $a^{r-1} = 1$, and hence $\psi(z) = a\varphi(z)$, as claimed.

Finally, (ii) follows because z^r and z^s are locally topologically conjugated if and only if $r = s$ (because the order is the number of preimages of points close to the origin). □

Therefore the one-dimensional local dynamics about a hyperbolic or superattracting fixed point is completely clear; let us now discuss what happens about a parabolic fixed point.

3 One Complex Variable: The Parabolic Case

Let $f \in \mathrm{End}(\mathbb{C}, 0)$ be a (non-linear) holomorphic local dynamical system with a parabolic fixed point at the origin. Then we can write

$$f(z) = e^{2i\pi p/q}z + a_{r+1}z^{r+1} + a_{r+2}z^{r+2} + \cdots, \tag{5}$$

with $a_{r+1} \neq 0$.

Definition 3.1. The rational number $p/q \in \mathbb{Q} \cap [0, 1)$ is the *rotation number* of f, and the number $r + 1 \geq 2$ is the *multiplicity* of f at the fixed point. If $p/q = 0$ (that is, if the multiplier is 1), we shall say that f is *tangent to the identity.*

The first observation is that such a dynamical system is never locally conjugated to its linear part, not even topologically, unless it is of finite order:

Proposition 3.2. *Let $f \in \mathrm{End}(\mathbb{C}, 0)$ be a holomorphic local dynamical system with multiplier λ, and assume that $\lambda = e^{2i\pi p/q}$ is a primitive root of the unity of order q. Then f is holomorphically (or topologically or formally) locally conjugated to $g(z) = \lambda z$ if and only if $f^q \equiv \mathrm{id}$.*

Proof. If $\varphi^{-1} \circ f \circ \varphi(z) = e^{2\pi i p/q}z$ then $\varphi^{-1} \circ f^q \circ \varphi = \mathrm{id}$, and hence $f^q = \mathrm{id}$.

Conversely, assume that $f^q \equiv \mathrm{id}$ and set

$$\varphi(z) = \frac{1}{q}\sum_{j=0}^{q-1}\frac{f^j(z)}{\lambda^j}.$$

Then it is easy to check that $\varphi'(0) = 1$ and $\varphi \circ f(z) = \lambda\varphi(z)$, and so f is holomorphically (and topologically and formally) locally conjugated to λz. □

In particular, if f is tangent to the identity then it *cannot* be locally conjugated to the identity (unless it was the identity to begin with, which is not a very interesting case dynamically speaking). More precisely, the stable set of such an f is never a neighbourhood of the origin. To understand why, let us first consider a map of the form

$$f(z) = z(1 + az^r)$$

for some $a \neq 0$. Let $v \in S^1 \subset \mathbb{C}$ be such that av^r is real and positive. Then for any $c > 0$ we have

$$f(cv) = c(1 + c^r av^r)v \in \mathbb{R}^+ v;$$

moreover, $|f(cv)| > |cv|$. In other words, the half-line $\mathbb{R}^+ v$ is f-invariant and repelled from the origin, that is $K_f \cap \mathbb{R}^+ v = \emptyset$. Conversely, if av^r is real and negative then the segment $[0, |a|^{-1/r}]v$ is f-invariant and attracted by the origin. So K_f neither is a neighbourhood of the origin nor reduces to $\{0\}$.

This example suggests the following definition:

Definition 3.3. Let $f \in \mathrm{End}(\mathbb{C}, 0)$ be tangent to the identity of multiplicity $r + 1 \geq 2$. Then a unit vector $v \in S^1$ is an *attracting* (respectively, *repelling*) *direction* for f at the origin if $a_{r+1}v^r$ is real and negative (respectively, positive).

Clearly, there are r equally spaced attracting directions, separated by r equally spaced repelling directions: if $a_{r+1} = |a_{r+1}|e^{i\alpha}$, then $v = e^{i\theta}$ is attracting (respectively, repelling) if and only if

$$\theta = \frac{2k+1}{r}\pi - \frac{\alpha}{r} \qquad \left(\text{respectively, } \theta = \frac{2k}{r}\pi - \frac{\alpha}{r}\right).$$

Furthermore, a repelling (attracting) direction for f is attracting (repelling) for f^{-1}, which is defined in a neighbourhood of the origin.

It turns out that to every attracting direction is associated a connected component of $K_f \setminus \{0\}$.

Definition 3.4. Let $v \in S^1$ be an attracting direction for an $f \in \mathrm{End}(\mathbb{C}, 0)$ tangent to the identity. The *basin* centered at v is the set of points $z \in K_f \setminus \{0\}$ such that $f^k(z) \to 0$ and $f^k(z)/|f^k(z)| \to v$ (notice that, up to shrinking the domain of f, we can assume that $f(z) \neq 0$ for all $z \in K_f \setminus \{0\}$). If z belongs to the basin centered at v, we shall say that the orbit of z *tends to* 0 *tangent to* v.

A slightly more specialized (but more useful) object is the following:

Definition 3.5. An *attracting petal* centered at an attracting direction v of an $f \in \mathrm{End}(\mathbb{C}, 0)$ tangent to the identity is an open simply connected f-invariant set $P \subseteq K_f \setminus \{0\}$ such that a point $z \in K_f \setminus \{0\}$ belongs to the basin centered at v if and only if its orbit intersects P. In other words, the orbit of a point tends to 0 tangent to v if and only if it is eventually contained in P. A *repelling petal* (centered at a repelling direction) is an attracting petal for the inverse of f.

It turns out that the basins centered at the attracting directions are exactly the connected components of $K_f \setminus \{0\}$, as shown in the *Leau-Fatou flower theorem:*

Theorem 3.6 (Leau, 1897 [L]; Fatou, 1919-20 [F1–3]). *Let $f \in \mathrm{End}(\mathbb{C},0)$ be a holomorphic local dynamical system tangent to the identity with multiplicity $r+1 \geq 2$ at the fixed point. Let $v_1^+,\dots,v_r^+ \in S^1$ be the r attracting directions of f at the origin, and $v_1^-,\dots,v_r^- \in S^1$ the r repelling directions. Then*

(i) *for each attracting (repelling) direction $v_j^\pm$ there exists an attracting (repelling) petal $P_j^\pm$, so that the union of these $2r$ petals together with the origin forms a neighbourhood of the origin. Furthermore, the $2r$ petals are arranged ciclically so that two petals intersect if and only if the angle between their central directions is π/r.*

(ii) *$K_f \setminus \{0\}$ is the (disjoint) union of the basins centered at the r attracting directions.*

(iii) *If B is a basin centered at one of the attracting directions, then there is a function $\varphi\colon B \to \mathbb{C}$ such that $\varphi \circ f(z) = \varphi(z)+1$ for all $z \in B$. Furthermore, if P is the corresponding petal constructed in part* (i), *then $\varphi|_P$ is a biholomorphism with an open subset of the complex plane containing a right half-plane — and so $f|_P$ is holomorphically conjugated to the translation $z \mapsto z+1$.*

Proof. Up to a linear conjugation, we can assume that $a_{r+1} = -1$, so that the attracting directions are the r-th roots of unity. For any $\delta > 0$, the set $\{z \in \mathbb{C} \mid |z^r - \delta| < \delta\}$ has exactly r connected components, each one symmetric with respect to a different r-th root of unity; it will turn out that, for δ small enough, these connected components are attracting petals of f, even though to get a pointed neighbourhood of the origin we shall need larger petals.

For $j = 1,\dots,r$ let $\Sigma_j \subset \mathbb{C}^*$ denote the sector centered about the attractive direction v_j^+ and bounded by two consecutive repelling directions, that is

$$\Sigma_j = \left\{ z \in \mathbb{C}^* \;\middle|\; \frac{2j-3}{r}\pi < \arg z < \frac{2j-1}{r}\pi \right\}.$$

Notice that each Σ_j contains a unique connected component $P_{j,\delta}$ of $\{z \in \mathbb{C} \mid |z^r - \delta| < \delta\}$; moreover, $P_{j,\delta}$ is tangent at the origin to the sector centered about v_j of amplitude π/r.

The main technical trick in this proof consists in transfering the setting to a neighbourhood of infinity in the Riemann sphere $\mathbb{P}^1(\mathbb{C})$. Let $\psi\colon \mathbb{C}^* \to \mathbb{C}^*$ be given by

$$\psi(z) = \frac{1}{rz^r};$$

it is a biholomorphism between Σ_j and $\mathbb{C}^* \setminus \mathbb{R}^-$, with inverse $\psi^{-1}(w) = (rw)^{-1/r}$, choosing suitably the r-th root. Furthermore, $\psi(P_{j,\delta})$ is the right half-plane $H_\delta = \{w \in \mathbb{C} \mid \mathrm{Re}\, w > 1/(2r\delta)\}$.

When $|w|$ is so large that $\psi^{-1}(w)$ belongs to the domain of definition of f, the composition $F = \psi \circ f \circ \psi^{-1}$ makes sense, and we have

$$F(w) = w + 1 + O(w^{-1/r}). \tag{6}$$

Thus to study the dynamics of f in a neighbourhood of the origin in Σ_j it suffices to study the dynamics of F in a neighbourhood of infinity.

The first observation is that when $\operatorname{Re} w$ is large enough then

$$\operatorname{Re} F(w) > \operatorname{Re} w + \frac{1}{2} ;$$

this implies that for δ small enough H_δ is F-invariant (and thus $P_{j,\delta}$ is f-invariant). Furthermore, by induction one has

$$\operatorname{Re} F^k(w) > \operatorname{Re} w + \frac{k}{2} \tag{7}$$

for all $w \in H_\delta$, which implies that $F^k(w) \to \infty$ in H_δ (and $f^k(z) \to 0$ in $P_{j,\delta}$) as $k \to \infty$.

Now we claim that the argument of $w_k = F^k(w)$ tends to zero. Indeed, (6) and (7) yield

$$\frac{w_k}{k} = \frac{w}{k} + 1 + \frac{1}{k} \sum_{l=0}^{k-1} O(w_l^{-1/r}) ;$$

hence Cesaro's theorem on the averages of a converging sequence implies

$$\frac{w_k}{k} \to 1, \tag{8}$$

and so $\arg w_k \to 0$ as $k \to \infty$. Going back to $P_{j,\delta}$, this implies that $f^k(z)/|f^k(z)| \to v_j^+$ for every $z \in P_{j,\delta}$. Since furthermore $P_{j,\delta}$ is centered about v_j^+, every orbit converging to 0 tangent to v_j^+ must intersect $P_{j,\delta}$, and thus we have proved that $P_{j,\delta}$ is an attracting petal.

Arguing in the same way with f^{-1} we get repelling petals; unfortunately, the petals obtained so far are too small to form a full pointed neighbourhood of the origin. In fact, as remarked before each $P_{j,\delta}$ is contained in a sector centered about v_j^+ of amplitude π/r; therefore the repelling and attracting petals obtained in this way do not intersect but are tangent to each other. We need larger petals.

So our aim is to find an f-invariant subset P_j^+ of Σ_j containing $P_{j,\delta}$ and which is tangent at the origin to a sector centered about v_j^+ of amplitude strictly greater than π/r. To do so, first of all remark that there are $R, C > 0$ such that

$$|F(w) - w - 1| \le \frac{C}{|w|^{1/r}} \tag{9}$$

as soon as $|w| > R$. Choose $\varepsilon \in (0,1)$ and select $\delta > 0$ so that $4r\delta < R^{-1}$ and $\varepsilon > 2C(4r\delta)^{1/r}$. Then $|w| > 1/(4r\delta)$ implies

$$|F(w) - w - 1| < \varepsilon/2.$$

Set $M_\varepsilon = (1+\varepsilon)/(2r\delta)$ and let

$$\tilde{H}_\varepsilon = \{w \in \mathbb{C} \mid |\mathrm{Im}\, w| > -\varepsilon \mathrm{Re}\, w + M_\varepsilon\} \cup H_\delta.$$

If $w \in \tilde{H}_\varepsilon$ we have $|w| > 1/(2r\delta)$ and hence

$$\mathrm{Re}\, F(w) > \mathrm{Re}\, w + 1 - \varepsilon/2 \qquad \text{and} \qquad |\mathrm{Im}\, F(w) - \mathrm{Im}\, w| < \varepsilon/2\,; \tag{10}$$

it is then easy to check that $F(\tilde{H}_\varepsilon) \subset \tilde{H}_\varepsilon$ and that every orbit starting in $\tilde{H}_\varepsilon$ must eventually enter H_δ. Thus $P_j^+ = \psi^{-1}(\tilde{H}_\varepsilon)$ is as required, and we have proved (i).

To prove (ii) we need a further property of $\tilde{H}_\varepsilon$. If $w \in \tilde{H}_\varepsilon$, arguing by induction on $k \geq 1$ using (10) we get

$$k\left(1 - \frac{\varepsilon}{2}\right) < \mathrm{Re}\, F^k(w) - \mathrm{Re}\, w$$

and

$$\frac{k\varepsilon(1-\varepsilon)}{2} < |\mathrm{Im}\, F^k(w)| + \varepsilon \mathrm{Re}\, F^k(w) - \big(|\mathrm{Im}\, w| + \varepsilon \mathrm{Re}\, w\big).$$

This implies that for every $w_0 \in \tilde{H}_\varepsilon$ there exists a $k_0 \geq 1$ so that we cannot have $F^{k_0}(w) = w_0$ for any $w \in \tilde{H}_\varepsilon$. Coming back to the z-plane, this says that any inverse orbit of f must eventually leave P_j^+. Thus every (forward) orbit of f must eventually leave any repelling petal. So if $z \in K_f \setminus \{O\}$, where the stable set is computed working in the neighborhood of the origin given by the union of repelling and attracting petals (together with the origin), the orbit of z must eventually land in an attracting petal, and thus z belongs to a basin centered at one of the r attracting directions — and (ii) is proved.

To prove (iii), first of all we notice that we have

$$|F'(w) - 1| \leq \frac{2^{1+1/r} C}{|w|^{1+1/r}} \tag{11}$$

in $\tilde{H}_\varepsilon$. Indeed, (9) says that if $|w| > 1/(2r\delta)$ then the function $w \mapsto F(w) - w - 1$ sends the disk of center w and radius $|w|/2$ into the disk of center the origin and radius $C/(|w|/2)^{1/r}$; inequality (11) then follows from the Cauchy estimates on the derivative.

Now choose $w_0 \in H_\delta$, and set $\tilde{\varphi}_k(w) = F^k(w) - F^k(w_0)$. Given $w \in \tilde{H}_\varepsilon$, as soon as $k \in \mathbb{N}$ is so large that $F^k(w) \in H_\delta$ we can apply Lagrange's theorem to the segment from $F^k(w_0)$ to $F^k(w)$ to get a $t_k \in [0,1]$ such that

$$\left|\frac{\tilde{\varphi}_{k+1}(w)}{\tilde{\varphi}_k(w)} - 1\right| = \left|\frac{F\big(F^k(w)\big) - F^k\big(F^k(w_0)\big)}{F^k(w) - F^k(w_0)} - 1\right| = \left|F'\big(t_k F^k(w) + (1-t_k)F^k(w_0)\big) - 1\right|$$
$$\leq \frac{2^{1+1/r} C}{\min\{|\mathrm{Re}\, F^k(w)|, |\mathrm{Re}\, F^k(w_0)|\}^{1+1/r}} \leq \frac{C'}{k^{1+1/r}},$$

where we used (11) and (8), and the constant C' is uniform on compact subsets of $\tilde{H}_\varepsilon$ (and it can be chosen uniform on H_δ).

As a consequence, the telescopic product $\prod_k \tilde{\varphi}_{k+1}/\tilde{\varphi}_k$ converges uniformly on compact subsets of $\tilde{H}_\varepsilon$ (and uniformly on H_δ), and thus the sequence $\tilde{\varphi}_k$ converges, uniformly on compact subsets, to a holomorphic function $\tilde{\varphi}\colon \tilde{H}_\varepsilon \to \mathbb{C}$. Since we have

$$\begin{aligned}\tilde{\varphi}_k \circ F(w) &= F^{k+1}(w) - F^k(w_0) = \tilde{\varphi}_{k+1}(w) + F\big(F^k(w_0)\big) - F^k(w_0)\\ &= \tilde{\varphi}_{k+1}(w) + 1 + O\big(|F^k(w_0)|^{-1/r}\big),\end{aligned}$$

it follows that

$$\tilde{\varphi} \circ F(w) = \tilde{\varphi}(w) + 1$$

on $\tilde{H}_\varepsilon$. In particular, $\tilde{\varphi}$ is not constant; being the limit of injective functions, by Hurwitz's theorem it is injective.

We now prove that the image of $\tilde{\varphi}$ contains a right half-plane. First of all, we claim that

$$\lim_{\substack{|w|\to+\infty\\ w\in H_\delta}} \frac{\tilde{\varphi}(w)}{w} = 1. \tag{12}$$

Indeed, choose $\eta > 0$. Since the convergence of the telescopic product is uniform on H_δ, we can find $k_0 \in \mathbb{N}$ such that

$$\left|\frac{\tilde{\varphi}(w) - \tilde{\varphi}_{k_0}(w)}{w - w_0}\right| < \frac{\eta}{3}$$

on H_δ. Furthermore, we have

$$\left|\frac{\tilde{\varphi}_{k_0}(w)}{w - w_0} - 1\right| = \left|\frac{k_0 + \sum_{j=0}^{k_0-1} O(|F^j(w)|^{-1/r}) + w_0 - F^{k_0}(w_0)}{w - w_0}\right| = O(|w|^{-1})$$

on H_δ; therefore we can find $R > 0$ such that

$$\left|\frac{\tilde{\varphi}(w)}{w - w_0} - 1\right| < \frac{\eta}{3}$$

as soon as $|w| > R$ in H_δ. Finally, if R is large enough we also have

$$\left|\frac{\tilde{\varphi}(w)}{w - w_0} - \frac{\tilde{\varphi}(w)}{w}\right| = \left|\frac{\tilde{\varphi}(w)}{w - w_0}\right| \left|\frac{w}{w_0}\right| < \frac{\eta}{3},$$

and (12) follows.

Equality (12) clearly implies that $(\tilde{\varphi}(w) - w^o)/(w - w^o) \to 1$ as $|w| \to +\infty$ in H_δ for any $w^o \in \mathbb{C}$. But this means that if $\operatorname{Re} w^o$ is large enough then the difference between the variation of the argument of $\tilde{\varphi} - w^o$ along a suitably small closed circle around w^o and the variation of the argument of $w - w^o$ along the same circle will be less than 2π — and thus it will be zero. Then the argument principle implies that $\tilde{\varphi} - w^o$ and $w - w^o$ have the same number of zeroes inside that circle, and thus $w^o \in \tilde{\varphi}(H_\delta)$, as required.

So setting $\varphi = \tilde{\varphi} \circ \psi$, we have defined a function φ with the required properties on P_j^+. To extend it to the whole basin B it suffices to put

$$\varphi(z) = \varphi\big(f^k(z)\big) - k, \tag{13}$$

where $k \in \mathbb{N}$ is the first integer such that $f^k(z) \in P_j^+$. □

A way to construct the conjugation φ as limit of hyperbolic linearizations given by Theorem 2.4 is described in [U3].

Remark 3.7. It is possible to construct petals that cannot be contained in any sector strictly smaller than Σ_j. To do so we need an F-invariant subset $\hat{H}_\varepsilon$ of $\mathbb{C}^* \setminus \mathbb{R}^-$ containing $\tilde{H}_\varepsilon$ and containing eventually every half-line issuing from the origin (but $\mathbb{R}^-$). For $M >> 1$ and $C > 0$ large enough, replace the straight lines bounding $\tilde{H}_\varepsilon$ on the left of $\mathrm{Re}\, w = -M$ by the curves

$$|\mathrm{Im}\, w| = \begin{cases} C \log|\mathrm{Re}\, w| & \text{if } r = 1\,, \\ C|\mathrm{Re}\, w|^{1-1/r} & \text{if } r > 1. \end{cases}$$

Then it is not too difficult to check that the domain $\hat{H}_\varepsilon$ so obtained is as desired (see [CG]).

So we have a complete description of the dynamics in the neighbourhood of the origin. Actually, Camacho has pushed this argument even further, obtaining a complete topological classification of one-dimensional holomorphic local dynamical systems tangent to the identity (see also [BH, Theorem 1.7]):

Theorem 3.8 (Camacho, 1978 [C]; Shcherbakov, 1982 [S]). *Let $f \in \mathrm{End}(\mathbb{C},0)$ be a holomorphic local dynamical system tangent to the identity with multiplicity $r+1$ at the fixed point. Then f is topologically locally conjugated to the map*

$$g(z) = z - z^{r+1}.$$

Remark 3.9. Camacho's proof ([C]; see also [Br2, J1]) shows that the topological conjugation can be taken smooth in a punctured neighbourhood of the origin. Jenkins [J1] also proved that if $f \in \mathrm{End}(\mathbb{C},0)$ is tangent to the identity with multiplicity 2 and the topological conjugation is actually real-analitic in a punctured neighbourhood of the origin, with real-analytic inverse, then f is locally holomorphically conjugated to $z - z^2$. Finally, Martinet and Ramis [MR] have proved that if a germ $f \in \mathrm{End}(\mathbb{C},0)$ tangent to the identity is C^1-conjugated (in a full neighbourhood of the origin) to $g(z) = z?z^{r+1}$, then the conjugation can be chosen holomorphic or antiholomorphic.

The formal classification is simple too, though different (see, e.g., Milnor [Mi]):

Proposition 3.10. *Let $f \in \mathrm{End}(\mathbb{C},0)$ be a holomorphic local dynamical system tangent to the identity with multiplicity $r+1$ at the fixed point. Then f is formally conjugated to the map*

$$g(z) = z - z^{r+1} + \beta z^{2r+1}, \tag{14}$$

where β is a formal (and holomorphic) invariant given by

$$\beta = \frac{1}{2\pi i}\int_\gamma \frac{dz}{z-f(z)}, \tag{15}$$

where the integral is taken over a small positive loop γ about the origin.

Proof. An easy computation shows that if f is given by (14) then (15) holds. Let us now show that the integral in (15) is a holomorphic invariant. Let φ be a local biholomorphism fixing the origin, and set $F = \varphi^{-1}\circ f\circ\varphi$. Then

$$\frac{1}{2\pi i}\int_\gamma \frac{dz}{z-f(z)} = \frac{1}{2\pi i}\int_{\varphi^{-1}\circ\gamma} \frac{\varphi'(w)\,dw}{\varphi(w)-f\big(\varphi(w)\big)} = \frac{1}{2\pi i}\int_{\varphi^{-1}\circ\gamma} \frac{\varphi'(w)\,dw}{\varphi(w)-\varphi\big(F(w)\big)}.$$

Now, we can clearly find $M, M_1 > 0$ such that

$$\begin{aligned}\left|\frac{1}{w-F(w)} - \frac{\varphi'(w)}{\varphi(w)-\varphi\big(F(w)\big)}\right| &= \frac{1}{\big|\varphi(w)-\varphi\big(F(w)\big)\big|}\left|\frac{\varphi(w)-\varphi\big(F(w)\big)}{w-F(w)} - \varphi'(w)\right| \\ &\le M\frac{|w-F(w)|}{\big|\varphi(w)-\varphi\big(F(w)\big)\big|} \le M_1,\end{aligned}$$

in a neighbourhood of the origin, where the last inequality follows from the fact that $\varphi'(0)\neq 0$. This means that the two meromorphic functions $1/\big(w-F(w)\big)$ and $\varphi'(w)/\big(\varphi(w)-\varphi((F(w))\big)$ differ by a holomorphic function; so they have the same integral along any small loop surrounding the origin, and

$$\frac{1}{2\pi i}\int_\gamma \frac{dz}{z-f(z)} = \frac{1}{2\pi i}\int_{\varphi^{-1}\circ\gamma} \frac{dw}{w-F(w)},$$

as claimed.

To prove that f is formally conjugated to g, let us first take a local formal change of coordinates φ of the form

$$\varphi(z) = z + \mu z^d + O_{d+1} \tag{16}$$

with $\mu\neq 0$, and where we are writing O_{d+1} instead of $O(z^{d+1})$. It follows that $\varphi^{-1}(z) = z - \mu z^d + O_{d+1}$, $(\varphi^{-1})'(z) = 1 - d\mu z^{d-1} + O_d$ and $(\varphi^{-1})^{(j)} = O_{d-j}$ for all $j\ge 2$. Then using the Taylor expansion of φ^{-1} we get

$$\begin{aligned}&\varphi^{-1}\circ f\circ\varphi(z)\\ &= \varphi^{-1}\left(\varphi(z) + \sum_{j\ge r+1} a_j\varphi(z)^j\right)\\ &= z + (\varphi^{-1})'(\varphi(z))\sum_{j\ge r+1} a_j z^j(1+\mu z^{d-1}+O_d)^j + O_{d+2r} \qquad (17)\\ &= z + [1 - d\mu z^{d-1} + O_d]\sum_{j\ge r+1} a_j z^j(1 + j\mu z^{d-1} + O_d) + O_{d+2r}\end{aligned}$$

$$= z + a_{r+1} z^{r+1} + \cdots + a_{r+d-1} z^{r+d-1} \\ + [a_{r+d} + (r+1-d)\mu a_{r+1}] z^{r+d} + O_{r+d+1}.$$

This means that if $d \neq r+1$ we can use a polynomial change of coordinates of the form $\varphi(z) = z + \mu z^d$ to remove the term of degree $r+d$ from the Taylor expansion of f without changing the lower degree terms.

So to conjugate f to g it suffices to use a linear change of coordinates to get $a_{r+1} = -1$, and then apply a sequence of change of coordinates of the form $\varphi(z) = z + \mu z^d$ to kill all the terms in the Taylor expansion of f but the term of degree z^{2r+1}.

Finally, formula (17) also shows that two maps of the form (14) with different β cannot be formally conjugated, and we are done. □

Definition 3.11. The number β given by (15) is called *index* of f at the fixed point. The *iterative residue* of f is then defined by

$$\mathrm{Resit}(f) = \frac{r+1}{2} - \beta.$$

The iterative residue has been introduced by Écalle [É1], and it behaves nicely under iteration; for instance, it is possible to prove (see [BH, Proposition 3.10]) that

$$\mathrm{Resit}(f^k) = \frac{1}{k}\mathrm{Resit}(f).$$

The holomorphic classification of maps tangent to the identity is much more complicated: as shown by Écalle [É2–3] and Voronin [Vo] in 1981, it depends on functional invariants. We shall now try and roughly describe it; see [I2, M1–2, Ki, BH, MR] and the original papers for details.

Let $f \in \mathrm{End}(\mathbb{C}, 0)$ be tangent to the identity with multiplicity $r+1$ at the fixed point; up to a linear change of coordinates we can assume that $a_{r+1} = -1$. Let $P_j^\pm$ be a set of petals as in Theorem 3.6.(i), ordered so that P_1^+ is centered on the positive real semiaxis, and the others are arranged cyclically counterclockwise. Denote by φ_j^+ (respectively, φ_j^-) the biholomorphism conjugating $f|_{P_j^+}$ (respectively, $f|_{P_j^-}$) to the shift $z \mapsto z+1$ in a right (respectively, left) half-plane given by Theorem 3.6.(iii) — applied to f^{-1} for the repelling petals. If we moreover require that

$$\varphi_j^\pm(z) = \frac{1}{rz^r} \pm \mathrm{Resit}(f) \cdot \log z + o(1), \tag{18}$$

then φ_j is uniquely determined.

Put now $U_j^+ = P_j^- \cap P_{j+1}^+$, $U_j^- = P_j^- \cap P_j^+$, and $S_j^\pm = \bigcup_{k \in \mathbb{Z}} U_j^\pm$. Using the dynamics as in (13) we can extend φ_j^- to $S_j^\pm$, and φ_j^+ to $S_{j-1}^+ \cup S_j^-$; put $V_j^\pm = \varphi_j^-(S_j^\pm)$, $W_j^- = \varphi_j^+(S_j^-)$ and $W_j^+ = \varphi_{j+1}^+(S_j^+)$. Then let $H_j^- : V_j^- \to W_j^-$ be the restriction of $\varphi_j^+ \circ (\varphi_j^-)^{-1}$ to V_j^-, and $H_j^+ : V_j^+ \to W_j^+$ the restriction of $\varphi_{j+1}^+ \circ (\varphi_j^-)^{-1}$ to V_j^+.

It is not difficult to see that $V_j^\pm$ and $W_j^\pm$ are invariant by translation by 1, and that V_j^+ and W_j^+ contain an upper half-plane while V_j^- and W_j^- contain a lower half-plane. Moreover, we have $H_j^{\pm}(z+1) = H_j^{\pm}(z) + 1$; therefore using the projection $\pi(z) = \exp(2\pi i z)$ we can induce holomorphic maps $h_j^{\pm}\colon \pi(V_j^{\pm}) \to \pi(W_j^{\pm})$, where $\pi(V_j^+)$ and $\pi(W_j^+)$ are pointed neighbourhoods of the origin, and $\pi(V_j^-)$ and $\pi(W_j^-)$ are pointed neighbourhoods of $\infty \in \mathbb{P}^1(\mathbb{C})$.

It is possible to show that setting $h_j^+(0) = 0$ one obtains a holomorphic germ $h_j^+ \in \mathrm{End}(\mathbb{C},0)$, and that setting $h_j^-(\infty) = \infty$ one obtains a holomorphic germ $h_j^+ \in \mathrm{End}\bigl(\mathbb{P}^1(C),\infty\bigr)$. Furthermore, denoting by λ_j^+ (respectively, λ_j^-) the multiplier of h_j^+ at 0 (respectively, of h_j^- at ∞), it turns out that

$$\prod_{j=1}^{r}(\lambda_j^+\lambda_j^-) = \exp\bigl[4\pi^2\mathrm{Resit}(f)\bigr]. \tag{19}$$

Now, if we replace f by a holomorphic local conjugate $\tilde{f} = \psi^{-1} \circ f \circ \psi$, and denote by $\tilde{h}_j^{\pm}$ the corresponding germs, it is not difficult to check that (up to a cyclic renumbering of the petals) there are constants $\alpha_j, \beta_j \in \mathbb{C}^*$ such that

$$\tilde{h}_j^-(z) = \alpha_j h_j^-\left(\frac{z}{\beta_j}\right) \qquad \text{and} \qquad \tilde{h}_j^+(z) = \alpha_{j+1} h_j^+\left(\frac{z}{\beta_j}\right). \tag{20}$$

This suggests the introduction of an equivalence relation on the set of $2r$-uple of germs $(h_1^{\pm},\dots,h_r^{\pm})$.

Definition 3.12. Let M_r denote the set of $2r$-uple of germs $\mathbf{h} = (h_1^{\pm},\dots,h_r^{\pm})$, with $h_j^+ \in \mathrm{End}(\mathbb{C},0)$, $h_j^- \in \mathrm{End}\bigl(\mathbb{P}^1(\mathbb{C}),\infty\bigr)$, and whose multipliers satisfy (19). We shall say that $\mathbf{h}, \tilde{\mathbf{h}} \in M_r$ are *equivalent* if up to a cyclic permutation of the indeces we have (20) for suitable $\alpha_j, \beta_j \in \mathbb{C}^*$. We denote by $\mathcal{M}_r$ the set of all equivalence classes.

The procedure described above allows then to associate to any $f \in \mathrm{End}(\mathbb{C},0)$ tangent to the identity with multiplicity $r+1$ an element $\mu_f \in \mathcal{M}_r$.

Definition 3.13. Let $f \in \mathrm{End}(\mathbb{C},0)$ be tangent to the identity. The element $\mu_f \in \mathcal{M}_r$ given by this procedure is the *sectorial invariant* of f.

Then the holomorphic classification proved by Écalle and Voronin is

Theorem 3.14 (Écalle, 1981 [É2–3]; Voronin, 1981 [Vo]). *Let $f, g \in \mathrm{End}(\mathbb{C},0)$ be two holomorphic local dynamical systems tangent to the identity. Then f and g are holomorphically locally conjugated if and only if they have the same multiplicity, the same index and the same sectorial invariant. Furthermore, for any $r \geq 1$, $\beta \in \mathbb{C}$ and $\mu \in \mathcal{M}_r$ there exists $f \in \mathrm{End}(\mathbb{C},0)$ tangent to the identity with multiplicity $r+1$, index β and sectorial invariant μ.*

Remark 3.15. In particular, holomorphic local dynamical systems tangent to the identity give examples of local dynamical systems that are topologically conjugated without being neither holomorphically nor formally conjugated, and of local dynamical systems that are formally conjugated without being holomorphically conjugated. See also [Na, Tr].

We would also like to mention a result of Ribón appeared in the appendix of [CGBM]. It is known (see, e.g., [Br2]) that any germ $f \in \mathrm{End}(\mathbb{C},0)$ tangent to the identity is the time-one map of a unique formal (not necessarily holomorphic) vector field X singular at the origin, the *infinitesimal generator* of f. It is not difficult to see that f is holomorphically locally conjugated to its formal normal form (given by Proposition 3.10) if and only if X is actually holomorphic; Ribón has shown that this is equivalent to the existence of a real-analytic foliation invariant under f. More precisely, he has proved the following

Theorem 3.16 (Ribón, 2008 [CGBM]). *Let $f \in \mathrm{End}(\mathbb{C},0) \setminus \{\mathrm{id}\}$ be a germ tangent to the identity. If there exists a germ of real-analytic foliation $\mathscr{F}$, having an isolated singularity at the origin, such that $f^*\mathscr{F} = \mathscr{F}$, then the formal infinitesimal generator of f is holomorphic at the origin. In particular, f is holomorphically conjugated to its formal normal form.*

We end this section recalling a few result on parabolic germs not tangent to the identity. If $f \in \mathrm{End}(\mathbb{C},0)$ satisfies $a_1 = e^{2\pi i p/q}$, then f^q is tangent to the identity. Therefore we can apply the previous results to f^q and then infer informations about the dynamics of the original f, because of the following

Lemma 3.17. *Let $f,\ g \in \mathrm{End}(\mathbb{C},0)$ be two holomorphic local dynamical systems with the same multiplier $e^{2\pi i p/q} \in S^1$. Then f and g are holomorphically locally conjugated if and only if f^q and g^q are.*

Proof. One direction is obvious. For the converse, let φ be a germ conjugating f^q and g^q; in particular,

$$g^q = \varphi^{-1} \circ f^q \circ \varphi = (\varphi^{-1} \circ f \circ \varphi)^q.$$

So, up to replacing f by $\varphi^{-1} \circ f \circ \varphi$, we can assume that $f^q = g^q$. Put

$$\psi = \sum_{k=0}^{q-1} g^{q-k} \circ f^k = \sum_{k=1}^{q} g^{q-k} \circ f^k.$$

The germ ψ is a local biholomorphism, because $\psi'(0) = q \neq 0$, and it is easy to check that $\psi \circ f = g \circ \psi$. □

We list here a few results; see [Mi, Ma, C, É2–3, Vo, MR, BH] for proofs and further details.

Proposition 3.18. *Let $f \in \mathrm{End}(\mathbb{C},0)$ be a holomorphic local dynamical system with multiplier $\lambda \in S^1$, and assume that λ is a primitive root of the unity of order q. Assume that $f^q \not\equiv \mathrm{id}$. Then there exist $n \geq 1$ and $\alpha \in \mathbb{C}$ such that f is formally conjugated to*

$$g(z) = \lambda z - z^{nq+1} + \alpha z^{2nq+1}.$$

Definition 3.19. The number n is the *parabolic multiplicity* of f, and $\alpha \in \mathbb{C}$ is the *index* of f; the *iterative residue* of f is then given by

$$\operatorname{Resit}(f) = \frac{nq+1}{2} - \alpha.$$

Proposition 3.20 (Camacho). *Let $f \in \operatorname{End}(\mathbb{C},0)$ be a holomorphic local dynamical system with multiplier $\lambda \in S^1$, and assume that λ is a primitive root of the unity of order q. Assume that $f^q \not\equiv \operatorname{id}$, and has parabolic multiplicity $n \geq 1$. Then f is topologically conjugated to*

$$g(z) = \lambda z - z^{nq+1}.$$

Theorem 3.21 (Leau-Fatou). *Let $f \in \operatorname{End}(\mathbb{C},0)$ be a holomorphic local dynamical system with multiplier $\lambda \in S^1$, and assume that λ is a primitive root of the unity of order q. Assume that $f^q \not\equiv \operatorname{id}$, and let $n \geq 1$ be the parabolic multiplicity of f. Then f^q has multiplicity $nq+1$, and f acts on the attracting (respectively, repelling) petals of f^q as a permutation composed by n disjoint cycles. Finally, $K_f = K_{f^q}$.*

Furthermore, it is possible to define the sectorial invariant of such a holomorphic local dynamical system, composed by $2nq$ germs whose multipliers still satisfy (19), and the analogue of Theorem 3.14 holds.

4 One Complex Variable: The Elliptic Case

We are left with the elliptic case:

$$f(z) = e^{2\pi i\theta} z + a_2 z^2 + \cdots \in \mathbb{C}_0\{z\}, \tag{21}$$

with $\theta \notin \mathbb{Q}$. It turns out that the local dynamics depends mostly on numerical properties of θ. The main question here is whether such a local dynamical system is holomorphically conjugated to its linear part. Let us introduce a bit of terminology.

Definition 4.1. We shall say that a holomorphic dynamical system of the form (21) is *holomorphically linearizable* if it is holomorphically locally conjugated to its linear part, the irrational rotation $z \mapsto e^{2\pi i\theta} z$. In this case, we shall say that 0 is a *Siegel point* for f; otherwise, we shall say that it is a *Cremer point*.

It turns out that for a full measure subset B of $\theta \in [0,1] \setminus \mathbb{Q}$ all holomorphic local dynamical systems of the form (21) are holomorphically linearizable. Conversely, the complement $[0,1] \setminus B$ is a G_δ-dense set, and for all $\theta \in [0,1] \setminus B$ the quadratic polynomial $z \mapsto z^2 + e^{2\pi i\theta} z$ is not holomorphically linearizable. This is the gist of the results due to Cremer, Siegel, Brjuno and Yoccoz we shall describe in this section.

The first worthwhile observation in this setting is that it is possible to give a topological characterization of holomorphically linearizable local dynamical systems.

Definition 4.2. We shall say that p is *stable* for $f \in \operatorname{End}(M,p)$ if it belongs to the interior of K_f.

Proposition 4.3. *Let $f \in \mathrm{End}(\mathbb{C},0)$ be a holomorphic local dynamical system with multiplier $\lambda \in S^1$. Then f is holomorphically linearizable if and only if it is topologically linearizable if and only if 0 is stable for f.*

Proof. If f is holomorphically linearizable it is topologically linearizable, and if it is topologically linearizable (and $|\lambda| = 1$) then it is stable. Assume that 0 is stable, and set

$$\varphi_k(z) = \frac{1}{k}\sum_{j=0}^{k-1}\frac{f^j(z)}{\lambda^j},$$

so that $\varphi_k'(0) = 1$ and

$$\varphi_k \circ f = \lambda\varphi_k + \frac{\lambda}{k}\left(\frac{f^k}{\lambda^k} - \mathrm{id}\right). \tag{22}$$

The stability of 0 implies that there are bounded open sets $V \subset U$ containing the origin such that $f^k(V) \subset U$ for all $k \in \mathbb{N}$. Since $|\lambda| = 1$, it follows that $\{\varphi_k\}$ is a uniformly bounded family on V, and hence, by Montel's theorem, it admits a converging subsequence. But (22) implies that a converging subsequence converges to a conjugation between f and the rotation $z \mapsto \lambda z$, and so f is holomorphically linearizable. □

The second important observation is that two elliptic holomorphic local dynamical systems with the same multiplier are always formally conjugated:

Proposition 4.4. *Let $f \in \mathrm{End}(\mathbb{C},0)$ be a holomorphic local dynamical system of multiplier $\lambda = e^{2\pi i\theta} \in S^1$ with $\theta \notin \mathbb{Q}$. Then f is formally conjugated to its linear part, by a unique formal power series tangent to the identity.*

Proof. We shall prove that there is a unique formal power series

$$h(z) = z + h_2 z^2 + \cdots \in \mathbb{C}[[z]]$$

such that $h(\lambda z) = f\big(h(z)\big)$. Indeed we have

$$\begin{aligned} h(\lambda z) - f\big(h(z)\big) &= \sum_{j\geq 2}\left\{[(\lambda^j - \lambda)h_j - a_j]z^j - a_j\sum_{\ell=1}^{j}\binom{j}{\ell}z^{\ell+j}\left(\sum_{k\geq 2}h_k z^{k-2}\right)^{\ell}\right\} \\ &= \sum_{j\geq 2}\big[(\lambda^j - \lambda)h_j - a_j - X_j(h_2,\dots,h_{j-1})\big]z^j, \end{aligned}$$

where X_j is a polynomial in $j-2$ variables with coefficients depending on $a_2,\dots,a_{j-1}$. It follows that the coefficients of h are uniquely determined by induction using the formula

$$h_j = \frac{a_j + X_j(h_2,\dots,h_{j-1})}{\lambda^j - \lambda}. \tag{23}$$

In particular, h_j depends only on λ, $a_2,\dots,a_j$. □

Remark 4.5. The same proof shows that any holomorphic local dynamical system with multiplier $\lambda \neq 0$ and not a root of unity is formally conjugated to its linear part.

The formal power series linearizing f is not converging if its coefficients grow too fast. Thus (23) links the radius of convergence of h to the behavior of $\lambda^j - \lambda$: if the latter becomes too small, the series defining h does not converge. This is known as the *small denominators problem* in this context.

It is then natural to introduce the following quantity:

$$\Omega_\lambda(m) = \min_{1 \leq k \leq m} |\lambda^k - \lambda|,$$

for $\lambda \in S^1$ and $m \geq 1$. Clearly, λ is a root of unity if and only if $\Omega_\lambda(m) = 0$ for all m greater or equal to some $m_0 \geq 1$; furthermore,

$$\lim_{m \to +\infty} \Omega_\lambda(m) = 0$$

for all $\lambda \in S^1$.

The first one to actually prove that there are non-linearizable elliptic holomorphic local dynamical systems has been Cremer, in 1927 [Cr1]. His more general result is the following:

Theorem 4.6 (Cremer, 1938 [Cr2]). *Let $\lambda \in S^1$ be such that*

$$\limsup_{m \to +\infty} \frac{1}{m} \log \frac{1}{\Omega_\lambda(m)} = +\infty. \tag{24}$$

Then there exists $f \in \mathrm{End}(\mathbb{C},0)$ with multiplier λ which is not holomorphically linearizable. Furthermore, the set of $\lambda \in S^1$ satisfying (24) *contains a G_δ-dense set.*

Proof. Choose inductively $a_j \in \{0,1\}$ so that $|a_j + X_j| \geq 1/2$ for all $j \geq 2$, where X_j is as in (23). Then

$$f(z) = \lambda z + a_2 z^2 + \cdots \in \mathbb{C}_0\{z\},$$

while (24) implies that the radius of convergence of the formal linearization h is 0, and thus f cannot be holomorphically linearizable, as required.

Finally, let $C(q_0) \subset S^1$ denote the set of $\lambda = e^{2\pi i\theta} \in S^1$ such that

$$\left| \theta - \frac{p}{q} \right| < \frac{1}{2^{q!}} \tag{25}$$

for some $p/q \in \mathbb{Q}$ in lowest terms, with $q \geq q_0$. Then it is not difficult to check that each $C(q_0)$ is a dense open set in S^1, and that all $\lambda \in \mathscr{C} = \bigcap_{q_0 \geq 1} C(q_0)$ satisfy (24). Indeed, if $\lambda = e^{2\pi i\theta} \in \mathscr{C}$ we can find $q \in \mathbb{N}$ arbitrarily large such that there is $p \in \mathbb{N}$ so that (25) holds. Now, it is easy to see that

$$|e^{2\pi i t} - 1| \leq 2\pi |t|$$

for all $t \in [-1/2, 1/2]$. Then let p_0 be the integer closest to $q\theta$, so that $|q\theta - p_0| \leq 1/2$. Then we have

$$|\lambda^q - 1| = |e^{2\pi i q\theta} - e^{2\pi i p_0}| = |e^{2\pi i(q\theta - p_0)} - 1| \leq 2\pi|q\theta - p_0| \leq 2\pi|q\theta - p| < \frac{2\pi}{2^{q!-1}}$$

for arbitrarily large q, and (24) follows. □

On the other hand, Siegel in 1942 gave a condition on the multiplier ensuring holomorphic linearizability:

Theorem 4.7 (Siegel, 1942 [Si]). *Let $\lambda \in S^1$ be such that there exists $\beta > 1$ and $\gamma > 0$ so that*

$$\frac{1}{\Omega_\lambda(m)} \leq \gamma m^\beta \tag{26}$$

for all $m \geq 2$. Then all $f \in \mathrm{End}(\mathbb{C}, 0)$ with multiplier λ are holomorphically linearizable. Furthermore, the set of $\lambda \in S^1$ satisfying (26) for some $\beta > 1$ and $\gamma > 0$ is of full Lebesgue measure in S^1.

Remark 4.8. If $\theta \in [0,1) \setminus \mathbb{Q}$ is algebraic then $\lambda = e^{2\pi i\theta}$ satisfies (26) for some $\beta > 1$ and $\gamma > 0$. However, the set of $\lambda \in S^1$ satisfying (26) is much larger, being of full measure.

Remark 4.9. It is interesting to notice that for generic (in a topological sense) $\lambda \in S^1$ there is a non-linearizable holomorphic local dynamical system with multiplier λ, while for almost all (in a measure-theoretic sense) $\lambda \in S^1$ every holomorphic local dynamical system with multiplier λ is holomorphically linearizable.

Theorem 4.7 suggests the existence of a number-theoretical condition on λ ensuring that the origin is a Siegel point for any holomorphic local dynamical system of multiplier λ. And indeed this is the content of the celebrated *Brjuno-Yoccoz theorem:*

Theorem 4.10 (Brjuno, 1965 [Brj1–3], Yoccoz, 1988 [Y1–2]). *Let $\lambda \in S^1$. Then the following statements are equivalent:*

(i) *the origin is a Siegel point for the quadratic polynomial $f_\lambda(z) = \lambda z + z^2$;*
(ii) *the origin is a Siegel point for all $f \in \mathrm{End}(\mathbb{C}, 0)$ with multiplier λ;*
(iii) *the number λ satisfies Brjuno's condition*

$$\sum_{k=0}^{+\infty} \frac{1}{2^k} \log \frac{1}{\Omega_\lambda(2^{k+1})} < +\infty. \tag{27}$$

Brjuno, using majorant series as in Siegel's proof of Theorem 4.7 (see also [He] and references therein) has proved that condition (iii) implies condition (ii). Yoccoz, using a more geometric approach based on conformal and quasi-conformal geometry, has proved that (i) is equivalent to (ii), and that (ii) implies (iii), that is that if λ does not satisfy (27) then the origin is a Cremer point for some $f \in \mathrm{End}(\mathbb{C}, 0)$ with multiplier λ — and hence it is a Cremer point for the quadratic polynomial $f_\lambda(z)$. See also [P9] for related results.

Remark 4.11. Condition (27) is usually expressed in a different way. Write $\lambda = e^{2\pi i\theta}$, and let $\{p_k/q_k\}$ be the sequence of rational numbers converging to θ given by the expansion in continued fractions. Then (27) is equivalent to

$$\sum_{k=0}^{+\infty} \frac{1}{q_k} \log q_{k+1} < +\infty,$$

while (26) is equivalent to

$$q_{n+1} = O(q_n^\beta),$$

and (24) is equivalent to

$$\limsup_{k\to+\infty} \frac{1}{q_k} \log q_{k+1} = +\infty.$$

See [He, Y2, Mi, Ma, K, P1] and references therein for details.

Remark 4.12. A clear obstruction to the holomorphic linearization of an elliptic $f \in \mathrm{End}(\mathbb{C},0)$ with multiplier $\lambda = e^{2\pi i\theta} \in S^1$ is the existence of *small cycles,* that is of periodic orbits contained in any neighbourhood of the origin. Perez-Marco [P2], using Yoccoz's techniques, has shown that when the series

$$\sum_{k=0}^{+\infty} \frac{\log\log q_{k+1}}{q_k}$$

converges then every germ with multiplier λ is either linearizable or has small cycles, and that when the series diverges there exists such germs with a Cremer point but without small cycles.

The complete proof (see [P1] and the original papers) of Theorem 4.10 is beyond the scope of this survey. We shall limit ourselves to describe a proof (adapted from [Pö]) of the implication (iii)$\Longrightarrow$(ii), to report two of the easiest results of [Y2], and to illustrate what is the connection between condition (27) and the radius of convergence of the formal linearizing map.

Let us begin with Brjuno's theorem:

Theorem 4.13 (Brjuno, 1965 [Brj1–3]). *Assume that $\lambda = e^{2\pi i\theta} \in S^1$ satisfies the Brjuno's condition*

$$\sum_{k=0}^{+\infty} \frac{1}{2^k} \log \frac{1}{\Omega_\lambda(2^{k+1})} < +\infty. \tag{28}$$

Then the origin is a Siegel point for all $f \in \mathrm{End}(\mathbb{C},0)$ with multiplier λ.

Proof. We already know, thanks to Proposition 4.4, that there exists a unique formal power series

$$h(z) = z + \sum_{k\geq 2} h_k z^k$$

such that $h^{-1} \circ f \circ h(z) = \lambda z$; we shall prove that h is actually converging. To do so it suffices to show that

$$\sup_k \frac{1}{k} \log |h_k| < \infty. \tag{29}$$

Since f is holomorphic in a neighbourhood of the origin, there exists a number $M > 0$ such that $|a_k| \le M^k$ for $k \ge 2$; up to a linear change of coordinates we can assume that $M = 1$, that is $|a_l| \le 1$ for all $k \ge 2$.

Now, $h(\lambda z) = f\big(h(z)\big)$ yields

$$\sum_{k\ge 2} (\lambda^k - \lambda) h_k z^k = \sum_{l\ge 2} a_l \left(\sum_{m\ge 1} h_m z^m \right)^l . \tag{30}$$

Therefore

$$|h_k| \le \varepsilon_k^{-1} \sum_{\substack{k_1+\cdots+k_\nu = k \\ \nu\ge 2}} |h_{k_1}| \cdots |h_{k_\nu}|,$$

where

$$\varepsilon_k = |\lambda^k - \lambda|.$$

Define inductively

$$\alpha_k = \begin{cases} 1 & \text{if } k = 1\,, \\ \displaystyle\sum_{\substack{k_1+\cdots+k_\nu = k \\ \nu\ge 2}} \alpha_{k_1} \cdots \alpha_{k_\nu} & \text{if } k \ge 2, \end{cases}$$

and

$$\delta_k = \begin{cases} 1 & \text{if } k = 1\,, \\ \varepsilon_k^{-1} \displaystyle\max_{\substack{k_1+\cdots+k_\nu = k \\ \nu\ge 2}} \delta_{k_1} \cdots \delta_{k_\nu}, & \text{if } k \ge 2. \end{cases}$$

Then it is easy to check by induction that

$$|h_k| \le \alpha_k \delta_k$$

for all $k \ge 2$. Therefore, to establish (29) it suffices to prove analogous estimates for α_k and δ_k.

To estimate α_k, let $\alpha = \sum_{k\ge 1} \alpha_k t^k$. We have

$$\alpha - t = \sum_{k\ge 2} \alpha_k t^k = \sum_{k\ge 2} \left(\sum_{j\ge 1} \alpha_j t^j \right)^k = \frac{\alpha^2}{1-\alpha}.$$

This equation has a unique holomorphic solution vanishing at zero

$$\alpha = \frac{t+1}{4} \left(1 - \sqrt{1 - \frac{8t}{(1+t)^2}} \right),$$

defined for $|t|$ small enough. Hence,

$$\sup_k \frac{1}{k} \log \alpha_k < \infty,$$

as we wanted.

To estimate δ_k we have to take care of small divisors. First of all, for each $k \geq 2$ we associate to δ_k a specific decomposition of the form

$$\delta_k = \varepsilon_k^{-1} \delta_{k_1} \cdots \delta_{k_\nu}, \tag{31}$$

with $k > k_1 \geq \cdots \geq k_\nu$, $k = k_1 + \cdots + k_\nu$ and $\nu \geq 2$, and hence, by induction, a specific decomposition of the form

$$\delta_k = \varepsilon_{l_0}^{-1} \varepsilon_{l_1}^{-1} \cdots \varepsilon_{l_q}^{-1}, \tag{32}$$

where $l_0 = k$ and $k > l_1 \geq \cdots \geq l_q \geq 2$. For $m \geq 2$ let $N_m(k)$ be the number of factors ε_l^{-1} in the expression (32) of δ_k satisfying

$$\varepsilon_l < \frac{1}{4} \Omega_\lambda(m).$$

Notice that $\Omega_\lambda(m)$ is non-increasing with respect to m and it tends to zero as m goes to infinity. The next lemma contains the key estimate.

Lemma 4.14. *For all $m \geq 2$ we have*

$$N_m(k) \leq \begin{cases} 0, & \text{if } k \leq m, \\ \frac{2k}{m} - 1, & \text{if } k > m. \end{cases}$$

Proof. We argue by induction on k. If $l \leq k \leq m$ we have $\varepsilon_l \geq \Omega_\lambda(m)$, and hence $N_m(k) = 0$.

Assume now $k > m$, so that $2k/m - 1 \geq 1$. Write δ_k as in (31); we have a few cases to consider.

Case 1: $\varepsilon_k \geq \frac{1}{4}\Omega_\lambda(m)$. Then

$$N(k) = N(k_1) + \cdots + N(k_\nu),$$

and applying the induction hypothesis to each term we get $N(k) \leq (2k/m) - 1$.

Case 2: $\varepsilon_k < \frac{1}{4}\Omega_\lambda(m)$. Then

$$N(k) = 1 + N(k_1) + \cdots + N(k_\nu),$$

and there are three subcases.

Case 2.1: $k_1 \leq m$. Then

$$N(k) = 1 \leq \frac{2k}{m} - 1,$$

and we are done.

Case 2.2: $k_1 \geq k_2 > m$. Then there is ν' such that $2 \leq \nu' \leq \nu$ and $k_{\nu'} > m \geq k_{\nu'+1}$, and we again have

$$N(k) = 1 + N(k_1) + \cdots + N(k_{\nu'}) \le 1 + \frac{2k}{m} - \nu' \le \frac{2k}{m} - 1.$$

Case 2.3: $k_1 > m \ge k_2$. Then

$$N(k) = 1 + N(k_1),$$

and we have two different subsubcases.

Case 2.3.1: $k_1 \le k - m$. Then

$$N(k) \le 1 + 2\frac{k-m}{m} - 1 < \frac{2k}{m} - 1,$$

and we are done in this case too.

Case 2.3.2: $k_1 > k - m$. The crucial remark here is that $\varepsilon_{k_1}^{-1}$ gives no contribution to $N(k_1)$. Indeed, assume by contradiction that $\varepsilon_{k_1} < \frac{1}{4}\Omega_\lambda(m)$. Then

$$|\lambda^{k_1}| > |\lambda| - \frac{1}{4}\Omega_\lambda(m) \ge 1 - \frac{1}{2} = \frac{1}{2},$$

because $\Omega_\lambda(m) \le 2$. Since $k - k_1 < m$, it follows that

$$\begin{aligned}\frac{1}{2}\Omega_\lambda(m) > \varepsilon_k + \varepsilon_{k_1} &= |\lambda^k - \lambda| + |\lambda^{k_1} - \lambda| \ge |\lambda^k - \lambda^{k_1}| \\ &= |\lambda^{k-k_1} - 1| \ge \Omega_\lambda(k - k_1 + 1) \ge \Omega_\lambda(m),\end{aligned}$$

contradiction.

Therefore Case 1 applies to δ_{k_1} and we have

$$N(k) = 1 + N(k_{1_1}) + \cdots + N(k_{1_{\nu_1}}),$$

with $k > k_1 > k_{1_1} \ge \cdots \ge k_{1_{\nu_1}}$ and $k_1 = k_{1_1} + \cdots + k_{1_{\nu_1}}$. We can repeat the argument for this decomposition, and we finish unless we run into case 2.3.2 again. However, this loop cannot happen more than $m+1$ times, and we eventually have to land into a different case. This completes the induction and the proof. □

Let us go back to the proof of Theorem 4.13. We have to estimate

$$\frac{1}{k}\log \delta_k = \sum_{j=0}^{q} \frac{1}{k}\log \varepsilon_{l_j}^{-1}.$$

By Lemma 4.14,

$$\operatorname{card}\left\{0 \le j \le q \,\middle|\, \frac{1}{4}\Omega_\lambda(2^{\nu+1}) \le \varepsilon_{l_j} < \frac{1}{4}\Omega_\lambda(2^\nu)\right\} \le N_{2^\nu}(k) \le \frac{2k}{2^\nu}$$

for $\nu \ge 1$. It is also easy to see from the definition of δ_k that the number of factors $\varepsilon_{l_j}^{-1}$ is bounded by $2k-1$. In particular,

$$\operatorname{card}\left\{0 \le j \le q \;\middle|\; \frac{1}{4}\Omega_\lambda(2) \le \varepsilon_{l_j}\right\} \le 2k = \frac{2k}{2^0}.$$

Then

$$\frac{1}{k}\log \delta_k \le 2\sum_{\nu\ge 0}\frac{1}{2^\nu}\log\left(4\Omega_\lambda(2^{\nu+1})^{-1}\right) = 2\log 4 + 2\sum_{\nu \ge 0}\frac{1}{2^\nu}\log\frac{1}{\Omega_\lambda(2^{\nu+1})},$$

and we are done. □

The second result we would like to present is Yoccoz's beautiful proof of the fact that almost every quadratic polynomial f_λ is holomorphically linearizable:

Proposition 4.15. *The origin is a Siegel point of $f_\lambda(z) = \lambda z + z^2$ for almost every $\lambda \in S^1$.*

Proof. (Yoccoz [Y2]) The idea is to study the radius of convergence of the inverse of the linearization of $f_\lambda(z) = \lambda z + z^2$ when $\lambda \in \Delta^*$. Theorem 2.4 says that there is a unique map φ_λ defined in some neighbourhood of the origin such that $\varphi_\lambda'(0) = 1$ and $\varphi_\lambda \circ f = \lambda \varphi_\lambda$. Let ρ_λ be the radius of convergence of φ_λ^{-1}; we want to prove that φ_λ is defined in a neighbourhood of the unique critical point $-\lambda/2$ of f_λ, and that $\rho_\lambda = |\varphi_\lambda(-\lambda/2)|$.

Let $\Omega_\lambda \subset\subset \mathbb{C}$ be the basin of attraction of the origin, that is the set of $z \in \mathbb{C}$ whose orbit converges to the origin. Notice that setting $\varphi_\lambda(z) = \lambda^{-k}\varphi_\lambda\big(f_\lambda(z)\big)$ we can extend φ_λ to the whole of Ω_λ. Moreover, since the image of φ_λ^{-1} is contained in Ω_λ, which is bounded, necessarily $\rho_\lambda < +\infty$. Let $U_\lambda = \varphi_\lambda^{-1}(\Delta_{\rho_\lambda})$. Since we have

$$(\varphi_\lambda' \circ f) f' = \lambda \varphi_\lambda' \tag{33}$$

and φ_λ is invertible in U_λ, the function f cannot have critical points in U_λ.

If $z = \varphi_\lambda^{-1}(w) \in U_\lambda$, we have $f(z) = \varphi_\lambda^{-1}(\lambda w) \in \varphi_\lambda^{-1}(\Delta_{|\lambda|\rho_\lambda}) \subset\subset U_\lambda$; therefore

$$f(\overline{U_\lambda}) \subseteq \overline{f(U_\lambda)} \subset\subset U_\lambda \subseteq \Omega_\lambda,$$

which implies that $\partial U \subset \Omega_\lambda$. So φ_λ is defined on ∂U_λ, and clearly $|\varphi_\lambda(z)| = \rho_\lambda$ for all $z \in \partial U_\lambda$.

If f had no critical points in ∂U_λ, (33) would imply that φ_λ has no critical points in ∂U_λ. But then φ_λ would be locally invertible in ∂U_λ, and thus φ_λ^{-1} would extend across $\partial \Delta_{\rho_\lambda}$, impossible. Therefore $-\lambda/2 \in \partial U_\lambda$, and $|\varphi_\lambda(-\lambda/2)| = \rho_\lambda$, as claimed.

(Up to here it was classic; let us now start Yoccoz's argument.) Put $\eta(\lambda) = \varphi_\lambda(-\lambda/2)$. From the proof of Theorem 2.4 one easily sees that φ_λ depends holomorphically on λ; so $\eta\colon \Delta^* \to \mathbb{C}$ is holomorphic. Furthermore, since $\Omega_\lambda \subseteq \Delta_2$, Schwarz's lemma applied to $\varphi_\lambda^{-1}\colon \Delta_{\rho_\lambda} \to \Delta_2$ yields

$$1 = |(\varphi_\lambda^{-1})'(0)| \le 2/\rho_\lambda,$$

that is $\rho_\lambda \le 2$. Thus η is bounded, and thus it extends holomorphically to the origin.

So $\eta: \Delta \to \Delta_2$ is a bounded holomorphic function not identically zero; Fatou's theorem on radial limits of bounded holomorphic functions then implies that

$$\rho(\lambda_0) := \limsup_{r\to 1^-} |\eta(r\lambda_0)| > 0$$

for almost every $\lambda_0 \in S^1$. This means that we can find $0 < \rho_0 < \rho(\lambda_0)$ and a sequence $\{\lambda_j\} \subset \Delta$ such that $\lambda_j \to \lambda_0$ and $|\eta(\lambda_j)| > \rho_0$. This means that $\varphi_{\lambda_j}^{-1}$ is defined in Δ_{ρ_0} for all $j \geq 1$; up to a subsequence, we can assume that $\varphi_{\lambda_j}^{-1} \to \psi\colon \Delta_{\rho_0} \to \Delta_2$. But then we have $\psi'(0) = 1$ and

$$f_{\lambda_0}\big(\psi(z)\big) = \psi(\lambda_0 z)$$

in Δ_{ρ_0}, and thus the origin is a Siegel point for f_{λ_0}. □

The third result we would like to present is the implication (i) $\Longrightarrow$ (ii) in Theorem 4.10. The proof depends on the following result of Douady and Hubbard, obtained using the theory of quasiconformal maps:

Theorem 4.16 (Douady-Hubbard, 1985 [DH]). *Given $\lambda \in \mathbb{C}^*$, let $f_\lambda(z) = \lambda z + z^2$ be a quadratic polynomial. Then there exists a universal constant $C > 0$ such that for every holomorphic function $\psi\colon \Delta_{3|\lambda|/2} \to \mathbb{C}$ with $\psi(0) = \psi'(0) = 0$ and $|\psi(z)| \leq C|\lambda|$ for all $z \in \Delta_{3|\lambda|/2}$ the function $f = f_\lambda + \psi$ is topologically conjugated to f_λ in $\Delta_{|\lambda|}$.*

Then

Theorem 4.17 (Yoccoz, 1988 [Y2]). *Let $\lambda \in S^1$ be such that the origin is a Siegel point for $f_\lambda(z) = \lambda z + z^2$. Then the origin is a Siegel point for every $f \in \mathrm{End}(\mathbb{C}, 0)$ with multiplier λ.*

Sketch of proof. Write

$$f(z) = \lambda z + a_2 z^2 + \sum_{k\geq 3} a_k z^k,$$

and let

$$f^a(z) = \lambda z + a z^2 + \sum_{k\geq 3} a_k z^k,$$

so that $f = f^{a_2}$. If $|a|$ is large enough then the germ

$$g^a(z) = a f^a(z/a) = \lambda z + z^2 + a \sum_{k\geq 3} a_k (z/a)^k = f_\lambda(z) + \psi^a(z)$$

is defined on $\Delta_{3/2}$ and $|\psi^a(z)| < C$ for all $z \in \Delta_{3/2}$, where C is the constant given by Theorem 4.16. It follows that g^a is topologically conjugated to f_λ. By assumption, f_λ is topologically linearizable; hence g^a is too. Proposition 4.3 then implies that g^a is holomorphically linearizable, and hence f^a is too. Furthermore, it is also possible to show (see, e.g., [BH, Lemma 2.3]) that if $|a|$ is large enough, say $|a| \geq R$, then the domain of linearization of g^a contains Δ_r, where $r > 0$ is such that Δ_{2r} is contained in the domain of linearization of f_λ.

So we have proven the assertion if $|a_2| \geq R$; assume then $|a_2| < R$. Since λ is not a root of unity, there exists (Proposition 4.4) a unique formal power series $\hat{h}^a \in \mathbb{C}[[z]]$ tangent to the identity such that $g^a \circ \hat{h}^a(z) = \hat{h}^a(\lambda z)$. If we write

$$\hat{h}^a(z) = z + \sum_{k \geq 2} h_k(a) z^k$$

then $h_k(a)$ is a polynomial in a of degree $k-1$, by (30). In particular, by the maximum principle we have

$$|h_k(a_2)| \leq \max_{|a|=R} |h_k(a)| \tag{34}$$

for all $k \geq 2$. Now, by what we have seen, if $|a| = R$ then $\hat{h}^a$ is convergent in a disk of radius $r(a) > 0$, and its image contains a disk of radius r. Applying Schwarz's lemma to $(\hat{h}^a)^{-1} \colon \Delta_r \to \Delta_{r(a)}$ we get $r(a) \geq r$. But then

$$\limsup_{k \to +\infty} |h_k(a_2)|^{1/k} \leq \max_{|a|=R} \limsup_{k \to +\infty} |h_k(a)|^{1/k} = \frac{1}{r(a)} \leq \frac{1}{r} < +\infty\,;$$

hence $\hat{h}^{a_2}$ is convergent, and we are done. □

Finally, we would like to describe the connection between condition (27) and linearization. From the function theoretical side, given $\theta \in [0,1)$ set

$$r(\theta) = \inf\{r(f) \mid f \in \mathrm{End}(\mathbb{C},0) \text{has multiplier } e^{2\pi i\theta} \text{and it is defined and injective in } \Delta\}$$

where $r(f) \geq 0$ is the radius of convergence of the unique formal linearization of f tangent to the identity.

From the number theoretical side, given an irrational number $\theta \in [0,1)$ let $\{p_k/q_k\}$ be the sequence of rational numbers converging to θ given by the expansion in continued fractions, and put

$$\alpha_n = -\frac{q_n\theta - p_n}{q_{n-1}\theta - p_{n-1}}, \qquad \alpha_0 = \theta,$$
$$\beta_n = (-1)^n (q_n\theta - p_n), \qquad \beta_{-1} = 1.$$

Definition 4.18. The *Brjuno function* $B \colon [0,1) \setminus \mathbb{Q} \to (0,+\infty]$ is defined by

$$B(\theta) = \sum_{n=0}^{\infty} \beta_{n-1} \log \frac{1}{\alpha_n}.$$

Then Theorem 4.10 is consequence of what we have seen and the following

Theorem 4.19 (Yoccoz, 1988 [Y2]).
(i) $B(\theta) < +\infty$ *if and only if* $\lambda = e^{2\pi i\theta}$ *satisfies Brjuno's condition* (27);
(ii) *there exists a universal constant* $C > 0$ *such that*

$$|\log r(\theta) + B(\theta)| \leq C$$

for all $\theta \in [0,1) \setminus \mathbb{Q}$ *such that* $B(\theta) < +\infty$;
(iii) *if* $B(\theta) = +\infty$ *then there exists a non-linearizable* $f \in \mathrm{End}(\mathbb{C},0)$ *with multiplier* $e^{2\pi i\theta}$.

If 0 is a Siegel point for $f \in \mathrm{End}(\mathbb{C},0)$, the local dynamics of f is completely clear, and simple enough. On the other hand, if 0 is a Cremer point of f, then the local dynamics of f is very complicated and not yet completely understood. Pérez-Marco (in [P3, 5–7]) and Biswas ([B1, 2]) have studied the topology and the dynamics of the stable set in this case. Some of their results are summarized in the following

Theorem 4.20 (Pérez-Marco, 1995 [P6, 7]). *Assume that* 0 *is a Cremer point for an elliptic holomorphic local dynamical system* $f \in \mathrm{End}(\mathbb{C},0)$. *Then:*

(i) *The stable set* K_f *is compact, connected, full (i.e.,* $\mathbb{C} \setminus K_f$ *is connected), it is not reduced to* $\{0\}$*, and it is not locally connected at any point distinct from the origin.*
(ii) *Any point of* $K_f \setminus \{0\}$ *is recurrent (that is, a limit point of its orbit).*
(iii) *There is an orbit in* K_f *which accumulates at the origin, but no non-trivial orbit converges to the origin.*

Theorem 4.21 (Biswas, 2007 [B2]). *The rotation number and the conformal class of* K_f *are a complete set of holomorphic invariants for Cremer points. In other words, two elliptic non-linearizable holomorphic local dynamical systems* f *and* g *are holomorphically locally conjugated if and only if they have the same rotation number and there is a biholomorphism of a neighbourhood of* K_f *with a neighbourhood of* K_g.

Remark 4.22. So, if $\lambda \in S^1$ is not a root of unity and does not satisfy Brjuno's condition (27), we can find f_1, $f_2 \in \mathrm{End}(\mathbb{C},0)$ with multiplier λ such that f_1 is holomorphically linearizable while f_2 is not. Then f_1 and f_2 are formally conjugated without being neither holomorphically nor topologically locally conjugated.

Remark 4.23. Yoccoz [Y2] has proved that if $\lambda \in S^1$ is not a root of unity and does not satisfy Brjuno's condition (27) then there is an uncountable family of germs in $\mathrm{End}(\mathbb{C},O)$ with multiplier λ which are not holomorphically conjugated to each other nor holomorphically conjugated to any entire function.

See also [P2, 4] for other results on the dynamics about a Cremer point. We end this section recalling the somewhat surprising fact that in the elliptic case the multiplier is a topological invariant (in the parabolic case this follows from Proposition 3.20):

Theorem 4.24 (Naishul, 1983 [N]). *Let $f, g \in \mathrm{End}(\mathbb{C}, O)$ be two holomorphic local dynamical systems with an elliptic fixed point at the origin. If f and g are topologically locally conjugated then $f'(0) = g'(0)$.*

See [P7] for another proof of this result.

5 Several Complex Variables: The Hyperbolic Case

Now we start the discussion of local dynamics in several complex variables. In this setting the theory is much less complete than its one-variable counterpart.

Definition 5.1. Let $f \in \mathrm{End}(\mathbb{C}^n, O)$ be a holomorphic local dynamical system at $O \in \mathbb{C}^n$, with $n \geq 2$. The *homogeneous expansion* of f is

$$f(z) = P_1(z) + P_2(z) + \cdots \in \mathbb{C}_0\{z_1, \ldots, z_n\}^n,$$

where P_j is an n-uple of homogeneous polynomials of degree j. In particular, P_1 is the differential df_O of f at the origin, and f is locally invertible if and only if P_1 is invertible.

We have seen that in dimension one the multiplier (i.e., the derivative at the origin) plays a main rôle. When $n > 1$, a similar rôle is played by the eigenvalues of the differential.

Definition 5.2. Let $f \in \mathrm{End}(\mathbb{C}^n, O)$ be a holomorphic local dynamical system at $O \in \mathbb{C}^n$, with $n \geq 2$. Then:

- if all eigenvalues of df_O have modulus less than 1, we say that the fixed point O is *attracting;*
- if all eigenvalues of df_O have modulus greater than 1, we say that the fixed point O is *repelling;*
- if all eigenvalues of df_O have modulus different from 1, we say that the fixed point O is *hyperbolic* (notice that we allow the eigenvalue zero);
- if O is attracting or repelling, and df_O is invertible, we say that f is in the *Poincaré domain;*
- if O is hyperbolic, df_O is invertible, and f is not in the Poincaré domain (and thus df_O has both eigenvalues inside the unit disk and outside the unit disk) we say that f is in the *Siegel domain;*
- if all eigenvalues of df_O are roots of unity, we say that the fixed point O is *parabolic;* in particular, if $df_O = \mathrm{id}$ we say that f is *tangent to the identity;*
- if all eigenvalues of df_O have modulus 1 but none is a root of unity, we say that the fixed point O is *elliptic;*
- if $df_O = O$, we say that the fixed point O is *superattracting.*

Other cases are clearly possible, but for our aims this list is enough. In this survey we shall be mainly concerned with hyperbolic and parabolic fixed points; however, in the last section we shall also present some results valid in other cases.

Remark 5.3. There are situations where one can use more or less directly the one-dimensional theory. For example, it is possible to study the so-called *semi-direct product* of germs, namely germs $f \in \mathrm{End}(\mathbb{C}^2, O)$ of the form

$$f(z_1, z_2) = \big(f_1(z_1), f_2(z_1, z_2)\big),$$

or the so-called *unfoldings*, i.e., germs $f \in \mathrm{End}(\mathbb{C}^n, O)$ of the form

$$f(z_1, \dots, z_n) = \big(f_1(z_1, \dots, z_n), z_2, \dots, z_n\big).$$

We refer to [J2] for the study of a particular class of semi-direct products, and to [Ri1–2] for interesting results on unfoldings.

Let us begin assuming that the origin is a hyperbolic fixed point for an $f \in \mathrm{End}(\mathbb{C}^n, O)$ not necessarily invertible. We then have a canonical splitting

$$\mathbb{C}^n = E^s \oplus E^u,$$

where E^s (respectively, E^u) is the direct sum of the generalized eigenspaces associated to the eigenvalues of df_O with modulus less (respectively, greater) than 1. Then the first main result in this subject is the famous *stable manifold theorem* (originally due to Perron [Pe] and Hadamard [H]; see [FHY, HK, HPS, Pes, Sh, AM] for proofs in the C^∞ category, Wu [Wu] for a proof in the holomorphic category, and [A3] for a proof in the non-invertible case):

Theorem 5.4 (Stable manifold theorem). *Let $f \in \mathrm{End}(\mathbb{C}^n, O)$ be a holomorphic local dynamical system with a hyperbolic fixed point at the origin. Then:*

(i) *the stable set K_f is an embedded complex submanifold of (a neighbourhood of the origin in) $\mathbb{C}^n$, tangent to E^s at the origin;*

(ii) *there is an embedded complex submanifold W_f of (a neighbourhood of the origin in) $\mathbb{C}^n$, called the* unstable set *of f, tangent to E^u at the origin, such that $f|_{W_f}$ is invertible, $f^{-1}(W_f) \subseteq W_f$, and $z \in W_f$ if and only if there is a sequence $\{z_{-k}\}_{k\in\mathbb{N}}$ in the domain of f such that $z_0 = z$ and $f(z_{-k}) = z_{-k+1}$ for all $k \geq 1$. Furthermore, if f is invertible then W_f is the stable set of f^{-1}.*

The proof is too involved to be summarized here; it suffices to say that both K_f and W_f can be recovered, for instance, as fixed points of a suitable contracting operator in an infinite dimensional space (see the references quoted above for details).

Remark 5.5. If the origin is an attracting fixed point, then $E^s = \mathbb{C}^n$, and K_f is an open neighbourhood of the origin, its *basin of attraction*. However, as we shall discuss below, this does not imply that f is holomorphically linearizable, not even when it is invertible. Conversely, if the origin is a repelling fixed point, then $E^u = \mathbb{C}^n$, and $K_f = \{O\}$. Again, not all holomorphic local dynamical systems with a repelling fixed point are holomorphically linearizable.

If a point in the domain U of a holomorphic local dynamical system with a hyperbolic fixed point does not belong either to the stable set or to the unstable set, it escapes both in forward time (that is, its orbit escapes) and in backward time (that is, it is not the end point of an infinite orbit contained in U). In some sense, we can think of the stable and unstable sets (or, as they are usually called in this setting, stable and unstable *manifolds*) as skewed coordinate planes at the origin, and the orbits outside these coordinate planes follow some sort of hyperbolic path, entering and leaving any neighbourhood of the origin in finite time.

Actually, this idea of straightening stable and unstable manifolds can be brought to fruition (at least in the invertible case), and it yields one of the possible proofs (see [HK, Sh, A3] and references therein) of the *Grobman-Hartman theorem:*

Theorem 5.6 (Grobman, 1959 [G1–2]; Hartman, 1960 [Har]). *Let $f \in \mathrm{End}(\mathbb{C}^n, O)$ be a locally invertible holomorphic local dynamical system with a hyperbolic fixed point. Then f is topologically locally conjugated to its differential df_O.*

Thus, at least from a topological point of view, the local dynamics about an invertible hyperbolic fixed point is completely clear. This is definitely not the case if the local dynamical system is not invertible in a neighbourhood of the fixed point. For instance, already Hubbard and Papadopol [HP] noticed that a Böttcher-type theorem for superattracting points in several complex variables is just not true: there are holomorphic local dynamical systems with a superattracting fixed point which are not even topologically locally conjugated to the first non-vanishing term of their homogeneous expansion. Recently, Favre and Jonsson (see, e.g., [Fa, FJ1, FJ2]) have begun a very detailed study of superattracting fixed points in $\mathbb{C}^2$, study that might lead to their topological classification. We shall limit ourselves to quote one result.

Definition 5.7. Given $f \in \mathrm{End}(\mathbb{C}^2, O)$, we shall denote by $\mathrm{Crit}(f)$ the set of critical points of f. Put

$$\mathrm{Crit}^\infty(f) = \bigcup_{k \geq 0} f^{-k}(\mathrm{Crit}(f));$$

we shall say that f is *rigid* if (as germ in the origin) $\mathrm{Crit}^\infty(f)$ is either empty, a smooth curve, or the union of two smooth curves crossing transversally at the origin. Finally, we shall say that f is *dominant* if $\det(df) \not\equiv 0$.

Rigid germs have been classified by Favre [Fa], which is the reason why next theorem can be useful for classifying superattracting dynamical systems:

Theorem 5.8 (Favre-Jonsson, 2007 [FJ2]). *Let $f \in \mathrm{End}(\mathbb{C}^2, O)$ be superattracting and dominant. Then there exist:*

(a) *a 2-dimensional complex manifold M (obtained by blowing-up a finite number of points; see next section);*
(b) *a surjective holomorphic map $\pi \colon M \to \mathbb{C}^2$ such that the restriction $\pi|_{M \setminus E} \colon M \setminus E \to \mathbb{C}^2 \setminus \{O\}$ is a biholomorphism, where $E = \pi^{-1}(O)$;*
(c) *a point $p \in E$; and*
(d) *a rigid holomorphic germ $\tilde{f} \in \mathrm{End}(M, p)$*

so that $\pi \circ \tilde{f} = f \circ \pi$.

See also Ruggiero [Ru] for similar results for semi-superattracting (one eigenvalue zero, one eigenvalue different from zero) germs in $\mathbb{C}^2$.

Coming back to hyperbolic dynamical systems, the holomorphic and even the formal classification are not as simple as the topological one. The main problem is caused by resonances.

Definition 5.9. Let $f \in \mathrm{End}(\mathbb{C}^n, O)$ be a holomorphic local dynamical system, and let denote by $\lambda_1, \ldots, \lambda_n \in \mathbb{C}$ the eigenvalues of df_O. A *resonance* for f is a relation of the form

$$\lambda_1^{k_1} \cdots \lambda_n^{k_n} - \lambda_j = 0 \tag{35}$$

for some $1 \le j \le n$ and some $k_1, \ldots, k_n \in \mathbb{N}$ with $k_1 + \cdots + k_n \ge 2$. When $n = 1$ there is a resonance if and only if the multiplier is a root of unity, or zero; but if $n > 1$ resonances may occur in the hyperbolic case too.

Resonances are the obstruction to formal linearization. Indeed, a computation completely analogous to the one yielding Proposition 4.4 shows that the coefficients of a formal linearization have in the denominators quantities of the form $\lambda_1^{k_1} \cdots \lambda_n^{k_n} - \lambda_j$; hence

Proposition 5.10. *Let $f \in \mathrm{End}(\mathbb{C}^n, O)$ be a locally invertible holomorphic local dynamical system with a hyperbolic fixed point and no resonances. Then f is formally conjugated to its differential df_O.*

In presence of resonances, even the formal classification is not that easy. Let us assume, for simplicity, that df_O is in Jordan form, that is

$$P_1(z) = (\lambda_1 z, \varepsilon_2 z_1 + \lambda_2 z_2, \ldots, \varepsilon_n z_{n-1} + \lambda_n z_n),$$

with $\varepsilon_1, \ldots, \varepsilon_{n-1} \in \{0, 1\}$.

Definition 5.11. We shall say that a monomial $z_1^{k_1} \cdots z_n^{k_n}$ in the j-th coordinate of f is *resonant* if $k_1 + \cdots + k_n \ge 2$ and $\lambda_1^{k_1} \cdots \lambda_n^{k_n} = \lambda_j$.

Then Proposition 5.10 can be generalized to (see [Ar, p. 194] or [IY, p. 53] for a proof):

Proposition 5.12 (Poincaré, 1893 [Po]; Dulac, 1904 [Du]). *Let $f \in \mathrm{End}(\mathbb{C}^n, O)$ be a locally invertible holomorphic local dynamical system with a hyperbolic fixed point. Then it is formally conjugated to a $g \in \mathbb{C}_O[[z_1, \ldots, z_n]]^n$ such that dg_O is in Jordan normal form, and g has only resonant monomials.*

Definition 5.13. The formal series g is called a *Poincaré-Dulac normal form* of f.

The problem with Poincaré-Dulac normal forms is that they are not unique. In particular, one may wonder whether it could be possible to have such a normal form including *finitely many* resonant monomials only (as happened, for instance, in Proposition 3.10).

This is indeed the case (see, e.g., Reich [Re1]) when f belongs to the Poincaré domain, that is when df_O is invertible and O is either attracting or repelling. As far as I know, the problem of finding canonical formal normal forms when f belongs to the Siegel domain is still open (see [J2] for some partial results).

It should be remarked that, in the hyperbolic case, the problem of formal linearization is equivalent to the problem of smooth linearization. This has been proved by Sternberg [St1–2] and Chaperon [Ch]:

Theorem 5.14 (Sternberg, 1957 [St1–2]; Chaperon, 1986 [Ch]). *Assume we have $f, g \in \mathrm{End}(\mathbb{C}^n, O)$ two holomorphic local dynamical systems, with f locally invertible and with a hyperbolic fixed point at the origin. Then f and g are formally conjugated if and only if they are smoothly locally conjugated. In particular, f is smoothly linearizable if and only if it is formally linearizable. Thus if there are no resonances then f is smoothly linearizable.*

Even without resonances, the holomorphic linearizability is not guaranteed. The easiest positive result is due to Poincaré [Po] who, using majorant series, proved the following

Theorem 5.15 (Poincaré, 1893 [Po]). *Let $f \in \mathrm{End}(\mathbb{C}^n, O)$ be a locally invertible holomorphic local dynamical system in the Poincaré domain. Then f is holomorphically linearizable if and only if it is formally linearizable. In particular, if there are no resonances then f is holomorphically linearizable.*

Reich [Re2] describes holomorphic normal forms when df_O belongs to the Poincaré domain and there are resonances (see also [ÉV]); Pérez-Marco [P8] discusses the problem of holomorphic linearization in the presence of resonances (see also Raissy [R1]).

When df_O belongs to the Siegel domain, even without resonances, the formal linearization might diverge. To describe the known results, let us introduce the following definition:

Definition 5.16. For $\lambda_1, \dots, \lambda_n \in \mathbb{C}$ and $m \geq 2$ set

$$\Omega_{\lambda_1,\dots,\lambda_n}(m) = \min\{|\lambda_1^{k_1} \cdots \lambda_n^{k_n} - \lambda_j| \mid k_1, \dots, k_n \in \mathbb{N}, 2 \leq k_1 + \cdots + k_n \leq m, 1 \leq j \leq n\}. \tag{36}$$

If $\lambda_1, \dots, \lambda_n$ are the eigenvalues of df_O, we shall write $\Omega_f(m)$ for $\Omega_{\lambda_1,\dots,\lambda_n}(m)$.

It is clear that $\Omega_f(m) \neq 0$ for all $m \geq 2$ if and only if there are no resonances. It is also not difficult to prove that if f belongs to the Siegel domain then

$$\lim_{m \to +\infty} \Omega_f(m) = 0,$$

which is the reason why, even without resonances, the formal linearization might be diverging, exactly as in the one-dimensional case. As far as I know, the best positive and negative results in this setting are due to Brjuno [Brj2–3] (see also [Rü] and [R4]), and are a natural generalization of their one-dimensional counterparts, whose proofs are obtained adapting the proofs of Theorems 4.13 and 4.6 respectively:

Theorem 5.17 (Brjuno, 1971 [Brj2–3]). *Let $f \in \mathrm{End}(\mathbb{C}^n, O)$ be a holomorphic local dynamical system such that f belongs to the Siegel domain, has no resonances, and df_O is diagonalizable. Assume moreover that*

$$\sum_{k=0}^{+\infty} \frac{1}{2^k} \log \frac{1}{\Omega_f(2^{k+1})} < +\infty. \tag{37}$$

Then f is holomorphically linearizable.

Theorem 5.18. *Let $\lambda_1, \dots, \lambda_n \in \mathbb{C}$ be without resonances and such that*

$$\limsup_{m \to +\infty} \frac{1}{m} \log \frac{1}{\Omega_{\lambda_1,\dots,\lambda_n}(m)} = +\infty.$$

Then there exists $f \in \mathrm{End}(\mathbb{C}^n, O)$, with $df_O = \mathrm{diag}(\lambda_1, \dots, \lambda_n)$, not holomorphically linearizable.

Remark 5.19. These theorems hold even without hyperbolicity assumptions.

Remark 5.20. It should be remarked that, contrarily to the one-dimensional case, it is not yet known whether condition (37) is necessary for the holomorphic linearizability of all holomorphic local dynamical systems with a given linear part belonging to the Siegel domain. However, it is easy to check that if $\lambda \in S^1$ does not satisfy the one-dimensional Brjuno condition then any $f \in \mathrm{End}(\mathbb{C}^n, O)$ of the form

$$f(z) = \left(\lambda z_1 + z_1^2, g(z)\right)$$

is not holomorphically linearizable: indeed, if $\varphi \in \mathrm{End}(\mathbb{C}^n, O)$ is a holomorphic linearization of f, then $\psi(\zeta) = \varphi(\zeta, O)$ is a holomorphic linearization of the quadratic polynomial $\lambda z + z^2$, against Theorem 4.10.

Pöschel [Pö] shows how to modify (36) and (37) to get partial linearization results along submanifolds, and Raissy [R1] (see also [Ro1] and [R2]) explains when it is possible to pass from a partial linearization to a complete linearization even in presence of resonances. Another kind of partial linearization results, namely "linearization modulo an ideal", can be found in [Sto]. Russmann [Rü] and Raissy [R4] proved that in Theorem 5.17 one can replace the hypothesis "no resonances" by the hypothesis "formally linearizable", up to define $\Omega_f(m)$ by taking the minimum only over the non-resonant multiindeces. See also and Il'yachenko [I1] for an important result related to Theorem 5.18. Raissy, in [R3], describes a completely different way of proving the convergence of Poincaré-Dulac normal forms, based on torus actions and allowing a detailed study of torsion phenomena. Finally, in [DG] results in the spirit of Theorem 5.17 are discussed without assuming that the differential is diagonalizable.

6 Several Complex Variables: The Parabolic Case

A first natural question in the several complex variables parabolic case is whether a result like the Leau-Fatou flower theorem holds, and, if so, in which form. To present what is known on this subject in this section we shall restrict our attention to holomorphic local dynamical systems tangent to the identity; consequences on dynamical systems with a more general parabolic fixed point can be deduced taking a suitable iterate (but see also the end of this section for results valid when the differential at the fixed point is not diagonalizable).

So we are interested in the local dynamics of a holomorphic local dynamical system $f \in \mathrm{End}(\mathbb{C}^n, O)$ of the form

$$f(z) = z + P_\nu(z) + P_{\nu+1}(z) + \cdots \in \mathbb{C}_0\{z_1, \dots, z_n\}^n, \tag{38}$$

where P_ν is the first non-zero term in the homogeneous expansion of f.

Definition 6.1. If $f \in \mathrm{End}(\mathbb{C}^n, O)$ is of the form (38), the number $\nu \geq 2$ is the *order* of f.

The two main ingredients in the statement of the Leau-Fatou flower theorem were the attracting directions and the petals. Let us first describe a several variables analogue of attracting directions.

Definition 6.2. Let $f \in \mathrm{End}(\mathbb{C}^n, O)$ be tangent at the identity and of order ν. A *characteristic direction* for f is a non-zero vector $v \in \mathbb{C}^n \setminus \{O\}$ such that $P_\nu(v) = \lambda v$ for some $\lambda \in \mathbb{C}$. If $P_\nu(v) = O$ (that is, $\lambda = 0$) we shall say that v is a *degenerate* characteristic direction; otherwise, (that is, if $\lambda \neq 0$) we shall say that v is *non-degenerate*. We shall say that f is *dicritical* if all directions are characteristic; *non-dicritical* otherwise.

Remark 6.3. It is easy to check that $f \in \mathrm{End}(\mathbb{C}^n, O)$ of the form (38) is dicritical if and only if $P_\nu \equiv \lambda\,\mathrm{id}$, where $\lambda\colon \mathbb{C}^n \to \mathbb{C}$ is a homogeneous polynomial of degree $\nu - 1$. In particular, generic germs tangent to the identity are non-dicritical.

Remark 6.4. There is an equivalent definition of characteristic directions that shall be useful later on. The n-uple of ν-homogeneous polynomials P_ν induces a meromorphic self-map of $\mathbb{P}^{n-1}(\mathbb{C})$, still denoted by P_ν. Then, under the canonical projection $\mathbb{C}^n \setminus \{O\} \to \mathbb{P}^{n-1}(\mathbb{C})$ non-degenerate characteristic directions correspond exactly to fixed points of P_ν, and degenerate characteristic directions correspond exactly to indeterminacy points of P_ν. In generic cases, there is only a finite number of characteristic directions, and using Bezout's theorem it is easy to prove (see, e.g., [AT1]) that this number, counting according to a suitable multiplicity, is given by $(\nu^n - 1)/(\nu - 1)$.

Remark 6.5. The characteristic directions are *complex* directions; in particular, it is easy to check that f and f^{-1} have the same characteristic directions. Later on we shall see how to associate to (most) characteristic directions $\nu - 1$ petals, each one in some sense centered about a *real* attracting direction corresponding to the same complex characteristic direction.

The notion of characteristic directions has a dynamical origin.

Definition 6.6. We shall say that an orbit $\{f^k(z_0)\}$ converges to the origin *tangentially* to a direction $[v]\in\mathbb{P}^{n-1}(\mathbb{C})$ if $f^k(z_0)\to O$ in $\mathbb{C}^n$ and $[f^k(z_0)]\to[v]$ in $\mathbb{P}^{n-1}(\mathbb{C})$, where $[\cdot]\colon\mathbb{C}^n\setminus\{O\}\to\mathbb{P}^{n-1}(\mathbb{C})$ denotes the canonical projection.

Then

Proposition 6.7. *Let $f\in\mathrm{End}(\mathbb{C}^n,O)$ be a holomorphic dynamical system tangent to the identity. If there is an orbit of f converging to the origin tangentially to a direction $[v]\in\mathbb{P}^{n-1}(\mathbb{C})$, then v is a characteristic direction of f.*

Sketch of proof. ([Ha2]) For simplicity let us assume $\nu=2$; a similar argument works for $\nu>2$.

If v is a degenerate characteristic direction, there is nothing to prove. If not, up to a linear change of coordinates we can assume $[v]=[1:v']$ and write

$$\begin{cases} f_1(z)=z_1+p_2^1(z_1,z')+p_3^1(z_1,z')+\cdots\,,\\ f'(z)=z'+p_2'(z_1,z')+p_3'(z_1,z')+\cdots\,,\end{cases}$$

where $z'=(z_2,\dots,z_n)\in\mathbb{C}^{n-1}$, $f=(f_1,f')$, $P_j=(p_j^1,p_j')$ and so on, with $p_2^1(1,v')\neq 0$. Making the substitution

$$\begin{cases} w_1=z_1\,,\\ z'=w'z_1\,,\end{cases}\tag{39}$$

which is a change of variable off the hyperplane $z_1=0$, the map f becomes

$$\begin{cases} \tilde f_1(w)=w_1+p_2^1(1,w')w_1^2+p_3^1(1,w')w_1^3+\cdots\,,\\ \tilde f'(w)=w'+r(w')w_1+O(w_1^2)\,,\end{cases}\tag{40}$$

where $r(w')$ is a polynomial such that $r(v')=O$ if and only if $[1:v']$ is a characteristic direction of f with $p_2^1(1,v')\neq 0$.

Now, the hypothesis is that there exists an orbit $\{f^k(z_0)\}$ converging to the origin and such that $[f^k(z_0)]\to[v]$. Writing $\tilde f^k(w_0)=\big(w_1^k,(w')^k\big)$, this implies that $w_1^k\to 0$ and $(w')^k\to v'$. Then, arguing as in the proof of (8), it is not difficult to prove that

$$\lim_{k\to+\infty}\frac{1}{kw_1^k}=-p_2^1(1,v'),$$

and then that $(w')^{k+1}-(w')^k$ is of order $r(v')/k$. This implies $r(v')=O$, as claimed, because otherwise the telescopic series

$$\sum_k\big((w')^{k+1}-(w')^k\big)$$

would not be convergent. □

Remark 6.8. There are examples of germs $f \in \mathrm{End}(\mathbb{C}^2, O)$ tangent to the identity with orbits converging to the origin without being tangent to any direction: for instance

$$f(z,w) = \big(z + \alpha zw, w + \beta w^2 + o(w^2)\big)$$

with $\alpha, \beta \in \mathbb{C}^*$, $\alpha \neq \beta$ and $\mathrm{Re}(\alpha/\beta) = 1$ (see [Riv1, AT3]).

The several variables analogue of a petal is given by the notion of parabolic curve.

Definition 6.9. A *parabolic curve* for $f \in \mathrm{End}(\mathbb{C}^n, O)$ tangent to the identity is an injective holomorphic map $\varphi\colon \Delta \to \mathbb{C}^n \setminus \{O\}$ satisfying the following properties:

(a) Δ is a simply connected domain in $\mathbb{C}$ with $0 \in \partial\Delta$;
(b) φ is continuous at the origin, and $\varphi(0) = O$;
(c) $\varphi(\Delta)$ is f-invariant, and $(f|_{\varphi(\Delta)})^k \to O$ uniformly on compact subsets as $k \to +\infty$.

Furthermore, if $[\varphi(\zeta)] \to [v]$ in $\mathbb{P}^{n-1}(\mathbb{C})$ as $\zeta \to 0$ in Δ, we shall say that the parabolic curve φ is *tangent* to the direction $[v] \in \mathbb{P}^{n-1}(\mathbb{C})$.

Then the first main generalization of the Leau-Fatou flower theorem to several complex variables is due to Écalle and Hakim (see also [W]):

Theorem 6.10 (Écalle, 1985 [É4]; Hakim, 1998 [Ha2]). *Let $f \in \mathrm{End}(\mathbb{C}^n, O)$ be a holomorphic local dynamical system tangent to the identity of order $\nu \geq 2$. Then for any non-degenerate characteristic direction $[v] \in \mathbb{P}^{n-1}(\mathbb{C})$ there exist (at least) $\nu - 1$ parabolic curves for f tangent to $[v]$.*

Sketch of proof. Écalle proof is based on his theory of resurgence of divergent series; we shall describe here the ideas behind Hakim's proof, which depends on more standard arguments.

For the sake of simplicity, let us assume $n = 2$; without loss of generality we can also assume $[v] = [1:0]$. Then after a linear change of variables and a transformation of the kind (39) we are reduced to prove the existence of a parabolic curve at the origin for a map of the form

$$\begin{cases} f_1(z) = z_1 - z_1^\nu + O(z_1^{\nu+1}, z_1^\nu z_2)\,, \\ f_2(z) = z_2\big(1 - \lambda z_1^{\nu-1} + O(z_1^\nu, z_1^{\nu-1} z_2)\big) + z_1^\nu \psi(z), \end{cases}$$

where ψ is holomorphic with $\psi(O) = 0$, and $\lambda \in \mathbb{C}$. Given $\delta > 0$, set $D_{\delta,\nu} = \{\zeta \in \mathbb{C} \mid |\zeta^{\nu-1} - \delta| < \delta\}$; this open set has $\nu - 1$ connected components, all of them satisfying condition (a) on the domain of a parabolic curve. Furthermore, if u is a holomorphic function defined on one of these connected components, of the form $u(\zeta) = \zeta^2 u_o(\zeta)$ for some bounded holomorphic function u_o, and such that

$$u\big(f_1(\zeta, u(\zeta))\big) = f_2\big(\zeta, u(\zeta)\big), \tag{41}$$

then it is not difficult to verify that $\varphi(\zeta) = \big(\zeta, u(\zeta)\big)$ is a parabolic curve for f tangent to $[v]$.

So we are reduced to finding a solution of (41) in each connected component of $D_{\delta,\nu}$, with δ small enough. For any holomorphic $u = \zeta^2 u_o$ defined in such a connected component, let $f_u(\zeta) = f_1\big(\zeta, u(\zeta)\big)$, put

$$H(z) = z_2 - \frac{z_1^\lambda}{f_1(z)^\lambda} f_2(z),$$

and define the operator T by setting

$$(Tu)(\zeta) = \zeta^\lambda \sum_{k=0}^{\infty} \frac{H\big(f_u^k(\zeta), u(f_u^k(\zeta))\big)}{f_u^k(\zeta)^\lambda}.$$

Then, if $\delta > 0$ is small enough, it is possible to prove that T is well-defined, that u is a fixed point of T if and only if it satisfies (41), and that T is a contraction of a closed convex set of a suitable complex Banach space — and thus it has a fixed point. In this way if $\delta > 0$ is small enough we get a unique solution of (41) for each connected component of $D_{\delta,\nu}$, and hence $\nu - 1$ parabolic curves tangent to $[v]$. □

Definition 6.11. A set of $\nu - 1$ parabolic curves obtained in this way is a *Fatou flower* for f tangent to $[v]$.

Remark 6.12. When there is a one-dimensional f-invariant complex submanifold passing through the origin tangent to a characteristic direction $[v]$, the previous theorem is just a consequence of the usual one-dimensional theory. But it turns out that in most cases such an f-invariant complex submanifold does not exist: see [Ha2] for a concrete example, and [É4] for a general discussion.

We can also have f-invariant complex submanifolds of dimension strictly greater than one attracted by the origin.

Definition 6.13. Given a holomorphic local dynamical system $f \in \mathrm{End}(\mathbb{C}^n, O)$ tangent to the identity of order $\nu \geq 2$, and a non-degenerate characteristic direction $[v] \in \mathbb{P}^{n-1}(\mathbb{C})$, the eigenvalues $\alpha_1, \ldots, \alpha_{n-1} \in \mathbb{C}$ of the linear operator

$$\frac{1}{\nu - 1}\big(d(P_\nu)_{[v]} - \mathrm{id}\big) : T_{[v]}\mathbb{P}^{n-1}(\mathbb{C}) \to T_{[v]}\mathbb{P}^{n-1}(\mathbb{C})$$

are the *directors* of $[v]$.

Then, using a more elaborate version of her proof of Theorem 6.10, Hakim has been able to prove the following:

Theorem 6.14 (Hakim, 1997 [Ha3]). *Let $f \in \mathrm{End}(\mathbb{C}^n, O)$ be a holomorphic local dynamical system tangent to the identity of order $\nu \geq 2$. Let $[v] \in \mathbb{P}^{n-1}(\mathbb{C})$ be a non-degenerate characteristic direction, with directors $\alpha_1, \ldots, \alpha_{n-1} \in \mathbb{C}$. Furthermore, assume that $\mathrm{Re}\,\alpha_1, \ldots, \mathrm{Re}\,\alpha_d > 0$ and $\mathrm{Re}\,\alpha_{d+1}, \ldots, \mathrm{Re}\,\alpha_{n-1} \leq 0$ for a suitable $d \geq 0$. Then:*

(i) *There exists an f-invariant $(d+1)$-dimensional complex submanifold M of $\mathbb{C}^n$, with the origin in its boundary, such that the orbit of every point of M converges to the origin tangentially to $[v]$;*

(ii) $f|_M$ *is holomorphically conjugated to the translation* $\tau(w_0, w_1, \ldots, w_d) = (w_0 + 1, w_1, \ldots, w_d)$ *defined on a suitable right half-space in* $\mathbb{C}^{d+1}$.

Remark 6.15. In particular, if all the directors of $[v]$ have positive real part, there is an open domain attracted by the origin. However, the condition given by Theorem 6.14 is not necessary for the existence of such an open domain; see Rivi [Riv1] for an easy example, or Ushiki [Us], Vivas [V] and [AT3] for more elaborate examples.

In his monumental work [É4] Écalle has given a complete set of formal invariants for holomorphic local dynamical systems tangent to the identity with at least one non-degenerate characteristic direction. For instance, he has proved the following

Theorem 6.16 (Écalle, 1985 [É4]). *Let* $f \in \mathrm{End}(\mathbb{C}^n, O)$ *be a holomorphic local dynamical system tangent to the identity of order* $\nu \geq 2$. *Assume that*

(a) f *has exactly* $(\nu^n - 1)/(\nu - 1)$ *distinct non-degenerate characteristic directions and no degenerate characteristic directions;*

(b) *the directors of any non-degenerate characteristic direction are irrational and mutually independent over* $\mathbb{Z}$.

Let $[v] \in \mathbb{P}^{n-1}(\mathbb{C})$ *be a non-degenerate characteristic direction, and denote by* $\alpha_1, \ldots, \alpha_{n-1} \in \mathbb{C}$ *its directors. Then there exist a unique* $\rho \in \mathbb{C}$ *and unique (up to dilations) formal series* $R_1, \ldots, R_n \in \mathbb{C}[[z_1, \ldots, z_n]]$, *where each* R_j *contains only monomial of total degree at least* $\nu + 1$ *and of partial degree in* z_j *at most* $\nu - 2$, *such that* f *is formally conjugated to the time-1 map of the formal vector field*

$$X = \frac{1}{(\nu-1)(1+\rho z_n^{\nu-1})}\left\{[-z_n^\nu + R_n(z)]\frac{\partial}{\partial z_n} + \sum_{j=1}^{n-1}[-\alpha_j z_n^{\nu-1} z_j + R_j(z)]\frac{\partial}{\partial z_j}\right\}.$$

Other approaches to the formal classification, at least in dimension 2, are described in [BM] and in [AT2].

Using his theory of resurgence, and always assuming the existence of at least one non-degenerate characteristic direction, Écalle has also provided a set of holomorphic invariants for holomorphic local dynamical systems tangent to the identity, in terms of differential operators with formal power series as coefficients. Moreover, if the directors of all non-degenerate characteristic directions are irrational and satisfy a suitable diophantine condition, then these invariants become a complete set of invariants. See [É5] for a description of his results, and [É4] for the details.

Now, all these results beg the question: what happens when there are no non-degenerate characteristic directions? For instance, this is the case for

$$\begin{cases} f_1(z) = z_1 + bz_1z_2 + z_2^2, \\ f_2(z) = z_2 - b^2z_1z_2 - bz_2^2 + z_1^3, \end{cases}$$

for any $b \in \mathbb{C}^*$, and it is easy to build similar examples of any order. At present, the theory in this case is satisfactorily developed for $n = 2$ only. In particular, in [A2] is proved the following

Theorem 6.17 (Abate, 2001 [A2]). *Every holomorphic local dynamical system $f \in \mathrm{End}(\mathbb{C}^2, O)$ tangent to the identity, with an isolated fixed point, admits at least one Fatou flower tangent to some direction.*

Remark 6.18. Bracci and Suwa have proved a version of Theorem 6.17 for $f \in \mathrm{End}(M, p)$ where M is a *singular* variety with not too bad a singularity at p; see [BrS] for details.

Let us describe the main ideas in the proof of Theorem 6.17, because they provide some insight on the dynamical structure of holomorphic local dynamical systems tangent to the identity, and on how to deal with it. A shorter proof, deriving this theorem directly from Camacho-Sad theorem [CS] on the existence of separatrices for holomorphic vector fields in $\mathbb{C}^2$, is presented in [BCL] (see also [D2]); however, such an approach provides fewer informations on the dynamical and geometrical structures of local dynamical systems tangent to the identity.

The first idea is to exploit in a systematic way the transformation (39), following a procedure borrowed from algebraic geometry.

Definition 6.19. If p is a point in a complex manifold M, there is a canonical way (see, e.g., [GH] or [A1]) to build a complex manifold $\tilde{M}$, called the *blow-up* of M at p, provided with a holomorphic projection $\pi\colon \tilde{M} \to M$, so that $E = \pi^{-1}(p)$, the *exceptional divisor* of the blow-up, is canonically biholomorphic to $\mathbb{P}(T_pM)$, and $\pi|_{\tilde{M}\setminus E}\colon \tilde{M} \setminus E \to M \setminus \{p\}$ is a biholomorphism. In suitable local coordinates, the map π is exactly given by (39). Furthermore, if $f \in \mathrm{End}(M, p)$ is tangent to the identity, there is a unique way to lift f to a map $\tilde{f} \in \mathrm{End}(\tilde{M}, E)$ such that $\pi \circ \tilde{f} = f \circ \pi$, where $\mathrm{End}(\tilde{M}, E)$ is the set of holomorphic maps defined in a neighbourhood of E with values in $\tilde{M}$ and which are the identity on E.

In particular, the characteristic directions of f become points in the domain of the lifted map $\tilde{f}$; and we shall see that this approach allows to determine which characteristic directions are dynamically meaningful.

The blow-up procedure reduces the study of the dynamics of local holomorphic dynamical systems tangent to the identity to the study of the dynamics of germs $f \in \mathrm{End}(M, E)$, where M is a complex n-dimensional manifold, and $E \subset M$ is a compact smooth complex hypersurface pointwise fixed by f. In [A2, BrT, ABT1] we discovered a rich geometrical structure associated to this situation.

Let $f \in \mathrm{End}(M, E)$ and take $p \in E$. Then for every $h \in \mathscr{O}_{M,p}$ (where $\mathscr{O}_M$ is the structure sheaf of M) the germ $h \circ f$ is well-defined, and we have $h \circ f - h \in \mathscr{I}_{E,p}$, where $\mathscr{I}_E$ is the ideal sheaf of E.

Definition 6.20. The *f-order of vanishing* at p of $h \in \mathscr{O}_{M,p}$ is

$$\nu_f(h; p) = \max\{\mu \in \mathbb{N} \mid h \circ f - h \in \mathscr{I}^{\mu}_{E,p}\},$$

and the *order of contact* ν_f of f with E is

$$\nu_f = \min\{\nu_f(h; p) \mid h \in \mathscr{O}_{M,p}\}.$$

In [ABT1] we proved that ν_f does not depend on p, and that

$$\nu_f = \min_{j=1,\dots,n} \nu_f(z_j;p),$$

where (U,z) is any local chart centered at $p \in E$ and $z=(z_1,\dots,z_n)$. In particular, if the local chart (U,z) is such that $E \cap U = \{z_1 = 0\}$ (and we shall say that the local chart is *adapted* to E) then setting $f_j = z_j \circ f$ we can write

$$f_j(z) = z_j + (z_1)^{\nu_f} g_j(z), \tag{42}$$

where at least one among $g_1,\dots,g_n$ does not vanish identically on $U \cap E$.

Definition 6.21. A map $f \in \mathrm{End}(M,E)$ is *tangential* to E if

$$\min\bigl\{\nu_f(h;p) \mid h \in \mathscr{I}_{E,p}\bigr\} > \nu_f$$

for some (and hence any) point $p \in E$.

Choosing a local chart (U,z) adapted to E so that we can express the coordinates of f in the form (42), it turns out that f is tangential if and only if $g^1|_{U\cap E} \equiv 0$.

The g^j's in (42) depend in general on the chosen chart; however, in [ABT1] we proved that setting

$$\mathscr{X}_f = \sum_{j=1}^n g_j \frac{\partial}{\partial z_j} \otimes (dz_1)^{\otimes \nu_f} \tag{43}$$

then $\mathscr{X}_f|_{U\cap E}$ defines a *global* section X_f of the bundle $TM|_E \otimes (N_E^*)^{\otimes \nu_f}$, where N_E^* is the conormal bundle of E into M. The bundle $TM|_E \otimes (N_E^*)^{\otimes \nu_f}$ is canonically isomorphic to the bundle $\mathrm{Hom}(N_E^{\otimes \nu_f}, TM|_E)$. Therefore the section X_f induces a morphism still denoted by $X_f\colon N_E^{\otimes \nu_f} \to TM|_E$.

Definition 6.22. The morphism $X_f\colon N_E^{\otimes \nu_f} \to TM|_E$ just defined is the *canonical morphism* associated to $f \in \mathrm{End}(M,E)$.

Remark 6.23. It is easy to check that f is tangential if and only if the image of X_f is contained in TE. Furthermore, if f is the lifting of a germ $f_o \in \mathrm{End}(\mathbb{C}^n,O)$ tangent to the identity, then (see [ABT1]) f is tangential if and only if f_o is non-dicritical; so in this case tangentiality is generic. Finally, in [A2] we used the term "non degenerate" instead of "tangential".

Definition 6.24. Assume that $f \in \mathrm{End}(M,E)$ is tangential. We shall say that $p \in E$ is a *singular point for* f if X_f vanishes at p.

By definition, $p \in E$ is a singular point for f if and only if

$$g_1(p) = \cdots = g_n(p) = 0$$

for any local chart adapted to E; so singular points are generically isolated.

In the tangential case, only singular points are dynamically meaningful. Indeed, a not too difficult computation (see [A2, AT1, ABT1]) yields the following

Proposition 6.25. *Let $f \in \mathrm{End}(M,E)$ be tangential, and take $p \in E$. If p is not singular, then the stable set of the germ of f at p coincides with E.*

The notion of singular point allows us to identify the dynamically meaningful characteristic directions.

Definition 6.26. Let M be the blow-up of $\mathbb{C}^n$ at the origin, and f the lift of a non-dicritical holomorphic local dynamical system $f_o \in \mathrm{End}(\mathbb{C}^n, O)$ tangent to the identity. We shall say that $[v] \in \mathbb{P}^{n-1}(\mathbb{C}) = E$ is a *singular direction* of f_o if it is a singular point of $\tilde{f}$.

It turns out that non-degenerate characteristic directions are always singular (but the converse does not necessarily hold), and that singular directions are always characteristic (but the converse does not necessarily hold). Furthermore, the singular directions are the dynamically interesting characteristic directions, because Propositions 6.7 and 6.25 imply that if f_o has a non-trivial orbit converging to the origin tangentially to $[v] \in \mathbb{P}^{n-1}(\mathbb{C})$ then $[v]$ must be a singular direction.

The important feature of the blow-up procedure is that, even though the underlying manifold becomes more complex, the lifted maps become simpler. Indeed, using an argument similar to one (described, for instance, in [MM]) used in the study of singular holomorphic foliations of 2-dimensional complex manifolds, in [A2] it is shown that after a finite number of blow-ups our original holomorphic local dynamical system $f \in \mathrm{End}(\mathbb{C}^2, O)$ tangent to the identity can be lifted to a map $\tilde{f}$ whose singular points (are finitely many and) satisfy one of the following conditions:

(o) they are dicritical; or,
($\star$) in suitable local coordinates centered at the singular point we can write

$$\begin{cases} \tilde{f}_1(z) = z_1 + \ell(z)\big(\lambda_1 z_1 + O(\|z\|^2)\big), \\ \tilde{f}_2(z) = z_2 + \ell(z)\big(\lambda_2 z_2 + O(\|z\|^2)\big), \end{cases}$$

with

- ($\star_1$) $\lambda_1, \lambda_2 \neq 0$ and $\lambda_1/\lambda_2, \lambda_2/\lambda_1 \notin \mathbb{N}$, or
- ($\star_2$) $\lambda_1 \neq 0$, $\lambda_2 = 0$.

Remark 6.27. This "reduction of the singularities" statement holds only in dimension 2, and it is not clear how to replace it in higher dimensions.

It is not too difficult to prove that Theorem 6.10 can be applied to both dicritical and ($\star_1$) singularities; therefore as soon as this blow-up procedure produces such a singularity, we get a Fatou flower for the original dynamical system f.

So to end the proof of Theorem 6.17 it suffices to prove that any such blow-up procedure *must* produce at least one dicritical or ($\star_1$) singularity. To get such a result, we need another ingredient.

Let again $f \in \mathrm{End}(M,E)$, where E is a smooth complex hypersurface in a complex manifold M, and assume that f is tangential; let E^o denote the complement in E of the singular points of f. For simplicity of exposition we shall assume $\dim M = 2$

and $\dim E = 1$; but this part of the argument works for any $n \geq 2$ (even when E has singularities, and it can also be adapted to non-tangential germs).

Since $\dim E = 1 = \mathrm{rk} N_E$, the restriction of the canonical morphism X_f to $N_{E^o}^{\otimes \nu_f}$ is an isomorphism between $N_{E^o}^{\otimes \nu_f}$ and TE^o. Then in [ABT1] we showed that it is possible to define a holomorphic connection ∇ on N_{E^o} by setting

$$\nabla_u(s) = \pi([\mathscr{X}_f(\tilde{u}), \tilde{s}]|_S), \tag{44}$$

where: s is a local section of N_{E^o}; $u \in TE^o$; $\pi \colon TM|_{E^o} \to N_{E^o}$ is the canonical projection; $\tilde{s}$ is any local section of $TM|_{E^o}$ such that $\pi(\tilde{s}|_{S^o}) = s$; $\tilde{u}$ is any local section of $TM^{\otimes \nu_f}$ such that $X_f\big(\pi(\tilde{u}|_{E^o})\big) = u$; and $\mathscr{X}_f$ is locally given by (43). In a chart (U,z) adapted to E, a local generator of N_{E^o} is $\partial_1 = \pi(\partial/\partial z_1)$, a local generator of $N_{E^o}^{\otimes \nu_f}$ is $\partial_1^{\otimes \nu_f} = \partial_1 \otimes \cdots \otimes \partial_1$, and we have

$$X_f(\partial_1^{\otimes \nu_f}) = g_2|_{U \cap E} \frac{\partial}{\partial z_2} ;$$

therefore

$$\nabla_{\partial/\partial z_2} \partial_1 = - \left. \frac{1}{g_2} \frac{\partial g_1}{\partial z_1} \right|_{U \cap E} \partial_1 .$$

In particular, ∇ is a meromorphic connection on N_E, with poles in the singular points of f.

Definition 6.28. The *index* $\iota_p(f,E)$ of f along E at a point $p \in E$ is by definition the opposite of the residue at p of the connection ∇ divided by ν_f:

$$\iota_p(f,E) = -\frac{1}{\nu_f} \mathrm{Res}_p(\nabla).$$

In particular, $\iota_p(f,E) = 0$ if p is not a singular point of f.

Remark 6.29. If $[v]$ is a non-degenerate characteristic direction of a non-dicritical $f_o \in \mathrm{End}(\mathbb{C}^2, O)$ with non-zero director $\alpha \in \mathbb{C}^*$, then it is not difficult to check that

$$\iota_{[v]}(f,E) = \frac{1}{\alpha},$$

where f is the lift of f_o to the blow-up of the origin.

Then in [A2] we proved the following *index theorem* (see [Br1, BrT, ABT1, ABT2] for multidimensional versions and far reaching generalizations):

Theorem 6.30 (Abate, Bracci, Tovena, 2004 [A2], [ABT1]). *Let E be a compact Riemann surface inside a 2-dimensional complex manifold M. Take $f \in \mathrm{End}(M,E)$, and assume that f is tangential to E. Then*

$$\sum_{q \in E} \iota_q(f,E) = c_1(N_E),$$

where $c_1(N_E)$ is the first Chern class of the normal bundle N_E of E in M.

Now, a combinatorial argument (inspired by Camacho and Sad [CS]; see also [Ca] and [T]) shows that if we have $f \in \mathrm{End}(\mathbb{C}^2, O)$ tangent to the identity with an isolated fixed point, and such that applying the reduction of singularities to the lifted map $\tilde{f}$ starting from a singular direction $[v] \in \mathbb{P}^1(\mathbb{C}) = E$ we end up only with $(\star_2)$ singularities, then the index of $\tilde{f}$ at $[v]$ along E must be a non-negative rational number. But the first Chern class of N_E is -1; so there must be at least one singular directions whose index is not a non-negative rational number. Therefore the reduction of singularities must yield at least one dicritical or $(\star_1)$ singularity, and hence a Fatou flower for our map f, completing the proof of Theorem 6.17.

Actually, we have proved the following slightly more precise result:

Theorem 6.31 (Abate, 2001 [A2]). *Let E be a (not necessarily compact) Riemann surface inside a 2-dimensional complex manifold M, and take $f \in \mathrm{End}(M,E)$ tangential to E. Let $p \in E$ be a singular point of f such that $\iota_p(f,E) \notin \mathbb{Q}^+$. Then there exist a Fatou flower for f at p. In particular, if $f_o \in \mathrm{End}(\mathbb{C}^2, O)$ is a non-dicritical holomorphic local dynamical system tangent to the identity with an isolated fixed point at the origin, and $[v] \in \mathbb{P}^1(\mathbb{C})$ is a singular direction such that $\iota_{[v]}\big(f, \mathbb{P}^1(\mathbb{C})\big) \notin \mathbb{Q}^+$, where f is the lift of f_o to the blow-up of the origin, then f_o has a Fatou flower tangent to $[v]$.*

Remark 6.32. This latter statement has been generalized in two ways. Degli Innocenti [D1] has proved that we can allow E to be singular at p (but irreducible; in the reducible case one has to impose conditions on the indeces of f along all irreducible components of E passing through p). Molino [Mo], on the other hand, has proved that the statement still holds assuming only $\iota_p(f,E) \neq 0$, at least for f of order 2 (and E smooth at p); it is natural to conjecture that this should be true for f of any order.

As already remarked, the reduction of singularities via blow-ups seem to work only in dimension 2. This leaves open the problem of the validity of something like Theorem 6.17 in dimension $n \geq 3$; see [AT1] and [Ro2] for some partial results.

As far as I know, it is widely open, even in dimension 2, the problem of describing the stable set of a holomorphic local dynamical system tangent to the identity, as well as the more general problem of the topological classification of such dynamical systems. Some results in the case of a dicritical singularity are presented in [BM]; for non-dicritical singularities a promising approach in dimension 2 is described in [AT3].

Let $f \in \mathrm{End}(M,E)$, where E is a smooth Riemann surface in a 2-dimensional complex manifold M, and assume that f is tangential; let E^o denote the complement in E of the singular points of f. The connection ∇ on N_{E^o} described above induces a connection (still denoted by ∇) on $N_{E^o}^{\otimes \nu_f}$.

Definition 6.33. In this setting, a *geodesic* is a curve $\sigma\colon I \to E^o$ such that

$$\nabla_{\sigma'} X_f^{-1}(\sigma') \equiv O.$$

It turns out that σ is a geodesic if and only if the curve $X_f^{-1}(\sigma')$ is an integral curves of a global holomorphic vector field G on the total space of $N_{E^o}^{\otimes\nu_f}$; furthermore, G extends holomorphically to the total space of $N_E^{\otimes\nu_f}$.

Now, assume that M is the blow-up of the origin in $\mathbb{C}^2$, and E is the exceptional divisor. Then there exists a canonical ν_f-to-1 holomorphic covering map $\chi_{\nu_f}\colon \mathbb{C}^2\setminus\{O\}\to N_E^{\otimes\nu_f}\setminus E$. Moreover, if f is the lift of a non-dicritical $f_o\in\mathrm{End}(\mathbb{C}^2,O)$ of the form (38) with $P_\nu=(P_\nu^1,P_\nu^2)$, then $\nu_f=\nu-1$ and it turns out that χ_{ν_f} maps integral curves of the homogeneous vector field

$$Q_\nu = P_\nu^1\frac{\partial}{\partial z_1}+P_\nu^2\frac{\partial}{\partial z_2}$$

onto integral curves of G.

Now, the time-1 map of Q_ν is tangent to the identity and of the form (38); the previous argument shows that to study its dynamics it suffices to study the dynamics of such a geodesic vector field G. This is done in [AT3], where we get: a complete formal classification of homogeneous vector fields nearby their characteristic directions; a complete holomorphic classification nearby generic characteristic directions (including, but not limited to, non-degenerate characteristic directions); and powerful tools for the study of the geodesic flow for meromorphic connections on $\mathbb{P}^1(\mathbb{C})$, yielding deep results on the global dynamics of real integral curves of homogeneous vector fields. For instance, we get the following Poincaré-Bendixson theorem:

Theorem 6.34 (Abate, Tovena, 2009 [AT3]). *] Let Q be a homogeneous holomorphic vector field on $\mathbb{C}^2$ of degree $\nu+1\geq 2$, and let $\gamma\colon [0,\varepsilon_0)\to\mathbb{C}^2$ be a recurrent maximal integral curve of Q. Then γ is periodic or $[\gamma]\colon [0,\varepsilon_0)\to\mathbb{P}^1(\mathbb{C})$ intersects itself infinitely many times, where $[\cdot]\colon \mathbb{C}^2\to\mathbb{P}^1(\mathbb{C})$ is the canonical projection.*

Furthermore, we also get examples of maps tangent to the identity with small cycles, that is periodic orbits of arbitrarily high period accumulating at the origin; and a complete description of the local dynamics in a full neighbourhood of the origin for a large class of holomorphic local dynamical systems tangent to the identity.

Since results like Theorem 3.8 seems to suggest that generic holomorphic local dynamical systems tangent to the identity might be topologically conjugated to the time-1 map of a homogeneous vector field, this approach might eventually lead to a complete topological description of the dynamics for generic holomorphic local dynamical systems tangent to the identity in dimension 2.

We end this section with a couple of words on holomorphic local dynamical systems with a parabolic fixed point where the differential is not diagonalizable. Particular examples are studied in detail in [CD, A4, GS]. In [A1] it is described a canonical procedure for lifting an $f\in\mathrm{End}(\mathbb{C}^n,O)$ whose differential at the origin is not diagonalizable to a map defined in a suitable iterated blow-up of the origin (obtained blowing-up not only points but more general submanifolds) with a canonical fixed point where the differential is diagonalizable. Using this procedure it is for instance possible to prove the following

Theorem 6.35 ([A2]). *Let $f \in \operatorname{End}(\mathbb{C}^2, O)$ be a holomorphic local dynamical system with $df_O = J_2$, the canonical Jordan matrix associated to the eigenvalue 1, and assume that the origin is an isolated fixed point. Then f admits at least one parabolic curve tangent to $[1:0]$ at the origin.*

7 Several Complex Variables: Other Cases

Outside the hyperbolic and parabolic cases, there are not that many general results. Theorems 5.17 and 5.18 apply to the elliptic case too, but, as already remarked, it is not known whether the Brjuno condition is still necessary for holomorphic linearizability. However, another result in the spirit of Theorem 5.18 is the following:

Theorem 7.1 (Yoccoz, 1995 [Y2]). *Let $A \in GL(n, \mathbb{C})$ be an invertible matrix such that its eigenvalues have no resonances and such that its Jordan normal form contains a non-trivial block associated to an eigenvalue of modulus one. Then there exists $f \in \operatorname{End}(\mathbb{C}^n, O)$ with $df_O = A$ which is not holomorphically linearizable.*

A case that has received some attention is the so-called semi-attractive case

Definition 7.2. A holomorphic local dynamical system $f \in \operatorname{End}(\mathbb{C}^n, O)$ is said *semi-attractive* if the eigenvalues of df_O are either equal to 1 or have modulus strictly less than 1.

The dynamics of semi-attractive dynamical systems has been studied by Fatou [F4], Nishimura [Ni], Ueda [U1–2], Hakim [Ha1] and Rivi [Riv1–2]. Their results more or less say that the eigenvalue 1 yields the existence of a "parabolic manifold" M — in the sense of Theorem 6.14.(ii) — of a suitable dimension, while the eigenvalues with modulus less than one ensure, roughly speaking, that the orbits of points in the normal bundle of M close enough to M are attracted to it. For instance, Rivi proved the following

Theorem 7.3 (Rivi, 1999 [Riv1–2]). *Let $f \in \operatorname{End}(\mathbb{C}^n, O)$ be a holomorphic local dynamical system. Assume that 1 is an eigenvalue of (algebraic and geometric) multiplicity $q \geq 1$ of df_O, and that the other eigenvalues of df_O have modulus less than 1. Then:*

(i) *We can choose local coordinates $(z, w) \in \mathbb{C}^q \times \mathbb{C}^{n-q}$ such that f expressed in these coordinates becomes*

$$\begin{cases} f_1(z,w) &= A(w)z + P_{2,w}(z) + P_{3,w}(z) + \cdots, \\ f_2(z,w) &= G(w) + B(z,w)z, \end{cases}$$

where: $A(w)$ is a $q \times q$ matrix with entries holomorphic in w and $A(O) = I_q$; the $P_{j,w}$ are q-uples of homogeneous polynomials in z of degree j whose coefficients are holomorphic in w; G is a holomorphic self-map of $\mathbb{C}^{n-q}$ such that $G(O) = O$

and the eigenvalues of dG_O are the eigenvalues of df_O with modulus strictly less than 1; and $B(z,w)$ is an $(n-q)\times q$ matrix of holomorphic functions vanishing at the origin. In particular, $f_1(z,O)$ is tangent to the identity.

(ii) *If $v\in\mathbb{C}^q\subset\mathbb{C}^m$ is a non-degenerate characteristic direction for $f_1(z,O)$, and the latter map has order ν, then there exist $\nu-1$ disjoint f-invariant $(n-q+1)$-dimensional complex submanifolds M_j of $\mathbb{C}^n$, with the origin in their boundary, such that the orbit of every point of M_j converges to the origin tangentially to $\mathbb{C}v\oplus E$, where $E\subset\mathbb{C}^n$ is the subspace generated by the generalized eigenspaces associated to the eigenvalues of df_O with modulus less than one.*

Rivi also has results in the spirit of Theorem 6.14, and results when the algebraic and geometric multiplicities of the eigenvalue 1 differ; see [Riv1, 2] for details.

Building on work done by Canille Martins [CM] in dimension 2, and using Theorem 3.8 and general results on normally hyperbolic dynamical systems due to Palis and Takens [PT], Di Giuseppe has obtained the topological classification when the eigenvalue 1 has multiplicity 1 and the other eigenvalues are not resonant:

Theorem 7.4 (Di Giuseppe, 2004 [Di]). *Let $f\in\mathrm{End}(\mathbb{C}^n,O)$ be a holomorphic local dynamical system such that df_O has eigenvalues $\lambda_1,\lambda_2,\ldots,\lambda_n\in\mathbb{C}$, where λ_1 is a primitive q-root of unity, and $|\lambda_j|\neq 0,\ 1$ for $j=2,\ldots,n$. Assume moreover that $\lambda_2^{r_2}\cdots\lambda_n^{r_n}\neq 1$ for all multi-indeces $(r_2,\ldots,r_n)\in\mathbb{N}^{n-1}$ such that $r_2+\cdots+r_n\geq 2$. Then f is topologically locally conjugated either to df_O or to the map*

$$z\mapsto(\lambda_1 z_1+z_1^{kq+1},\lambda_2 z_2,\ldots,\lambda_n z_n)$$

for a suitable $k\in\mathbb{N}^*$.

We end this survey by recalling results by Bracci and Molino, and by Rong. Assume that $f\in\mathrm{End}(\mathbb{C}^2,O)$ is a holomorphic local dynamical system such that the eigenvalues of df_O are 1 and $e^{2\pi i\theta}\neq 1$. If $e^{2\pi i\theta}$ satisfies the Brjuno condition, Pöschel [Pö] proved the existence of a 1-dimensional f-invariant holomorphic disk containing the origin where f is conjugated to the irrational rotation of angle θ. On the other hand, Bracci and Molino give sufficient conditions (depending on f but not on $e^{2\pi i\theta}$, expressed in terms of two new holomorphic invariants, and satisfied by generic maps) for the existence of parabolic curves tangent to the eigenspace of the eigenvalue 1; see [BrM] for details, and [Ro3] for generalizations to $n\geq 3$.

References

[A1] Abate, M.: Diagonalization of non-diagonalizable discrete holomorphic dynamical systems. Am. J. Math. **122**, 757–781 (2000)

[A2] Abate, M.: The residual index and the dynamics of holomorphic maps tangent to the identity. Duke Math. J. **107**, 173–207 (2001a)

[A3] Abate, M.: An introduction to hyperbolic dynamical systems. I.E.P.I. Pisa (2001b)

[A4] Abate, M.: Basins of attraction in quadratic dynamical systems with a Jordan fixed point. Nonlinear Anal. **51**, 271–282 (2002)

[AT1] Abate, M., Tovena, F.: Parabolic curves in $\mathbb{C}^3$. Abstr. Appl. Anal. **2003**, 275–294 (2003)

[AT2] Abate, M., Tovena, F.: Formal classification of holomorphic maps tangent to the identity. Disc. Cont. Dyn. Sys. **Suppl.**, 1–10 (2005)

[AT3] Abate, M., Tovena, F.: Poincaré-Bendixson theorems for meromorphic connections and homogeneous vector fields, Preprint, arXiv:0903.3485, 2009

[ABT1] Abate, M., Bracci, F., Tovena: Index theorems for holomorphic self-maps. Ann. Math. **159**, 819–864 (2004)

[ABT2] Abate, M., Bracci, F., Tovena, F.: Index theorems for holomorphic maps and foliations. Indiana Univ. Math. J. **57**, 2999–3048 (2008)

[AM] Abbondandolo, A., Majer, P.: On the global stable manifold. Studia Math. **177**, 113–131 (2006)

[Ar] Arnold, V.I.: Geometrical Methods in the Theory of Ordinary Differential Equations. Springer-Verlag, Berlin (1988)

[B1] Biswas, K.: Smooth combs inside hedgehogs. Discrete Contin. Dyn. Syst. **12**, 853–880 (2005)

[B2] Biswas, K.: Complete conjugacy invariants of nonlinearizable holomorphic dynamics, Preprint, arXiv:0903.2394, 2007

[Bö] Böttcher, L.E.: The principal laws of convergence of iterates and their application to analysis. Izv. Kazan. Fiz.-Mat. Obshch. **14**, 155–234 (1904)

[Br1] Bracci, F.: The dynamics of holomorphic maps near curves of fixed points. Ann. Scuola Norm. Sup. Pisa **2**, 493–520 (2003)

[Br2] Bracci, F.: Local dynamics of holomorphic diffeomorphisms. Boll. Unione Mat. Ital. **7–B**, 609–636 (2004)

[BrM] Bracci, F., Molino, L.: The dynamics near quasi-parabolic fixed points of holomorphic diffeomorphisms in $\mathbb{C}^2$. Am. J. Math. **126**, 671–686 (2004)

[BrS] Bracci, F., Suwa, T.: Residues for singular pairs and dynamics of biholomorphic maps of singular surfaces. Internat. J. Math. **15**, 443–466 (2004)

[BrT] Bracci, F., Tovena, F.: Residual indices of holomorphic maps relative to singular curves of fixed points on surfaces. Math. Z. **242**, 481–490 (2002)

[BM] Brochero Martínez, F.E.: Groups of germs of analytic diffeomorphisms in $(\mathbb{C}^2, O)$. J. Dynamic. Contrl. Syst. **9**, 1–32 (2003)

[BCL] Brochero Martínez, F.E., Cano, F., López-Hernanz, L.: Parabolic curves for diffeomorphisms in $\mathbb{C}^2$. Publ. Mat. **52**, 189–194 (2008)

[Brj1] Brjuno, A.D.: Convergence of transformations of differential equations to normal forms. Dokl. Akad. Nauk. USSR, **165**, 987–989 (1965)

[Brj2] Brjuno, A.D.: Analytical form of differential equations, I. Trans. Moscow Math. Soc. **25**, 131–288 (1971)

[Brj3] Brjuno, A.D.: Analytical form of differential equations, II. Trans. Moscow Math. Soc. **26**, 199–239 (1972)

[BH] Buff, X., Hubbard, J.H.: Dynamics in One Complex Variable. To be published by Matrix Edition, Ithaca, NY

[C] Camacho, C.: On the local structure of conformal mappings and holomorphic vector fields. Astérisque **59–60**, 83–94 (1978)

[CS] Camacho, C., Sad, P.: Invariant varieties through singularities of holomorphic vector fields. Ann. Math. **115**, 579–595 (1982)

[CM] Canille Martins, J.C.: Holomorphic flows in $(\mathbb{C}^3, O)$ with resonances. Trans. Am. Math. Soc. **329**, 825–837 (1992)

[Ca] Cano, J.: Construction of invariant curves for singular holomorphic vector fields. Proc. Am. Math. Soc. **125**, 2649–2650 (1997)

[CG] Carleson, S., Gamelin, F.: Complex Dynamics. Springer, Berlin (1994)

[CGBM] Cerveau, D., Garba Belko, D., Meziani, R.: Techniques complexes détude dE.D.O. (avec un appendice de J. Ribón). Publ. Mat. **52**, 473–502 (2008)

[Ch] Chaperon, M.: Géométrie différentielle et singularités des systèmes dynamiques. Astérisque **138–139** (1986)

[CD] Coman, D., Dabija, M.: On the dynamics of some diffeomorphisms of $\mathbb{C}^2$ near parabolic fixed points. Houston J. Math. **24**, 85–96 (1998)

[Cr1] Cremer, H.: Zum Zentrumproblem. Math. Ann. **98**, 151–163 (1927)

[Cr2] Cremer, H.: Über die Häufigkeit der Nichtzentren. Math. Ann. **115**, 573–580 (1938)

[D1] Degli Innocenti, F.: Holomorphic dynamics near germs of singular curves. Math. Z. **251**, 943–958 (2005)

[D2] Degli Innocenti, F.: On the relations between discrete and continuous dynamics in $\mathbb{C}^2$. Ph.D. Thesis, Università di Pisa (2007)

[DG] DeLatte, D., Gramchev, T.: Biholomorphic maps with linear parts having Jordan blocks: linearization and resonance type phenomena. Math. Phys. El. J. **8**, 1–27 (2002)

[Di] Di Giuseppe, P.: Topological classification of holomorphic, semi-hyperbolic germs in quasi-absence of resonances. Math. Ann. **334**, 609–625 (2006)

[DH] Douady, A., Hubbard, J.H.: On the dynamics of polynomial-like mappings. Ann. Sc. Éc. Norm. Sup. **18**, 287–343 (1985)

[Du] Dulac, H.: Recherches sur les points singuliers des équationes différentielles. J. Éc. Polytech. **IX**, 1–125 (1904)

[É1] Écalle, J.: Théorie itérative: introduction à la théorie des invariants holomorphes. J. Math. Pures Appl. **54**, 183–258 (1975)

[É2] Écalle, J.: Les fonctions résurgentes. Tome I: Les algèbres de fonctions résurgentes, Prépublications Math. Orsay **81-05,** Université de Paris-Sud, Orsay, 1981. Download site: `http://www.math.u-psud.fr/~biblio/numerisation/docs/E_ECALLE_81_05/pdf/E_ECALLE_81_05.pdf`

[É3] Écalle, J.: Les fonctions résurgentes. Tome II: Les fonctions résurgentes appliquées à l'itération, Prépublications Math. Orsay **81-06,** Université de Paris-Sud, Orsay, 1981. Download site: `http://www.math.u-psud.fr/~biblio/numerisation/docs/E_ECALLE_81_06/pdf/E_ECALLE_81_06.pdf`

[É4] Écalle, J.: Les fonctions résurgentes. Tome III: L'équation du pont et la classification analytique des objects locaux, Prépublications Math. Orsay **85-05,** Université de Paris-Sud, Orsay, 1985. Download site: `http://www.math.u-psud.fr/~biblio/numerisation/docs/E_ECALLE_85_05/pdf/E_ECALLE_85_05.pdf`

[É5] Écalle, J.: Iteration and analytic classification of local diffeomorphisms of $\mathbb{C}^\nu$, in Iteration theory and its functional equations. Lecture Notes in Mathematics, vol. 1163, pp. 41–48. Springer-Verlag, Berlin (1985)

[ÉV] Écalle, J., Vallet, B.: Correction and linearization of resonant vector fields and diffeomorphisms. Math. Z. **229**, 249–318 (1998)

[FHY] Fathi, A., Herman, M., Yoccoz, J.-C.: A proof of Pesin's stable manifold theorem, in Geometric Dynamics. Lecture Notes in Mathematics, vol. 1007, pp. 177–216. Springer Verlag, Berlin (1983)

[F1] Fatou, P.: Sur les équations fonctionnelles, I. Bull. Soc. Math. France, **47**, 161–271 (1919)

[F2] Fatou, P.: Sur les équations fonctionnelles, II. Bull. Soc. Math. France, **48**, 33–94 (1920a)

[F3] Fatou, P.: Sur les équations fonctionnelles, III. Bull. Soc. Math. France, **48**, 208–314 (1920b)

[F4] Fatou, P.: Substitutions analytiques et équations fonctionnelles à deux variables. Ann. Sci. École Norm. Sup. **40**, 67–142 (1924)

[Fa] Favre, C.: Classification of 2-dimensional contracting rigid germs and Kato surfaces. I. J. Math. Pures Appl. **79**, 475–514 (2000)

[FJ1] Favre, C., Jonsson, M.: The valuative tree. Lecture Notes in Mathematics, vol. 1853. Springer Verlag, Berlin (2004)

[FJ2] Favre, C., Jonsson, M.: Eigenvaluations. Ann. Sci. École Norm. Sup. **40**, 309–349 (2007)

[GS] Gelfreich, V., Sauzin, D.: Borel summation and splitting of separatrices for the Hénon map. Ann. Inst. Fourier Grenoble, **51**, 513–567 (2001)

[GH] Griffiths, P., Harris, J.: Principles of algebraic geometry. Wyley, New York (1978)

[G1] Grobman, D.M.: Homeomorphism of systems of differential equations. Dokl. Akad. Nauk. USSR, **128**, 880–881 (1959)

[G2] Grobman, D.M.: Topological classification of neighbourhoods of a singularity in n-space. Math. Sbornik, **56**, 77–94 (1962)

[H] Hadamard, J.S.: Sur l'itération et les solutions asymptotyques des équations différentielles. Bull. Soc. Math. France, **29**, 224–228 (1901)

[Ha1] Hakim, M.: Attracting domains for semi-attractive transformations of $\mathbb{C}^p$. Publ. Matem. **38**, 479–499 (1994)

[Ha2] Hakim, M.: Analytic transformations of $(\mathbb{C}^p, 0)$ tangent to the identity. Duke Math. J. **92**, 403–428 (1998)

[Ha3] Hakim, M.: Transformations tangent to the identity. Stable pieces of manifolds. Prépublications Math. Orsay **97-30,** Université de Paris-Sud, Orsay (1997)

[Har] Hartman, P.: A lemma in the theory of structural stability of differential equations. Proc. Am. Math. Soc. **11**, 610–620 (1960)

[HK] Hasselblatt, B., Katok, A.: Introduction to the modern theory of dynamical systems. Cambridge University Press, Cambridge (1995)

[He] Herman, M.: Recent results and some open questions on Siegel's linearization theorem of germs of complex analytic diffeomorphisms of $\mathbb{C}^n$ near a fixed point, in **Proc.** 8th **Int. Cong. Math. Phys.** World Scientific, Singapore, pp. 138–198 (1986)

[HPS] Hirsch, M., Pugh, C.C., Shub, M.: Invariant manifolds. Lecture Notes Mathematics, vol. 583. Springer-Verlag, Berlin (1977)

[HP] Hubbard, J.H., Papadopol, P.: Superattractive fixed points in $\mathbb{C}^n$. Indiana Univ. Math. J. **43**, 321–365 (1994)

[I1] Il'yashenko, Yu.S.: Divergence of series reducing an analytic differential equation to linear normal form at a singular point. Funct. Anal. Appl. **13**, 227–229 (1979)

[I2] Il'yashenko, Yu.S.: Nonlinear Stokes phenomena, in Nonlinear Stokes phenomena, Adv. Soviet Math. **14,** Am. Math. Soc. Providence, pp. 1–55 (1993)

[IY] Il'yashenko, Y.S., Yakovenko, S.: Lectures on analytic differential equations, Graduate Studies in Mathematics 86, American Mathematical Society, Providence, RI (2008)

[J1] Jenkins, A.: Holomorphic germs and the problem of smooth conjugacy in a punctured neighborhood of the origin. Trans. Am. Math. Soc. **360**, 331–346 (2008)

[J2] Jenkins, A.: Further reductions of Poincaré-Dulac normal forms in $\mathbb{C}^{n+1}$. Proc. Am. Math. Soc. **136**, 1671–1680 (2008)

[K] Ya. Khinchin, A.: Continued fractions. Dover Publications, Inc., Mineola, NY (1997)

[Ki] Kimura, T.: On the iteration of analytic functions. Funk. Eqvacioj, **14**, 197–238 (1971)

[Kœ] Kœnigs, G.: Recherches sur les integrals de certain équations fonctionelles. Ann. Sci. Éc. Norm. Sup. **1**, 1–41 (1884)

[L] Leau, L.: Étude sur les équations fonctionelles à une ou plusieurs variables. Ann. Fac. Sci. Toulouse, **11**, E1–E110 (1897)

[M1] Malgrange, B.: Travaux d'Écalle et de Martinet-Ramis sur les systèmes dynamiques. Astérisque, **92–93**, 59–73 (1981/82)

[M2] Malgrange, B.: Introduction aux travaux de J. Écalle. Ens. Math. **31**, 261–282 (1985)

[Ma] Marmi, S.: An introduction to small divisors problems, I.E.P.I., Pisa (2000)

[MR] Martinet, J., Ramis, J.-P.: Classification analytique des équations différentielles non linéaires résonnantes du premier ordre. Ann. Sci. École Norm. Sup. **4**, 571–621 (1984)

[MM] Mattei, J.F., Moussu, R.: Holonomie et intégrales premières. Ann. Sci. Éc. Norm. Sup. **13**, 469–523 (1980)

[Mi] Milnor, J.: Dynamics in one complex variable. Annals of Mathematics Studies, vol. 160. Princeton University Press, Princeton, NJ (2006)

[Mo] Molino, L.: The dynamics of maps tangent to the identity and with nonvanishing index. Trans. Am. Math. Soc. **361**, 1597–1623 (2009)

[N] Naishul, V.I.: Topological invariants of analytic and area preserving mappings and their application to analytic differential equations in $\mathbb{C}^1$ and $\mathbb{CP}^2$. Trans. Moscow Math. Soc. **42**, 239–250 (1983)

[Na] Nakai, I.: The classication of curvilinear angles in the complex plane and the groups of $\pm$ holomorphic diffeomorphisms. Ann. Fac. Sci. Toulouse Math. **6,** 313–334 (1998)

[Ni] Nishimura, Y.: Automorphismes analytiques admettant des sousvariétés de point fixes attractives dans la direction transversale. J. Math. Kyoto Univ. **23**, 289–299 (1983)

[PT] Palis, J., Takens, F.: Topological equivalence of normally hyperbolic dynamical systems. Topology **16**, 335–345 (1977)

[P1] Pérez-Marco, R.: Solution complète au problème de Siegel de linéarisation d'une application holomorphe au voisinage d'un point fixe (d'après J.-C. Yoccoz). Astérisque **206**, 273–310 (1992)

[P2] Pérez-Marco, R.: Sur les dynamiques holomorphes non linéarisables et une conjecture de V.I. Arnold. Ann. Sci. École Norm. Sup. **26**, 565–644 (1993)

[P3] Pérez-Marco, R.: Topology of Julia sets and hedgehogs. Prépublications Math. Orsay, **94-48,** Université de Paris-Sud, Orsay (1994)

[P4] Pérez-Marco, R.: Non-linearizable holomorphic dynamics having an uncountable number of symmetries. Invent. Math. **199**, 67–127 (1995a)

[P5] Pérez-Marco, R.: Hedgehogs Dynamics, Preprint, UCLA (1995b)

[P6] Pérez-Marco, R.: Sur une question de Dulac et Fatou. C.R. Acad. Sci. Paris **321**, 1045–1048 (1995c)

[P7] Pérez-Marco, R.: Fixed points and circle maps. Acta Math. **179**, 243–294 (1997)

[P8] Pérez-Marco, R.: Linearization of holomorphic germs with resonant linear part. Preprint, arXiv: math.DS/0009030 (2000)

[P9] Pérez-Marco, R.: Total convergence or general divergence in small divisors. Comm. Math. Phys. **223**, 451–464 (2001)

[Pe] Perron, O.: Über Stabilität und asymptotisches Verhalten der Integrale von Differentialgleichungssystemen. Math. Z. **29**, 129–160 (1928)

[Pes] Pesin, Ja.B.: Families of invariant manifolds corresponding to non-zero characteristic exponents. Math. USSR Izv. **10**, 1261–1305 (1976)

[Po] Poincaré, H.: Œuvres, Tome I. Gauthier-Villars, Paris, pp. XXXVI–CXXIX (1928)

[Pö] Pöschel, J.: On invariant manifolds of complex analytic mappings near fixed points. Exp. Math. **4**, 97–109 (1986)

[R1] Raissy, J.: Linearization of holomorphic germs with quasi-Brjuno xed points. Math. Z., Online First, (2009a). Download site: `http://www.springerlink.com/content/3853667627008057/fulltext.pdf`

[R2] Raissy, J.: Simultaneous linearization of holomorphic germs in presence of resonances. Conform. Geom. Dyn. **13**, 217–224 (2009b)

[R3] Raissy, J.: Torus actions in the normalization problem. J. Geometric Anal., Online First, (2009c). Download site: `http://springerlink.com/content/r20323122p74h06j/fulltext.pdf`

[R4] Raissy, J.: Brjuno conditions for linearization in presence of resonances. Preprint, arXiv:0911.4341 (2009d)

[Re1] Reich, L.: Das Typenproblem bei formal-biholomorphien Abbildungen mit anziehendem Fixpunkt. Math. Ann. **179**, 227–250 (1969)

[Re2] Reich, L.: Normalformen biholomorpher Abbildungen mit anziehendem Fixpunkt. Math. Ann. **180**, 233–255 (1969)

[Ri1] Ribón, J.: Formal classication of unfoldings of parabolic diffeomorphisms. Ergod. Theor. Dyn. Syst. **28**, 1323–1365 (2008)

[Ri2] Ribón, J.: Modulus of analytic classication for unfoldings of resonant diffeomorphisms. Mosc. Math. J. **8**, 319–395, 400 (2008)

[Riv1] Rivi, M.: Local behaviour of discrete dynamical systems. Ph.D. Thesis, Università di Firenze (1999)

[Riv2] Rivi, M.: Parabolic manifolds for semi-attractive holomorphic germs. Mich. Math. J. **49**, 211–241 (2001)

[Ro1] Rong, F.: Linearization of holomorphic germs with quasi-parabolic fixed points. Ergod Theor Dyn. Syst. **28**, 979–986 (2008)

[Ro2] Rong, F.: Robust parabolic curves in $\mathbb{C}^m$ ($m \geqslant 3$). Houston J. Math. **36**, 147–155 (2010)

[Ro3] Rong, F.: Quasi-parabolic analytic transformations of $\mathbb{C}^n$. J. Math. Anal. Appl. **343**, 99–109 (2008)

[Ru] Ruggiero, M.: Rigidification of holomorphic germs with non-invertible differential. Preprint, arXiv:0911.4023 (2009)

[Rü] Rüssmann, H.: Stability of elliptic fixed points of analytic area-preserving mappings under the Brjuno condition. Ergod. Theor. Dyn. Syst. **22**, 1551–1573 (2002)

[S] Shcherbakov, A.A.: Topological classification of germs of conformal mappings with identity linear part. Moscow Univ. Math. Bull. **37**, 60–65 (1982)

[Sh] Shub, M.: Global Stability of Dynamical Systems. Springer, Berlin (1987)

[Si] Siegel, C.L.: Iteration of analytic functions. Ann. Math. **43**, 607–612 (1942)

[St1] Sternberg, S.: Local contractions and a theorem of Poincaré. Am. J. Math. **79**, 809–824 (1957)

[St2] Sternberg, S.: The structure of local homomorphisms, II. Am. J. Math. **80**, 623–631 (1958)

[Sto] Stolovitch, L.: Family of intersecting totally real manifolds of $(\mathbb{C}^n, 0)$ and CR-singularities. Preprint, arXiv: math/0506052v2 (2005)

[T] Toma, M.: A short proof of a theorem of Camacho and Sad. Enseign. Math. **45**, 311–316 (1999)

[Tr] Trépreau, J.-M.: Discrimination analytique des difféomorphismes résonnants de $(\mathbb{C}, 0)$ et r´eexion de Schwarz. Ast´erisque **284**, 271–319 (2003)

[U1] Ueda, T.: Local structure of analytic transformations of two complex variables, I. J. Math. Kyoto Univ. **26**, 233–261 (1986)

[U2] Ueda, T.: Local structure of analytic transformations of two complex variables, II. J. Math. Kyoto Univ. **31**, 695–711 (1991)

[U3] Ueda, T.: Simultaneous linearization of holomorphic maps with hyperbolic and parabolic fixed points. Publ. Res. Inst. Math. Sci. **44**, 91–105 (2008)

[Us] Ushiki, S.: Parabolic fixed points of two-dimensional complex dynamical systems. Sūrikaisekikenkyūsho Kōkyūroku **959**, 168–180 (1996)

[V] Vivas, L.R.: Fatou-Bieberbach domains as basins of attraction of automorphisms tangent to the identity. Preprint, arXiv: 0907.2061v1 (2009)

[Vo] Voronin, S.M.: Analytic classification of germs of conformal maps $(\mathbb{C}, 0) \to (\mathbb{C}, 0)$ with identity linear part. Func. Anal. Appl. **15**, 1–17 (1981)

[W] Weickert, B.J.: Attracting basins for automorphisms of $\mathbb{C}^2$. Invent. Math. **132**, 581–605 (1998)

[Wu] Wu, H.: Complex stable manifolds of holomorphic diffeomorphisms. Indiana Univ. Math. J. **42**, 1349–1358 (1993)

[Y1] Yoccoz, J.-C.: Linéarisation des germes de difféomorphismes holomorphes de $(\mathbb{C}, 0)$. C.R. Acad. Sci. Paris **306**, 55–58 (1988)

[Y2] Yoccoz, J.-C.: Théorème de Siegel, nombres de Brjuno et polynômes quadratiques. Astérisque **231**, 3–88 (1995)

Dynamics of Rational Surface Automorphisms

Eric Bedford

Abstract This is a 2-part introduction to the dynamics of rational surface automorphisms. Such maps can be written in coordinates as rational functions or polynomials. The first part concerns polynomial automorphisms of complex 2-space and includes the complex Henon family.

The second part concerns compact (complex) rational surfaces. The basic properties of automorphisms of positive entropy are given, as well as the construction of invariant currents and measures. This is illustrated by a number of examples.

1 Polynomial Automorphisms of $\mathbf{C}^2$

1.1 Hénon Maps

A surface is said to be rational if it is birationally equivalent to the plane. The purpose of these notes is to give an entry into the dynamics of the automorphisms of rational surfaces. The first part is devoted to the complex Hénon family of maps, which has been the most heavily studied family of invertible holomorphic maps. Up to this point, the investigations of the Hénon maps have been guided by the study of polynomial maps of one variable. The dynamics of polynomials in the one-dimensional case has developed into a very rich topic, and these maps are understood in considerable detail. Although the Hénon family is only partially understood, its methods and results should provide motivation and guidance for the understanding of other automorphisms in dimension 2. We have selected for discussion only a part of what is known on the subject, and the reader is recommended to consult the expository treatments in [MNTU] and [S], as well as the original works [HOV1,2, FS1] and the series of papers [BS1,2,...], [BLS].

E. Bedford
Department of Mathematics, Indiana University, 47405 Bloomington, IN, USA
e-mail: bedford@indiana.edu

G. Gentili et al. (eds.), *Holomorphic Dynamical Systems*,
Lecture Notes in Mathematics 1998, DOI 10.1007/978-3-642-13171-4_2,

Only certain compact complex manifolds carry automorphisms of positive entropy (see [C1]), and the majority of these are rational surfaces. The second part of these notes considers the geometry of compact rational surfaces, and then presents some examples of nontrivial automorphisms. In contrast with the case of the polynomial automorphisms of $\mathbf{C}^2$, not much is known about the set of all rational surface automorphisms — neither a description of all the automorphisms are nor a dynamical classification of them. As will be seen from the second part, the rational surface automorphisms have a close connection with certain birational maps of the plane. The reader is referred to [CD] for further discussion of the Cremona group of all birational transformations of the plane.

We say that a map $f = (f_1, f_2)$ of $\mathbf{C}^2$ to itself is *polynomial* if both coordinates are given by polynomials. We let $PolyAut(\mathbf{C}^2)$ denote the set of polynomial automorphisms of $\mathbf{C}^2$. That is, these are invertible maps of $\mathbf{C}^2$; it is a Theorem that if a polynomial map is one-to-one and onto, then its inverse must be polynomial. Our goal here is to study dynamical properties of such automorphisms, by which we mean the behavior of the iterates $f^n := f \circ \cdots \circ f$ as $n \to \infty$. If $f, g \in PolyAut(\mathbf{C}^2)$, then we may consider the conjugate $g^{-1}fg$ of f. However, we note that $(g^{-1}fg)^n = g^{-1}f^n g$, and so the dynamics of f and its conjugate are essentially the same. Thus in the study of $PolyAut(\mathbf{C}^2)$, we need to understand representatives of the various conjugacy classes. Friedland and Milnor [FM] have done this and have shown that, in order to understand the whole family $PolyAut(\mathbf{C}^2)$, it suffices to consider the family of *generalized Hénon mappings*, which are the class

$$\mathscr{H} = \{f(x,y) = (y, p(y) - \delta x), p(y) = y^d + a_{d-2}y^{d-2} + \cdots + a_0,\ d \geq 2, \delta \in \mathbf{C}, \delta \neq 0\}$$

In fact, much of what we do will remain valid if we replace f^n by certain more general compositions $f_n \circ \cdots \circ f_1$, where $f_j \in \mathscr{H}$. While these mappings may appear to be somewhat special, in fact they are the key to understanding invertible polynomial dynamics in $\mathbf{C}^2$.

Exercises:

1. Each map $f \in \mathscr{H}$ is invertible, and f^{-1} is polynomial.
2. The Jacobian of f is δ, and in particular it is constant.
3. $\deg(f^n) = (d^{n-1}, d^n)$.
4. Consider the dilation by t: $D(x,y) = (tx, ty)$, and the translation by (s,s): $T(x,y) = (x+s, y+s)$. Show that if $f \in \mathscr{H}$, then the conjugate $T \circ D \circ f \circ D^{-1} \circ T^{-1}$ has the form of an element of $\mathscr{H}$, except that the coefficient of y^d is an arbitrary nonzero number, and the coefficient of y^{d-1} is arbitrary.
5. Define the involution $\tau(x,y) = (y,x) = \tau^{-1}$. Then $f^{-1} \notin \mathscr{H}$, but $\tau \circ f^{-1} \circ \tau$ has the form of an element of $\mathscr{H}$, except that p is not monic.

We will use the notation $(x_n, y_n) = f^n(x,y)$ for $n \in \mathbf{Z}$. Thus we have $x_n = y_{n-1}$, and the whole orbit $\{(x_n, y_n) : n \in \mathbf{Z}\}$ is essentially given by the bi-infinite scalar sequence

$$\ldots, y_{-2}, y_{-1}, y_0, y_1, y_2, \ldots.$$

One advantage of this reduction is that the action of f may be replaced by the shift. That is, if the f-orbit of the point $(x',y') = f(x,y)$ corresponds to the sequence

$$\dots, y'_{-2}, y'_{-1}, y'_0, y'_1, y'_2, \dots,$$

then $y'_n = y_{n+1}$.

1.2 Filtration

We will find it useful to look at Hénon maps in terms of their escape to infinity. Let us define

$$V = \{|x|,|y| \le R\},\ \ V^+ = \{|y| \ge \max\{|x|,R\}\},\ \ V^- = \{|x| \ge \max\{|y|,R\}\}.$$

Thus V is a bi-disk, and the sets $V^{\pm}$ are (topologically) the product of a disk and an annulus. Let us choose R to be sufficiently large that

$$|p(t)| - |\delta t| \ge \max\left\{\frac{1}{2}|t^d|, 2|t|\right\},\ \text{for all } |t| \ge R. \tag{1}$$

It follows that:

$$(x,y) \in V^+ \ \Rightarrow\ |y_1| \ge 2|y_0| = 2|x_1| \ \Rightarrow\ (x_1,y_1) \in V^+$$

and thus

$$(x,y) \in V^+ \ \Rightarrow\ |y_n| \ge 2^n|y| \ge 2^n R \tag{2}$$

for $n = 1,2,3\dots$ In fact, it is possible to show that for any $\varepsilon > 0$ we may choose R sufficiently large that for $(x_0,y_0) \in V^+$ we have

$$(1-\varepsilon)|y_0|^{d^n} \le |y_n| \le (1+\varepsilon)|y_0|^{d^n}. \tag{3}$$

Theorem 1.1. *If $(x,y) \in V^-$, then there is a number N such that $f^N(x,y) \in V \cup V^+$.*

Proof. Suppose not. Then we have $(x_n,y_n) \in V^-$ for all $n \ge 0$. Thus $|y_n| \le |x_n|$. Since $x_n = y_{n-1}$, we have $|x_1| \ge |x_2| \ge \cdots \ge R$, and $|y_1| \ge |y_2| \ge \cdots$, and the both approach the same limit M. Thus for N sufficiently large, we will have $|x_N| \sim |y_N| \sim M \ge R$. Thus by (*), we will have $|y_{N+1}| \ge 2M$, which is a contradiction. □

We summarize this discussion in the following Theorem:

Theorem 1.2. *1. $fV^+ \subset V^+$, and for each $(x,y) \in V^+$, the forward orbit escapes to infinity.*
2. $f(V \cup V^+) \subset V \cup V^+$.

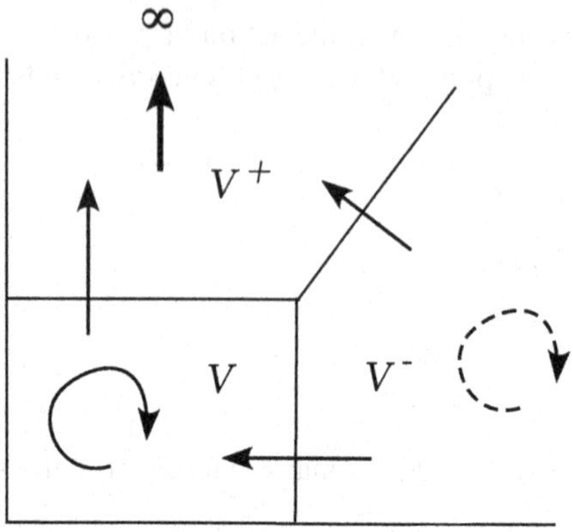

Fig. 1 Filtration behavior of Hénon maps

3. *If $(x,y) \in V^-$, then the orbit $f^n(x,y)$, $n \geq 0$, can stay in V^- for only finite time. After $f^n(x,y)$ leaves V^-, it cannot reenter, which means that it stays in $V \cup V^+$. In particular, if an orbit is bounded in forward time, then it must enter V and remain there.*

This Theorem is illustrated in Figure 1. The arrows give the possibilities where a point might map; the dashed circular arrow indicates that an orbit can remain in V^- for only finite time.

$$K^{\pm} = \{(x,y) \in C^2 : \{f^{\pm n}(x,y), n \geq 0\} \text{ is bounded}\}$$
$$= \{(x,y) \in C^2 : \text{with bounded forward/backward orbit}\}$$

$$K := K^+ \cap K^-, \quad J^{\pm} = \partial K^{\pm}, \quad J = J^+ \cap J^-, \quad U^+ = \mathbf{C}^2 - K^+$$

Proposition 1.3. *The iterates $\{f^n, n \geq 0\}$ are a normal family on the interior of K^+, and for $(x_0, y_0) \in J^+$, there is no neighborhood $U \ni (x_0, y_0)$ on which this family is normal.*

Proof. If $p \in int(K^+)$, then the forward orbit cannot enter V^+. It can remain in V^- for only finite time, so a neighborhood of p must ultimately be in V. Thus the forward iterates of a neighborhood of p are ultimately bounded by R in a neighborhood of p, so the iterates are a normal family in a neighborhood of p. If U is an open set that intersects J^+, then $U \cap K^+ \neq \emptyset$ and $U \cap (\mathbf{C}^2 - K^+) \neq \emptyset$. Thus U contains points where the forward iterates are bounded and points whose orbits tend to ∞. Thus f^n cannot be normal on U. □

Exercise: Show that if $p \in int(K^+)$ is fixed, then the eigenvalues of Df_p have modulus ≤ 1.

Proposition 1.4. *We have $K \subset V$ and $K^+ \subset V \cup V^-$. Further, $U^+ = \bigcup_{n \geq 0} f^{-n} V^+$, where the union $V^+ \subset f^{-1} V^+ \subset \cdots$ is increasing.*

Proof. This is a consequence of the previous Theorem. □

A point p is *periodic* if $f^N p = p$ for some N. The minimal such $N > 0$ is called the *period* of p.

Proposition 1.5. *For each N there are only finitely many periodic points of period N.*

Proof. The set $P_N := \{(x,y) \in \mathbf{C}^2 : f^N(x,y) = (x,y)\}$ is a subvariety. Further, we have $P_N \subset K$, so it is bounded. But any bounded subvariety of $\mathbf{C}^2$ must be a finite set. □

A periodic point p of period N is a *saddle* if Df_p^N has one eigenvalue with modulus <1 and one eigenvalue with modulus >1. We use the notation *SPer* for the set of saddle (periodic) points and $J^* = \overline{SPer}$ for its closure. By an exercise above, we have:

Proposition 1.6. *If p is a saddle (periodic) point, then $p \in J$. Thus $J^* \subset J$.*

Problem: It is an interesting and basic question to determine whether $J^* = J$ holds for all Hénon maps. However, this is not yet known.

Exercise: Suppose that p is a fixed point, and there is a neighborhood U of p such that $\overline{fU}$ is a compact subset of U. Show that p is a sink, and $\bigcap f^n U = p$.

Theorem 1.7. *If $|a| = 1$, then the volume of $(K^+ \cup K^-) - K$ is zero. If $|a| < 1$, then the volume of K^- is zero.*

Proof. By the previous Corollary, the sets $S_n = K^- - f^n V^+$ are increasing in n. But the volume is $|S_{n+1}| = |a|^2 |S_n|$. If $|a| \le 1$, we have $|S_{n+1}| = |S_n|$, or $|S_{n+1} - S_n| = 0$. Their union is $K^- - V^+ = \bigcup S_n$, so we see that $|K^- - V^+| = 0$. By Theorem, then, we have $|K^- - K| = 0$. If $|a| = 1$, then we apply the argument to f^{-1} to obtain the first statement of the Theorem. If $|a| < 1$, then the volumes are $|K| = |fK| = |a|^2 |K|$, so $|K| = 0$. □

Problem: Suppose that f is a Hénon map with real coefficients, so f is a diffeomorphism of $\mathbf{R}^2$. If $\delta = \pm 1$, then f preserves area. Is it possible for K to have positive volume in $\mathbb{C}^2$? Or if $K \subset \mathbb{R}^2$ can it have positive area?

For a point p, we define the *ω-limit set* $\omega(p)$ to be the set of all limit points $f^{n_j} p \to q$ for subsequences $n_j \to +\infty$. A property of $\omega(p)$ is that it is invariant under both f and f^{-1}.

For a compact set X, define its *stable set* as

$$W^s(X) = \{y : \lim_{n \to \infty} dist(f^n y, f^n X) = 0\},$$

and the unstable set is defined to be the stable set in backward time:

$$W^u(X) = \{y : \lim_{n \to -\infty} dist(f^{-n} y, f^{-n} X) = 0\}.$$

Proposition 1.8. $W^s(K) = K^+$.

Proof. Since K is bounded, it is clear from the definition that $W^s(K) \subset K^+$. On the other hand, let $p \in K^+$ be given. Then by the filtration properties, we have $\omega(p) \subset V$. Since $\omega(p)$ is invariant under f^{-1}, is is contained in K^-. We conclude that $\omega(p) \subset K$, so $dist(f^n p, K) \to 0$ as $n \to \infty$. □

1.3 Fatou Components

We define the *Fatou set* to be the points p for which there is a neighborhood U such that the restrictions of the forward iterates $\{f^n|U : n \geq 0\}$ are equicontinuous. If we consider $\mathbf{C}^2$ as imbedded in the projective plane $\mathbf{P}^2$ (this is defined in 2nd part), then the iterates $f^n : \mathbf{C}^2 \to \mathbf{P}^2$ are locally equicontinuous on $\mathbf{C}^2 - K^+$. That is, they converge locally uniformly to the point $[0:0:1] \in \mathbf{P}^2$, there the y-axis intersects the line at infinity. By the previous section on Filtration, then, we have:

Proposition 1.9. *The Fatou set of a complex Hénon map is* $\mathbf{C}^2 - J^+$.

A domain $U \subset \mathbf{C}^2$ is said to be a *Runge domain* if for each compact subset $X \subset U$, the polynomial hull

$$\hat{X} := \{q \in \mathbf{C}^2 : |P(q)| \leq \max_{x \in X} |P(x)| \text{ for all polynomials } P\}$$

is also contained in U.

Proposition 1.10. *$int(K^+)$ is Runge.*

Proof. Suppose that X is a compact subset of $int(K^+)$. By the filtration properties of the previous Section, we know that $\sup_{n \geq 0} \sup_{x \in X} ||f^n(x)|| = C < \infty$. Since the components of f^n are polynomials, it follows from the definition of the polynomial hull that $\sup_{x \in \hat{X}} ||f^n(x)|| \leq C$. Thus $\hat{X}$ is contained in K^+. Further, since X is compact, we know that the translates $X + v$ are contained in $int(K^+)$ if $||v||$ is sufficiently small. Thus we have $\widehat{X+v} = \hat{X} + v \subset K^+$. It follows that $dist(X, \partial K^+) = dist(\hat{X}, \partial K^+)$ and thus $\hat{X}$ is in the interior of K^+. □

If Ω is a Fatou component (i.e., a connected component of the Fatou set), then $f\Omega$ is again a Fatou component. Let δ denote the (constant) Jacobian of f. There are two distinct cases: $|\delta| = 1$ and $|\delta| \neq 1$. We will first consider the case $|\delta| \neq 1$. In this case, after passing to f^{-1} if necessary, we may suppose that $|\delta| < 1$. If $|\delta| < 1$, then the Fatou components consist of $\mathbf{C}^2 - K^+$ and the connected components of $int(K^+)$.

Problem: Is every Fatou component periodic, i.e., is $f^n\Omega = \Omega$ for some $n > 0$? A non-periodic Fatou component (if it existed) would be said to be *wandering*.

A Fatou component Ω is said to be *recurrent* if there is a point $p \in \Omega$ such that $\Omega \cap \omega(p) \neq \emptyset$. Equivalently, Ω is recurrent if there are a compact set $C \subset \Omega$ and a

point $p \in \Omega$ such that $f^{n_j} p \in C$ for infinitely many $n_j \to +\infty$. Note that a recurrent Fatou component is necessarily periodic.
Exercise: Show that a Fatou component is recurrent if it is a periodic domain and not all points tend to $\bigcup_j f^j \partial \Omega$.

We say that a fixed point p is a *sink* if $p \in Int(W^s(p))$. In this case, we refer to $W^s(p)$ as the *basin* of p. If $\mathcal{O} = \{p, fp, \dots, f^N p = p\}$ is a periodic orbit, we say that the orbit is a sink if $\mathcal{O} \subset W^s(\mathcal{O})$. In other words, a periodic orbit is a sink orbit for f is p is a sink fixed point for f^N. The following show that every sink is hyperbolic.

Proposition 1.11. *If p is a sink, then the eigenvalues of Df_p all have modulus less than 1.*

Proof. Let B be a ball containing p such that $\bar{B} \subset W^s(p)$. Thus the iterates f^n converge uniformly to p on $\bar{B}$. We may assume that $f^n \bar{B} \subset B$ for $n \geq N$. Let I_ε denote the affine map which fixes p and which dilates $\mathbf{C}^2$ by a factor of $1+\varepsilon$. For $\varepsilon > 0$ small the new map $g := I_\varepsilon \circ f$ maps $\bar{B}$ inside B. Thus the derivatives of g^n are bounded at p. This proves that the derivatives of f^n must tend to 0 as $n \to \infty$. Thus the eigenvalues of Df at p must be less than 1 in modulus. □

Exercise: If p is a periodic point such that $f^n p = p$ and the eigenvalues of $Df^n p$ have modulus < 1, then the basin $W^s(p)$ is a Fatou component. (It is clear that $W^s(p)$ is contained in K^+. The issue is to show that $W^s(p)$ is exactly a connected component of $int(K^+)$.)

Now we give other examples of Fatou components. By a *Siegel disk* we mean a set of the form $\mathcal{D} = \varphi(\Delta)$, where $\varphi : \Delta \to \mathbf{C}^2$ is an injective map of the disk $\Delta \subset \mathbf{C}$ with the property that

$$f(\varphi(\zeta)) = \varphi(\alpha\zeta) \tag{4}$$

holds for $\alpha = e^{\pi i a}$ with fixed irrational a and all $\zeta \in \Delta$. We also call $\mathcal{D}$ a Siegel disk if it is periodic: $f^N \mathcal{D} = \mathcal{D}$ for some $N > 1$, and $\mathcal{D}$ is a Siegel disk for f^N. A *Herman ring* is similar; it is the image of a maximal injective holomorphic mapping of an annulus $\{\zeta \in \mathbf{C} : r_1 < |\zeta| < r_2\}$ such that (4) holds. "Maximal" here means that it is not contained in any larger such annulus.

Theorem 1.12. *Suppose that $|\delta| < 1$ and Ω is a recurrent Fatou component. Then Ω is the basin of either (1) a sink, (2) a Siegel disk, or (3) a Herman ring. In cases (2) and (3), the Siegel disk (or Herman ring) is a subvariety of Ω.*

Proof. Passing to an iterate of f, we may suppose that $f\Omega = \Omega$. Let $p \in \Omega$ be a point with $\omega(p) \cap \Omega \neq \emptyset$. We may choose a sequence $n_j \to \infty$ such that f^{n_j} converges normally to a map $g : \Omega \to \bar{\Omega}$, and we may choose the sequence such that $g(p) \in \Omega$. Since $f \circ f^{n_j} = f^{n_j} \circ f$, it follows that $f \circ g = g \circ f$ on Ω. Thus if $g(\Omega)$ is a single point, then it must be a fixed sink, and Ω is the basin of that sink.

If g is not constant, we set $\Sigma := \Omega \cap g(\Omega) \neq \emptyset$. We may pass to a subsequence so that $n_{j+1} - n_j \to \infty$, and $f^{n_{j+1}-n_j}$ converges normally to a holomorphic map $h : \Omega \to \bar{\Omega}$. Now choose a point $w = g(z) \in \Sigma$. We see that

$$hw = \lim_{j\to\infty} f^{n_{j+1}-n_j}(f^{n_j} z) = \lim_{j\to\infty} f^{n_{j+1}} z = gz = w.$$

Thus, setting $M := \{w \in \Omega : hw = w\}$, we see that $\Sigma \subset M$. Conversely, if $w \in M$, then $gw \in M$ because $h(gw) = g(hw) = g(w)$. Thus $\Sigma = M$.

Again, since $f \circ g = g \circ f$, we have $f\Sigma = \Sigma$. Further, since Σ consists of infinite forward limits of points of K^+, we have $\Sigma \subset K$. Thus Σ is a bounded Riemann surface, so it is covered by the disk. Finally, since the fixed points of f^N are isolated, the restriction $f^N|\Sigma$ is not the identity. Since there is a point of $\omega(p) \cap \Sigma$, we know that the iterates $f^n|\Sigma$ cannot diverge to the boundary of Σ as $n \to \infty$. We conclude that Σ must be uniformized by either the disk or an annulus, and f must be an irrational rotation. Thus Σ must be either a Siegel disk or a Herman ring. □

It is clear that Hénon maps can have sinks. For a Siegel disk, let us consider the diagonal linear map L of $\Delta \times \mathbf{C}$ to itself which is given by $(\zeta, \eta) \mapsto (\alpha\zeta, \delta\eta/\alpha)$, and $\alpha = e^{\pi i a}$ with a irrational. If $\Phi : \Delta \times \mathbf{C} \to \mathbf{C}^2$ is a holomorphic imbedding such that $f \circ \Phi = \Phi \circ L$, then $\mathscr{D} = \Phi(\Delta \times \{0\})$ will be contained in a Siegel disk. Further, if $|\delta| < 1$, we will have that $\Omega := \Phi(\Delta \times \mathbf{C}^2)$ is equal to $W^s(\mathscr{D})$.

Such a map occurs for Hénon maps which have the form

$$f(x,y) = (\alpha x + \cdots, \delta y/\alpha + \cdots)$$

where the $\cdots$ indicate arbitrary terms of higher order. (Or rather we conjugate a Hénon map by an affine map so that the origin is fixed, and the linear part is diagonal.) If $|\delta| < 1$ and if α satisfies a suitable Diophantine condition (see [Z]), then there will exist a linearizing map Φ as in the previous paragraph.

Problem: Is it possible for a Fatou component to be the basin of a Herman ring?

Another sort of Fatou component can arise as follows. Suppose that f is a Hénon map fixing $(0,0)$ and taking the form

$$f(x,y) = (x + x^2 + \cdots, by + \cdots),$$

where $0 < |b| < 1$, and the terms $\cdots$ involve both x and y, but they are "smaller." This is like a parabolic fixed point in the x-direction and an attracting fixed point in the y-direction. We may choose r small so that $D := \{|x+r| < r\} \times \{|y| < r\}$ is mapped inside itself. In fact, for each $p \in D$ there is a neighborhood U such that $f^n \to (0,0)$ uniformly on U as $n \to +\infty$. The forward basin $\mathscr{B} = \{(x,y) \in \mathbf{C}^2 : f^n \to (0,0)$ uniformly in a neighborhood of $(x,y)\}$ is a non-recurrent Fatou component.

Problem: Is such a semi-attracting basin the ony possible sort of non-recurrent Fatou component? More precisely, suppose that $|\delta| \neq 1$ and that Ω is a periodic Fatou component which is not recurrent. Is there necessarily a point $p \in \partial\Omega$ with $f^n p = p$ and such that the multipliers of $Df^n p$ are α and δ^n/α, where α is a root of unity?

The volume preserving case: $|\delta| = 1$.

Except for $\mathbf{C}^2 - K^+$, all Fatou components are contained in the bounded set $int(K^+) = int(K^-) = int(K)$. If Ω is such a bounded component, then there are only finitely many other components with the same volume, so Ω is necessarily periodic.

If U is a Fatou component, we may pass to an iterate of f and assume that $fU = U$. Since $\{f^n|U : n \geq 0\}$ is a normal family, we let $\mathscr{G} = \mathscr{G}(U)$ denote the set of normal limits $g = \lim_{j\to\infty} f^{n_j}$ on U. When a sequence of analytic functions converges, the derivatives converge, too. So it follows that each $g \in \mathscr{G}$ preserves volume, and $g(U)$ is an open set. Thus $g : U \to U$, and not just $g : U \to \bar{U}$. Since the iterates of f commute with each other, so do the elements of $\mathscr{G}$. Finally, we may pass to a subsequence of the $\{n_j\}$ so that $n_{j+1} - 2n_j \to +\infty$. Passing to a further subsequence, we find that $f^{n_{j+1}-2n_j} \to g^{-1}$. Thus $\mathscr{G}$ is a group. By a theorem of H. Cartan, the full automorphism group $Aut(U)$ of a bounded domain is a Lie group (see [N]). Thus $\mathscr{G}$ is a compact abelian subgroup of a Lie group. Since $\mathscr{G}$ is compact, it can have only finitely many connected components. Let $\mathscr{G}_0$ denote the connected component of the identity in $\mathscr{G}$. It follows that $\mathscr{G}_0$ is a torus $\mathbf{T}^d$. Since $f \in \mathscr{G}$, it follows that $\mathscr{G}$ is infinite, so $d \geq 1$. Further, we may choose $m \neq 0$ so that $f^m \in \mathscr{G}_0$. Since $\mathscr{G}_0$ is compact, we have the following:

Proposition 1.13. *For volume-preserving Hénon maps, all Fatou components are periodic and recurrent.*

Since $U \subset \mathbf{C}^2$, it follows that $1 \leq d \leq 2$. Examples of both cases $d = 1$ and $d = 2$ can be obtained by starting with maps of the form $f(x,y) = (\alpha x + \cdots, \beta y + \cdots)$ with α and β jointly Diophantine, so that the map can be linearized in a neighborhood of the fixed points. If $d = 2$, then the action of $\mathscr{G}_0$ on U is equivalent to the standard $\mathbf{T}^2$-action on a Reinhardt domain in $\mathbf{C}^2$.

Theorem 1.14. *Suppose that U is a Fatou component, and suppose that f has a fixed point $a \in U$ such that $A = Df_a$ is unitary. Then there is a holomorphic semi-conjugacy $\Phi : U \to \mathbf{C}^2$, $\Phi(a) = 0$ and taking (f,U) to $(A, \Phi(U))$.*

Proof. Conjugating by a translation, we may assume that $a = 0$. For each compact subset $S \subset U$, the sets $f^j(S)$ are bounded independently of j. Thus $\Phi_j := N^{-1}\sum_{j=0}^{N-1} A^{-j} f^j$ is a bounded sequence of analytic functions. Now let Φ be any limit point of a subsequence Φ_{j_k}, and observe that Φ has the desired property. □

Problem: What $\mathbf{T}^1$ actions can appear as Fatou components? What Reinhardt domains can appear as Fatou components? Is a Fatou component necessarily simply connected?

Notes: Recurrent Fatou components are discussed in [BS2, FS1, BSn]. For a deeper discussion of semi-attracting basins, see [U1,2].

1.4 Hyperbolicity

A compact set $X \subset \mathbf{C}^2$ is said to be *hyperbolic* if there is a continuous splitting $X \ni x \mapsto E^s_x \oplus E^u_x = T_x\mathbf{C}^2$ of the complex tangent space such that $Df_x E^{s/u}_x = E^{s/u}_{fx}$, and if there are constants $c < \infty$ and $\lambda < 1$ such that

$$||Df^n_x|_{E^s_x}|| \leq c\lambda^n, \quad ||Df^{-n}_x|_{E^u_x}|| \leq c\lambda^n.$$

An alternative definition which is sometimes useful is that there are invariant cone fields which are expanding and contracting. For an expanding cone field, there is a neighborhood $U_0 \supset X$, and for each $p \in U_0$ there is an open cone $\mathscr{C}_p \subset T_p\mathbf{C}^2$. This cone field has the property that there is an n such that for each $p \in U_0$ and nonzero $\xi \in \mathscr{C}_p$, we have $Df^n\mathscr{C}_p \subset int(\mathscr{C}_{f^n p})$, and $||Df^n\xi|| \geq 2||\xi||$. A contracting cone field is defined as an expanding cone field for f^{-1}.

Exercise: A finite set is hyperbolic if and only if it is the union of orbits of periodic points, each of which is of saddle, attracting, or repelling type.

Exercise: We say that a hyperbolic set X has *index i* if $\dim E^u_x = i$ for all $x \in X$. Show that if J is a hyperbolic set for f, then J has index 1. (Hint: First subdivide $J = \bigcup_{0 \leq i \leq 2} \Lambda_i$, where Λ_i has index i. Then show that Λ_0 is contained in the interior of K^+.)

Stable Manifold Theorem: Let f be a diffeomorphism of a smooth manifold, and let $X \subset M$ be a hyperbolic set for f. For each $x \in X$, the stable set $W^s(x)$ is a submanifold of M, with $T_xW^s(x) = E^s_x$. Further $W^s(X) = \bigcup_{x \in X} W^s(x)$.
From the proof of the Stable Manifold Theorem, it is clear that if f is also holomorphic, then each $W^s(x)$ is a complex submanifold. However, $W^s(x)$ is not necessarily a closed submanifold.

Proposition 1.15. *If f is a complex Hénon map, and if J is a hyperbolic set, then $W^s(J) \subset J^+$ and $W^u(J) \subset J^-$.*

Proof. We will prove that $W^s(J) \subset J^+$; the statement that $W^u(J) \subset J^-$ follows by considering f^{-1}. It is clear that $W^s(J) \subset K^+$, so it will be sufficient to show that $W^s(J)$ is disjoint from the interior of $int(K^+)$. Now the iterates $\{f^n : n \geq 0\}$ are a normal family on the interior of K^+, so if $p \in int(K^+)$, $||Df^n p||$ is bounded for $n \geq 0$. On the other hand, but the hyperbolicity of f, there is an expanding cone field $\mathscr{C}_q$ which is defined for $q \in U_0$. If $p \in W^s(J)$, then $f^m p \in U_0$ for m sufficiently large. Thus there is a vector $\xi \in \mathscr{C}_{f^m p}$ which is uniformly expanded by Df^n. This means that $||Df^{m+kn}\xi||$ is unbounded for k large. The contradiction shows that p, and thus $W^s(J)$, is disjoint from the interior of K^+. □

We say that X has *local product structure* for a map f if $W^s(p_1) \cap W^u(p_2) \subset X$ whenever $p_1, p_2 \in X$.

Proposition 1.16. *If f is hyperbolic on J, then J has local product structure.*

Proof. By the previous Proposition, $W^s(p_1) \subset J^+$, and $W^u(p_2) \subset J^-$. Thus the intersection is contained in $J = J^+ \cap J^-$. □

A consequence of the local product structure and hyperbolicity (see [S, Prop. 8.22]) is the following:

Corollary 1.17. *The set J is locally maximal, which means that there is a neighborhood U of J so that every invariant contained in U is also contained in J.*

Proposition 1.18. *If f is a complex Hénon map, and if J is a hyperbolic set, then each stable manifold $W^s(p)$, $p \in J$ is conformally (biholomorphically) equivalent to $\mathbf{C}$.*

Proof. By the stable manifold theorem, $W^s(p)$ is homeomorphic to a disk, so as a Riemann surface it is conformally equivalent either to the disk or to **C**. In order to show it is equivalent to **C**, it suffices to show that $W^s(p)$ contains an infinite increasing family of disjoint annuli $A(k)$ so that $A(1)$ surrounds p, $A(k+1)$ surrounds $A(k)$, and the moduli of the $A(k)$ are bounded below.

Let $B_p(r) = \{q : ||p-q|| < r\}$ be the ball with radius r and center p. Then for r_2 small, the connected component of $B_p(r_2) \cap W^s(p)$ containing p will be a disk, and for $0 < r_1 < r_2$, $A_p := (B_p(r_2) - B_p(r_1)) \cap W^s(p)$ will be an annulus in $W^s(p)$ which goes around p. By compactness, there is a lower bound on the moduli of the annuli A_p for $p \in J$. By the uniform contraction of hyperbolicity, there is an n such that for every p, $A_p(1) := f^{-n}A_{f^n p}$ contains A_p in its "hole". Thus $A(k) := f^{-nk}A_{f^{nk}p}$ is the desired increasing sequence of annuli in $W^s(p)$. □

Theorem 1.19. *If f is a complex Hénon map with $|\delta| \neq 1$, and if J is a hyperbolic set, then the interior of K^+ consists of the basins of finitely may hyperbolic sink orbits.*

Proof. We may assume that $|\delta| \leq 1$, for if $|\delta| > 1$, then $int(K^+)$ is empty, and there is nothing to prove. The theorem will be a consequence of the following assertions:

(1) Each component is periodic and recurent.
(2) Each component is the basin of a sink.
(3) There are finitely many sink orbits.

Proof of (1). So assume that $|\delta| < 1$. Let C be a wandering component of $int(K^+)$, and let p be a point of C. For n sufficiently large, $f^n p$ will be a point of V. Since V is compact, then, the ω-limit set $\omega(p)$ is a nonempty compact subset of V. Since $\omega(p)$ is invariant under f^{-1} and bounded, it follows that $\omega(p) \subset K \subset K^-$. Since $|\delta| < 1$, we have $J^- = K^-$. Thus $\omega(p) \subset J^-$.

Now we claim that $\omega(p)$ intersects $int(K^+)$. For otherwise, $\omega(p) \subset J^+ = \partial K^+$, which means that $\omega(p) \subset J$. Thus $p \in W^s(J)$, and by the Proposition above, we conclude that $p \in J^+$, which is not the case.

Now choose $q_0 \in \omega(p) \cap int(K^+)$, and let C_0 denote the component of $int(K^+)$ containing q_0. There are n_1 and n_2 so that $f^{n_1}p$ and $f^{n_2}p$ are arbitrarily close to q_0, and so they both belong to C_0. We conclude, then, that $f^{n_1}C = f^{n_2}C$, from which we conclude that C must be periodic, and since it contains a point of $\omega(p)$ it is recurrent.

Proof of (2) Assume $|\delta| < 1$. We Know By Theorem 1.12 that each recurrent component is the basin of either a sink or a Siegel disk or Herman ring. Suppose that is the basin of Σ, which is a Siegel disk or a Herman ring. By Theorem 1.12 again, Σ is a subvariety of C. Since Σ is bounded and invariant under f^{-1}, we have $\Sigma \subset K^-$. Finally, since $\partial\Sigma \subset \partial C \subset \partial K^+$, we have $\partial\Sigma \subset J$. By the rotation property of f on Σ, the closure of the orbit of a point of Σ is a compact curve. If we take $q \in \Sigma$ close to J, then the closure of the orbit will be contained in a small neighborhood of J. However, this contradicts the maximality property of J. We conclude that the Siegel disk Herman ring cases cannot occur, so C must be the basin of a sink.

Proof of 3. The sink orbits are all contained in V, which is compact. Define the set L to be the limit points of sequences of (distinct) sinks. If there are infinitely many sink orbits $\{p_j\}$, then $L \neq \emptyset$. Since $int(K^+)$ is the union of basins of sinks, we see that $L \cap int(K^+) = \emptyset$. Thus we must have $L \subset J$. On the other hand, this means that there are sink orbits which are arbitrarily close to J, which contradicts the maximality of J. □

Corollary 1.20. *If f is hyperbolic on J, and if $|\delta| = 1$, then $int(K^+) = int(K^-) = int(K) = \emptyset$.*

Proof. In the volume-preserving case, we have $int(K^+) = int(K^-)$. We saw in the previous section that all components of $int(K^+)$ are periodic, and if C is such a component, we may pass to f^N and suppose that it is fixed. The iterates of f^N then induce a torus action on C. There will be (compact) orbits if this torus action inside any neighborhood of J, which will contradict the maximality of J. But we saw that in he hyperbolic case, J is always maximal. □

In the following, we assume that the Jacobian δ satisfies $|\delta| \leq 1$, i.e., f decreases volume. Since we may replace f by f^{-1} if necessary, we may always make this assumption.

Theorem 1.21. *Let f be a complex Hénon with $|\delta| \leq 1$, and let J be a hyperbolic set. Then $W^s(J) = J^+$, and if $s_1, \ldots, s_k$ are the sinks of f, then $W^u(J) = J^- - \{s_1, \ldots, s_k\}$.*

Proof. By Proposition 1.15 we have $W^s(J) \subset J^+$. We also know that $J^+ \subset K^+ = W^s(K)$. By Theorem 1.7 and previous corollary we have $int(K^-) = \emptyset$, that is $K^- = J^-$. Since J^+ is invariant $J^+ \subset W^s(K \cap J^+) = W^s(K^- \cap J^+) = W^s(J)$. Thus, we have $W^s(J) = J^+$.

For the other statement, let us note that $J^- = K^-$. We have seen in §2 that this holds for all Hénon maps when $|\delta| < 1$. And the case $|\delta| = 1$ follows from the previous Corollary.

For the second statement, we have $K^- = W^u(K)$, so $J^- \subset W^u(K)$. Suppose first that $p \in J^- - int(K^+)$. Then $p \in W^u(K - int(K^+))$. Since $K^- = J^-$, we have $K - int(K^+) = J$, so $J^- - int(K^+) \subset W^u(J)$.

Now if $p \in J^- \cap int(K^+)$, then by Theorem 1.19, $p \in W^s(s_j)$ is in the basin of some sink s_j. If p is not the sink itself, then the backward iterates converge to the boundary, so $f^{-n}p \to \partial K^+ = J^+$ as $n \to +\infty$. We conclude that if p is not a sink, then $p \in W^u(J^-)$. Since $p \in W^u(J^+)$, we have $p \in W^u(J)$. □

A point p is *wandering* if there is a neighborhood U containing p such that $U \cap f^n U = \emptyset$ for all $n \neq 0$. And p is *nonwandering* if it is not wandering. We let Ω_f denote the set of nonwandering points.

A point p is *chain recurrent* if for each $\varepsilon > 0$ there is a sequence of points $p_0 = p$, $p_1, \ldots, p_{N-1}$ and $p_N = p$ such that $dist(fp_j, p_{j+1}) < \varepsilon$ for $0 \leq j \leq N-1$. The number N and the sequence of points depend on ε. The set of all chain recurrent points is denoted $R(f)$.

Exercise: Show that $\Omega_f \subset R(f)$.

We will use the following transitivity property which will be proved using the convergence theorems in 1.9:

Theorem 1.22. *Let $p_+ \in J^+$ and $p_- \in J^-$ be given. Then for any neighborhoods $U_- \ni p_-$ and $U_+ \ni p_+$, there is an n such that $U_+ \cap f^n U_- \neq \emptyset$.*

Let us note that an immediate consequence of this Theorem is that the restriction $f|_J$ is *topologically mixing*, which means that for any open sets U_1 and U_2 with $U_i \cap J \neq \emptyset$ for $i = 1,2$, there is an $n \neq 0$ such that $U_i \cap f^n U_i \neq \emptyset$. In particular, it follows that $J \subset \Omega_f$.

We will call a Hénon map *hyperbolic* if any of the following sets is a hyperbolic set:

Theorem 1.23. *The following are equivalent:*

(i) f has a hyperbolic splitting over the chain recurrent set.
(ii) f has a hyperbolic splitting over the nonwandering set.
(iii) f has a hyperbolic splitting over J.

Proof. The implication $(i) \Rightarrow (ii)$ follows because $\Omega_f \subset R(f)$. The implication $(ii) \Rightarrow (iii)$ follows because $J \subset \Omega_f$. Now we prove $(iii) \Rightarrow (i)$. Without loss of generality, we may assume that $|\delta| \leq 1$. It will be sufficient to show the claim that $R(f)$ is contained in the union of J with finitely many sink orbits. It is clear that $R(f)$ is contained in K^+ and K^-. Since $|\delta| \leq 1$, we have $J^- = K^-$, so $R(f) \subset J^- \cap K^+$. Now by Theorem 1.19, the interior of K^+ consists of the basins of sink orbits, so the only points of $R(f) \cap int(K^+)$ are the sink orbits themselves. This proves the claim. □

1.5 Rate of Escape

In the second part of these notes, we will look at the Hénon family as maps in projective space. Seen from this point of view, there is an invariant line (the line at infinity), and a superattracting fixed point (the intersection of the line at infinity with the y-axis). Orbits in the basin approach this superattracting fixed point at a super-exponential rate. This rate of escape function is very useful for understanding K^+ (the set of non-escaping points, or the complement of the basin). For the moment, we are able to look at this situation very adequately from within $\mathbf{C}^2$. Let us define

$$q(x,y) = p(y) - y^d = a_{d-2}y^{d-2} + \cdots + a_0 - \delta x$$

Thus we have

$$\frac{y_{n+1}}{y_n^d} = \frac{y_n^d + q(x_n,y_n)}{y_n^d} = 1 + \frac{q(x_n,y_n)}{y_n^d}$$

For $(x,y) \in V^+$, then, we have

$$\left| \frac{y_{n+1}}{y_n^d} - 1 \right| \leq \frac{\kappa}{|y_n|^2}$$

for some $\kappa > 0$. Let us set

$$G_n^+(x,y) = \frac{1}{d^n} \log^+ |y_n|$$

Definition 1.24. A function ψ is *upper-semicontinuous* if $\psi(z_0) \geq \limsup_{z \to z_0} \psi(z)$ at all points z_0. We say that a function ψ is *pluri-subharmonic*, or simply *psh* on a domain $\Omega \subset \mathbf{C}^2$ if ψ is upper-semicontinuous, and if for all $\alpha, \beta \in \mathbf{C}^2$, the function $\zeta \mapsto \psi(\alpha\zeta + \beta)$ is subharmonic in ζ, wherever it is defined. ψ is said to be *pluri-harmonic* if both ψ and $-\psi$ are psh. If ψ is pluri-harmonic, then it is locally the real part of a holomorphic function.

Recall that $\log^+ |t| := \max\{\log|t|, 0\} = \log(\max\{|t|, 1\})$ is continuous on all of $\mathbf{C}$.

Proposition 1.25. $G^+(x,y) := \lim_{n\to\infty} G_n^+(x,y)$ *exists uniformly on* V^+. *Further* $G^+ > 0$ *on* V^+ *and is pluri-harmonic there.*

Proof. We have $y_n \neq 0$ on V^+, so $\log^+ |y_n| = \log|y_n|$ is pluriharmonic there. It suffices to show uniform convergence of the limit. For this we rewrite it as a telescoping sum

$$\begin{aligned} G_N^+(x,y) &= G_0(x,y) + \sum_{n=1}^{N-1} (G_{n+1}(x,y) - G_n(x,y)) \\ &= \log|y| + \sum_{n=1}^{N-1} \frac{1}{d^n} \left(\frac{1}{d} \log|y_{n+1}| - \log|y_n| \right) \\ &= \log|y| + \sum \frac{1}{d^n} \log\left| \frac{y_{n+1}}{y_n^d} \right|. \end{aligned}$$

Now on V^+ we have

$$\left| \frac{y_{n+1}}{y_n^d} - 1 \right| < \frac{\kappa}{|y_n|^2} \leq \frac{\kappa}{R^2}$$

so the series converges uniformly. □

The formula $G^+ \circ f^n = d^n G^+$ holds on V^+, and this formula may be used to extend G^+ to $U^+ = \bigcup_{n\geq 0} f^{-n}V^+ = \mathbf{C}^2 - K^+$. Recall that $H_1(V^+;\mathbf{Z}) \cong \mathbf{Z}$. Since $f(x,y) = (y, y^d) + \cdots$, we see that the action of f_* on $H_1(V^+;\mathbf{Z})$ to itself is multiplication by d. We may use G^+ to describe the homology of U^+.

Theorem 1.26. *The map defined by* $\gamma \mapsto \omega(\gamma) = \frac{1}{\pi i} \int_\gamma \partial G^+$ *yields an isomorphism* $\omega : H_1(U^+;\mathbf{Z}) \to \mathbf{Z}[\frac{1}{d}]$.

Proof. Let τ denote the circle inside V^+ which is given by $t \mapsto (0, \rho e^{2\pi i t})$. First we show that τ is nonzero in $H_1(U^+;\mathbf{Z})$. For this we recall that on V^+ we have $G^+(x,y) = \log|y| + O(y^{-1})$. Thus we have

$$\int_\tau \partial G^+ = \int_\tau \partial \log|y| + O(y^{-2}) = \int \frac{dy}{2y} + O(y^{-2}) = i\pi + O(\rho^{-1}).$$

Letting $\rho \to \infty$, we see that the integral is $i\pi \neq 0$, so τ is nonzero in $H_1(U^+)$. Further, τ generates $H_1(V^+;\mathbf{Z})$.

Now if $\gamma \in H_1(U^+;\mathbf{Z})$ is arbitrary, there exists $m \geq 0$ such that $f_*^m \gamma$ is supported in V^+. Thus $f_m^* \gamma \sim k\tau$ for some $k \in \mathbf{Z}$. Thus $\gamma \sim kd^{-m}\tau$. □

Two global convergence theorems will be useful to us. The first concerns locally uniform convergence on $\mathbf{C}^2$.

Theorem 1.27. *The limit $G^+(x,y) = \lim_{n\to\infty} \frac{1}{d^n} \log^+ |y_n|$ exists uniformly on compact subsets of $\mathbf{C}^2$, and the following hold:*

(i) G^+ is continuous and psh on $\mathbf{C}^2$;
(ii) $\{G^+ = 0\} = K^+$;
(iii) G^+ is pluri-harmonic on U^+;
(iv) $G^+ \circ f = (d) \cdot G^+$.

Proof. The two things that we must prove are the continuity of G^+ and the uniform convergence. Everything else should be clear. We first show that G^+ is continuous on $\mathbf{C}^2$. By Proposition 1.25 we know that G^+ is continuous (and even pluri-harmonic) on U^+, and G^+ is equal to 0 on K^+. Thus we need to show that $\lim_{\zeta\to z} G^+(\zeta) = 0$ for all $z \in \partial K^+$. Let $M = \max_{V^+\cap fV} G^+$. Without loss of generality, we may assume that $z \in V$. Let $\zeta_j \to z$ be any sequence. If $\zeta_j \notin K^+$, then there is a smallest number n_j such that $f^{n_j}\zeta_j \notin V$. As $\zeta_j \to z \in K^+$, we must have $n_j \to \infty$. Since $f^{n_j}\zeta_j \in V +^{\cap} fV$, we have

$$G^+(\zeta_j) = \frac{1}{d^{n_j}} G^+(f^{n_j}\zeta_{n_j}) \le \frac{M}{d^{n_j}},$$

from which we conclude that $G^+(\zeta_j) \to 0$ as $j \to \infty$.

Now we show uniform convergence on a set of the form $D = \{|x| < \rho', |y| < \rho''\}$. Choosing ρ' large, we may assume that $K^+ \cap \{|x| = \rho', |y| \le \rho''\} = \emptyset$. Let $\varepsilon > 0$ be given, and choose $U \subset D$ such that for any $|y_0| < \rho''$, U contains a neighborhood of $K^+ \cap \{y = y_0\}$ inside $\{y = y_0\}$. We may choose U sufficiently large that $G^+ < \varepsilon$ on $\{|y| < \rho''\} \cap \partial U$. By the Proposition, we may choose n such that $|G_n^+ - G^+| < \varepsilon$ on $D - U$. Thus $G_n^+ \le 2\varepsilon$ on $\partial U \cap D$, so by the maximum principle, G^+ and G_n^+ are both bounded above by 2ε on U. We conclude that $|G^+ - G_n^+|$ is uniformly small on D. □

Theorem 1.28. *$d^{-n} \log |y_n|$ converges to G^+ in L^1_{loc} as $n \to \infty$.*

Proof. Since $\log^+ |y| = \log |y|$ for $|y| > 1$, it follows from the previous Theorem that $d^{-n} \log |y_n|$ converges locally uniformly to G^+ on $\mathbf{C}^2 - K^+ = \{G^+ > 0\}$. By Hartogs' Theorem, we know that for any sequence $d^{-n_j} \log |y_{n_j}|$ there is a further subsequence which will converge in L^1_{loc} to a psh limit v. If we can show that $v = 0$ on the interior of K^+, then by the upper semicontinuity of G^+, it will follow that $v = G^+$. Since the limit is independent of the subsequence, we will conclude that $d^{-n} \log |y_n|$ converges in L^1_{loc} to G^+.

Now if v is not identically zero on $int(K^+)$, there will be a $\delta > 0$ such that $\{v < -2\delta\}$ is a nonempty open set W. By Hartogs' Theorem again, we may choose a relatively compact open subset $W_0 \subset W$ such that $d^{n_j} \log |y_{n_j}| < -\delta$ holds on W_0 for large j. This means that $f^{n_j} W_0$ is contained in the set $\{\log |y| < -\delta d^{n_j}\}$. Further, since $W_0 \subset int(K^+)$ we may assume that $f^{n_j} W_0 \subset V$. Now the standard comparison between volume and capacity gives us that:

$$Vol(V \cap \{|y| < e^{-\delta d^{n_j}}\}) < Ce^{-\delta d^{n_j}}$$

for some constant C. On the other hand, the Jacobian δ is constant, so we have

$$Vol(f^{n_j}W_0) = |\delta|^{2n_j}Vol(W_0).$$

Since $f^{n_j}W_0 \subset V \cap \{|y| < e^{\delta d^{n_j}}\}$ the first estimate needs to dominate the second estimate. This is not possible when n_j is large, so we conclude that we must have $v = 0$ on $int(K^+)$. □

1.6 Böttcher Coordinate

We would like to define a function $\varphi^+ = \lim_{n\to\infty} y_n^{1/d^n}$, because if we have such a function, then we will have $\varphi^+ \circ f = (\varphi^+)^d$. In order to make this work, we consider the product

$$y_n^{1/d^n} = y_0 \left(\frac{y_1}{y_0^d}\right)^{1/d} \left(\frac{y_2}{y_1^d}\right)^{1/d^2} \cdots \left(\frac{y_n}{y_{n-1}^d}\right)^{1/d^n}.$$

By the estimate (2), $y_n/y_{n-1}^d = 1 + \frac{q(x_j,y_j)}{y_j^d}$ is close to 1 for $(x,y) \in V^+$, so we have the following:

Theorem 1.29. *The limit*

$$\varphi^+(x,y) := y \lim_{n\to\infty} \prod_{j=0}^{n-1} \left(1 + \frac{q(x_j,y_j)}{y_j^d}\right)^{\frac{1}{d^j}}$$

exists uniformly on V^+ and defines a holomorphic function there. Further, we have $\varphi^+ \circ f = (\varphi^+)^d$.

Theorem 1.30.
(1) If γ is a path in U^+ which starts at a point of V^+, then φ^+ has an analytic continuation along γ.
(2) There exists no function ψ which is holomorphic on $f^{-1}V^+$ and which is equal to φ^+ on V^+.

Proof. It is clear from the definitions that $G^+ = \log|\varphi^+|$ on V^+. Now let G^* be a pluriharmonic conjugate for G^+ in a neighborhood of the starting point of γ, such that $\varphi^+ = exp(G^+ + iG^*)$. Now we may continue G^* pluri-harmonically along γ, which gives us the desired analytic continuation of φ^+.

The function $\psi \circ f^{-1}$ is defined on V^+, and it is equal to $\varphi^+ \circ f^{-1}$ on fV^+. Thus $(\psi \circ f^{-1})^d = (\varphi \circ f^{-1})^d = \varphi^+$ on fV^+. Thus $\varphi \circ f$ is a dth root of φ^+ on V^+. On the other hand, by the formula, we see that $\varphi^+ \sim y$ for $(x,y) \in V^+$ for $|y|$ large. Since y does not have a dth root on $\{|y| > M\}$, this is a contradiction. □

Exercises:

1. Re-do the previous discussions in the case where p is not monic. How does the factor $a_d \neq 1$ change things? In particular, if we write
$$G^+(x,y) = \log|y| + c_0 + \frac{c_1}{y} + O\left(\frac{1}{y^2}\right)$$
then what are c_0 and c_1?
2. Re-do the previous discussion for backward time. In particular, construct $G^- := \lim_{n\to\infty} \frac{1}{d^n} \log^+ |x_{-n}|$, as well as a function φ^- on V^- which has the property that $\varphi^- \circ f^{-1} = (\varphi^-)^d$.
3. Let $||\cdot||$ be any norm on $\mathbf{C}^2$. Show that $G^+(x,y) = \lim_{n\to\infty} \frac{1}{d^n} \log^+ ||(x_n,y_n)||$.
4. Show that for all $(x,y) \in V^+$, we have $|y_n| \geq c\, e^{d^n G^+(x,y)}$. Give a value for c.

In dimension 1, we consider a polynomial map $p(z) = z^d + \cdots : \mathbf{C} \to \mathbf{C}$. In this case, the Böttcher coordinate has been fundamental in the description of the dynamics. The dynamical systems that serves as a 1-D model domain is $(\sigma, \mathbf{C} - \bar{D})$, where $\sigma(w) = w^d$, and $\mathbf{C} - \bar{D}$ is the complement of the closed unit disk. The dynamics in the model system is seen in terms of polar coordinates (r,θ): orbits are escaping to infinity by $r \mapsto r^d$, and the recurrent part of the dynamics is given by $\theta \mapsto d\cdot\theta$ (modulo 2π). The Böttcher coordinate φ^+ may be analytically continued to $\mathbf{C} - K$ if and only if K is connected. In this case, it gives a conjugacy $\varphi^+ : (p, \mathbf{C} - K) \to (\sigma, \mathbf{C} - \bar{D})$. The radial lines $\{\theta = c\}$ in the image correspond to the gradient lines of G^+ in $\mathbf{C} - K$. When $d = 2$, the most natural picture represents the level sets $\{G^+ = c\}$ for $c = 2^{-n}$ and gradient lines for G^+, i.e., $(\varphi^+)^{-1}\{\theta = c\}$ for $\theta = j2^{-n}$.

This is shown for the map $p(z) = z^2 - 1$ in Figure 2; the set K is black, and its complement is colored in white/gray. A starting point z is iterated until some iterate $z_n = p^n(z)$ has modulus bigger than 100. The color "white" means that $Im(z_n) > 0$. This is equivalent to the statement on the argument for the bottcher coordinate $\varphi(z)$: we have $2\pi j 2^{-n} < Arg(\varphi(z)) < 2\pi(j+1)2^{-n}$ (mod 2π), with j being even. If n is odd, then the color is "gray". The binary (white/gray) color scheme corresponds to the fact that we may identify the circle with binary expansions $\theta = .b_1b_2b_3\ldots$ in base 2. That is, we may represent the circle as infinite binary sequences $\{0,1\}^{\mathbf{N}}/\sim$, where the equivalence relation $\sim$ identifies two binary expansions $.*1\bar{0} = .*0\bar{1}$ for the same real number.

Looking at Figure 2, the eye of the observer can fill in the curves of the level sets $\{G = 2^{-n}\}$ of the Green function, as well as arcs of the form $\{Arg(\varphi) = j2^{-n} = \theta\}$. Looking more carefully at one of these arcs, we may observe the alternating white/gray sequences as the arc lands, and this lets us deduce part of the binary expansion of θ. If we make a zoom of this picture near the landing point, we would see more binary digits of θ.

The two fixed points $\{fix_\alpha, fix_\beta\}$ of p are repelling; fix_α is the right hand tip of of J, and fix_β is the point of $J \cap [-1,0]$, which is the intersection of the immediate basin of -1 and the basin of 0.

In dimension 1, the subject of representing a Julia set as the quotient of a circle is a very well developed subject, so we would hope to develop something analogous

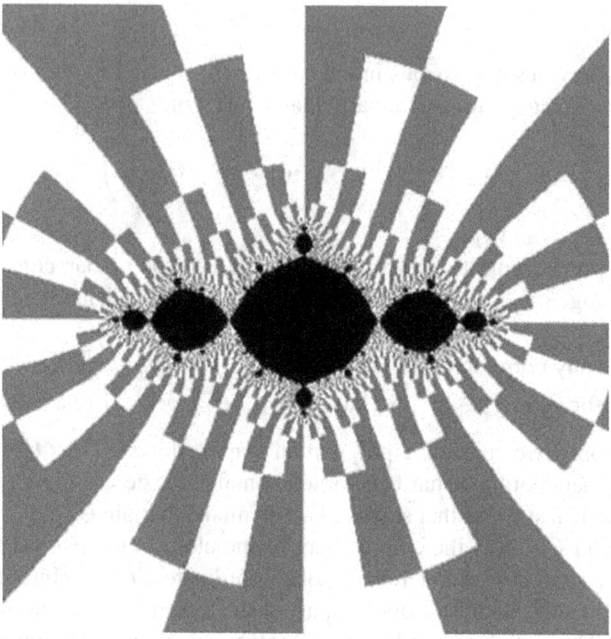

Fig. 2 Julia set for $p(z) = z^2 - 1$

in 2D. It is natural to take the corresponding 2-D model to be the complex solenoid. We set

$$\Sigma_* = \{\zeta = (\zeta_n)_{n\in\mathbf{Z}} : \zeta_n \in \mathbf{C}_*, \zeta_{n+1} = \zeta_n^d\}$$

and we give Σ_* the infinite product topology induced by $\Sigma_* \subset \mathbf{C}_*^\infty$. Also, Σ_* is a group under the operation $\zeta\cdot\eta = (\zeta_n\eta_n)_{n\in\mathbf{Z}}$. The map $\sigma(\zeta) = \zeta^d$ is the same as the (bilateral) shift map, and σ defines a homeomorphism of Σ_*.

Let us write $\pi : \Sigma_* \to \mathbf{C}_*$ for the projection $\pi(\zeta) = \zeta_0$ and set $\Sigma_+ = \pi^{-1}(\mathbf{C} - \bar{D})$. Our model dynamical system is now (σ, Σ_+); in fact this is the projective limit of the dynamical system $(\sigma, \mathbf{C} - \bar{D})$. The 2-D analogue of $\mathbf{C} - \bar{D}$ is $J_+^- := J^- - K$. The following should be clear from the definitions.

Proposition 1.31. *If φ^+ extends holomorphically to a neighborhood of J_+^-, then it yields a semiconjugacy $\Phi^+ : J_+^- \to \Sigma_+$ defined at $p \in J_+^-$ by $\Phi^+(p) = (\varphi^+(f^n p))_{n\in\mathbf{Z}}$.*

Problem: Suppose that f is hyperbolic and that φ^+ extends to J_+^-. Is the map Φ^+ injective, i.e., does it give a conjugacy between (f, J_+^-) and the model (σ, Σ_*)?

The 2-D version of the circle variable θ is the real solenoid $\Sigma_0 := \{\zeta \in \Sigma_* : |\zeta_0| = 1\}$. The path components of Σ_0 are homeomorphic to $\mathbf{R}$. The "natural" picture to draw for a complex Hénon map is to start with a saddle point p. The stable and unstable manifolds $W^{s/u}(p)$ are conformally equivalent to $\mathbf{C}$. Let $\psi : \mathbf{C} \to W^u(p)$ denote a uniformization such that $\psi(0) = p$. Any other such uniformization $\hat{\psi}$ is given by $\hat{\psi}(\zeta) = \psi(\alpha\zeta)$ for some $\alpha \in \mathbf{C}_*$. A useful picture, then, is to draw the

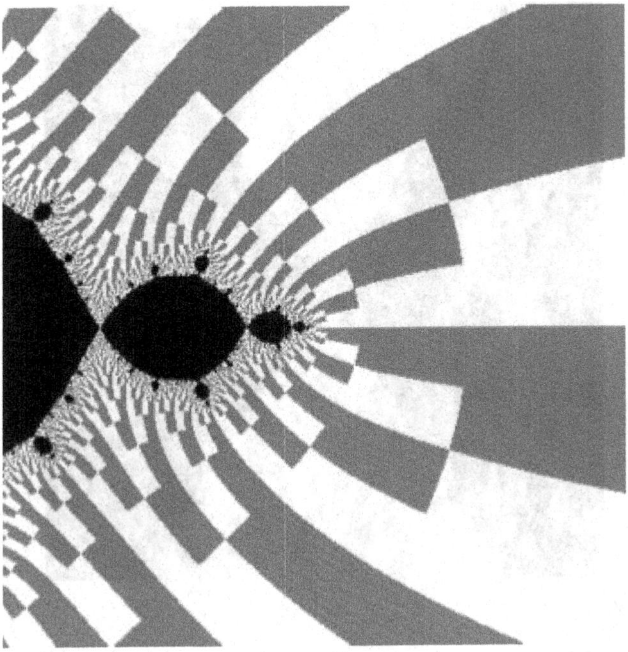

Fig. 3 The map $f(x,y) = (y, y^2 - 1 - .01x)$, slice $W^u(q_\alpha) \cap K^+$

level sets of $G^+ \circ \psi$ and the gradient lines, which are parametrized by a solenoidal variable $s \in \Sigma_0$. If λ is the unstable multiplier for Df at p, then we will have the relationship $d \cdot G^+(\psi(\zeta)) = G^+(\psi(\lambda\zeta))$, which means that the picture will be self-similar about the origin with a factor of λ.

Figures 3 and 4 show this phenomenon for the map $f(x,y) = (y, y^2 - 1 - .01x)$. There are saddle points $\{q_\alpha, q_\beta\}$ which are "close" to fix_α and fix_β. Figure 3 shows the slice through $W^s(q_\alpha)$, and the appearance is consistent with the idea that Figure 3 is obtained by expanding Figure 2, centered at fix_α, by the unstable multiplier infinitely many times until it becomes self-similar. Similarly, Figure 4 looks like Figure 2 made self-similar by expanding about fix_β. The black in Figures 3 and 4 is $W^u(q_\alpha) \cap K^+$; the white/gray parts are the intersection of $W^u(q_\alpha)$ with the basin of infinity, i.e., $W^u(q_\alpha) \cap U^+$. A starting point $(x,y) = \psi(\zeta)$ is iterated until $|y_n| > 1000$. Then the picture is colored white/gray depending on whether $Im(y_n) > 0$ or not. If we want to interpret this in terms of an angular or "argument" variable, we must consider it as an "argument" of Φ^+; this so-called argument, then, would be interpreted as an element of the real solenoid Σ_0. The slices $W^u(q) \cap U^+$ appear to be simply connected in Figures 3 and 4, and the pictures are self-similar with respect to the origin, so they would also be simply connected in the large. Thus we see that it would not work to have an argument (mod 2π); working in the solenoid, the "argument" is obtained as a value in $\mathbf{R}$. The binary nature of the color scheme corresponds to the fact that we may also represent the real solenoid Σ_0 in terms of binary sequences: $\Sigma_0 \cong \{0,1\}^{\mathbf{Z}}/\sim$.

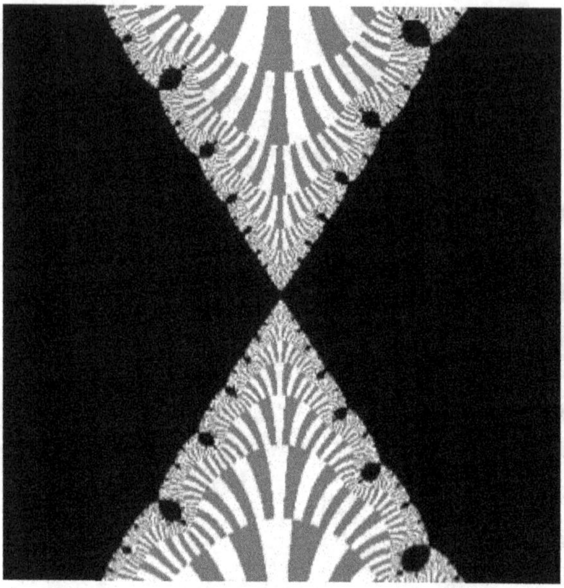

Fig. 4 Slice $W^u(q_\beta) \cap K^+$

Problem Suppose that J is connected, and f is hyperbolic. How do you write J as a quotient of the solenoid? What quotients can appear?

Notes. The geometry of U^+ is discussed further in [HO]. The problem of writing J as a quotient of the solenoid is discussed in [BS7]. The thesis of Oliva [O] gives a number of computer experiments that are helpful in understanding what identifications can appear when writing J as a quotient. The papers [I] and [IS] give a combinatorial/topological approach to this problem.

1.7 Currents on $\mathbf{R}^N$

Let $\mathscr{D}^k$ be the set of smooth k-forms on $\mathbf{R}^N$ with compact support. We define the space of k-currents to be the dual: $\mathscr{D}'_k := (\mathscr{D}^k)'$. A basic example is the *current of integration*. Let $M \subset \mathbf{R}^N$ be a k-dimensional submanifold which is oriented and has locally bounded (k-dimensional) area. Then the current of integration $[M]$ is defined by

$$\mathscr{D}^k \ni \varphi \mapsto \langle [M], \varphi \rangle = \int_M \varphi.$$

If φ is a k-form, and if $\mathbf{t}$ is a k-vector, then $x \mapsto \varphi(x) \cdot \mathbf{t}$ is a scalar-valued function. Now let ν be a Borel measure on $\mathbf{R}^N$, and let $\mathbf{t}(x)$ denote a field of k-vectors on $\mathbf{R}^N$ which is Borel measurable, and which is locally integrable with respect to ν. Then the current $T = \mathbf{t}\nu$ is defined by the action

$$\mathscr{D}^k \ni \varphi \mapsto \langle T, \varphi \rangle := \int \varphi(x) \cdot \mathbf{t}(x)\, \nu(x).$$

We see that currents of the form $T = \mathbf{t}\,\nu$ have the property that for each compact set K, there is a constant c_K such that

$$|\langle T, \varphi \rangle| \le c_K \sup_{x\in K} |\varphi(x)|$$

for every $\varphi \in \mathscr{D}^k$ with support in K. Currents with this property are said to be *represented by integration*. Recall that the dual space of the continuous functions may be identified with a space of measures. In a similar way, the currents represented by integration may be identified with polar representations $\mathbf{t}\,\nu$.

Currents of integration may be represented in this form, which is called the *polar representation*. Namely, we let dS_M denote the Euclidean k-dimensional surface measure on M, and at each $x \in M$, we let $\mathbf{t}(x)$ denote the k-vector which has unit Euclidean length, and which gives the orientation of the tangent space of M at x. Thus we have

$$\int_M \varphi = \int_M \mathbf{t}(x) \cdot \varphi(x)\, dS_M(x)$$

Examples. Let $\mathbf{C}^2_{x,y}$, and write $x = t + is$ and $y = u + iv$. Let dS denote the 3-dimensional surface measure on the set $\mathbf{R} \times \mathbf{C} = \{s = 0\}$. We give some examples of currents which correspond to different ways in which we might "laminate" or "stratify" $\mathbf{R} \times \mathbf{C} = \{s = 0\}$. Let us define

$T_0 := dS$. This is a measure, and as a current it has dimension 0.

$T_1 := \partial_x dS$. Here ∂_x denotes the 1-vector dual to dx. We may fill the set $\mathbf{R} \times \mathbf{C}$ by the disjoint family of lines $\mathbf{R} \times \{y_0\}$ for all $y_0 \in \mathbf{C}$. The current T_1 may be described as the family of 1-dimensional currents of integration $[\mathbf{R} \times \{y_0\}]$, averaged with respect to area measure on $\mathbf{C}_y$. Thus, as a current, T_1 has dimension 1. We may write $T_1 = \int_{\mathbf{C}} du\,dv\,[\mathbf{R} \times \{y = u + iv\}]$, where we interpret the "current-valued" integral as follows:

$$\langle T_1, \xi \rangle = \int_{\mathbf{C}} du\,dv \left(\int_{\mathbf{R}_x \times \{y=u+iv\}} \xi \right).$$

$T_2 := \partial_u \wedge \partial_v\, dS$. This is a 2-dimensional current which is in some sense dual to T_1. Namely, we consider $\mathbf{R} \times \mathbf{C}$ to be filled by the disjoint family of complex lines $\{t_0\} \times \mathbf{C}$. The current T_2 acts by averaging the currents of integration $[\{t_0\} \times \mathbf{C}]$ with respect to dt. In notation analogous to what we used for T_1, we may write

$$T_2 = \int_{t\in\mathbf{R}} dt\,[\{x = t\} \times \mathbf{C}].$$

$T_3 := \partial_t \wedge \partial_u \wedge \partial_v\, dS = [\mathbf{R} \times \mathbf{C}]$ is the (3-dimensional) current of integration. We may interpret the space of $N-k$-forms, $\mathscr{A}^{N-k}$, as currents. We define $\iota : \mathscr{A}^{N-k} \to \mathscr{D}'_k$ from $N-k$-forms to currents of *degree* $N-k$ or *dimension* k by considering them as densities with respect to the current of integration:

$$\mathscr{A}^{N-k} \ni \eta \mapsto \iota\eta := \eta\,[\mathbf{R}^N] \in \mathscr{D}'_k,$$

which acts on k-forms as

$$\varphi \mapsto \langle \varphi\eta, [\mathbf{R}^N] \rangle = \int_{\mathbf{R}^N} \varphi \wedge \eta.$$

If we say that a current T is *smooth*, we mean that there is a smooth form η such that $T = \iota\eta$. Note that if f is a diffeomorphism, then we may pull back in two different ways: $f^* : \mathscr{D}^k \to \mathscr{D}^k$ or $(f^{-1})^* : \mathscr{D}^k \to \mathscr{D}^k$. The second operation is equivalent to a push-forward f_*. Taking adjoints, we may push forward $f_* : \mathscr{D}'_k \to \mathscr{D}'_k$. And pushing currents forward by f^{-1} is equivalent to pulling them back by f. These operations are natural with respect to taking currents of integration:

$$f^*[M] = [f^{-1}M], \quad f_*[M] = [fM].$$

There is also the exterior derivative $d : \mathscr{D}^k \to \mathscr{D}^{k+1}$, and its adjoint $d : \mathscr{D}'_{k+1} \to \mathscr{D}'_k$. The operator d on currents is a natural generalization of the boundary operator, since if M is a compact manifold-with-boundary, the statement of Stokes theorem becomes simply $d[M] = [\partial M]$.

1.8 Currents on $\mathbf{C}^N$ *and Especially* $\mu^\pm$

The forms $dz_{i_1} \wedge \cdots \wedge dz_{i_p} \wedge d\bar{z}_{j_1} \wedge \cdots \wedge d\bar{z}_{j_q}$ are said to have bidegree p,q. We may grade the k-forms according to their bidegrees: $\mathscr{D}^k = \bigoplus_{p+q=k} \mathscr{D}^{p,q}$. We may apply the complex conjugation operator to p,q-forms and obtain forms of bidegree q,p. A form α is *real* if $\bar{\alpha} = \alpha$. If we write $z = x + iy$, then $\frac{i}{2} dz \wedge d\bar{z} = dx \wedge dy$, so this is a real form. On the other hand, if $\beta = dz \wedge d\bar{z}$, then $\bar{\beta} = -\beta$.

A current T is *positive* if it is of type (p,p), and if $\langle T, \bigwedge_{j=1}^p \frac{i}{2} \alpha_j \wedge \bar{\alpha}_j \rangle \geq 0$ for all $\alpha_j \in \mathscr{D}^{1,0}$.

Theorem 1.32. *A current of integration $[M]$ over an oriented, 2-dimensional submanifold of $\mathbf{C}^2$ is real if and only if M is complex. In this case $[M]$ is positive if and only if it is real.*

Exercise: Let M be a real, oriented submanifold of $\mathbf{C}^N$. Under what conditions on M is the current $[M]$ real?

Theorem 1.33. *If T is a positive current, then T is represented by integration.*

We may split the operator d into its parts of type 1,0 and 0,1 respectively, which produces the splitting $d = \partial + \bar{\partial}$. Thus we have $\partial = \sum_j \partial_{z_j} dz_j$. We define $d^c = i(\bar{\partial} - \partial)$. In dimension 1, we write $z = x + iy$, and we have

$$dd^c = 2i\partial\bar{\partial} = 2i dz \wedge d\bar{z} \partial_z \partial_{\bar{z}} = 4 dx \wedge dy \partial_z \partial_{\bar{z}} = dx \wedge dy \Delta$$

where $\Delta = \partial_x^2 + \partial_y^2$ is the Laplacian. In dimension $N > 1$, the operator dd^c is not a scalar object, and when it acts on functions, it is essentially equal to the $n \times n$ hermitian matrix of second derivatives $(\partial_{z_j} \partial_{\bar{z}_k})$.

The currents we are interested are

$$\mu^+ := \frac{1}{2\pi} dd^c G^+, \quad \mu^- := \frac{1}{2\pi} dd^c G^-.$$

These are positive, closed currents which satisfy $f^*\mu^+ = d\cdot\mu^+$ and $f^*\mu^- = d^{-1}\cdot\mu^-$. Further, we have

Proposition 1.34. *The support of μ^+ is J^+.*

Proof. We have seen that G^+ is pluri-harmonic on the complement of J^+, so the support of μ^+ is contained in J^+. On the other hand, if $p\in J^+$, then we have $G^+(p)=0$, but $G^+>0$ at points of $\mathbf{C}^2-K^+$ arbitrarily close to p. Since $G^+\geq 0$, it follows from the maximum principle that G^+ cannot be pluriharmonic in a neighborhood of p. On the other hand, if p is not in the support of dd^cG^+, then G^+ must be pluriharmonic there. □

Of course, the defining property of currents is that they act on test forms. Another operation that is basic for μ^+ is taking slice measures. That is, suppose that $M\subset\mathbf{C}^2$ is a 1-dimensional complex submanifold. We may define the slice measure

$$\mu^+\Big|_M := \frac{1}{2\pi}dd^c_M G^+\Big|_M$$

That is, we restrict G^+ to the manifold M and we take dd^c intrinsically to M. The continuity of G^+ means that the slice measure depends continuously on M.

If f is a hyperbolic Hénon map, we can present a heuristic picture to show the connection between the geometry of the stable manifolds and the current μ^+. By the Stable Manifold Theorem, the stable manifolds $\mathcal{W}^s=\mathcal{W}^s(J)$ form a lamination in a neighborhood of J. That is there are local charts U for which $\mathcal{W}^s\cap U$ is homeomorphic to $A\times D$, where A is closed, and D is a disk. Figure 5 shows two complex manifold M_1 and M_2 which are transverse to $\mathcal{W}^s$. The set A_j is $M_j\cap\mathcal{W}^s$. For $\alpha\in A$, we define the local stable disk D^s_α to be the component of $W^s(\alpha)\cap U$ containing α. There is also a *holonomy* map χ which takes the intersection points $D^s_\alpha\cap M_1$ to $D^s_\alpha\cap M_2$. A property of the current μ^+ is that the holonomy map transports the slice measure $\mu^+|_{M_1}$ to the slice measure $\mu^+|_{M_2}$. So, within this flow box, all slice measures are equivalent via the holonomy map.

We can use this to generate the local pieces of μ^+. Since D^s_α has locally finite area it defines a current $[D^s_\alpha]$. We define a current by integrating these currents of integration with respect to the slice measure:

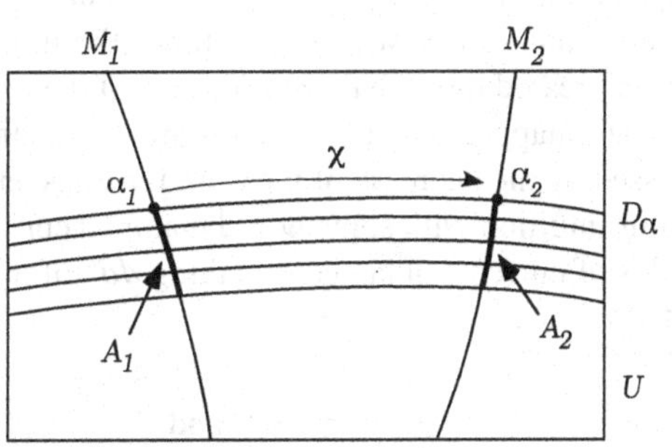

Fig. 5 Holonomy map χ

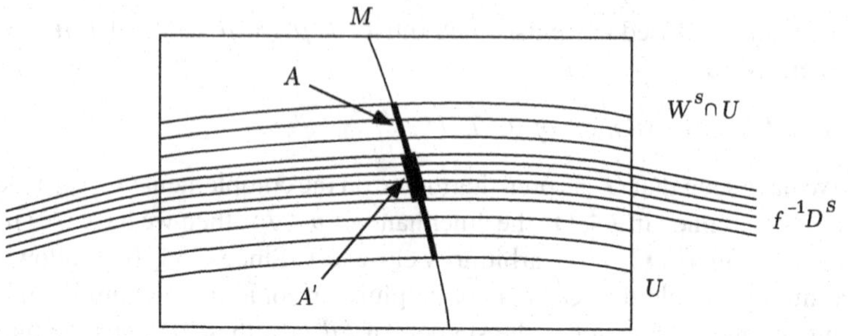

Fig. 6 Expansion in terms of transversal measure

$$\int_{\alpha\in A}[D^s_\alpha]\ \mu^+|_M(\alpha)$$

where we use the notation $\mu^+|_M(\alpha)$ in the integral to mean that we are integrating with respect to the variable α and the measure $\mu^+|_M$. It is shown in [BS1] that the restriction $\mu^+\llcorner U$ can be written in this manner.

Now let us interpret the relationship between "expansion" of the current μ^+ and a more directly geometrical sort of expansion. First, what is the effect of pulling back the restriction $\mu^+\llcorner U$ under f^*? If we write $\mu^+\llcorner U$ in the laminar form above, we may pull the currents of integration to obtain

$$f^*(\mu^+\llcorner U) = \int_{\alpha\in A}[f^{-1}D^s_\alpha]\ \mu^+|_M(\alpha)$$

Now let us restrict the pullback to U again, and let us suppose that the box U maps across itself as in Figure 6. Then we find that the pullbacks $f^{-1}D^s_\alpha$ will again be disks of $\mathcal{W}^s\cap U$, and these will correspond to a subset of the transversal $A'\subset M$. Thus we may write

$$(f^*\mu^+\llcorner U)\llcorner U = \int_{\alpha'\in A'}[f^{-1}D^s_{\alpha'}]\ \lambda'(\alpha')$$

where λ' is a measure on A' determined by the pullback. The total mass of λ' on A' is the same as the total mass of $\mu^+|_M$ on A. The fact that $f^*\mu^+ = d\cdot\mu^+$ means the amount of transversal mass of A' will be d^{-1} times the transversal mass of A. Thus the expansion is expressed through the contraction of the "thickness" of bands of stable disks as they are mapped under f^{-1}. Thus the contraction is not precisely geometric, it is expressed by the decrease of measure for transversals.

Now we develop our intuition with some non-dynamical currents. Let us look at some simpler examples of currents obtained by taking dd^c of a psh function. First let us define for $z\in\mathbf{C}$:

$$L(z) = \log|z|,\ \ L_\varepsilon(z) = \log|z|,\ \text{if } |z|\ge\varepsilon,\ \text{and } \frac{|z|^2}{2\varepsilon^2}+C_\varepsilon,\ \text{if } |z|\le\varepsilon.$$

Thus L_ε is globally C^1 and piecewise C^2. Thus we may compute:

$$dd^c L_\varepsilon = \frac{2}{\varepsilon^2} \cdot \text{area measure}|_{|z|<\varepsilon}.$$

Since L_ε decreases to L, it converges in the topology of $\mathscr{D}'_2$, so $dd^c L_\varepsilon \to dd^c L$ as $\varepsilon \to 0$. Since $dd^c L_\varepsilon$ is a measure with total mass 2π, we see that we obtain

$$dd^c L = 2\pi\delta_0.$$

Now let us apply a similar argument to the function $L_\varepsilon(x)$ on $\mathbf{C}^2$. We find, this time, that

$$dd^c L_\varepsilon = \frac{2}{\varepsilon^2}\int_{|x|<\varepsilon} dA(x)\,[\{x\}\times\mathbf{C}].$$

That is, we obtain an average of the currents of integration over $\{x\}\times\mathbf{C}$. Letting $\varepsilon \to 0$, we obtain the Poincaré-Lelong formula in the case of a line:

$$\frac{1}{2\pi} dd^c \log|x| = [\{x=0\}].$$

Now we may interpret this for submanifolds which are not necessarily linear. Suppose that $M = \{h=0\}$, where h is holomorphic, and the gradient of h does not vanish on M. Observe that if $\tilde{h}$ is any other such function, then we have $\tilde{h} = \alpha h$, where α is holomorphic and nonvanishing in a neighborhood of M. It follows that $\log|\tilde{h}| = \log|h| + \log|\alpha|$, and $dd^c \log|\alpha| = 0$. Thus we have $dd^c \log|h| = dd^c \log|\tilde{h}|$.

At each point of $\{x=0\}$, there is a (locally) biholomorphic map φ of $\mathbf{C}^2$ with $\varphi : \{x=0\} \to M$. Thus $\varphi^* h = h\circ\varphi$ is a defining function for $\{x=0\}$, so we have $\varphi^* h = x\cdot\alpha$ for some invertible holomorphic function α. Thus we have $dd^c \log|h| = \varphi_* dd^c \log|\varphi^* h| = \varphi_* dd^c \log|x\cdot\alpha| = \varphi_* dd^c \log|x| = [M]$. We summarize this as:

Theorem 1.35. *(Poincaré-Lelong formula)*

$$\frac{1}{2\pi} dd^c \log|h| = [\{h=0\}].$$

Exercises: (1) Let $\chi(x,y) = \max\{Re\, x, 0\}$. Show that $\frac{1}{2\pi} dd^c \chi = T_2$. (Hint: You may deal with this like the case with L above. Consider the maps $\chi_\varepsilon(t) = 0$ for $t \le 0$, $\chi_\varepsilon(t) = t^2/(2\varepsilon)$ for $0 < t < \varepsilon$, and $\chi_\varepsilon(t) = t - \varepsilon/2$ for $t \ge \varepsilon$.)
(2) Use an exponential change of variable on (1) to obtain:

$$dd^c \log^+|x| = \int_0^{2\pi} d\theta\,[\{x = e^{i\theta}\}\times\mathbf{C}].$$

Theorem 1.36. *The currents $d^{-n}[f^{-n}\{y=0\}]$ converge to μ^+ as $n\to+\infty$.*

Proof. By Theorem 1.28, we know that $\lim_{n\to\infty} d^{-n}\log|y_n| = G^+$. If we apply $\frac{1}{2\pi} dd^c$ to both sides of this equation, we obtain the Theorem from the Poincaré-Lelong formula. □

This Theorem gives us a measure-theoretic sense of the location of $f^{-n}\{y=0\}$. Namely, if B is a ball that is disjoint from J^+, we can let M_n denote the number of connected components of $B\cap f^{-n}\{y=0\}$. From the Theorem, we conclude that $d^{-n}M_n \to 0$ as $n\to\infty$.

Notes: Many of the important properties of μ^+ come from its laminar structure, which must be interpreted in a measure-theoretic sense. A good starting place for entering the literature is to look at [Du, §2].

1.9 Convergence to μ^+

Much of the utility of the invariant current μ^+ comes from various convergence theorems. The proofs of these results use some potential theory; the relevant tools have been well developed in [S]. So we will state two results below and refer the reader to [S] for the proofs.

First we consider a pair of complex disks $D\subset\tilde{D}$ such that $\tilde{D}$ contains the closure of D. As we saw in the previous section, we may slice the invariant current μ^- along $\tilde{D}$ to obtain a slice measure $\mu^-|_{\tilde{D}}$. The main convergence theorem is:

Theorem 1.37. *Let D and $\tilde{D}$ be complex disks such that the closure of D is contained in $\tilde{D}$. Suppose that the slice measure $\mu^-|_{\tilde{D}}$ puts no mass on ∂D, and let c be the measure $\mu^-|_{\tilde{D}}(D)$. Then the normalized pullbacks $d^{-n}f^{*n}[D]$ converge in the weak sense of currents to $c\mu^+$.*

For an alternative formulation: if we replace $[D]$ by $\chi[D]$, where χ is a test function with support in D, then we may omit the hypothesis that ∂D has no mass for the slice measure, and we use the value $c=\int_D\chi\,\mu^-|_{\tilde{D}}$.

We get several consequences from this theorem.

Corollary 1.38. *Let p be a saddle point, and let $W^s(p)$ be its stable manifold. Then J^+ is the closure of $W^s(p)$. In particular, J^+ is connected.*

Proof. We have seen already that $W^s(p)\subset J^+$. Let D^s be a disk inside $W^s(p)$. We may assume that $p\in D^s$. It follows that $\mu^-|_{D^s}$ is not the zero measure, so we may choose χ so that $\int_{D^s}\chi\,\mu^-|_{\tilde{D}}>0$. Each pullback $d^{-n}f^{n*}(\chi[D^s])$ has support in $W^s(p)$. Since the sequence of currents converges to μ^+, and J^+ is the support of μ^+, the Theorem follows. □

Proof of Theorem 1.22. Let D denote a linear complex disk inside U_-. We may choose D so that it intersects both J^- and the complement of K^-. Thus the slice $\mu^-|_D$ is not the zero measure, and we may choose a test function χ such that $c>0$. Now since the pullbacks converge to μ^+, their supports must be dense in J^+, so we see that for some large n, we will have $f^{-n}D\cap U_+\neq\emptyset$. This proves the theorem. □

We may wedge the currents μ^+ and μ^- together to obtain a measure μ. In order to define a measure, it suffices to show how to integrate a function χ with respect to μ. If χ is smooth, then we can define

$$\int\chi\mu := \int(dd^c\chi)\wedge G^+\mu^- \tag{5}$$

which would be exactly what you would obtain G^+ and G^- were smooth and you integrate by parts: $\int \chi dd^c G^+ \wedge dd^c G^- = \int \chi dd^c(G^+ \wedge \mu^-) = \int (dd^c \chi) \wedge (G^+ \wedge \mu^-)$. This defines μ as a distribution. But since $\mu^{\pm}$ are a positive currents, we see that (*) defines μ as a positive distribution, and thus a measure.

A variant of the convergence theorem above deals with the normalized pullbacks $d^{-n} f^{n*}(\chi \mu^+) = \chi(f^n)\mu^+$:

Theorem 1.39. *Let χ be a test function, and let $c = \int \chi \mu$. Then*

$$\lim_{n\to\infty} d^{-n} f^{*n}(\chi \mu^+) = \lim_{n\to\infty} \chi(f^n)\mu^+ = c\mu^+.$$

A measure ν is said to be *mixing* with respect to a transformation f if $\lim_{n\to\infty} \nu(A \cap f^{-n}B) = \nu(A)\nu(B)$ for all Borel sets A and B. For this, it suffices to show that for every pair of smooth functions φ and ψ, we have

$$\lim_{n\to\infty} \int \varphi(f^n)\psi \mu = \int \varphi \mu \int \psi \mu$$

Corollary 1.40. *The measure μ is mixing (and thus ergodic).*

Proof. We have $\int \varphi(f^n)\psi \mu = \int (\varphi(f^n)\mu^+) \wedge (\psi \mu^-)$, so as $n \to \infty$, the right hand integral converges to $\int (c\mu^+) \wedge \psi \mu^- = c \int \psi \mu^+ \wedge \mu^- = c \int \psi \mu$, and the constant is $c = \int \varphi \mu$. □

2 Rational Surfaces

2.1 Blowing Up

In the second section of these notes, we will consider automorphisms of a compact, rational surface M. Given an automorphism $f \in Aut(M)$, we can look at its pullback on cohomology $f^* \in GL(H^2(M))$. The dynamical degree is then defined as

$$\lambda(f) := \lim_{n\to\infty} ||f^{n*}||^{1/n}.$$

A surface M' that is birationally equivalent to M may be topologically different: for instance, $H^2(M')$ and $H^2(M)$ can have different dimensions, and so the induced maps on cohomology will not be conjugate. However, in [DF] it is shown that $\lambda(f)$ is the same for all birationally equivalent maps.

Here we will focus on automorphisms for which $\lambda(f) > 1$. Although we will not discuss entropy here, we note that in this case the entropy is $\log(\lambda(f)) > 0$. Much of the theory for Hénon maps can be carried over to the case of automorphisms of compact surfaces. However, the Hénon family of diffeomorphisms themselves will not be part of this section: for a Hénon map H, there is no compact, complex manifold M which compactifies $\mathbf{C}^2$ in such a way that H becomes a homeomorphism of M.

(Exercise: The most obvious first attempt, the one point compactification $\hat{\mathbf{C}}^2$ of $\mathbf{C}^2$, is not a complex manifold.) We note, too, that we do not discuss are the complex 2-tori or the $K3$ surfaces, which are the other possibilities for projective surfaces with automorphisms for which $\lambda(f) > 1$ (see [C1]). In fact, rational surface automorphisms are more abundant than these other two cases, so we will be drawing from a rich family of dynamical systems.

In §1 we used the dynamical classification of $PolyAut(\mathbf{C}^2)$ which shows that the Hénon diffeomorphisms represent the conjugacy classes with positive entropy. On the other hand, it is not easy to determine the set of all rational surfaces that admit nontrivial automorphisms; and given such a surface, it is not easy to determine its automorphisms. In other words, an analogous dynamical classification of rational surface automorphisms is not yet known.

Here we focus on the surfaces that are obtained from $\mathbf{P}^2$ by blow-ups; this focus is justified by a Theorem of Nagata which is stated in §2.7. However, we note that a surface constructed from $\mathbf{P}^2$ by making "generic" blowups will not have any automorphisms except the identity (see [H, K]). Our starting place will be complex projective space $\mathbf{P}^n$, which is $\mathbf{C}^{n+1} - 0$, modulo the equivalence $(x_0,\dots,x_n) \sim (\lambda x_0,\dots,\lambda x_n)$ for any $\lambda \in \mathbf{C}$ with $\lambda \neq 0$. We write $[x_0 : \cdots : x_n]$ (square brackets to denote homogeneous coordinates) for the equivalence class. It is classical that $Aut(\mathbf{P}^n) = PGL(n+1,\mathbf{C})$, and $Aut(\mathbf{P}^1 \times \mathbf{P}^1)$ is the group generated by $PGL(2,\mathbf{C}) \times PGL(2,\mathbf{C})$, together with the map $\tau(x,y) = (y,x)$.

The blow-up of a manifold X at a point $p \in X$ is a manifold $\tilde{X}$, together with a projection $\pi : \tilde{X} \to X$ such that the *exceptional fiber* $E := \pi^{-1}p$ is equivalent to $\mathbf{P}^1$, and $\pi : \tilde{X} - E \to X - p$ is biholomorphic. A concrete presentation of this is the space

$$\Gamma = \{(x,y);[\xi : \eta] \in \mathbf{C}^2 \times \mathbf{P}^1 : x\eta = y\xi\}$$

together with the projection π to the first coordinate, so we define the pair (Γ,π) to be the blow-up of $\mathbf{C}^2$ at 0. This construction is essentially local at the center of blowup. Thus we may use this construction of Γ to define the blowup of X at the point p.

Observe that $\pi^{-1} : \mathbf{C}^2 - 0 \to \Gamma$ is the tautological map $\pi^{-1}(x,y) = (x,y);[x : y]$. We may also write $\pi^{-1}(x,y) = (x,y);[1 : y/x] = (x,y);[x/y : 1]$, and we may use these two representations to define local coordinate charts $U' = \mathbf{C}^2_{s,\eta}$ and $U'' = \mathbf{C}^2_{\xi,t}$, where the coordinates are defined via the form that the projection π takes:

$$\pi' : (s,\eta) \mapsto (s,s\eta) = (x,y), \quad \pi'' : (\xi,t) \mapsto (\xi t,t) = (x,y).$$

It follows that $U' \cup U'' = \Gamma$, and $E' := E \cap U' = \{s = 0\}$ is equivalent to $\mathbf{C}$.

We may pull the 2-form $dx \wedge dy$ back to Γ; in the U' coordinate chart, for instance, we have $(\pi')^* dx \wedge dy = ds \wedge d(s\eta) = s\, ds \wedge d\eta$.

We will use the following notational convention: if C is a curve in X, then C will also denote the *strict transform* which is the curve in $\tilde{X}$ which is obtained by taking the closure of $\pi^{-1}(C-p)$ inside $\tilde{X}$. In the blowup, the strict transform of the x-axis $\{y = 0\}$ inside Γ is given by $\{\eta = 0\}$. Thus we use the (s,η) coordinate system if we want to work in a neighborhood of the x-axis, and we use the (ξ,t) coordinate system if we want to work in a neighborhood of the y-axis.

Exercises:

1. Show that Γ is smooth.
2. Let $f : \mathbf{C}^2,0 \to \mathbf{C}^2,0$ be a local biholomorphism. Show that f lifts to a locally invertible holomorphic map f_Γ of Γ, and f_Γ maps E to itself E. Show that if f is not locally invertible at 0, then f_Γ is not everywhere holomorphic on E.
3. Let X_1 denote the space obtained by blowing up $\mathbf{P}^2$ at two points, and let X_2 denote the space obtained by blowing up $\mathbf{P}^1 \times \mathbf{P}^1$ at one point. Show that X_1 and X_2 are biholomorphic, and thus $\mathbf{P}^2$ is birationally equivalent to $\mathbf{P}^1 \times \mathbf{P}^1$. (The easiest way to do this is to let the two points of blowup of $\mathbf{P}^2$ be the points where the x- and y- axes cross the line at infinity; and blowup $\mathbf{P}^1 \times \mathbf{P}^1$ at (∞,∞).)

We define a divisor to be a linear combination $D = \sum c_j D_j$, where D_j is a hypersurface in X. We say that divisors D' and D'' are linearly equivalent if $D' - D''$ is the divisor of a rational function r. That is, $D' - D''$ is equal to the zero set (with multiplicity) of r minus its pole set. We define the Picard group $Pic(X)$ to be the set of divisors modulo linear equivalence.

Proposition 2.1. $Pic(\mathbf{P}^2) \cong \mathbf{Z}$.

Proof. If D is a hypersurface, then $D = \{p([x_0 : x_1 : x_2]) = 0\}$ is defined by a homogeneous polynomial of some degree d. Thus $r = p/x_0^d$ is a well defined rational function on $\mathbf{P}^2$, so we see that as an element of $Pic(\mathbf{P}^2)$, D is equivalent to d times the line $\{x_0 = 0\}$. Similarly, we see that all lines define the same element, and this generates $Pic(\mathbf{P}^2)$. □

Remark: In the proof of the Proposition, we saw that if $H \in Pic(\mathbf{P}^2)$ is the class of a line, then $C \sim d \cdot H \in Pic(\mathbf{P}^2)$, where d is the degree of C.

Theorem 2.2. *If f is an automorphism of $\mathbf{P}^2$, then f must have degree 1.*

Proof. The action f^* on $Pic(\mathbf{P}^2)$ is multiplication by d, where d is the degree of f. If f is invertible, then the action of ${f^{-1}}^*$ must be multiplication by d_1, where d_1 is the degree of f^{-1}. On the other hand, we must have $1 = id^* = (f \circ f^{-1})^* = dd_1$, so $d = d_1 = 1$, since d and d_1 are both positive integers. □

Now let $\tilde{\mathbf{P}}^2$ denote $\mathbf{P}^2$ blown up at a point p, and let P denote the exceptional fiber. The rational functions on $\tilde{\mathbf{P}}^2$ are defined to be the pullbacks under π of rational functions on $\mathbf{P}^2$, i.e., $Rat(\tilde{\mathbf{P}}^2) := \pi^* Rat(\mathbf{P}^2)$. It follows that $Pic(\tilde{\mathbf{P}}^2)$ is generated by $\tilde{L}$ and P, where $L = \{\ell = \sum a_j x_j = 0\}$ is a line, and $\tilde{L} := \pi^* L = \{\ell \circ \pi = 0\}$ is the pullback. More generally, we consider N distinct points $p_1, \dots, p_N$ and let X denote the complex manifold obtained by blowing up $\mathbf{P}^2$ at the points p_j, $1 \le j \le N$. We will let $H_X = \pi^* H$ denote the class of a line; if $H \subset \mathbf{P}^2$ which does not contain any p_j, $1 \le j \le N$, then H_X is represented by the strict transform $\tilde{H}$. We let $P_j = \pi^{-1} p_j$ denote the class in $Pic(X)$ of the exceptional fiber. In general, if $H \subset \mathbf{P}^2$ is a line, then $H_X = \tilde{H} + \sum' P_j$, where the sum is taken over the indices j for which $p_j \in H$. By the discussion above, we see that:

Theorem 2.3. *Suppose that $\pi : X \to \mathbf{P}^2$ is obtained by blowing up N distinct points $p_1,\dots,p_N \in \mathbf{P}^2$. If $P_j = \pi^{-1}p_j$ denotes the exceptional fiber, then $Pic(X)$ is generated by H_X and P_j, $1 \le j \le N$.*

Theorem 2.4. *Let X denote $\mathbf{P}^2$ blown up at distinct points $p_1,\dots,p_N$, and consider two elements $T' = D' + \sum a'_j P_j$ and $T'' = D'' + \sum a''_j P_j$ in $Pic(X)$, where D' and D'' denote strict transforms of divisors in $\mathbf{P}^2$. It follows that $T' \sim T'' \in Pic(X)$ if and only if D' and D'' have the same degrees, and $a'_j = a''_j$ for all j.*

Proof. We may replace D' and D'' by $d' \cdot H$ and $d'' \cdot H$, and we may suppose that H is disjoint from the centers of blowup. If $T' \sim T''$, then there is a rational function $r_X = \pi^* r$ whose divisor is $T' - T''$. It follows that the divisor of r is $(d' - d'')H$. Thus we must have $d' - d'' = 0$ and $a'_j - a''_j = 0$. □

2.2 Cohomology

A basic result is that $H^2(\mathbf{P}^2;\mathbf{Z}) \cong \mathbf{Z}$. By the DeRham theorem, we may represent cohomology classes by closed 2-forms. An example of a global closed 2-form may be written loosely as $\omega := dd^c \log(|x_0|^2 + |x_1|^2 + |x_2|^2)$, which we interpret as follows. On the coordinate chart $\mathbf{C}^2_{x,y} \ni (x,y) \mapsto [1:x:y] \in \mathbf{P}^2$, we write $\omega' = dd^c \log(1+|x|^2+|y|^2)$. On the coordinate chart $\mathbf{C}^2_{t,v} \ni (t,v) \mapsto [t:1:v] \in \mathbf{P}^2$, we have $\omega'' = \log(|t|^2 + 1 + |v|^2)$. These coordinates are related by $[1:x:y] = [t:1:v]$, so $t = 1/x$ and $v = y/x$. Thus $\omega' - \omega'' = dd^c \log|x|^2 = dd^c \log|t|^{-2} = 0$ on the overlap $\{x_0 x_1 \ne 0\}$ of these two coordinate charts, so this definition is well-defined globally. *Exercise:* $\log(1+|x|^2+|y|^2)$ is strictly psh on $\mathbf{C}^2$, and thus $\omega > 0$. Find $c > 0$ such that $\int_L c\omega = 1$ for some (or equivalently, every) line $L \subset \mathbf{P}^2$.

In complex dimension 2, there is a duality on $H^2(X;\mathbf{C})$ which is given by $(\alpha,\beta) = \int \alpha \wedge \bar{\beta}$. We will sometimes write this as $\alpha \cdot \beta$ and call it the *intersection product*. The Poincaré duality theorem says that H^2 is self-dual under this pairing. In the case of Kähler manifolds, we have the following Signature Theorem (see Theorem IV.2.14 of [BHPV]).

Theorem 2.5. *Let X be a compact Kähler surface, and let $h_{1,1}$ denote the dimension of $H^{1,1}(X;\mathbf{R}) \subset H^2(X;\mathbf{R})$. Then the signature of the restriction of the intersection product to $H^{1,1}$ is $(1, h_{1,1} - 1)$. In particular, there is no 2-dimensional linear subspace $L \subset H^{1,1}(X;\mathbf{R})$ with the property that $\omega \cdot \omega = 0$ for all $\omega \in L$.*

For a complex curve C, the current of integration $[C]$ defines an element in the dual of H^2, and thus we may consider C as an element of H^2 as well. With c as in the preceding exercise, we see that $c\omega$ represents the class $\langle L \rangle$, where L is any complex line in $\mathbf{P}^2$. Let $C \subset \mathbf{P}^2$ be a curve of degree μ. Then its class in $H^2(\mathbf{P}^2)$ is $C = \mu H$. We see that if $\tilde{\mathbf{P}}^2$ is the blowup of $\mathbf{P}^2$ at a point p, then we may identify $H^2(\tilde{\mathbf{P}}^2)$ and $Pic(\tilde{\mathbf{P}}^2)$. In the sequel, we will find it convenient to use H^2 and Pic interchangeably. But note that this is a special property of blowups of $\mathbf{P}^2$; for other complex manifolds, it is possible for H^2 and Pic to be very different.

Let C_1 and C_2 be complex curves which intersect transversally. A basic result of intersection theory is that the number of intersection points $C_1 \cap C_2$ is equal to the intersection product (C_1, C_2) of their cohomology classes. We note that for general 2-manifolds, the intersection multiplicity is determined by the orientation. That is, if E_1 and E_2 are smooth 2-manifolds in X which intersect transversally, then $E_1 \cdot E_2$ is equal to the total number of intersection points, counted with multiplicity, which is ± 1, depending on the orientation.

Let us consider the case of the exceptional blowup fiber E in Γ, defined in the previous section. We would like to determine $E \cdot E$, but of course E does not intersect itself transversally. We will perturb E to obtain a new surface $\tilde{E}$ which intersects E transversally. We define $\tilde{E}$ in the U' coordinate system as

$$\tilde{E} \cap U' = \{s = \varepsilon \bar{\eta} : |\eta| \leq 1\} \cup \{s = 1/\eta : |\eta| \geq 1\}$$
$$\tilde{E} \cap U'' = \{t = \varepsilon : |\xi| \leq 1\} \cup \{t = \varepsilon/|\xi|^2 : |\xi| \geq 1\}$$

By the relation $(x,y) = (s, \eta s)$, for instance, we see that

$$\pi\tilde{E} = \{y = \varepsilon |x|^2 : |x| \leq 1\} \cup \{(x, \varepsilon) : |x| \leq 1\}.$$

We see that this surface is the union of two smooth surfaces, and it is not hard to smoothen $\tilde{E}$ along the circle where the two surfaces intersect. We see that the intersection point of $E \cap \tilde{E}$ occurs at $(0,0) \in U'$. The canonical 2-form which orients E is $i ds \wedge d\bar{s}$. The canonical 2-form which orients $\tilde{E}$ at $\{s - \bar{\eta} = 0\}$ is $i d(s - \bar{\eta}) \wedge \overline{d(s - \bar{\eta})}$. We wedge these together and find $i ds \wedge d\bar{s} \wedge i d\bar{\eta} \wedge d\eta$, which is a negative multiple of the canonical orientation 4-form on X. Thus we have the intersection number $E \cdot \tilde{E} = -1$.

The cohomology class of $E \in H^2$ is represented by the current of integration $[E]$, which is a positive, closed current. However, if ω is a smooth 2-form which represents the class of $E \in H^2$, then $\int \omega^2 = -1$, so we cannot have $\omega \geq 0$ everywhere. In other words, $[E]$ cannot be approximated by a positive, smooth form.

Theorem 2.6. *Suppose that $\pi : \tilde{X} \to X$ is the blowup of X at the center p. Suppose that C is a smooth curve in X containing p, and suppose that $\tilde{C}$ represents the strict transformation of C inside $\tilde{X}$. Then $\tilde{C}^2 = C^2 - 1$.*

Proof. Since C is smooth, the curves $\tilde{C}$ and P are smooth and intersect transversally, with $\tilde{C} \cdot P = 1$. Thus

$$C^2 = (\pi^* C)^2 = (\tilde{C} + P)^2 = \tilde{C}^2 + 2\tilde{C} \cdot P + P^2 = \tilde{C}^2 + 2 - 1,$$

which gives us what we wanted. □

Now let us suppose that C is a curve of degree μ, that $C \cap \{p_1, \ldots, p_N\} = p_1$ and that C is smooth at p_1. Now $\pi^* C = C + P_1 \in H^2(X)$. We conclude, then, that $C = \mu H - P_1$. Conversely, if C is any curve whose cohomology class is $\mu H - P_1$, then C is a curve of degree μ, which contains p_1 with multiplicity 1. "Containing p_1

with multiplicity 1" means, in particular, that C is regular at p_1. In a similar manner, we see that if C is a curve with the cohomology class $H - P_1 - P_2$, then C must be the line $p_1 p_2$.

If X is a surface obtained by repeated blowups, the total cohomology is given by $H^*(X;\mathbf{C}) = H^0(X;\mathbf{C}) \oplus H^{1,1}(X;\mathbf{C}) \oplus H^4(X;\mathbf{C})$. If f is a holomorphic map, then the total map f^* acts on each of these factors. We have $H^0(X;\mathbf{C}) \cong \mathbf{C}$, and $f^*|_{H^0} = 1$. Similarly, the dimension of H^4 is the number of connected components (which is equal to 1), and $f^*|_{H^4}$ is multiplication the mapping degree of f, which is 1 in the case of an automorphism. Thus the Lefschetz Fixed Point Formula takes the form:

Theorem 2.7. *If X is obtained from $\mathbf{P}^2$ by iterated blowups and if each f^n has isolated fixed points, then*

$$Per_n = 2 + trace(f^{*n})$$

where f^* denotes the restriction of f^* to $H^{1,1}$, and Per_n denotes the number of solutions of $\{p \in X : f^n(p) = p\}$, counted with multiplicity.

2.3 Invariant Currents and Measures

Let f be an automorphism of a compact Kähler surface X. Since $f^* \in GL(H^{1,1};\mathbf{Z})$, the determinant of f^* must be ± 1. The pullback and push-forward preserve the intersection product: $(\omega, \eta) = (f^*\omega, f^*\eta) = (f_*\omega, f_*\eta)$. In fact, the push-forward and pullback are adjoint: $f^*\alpha \cdot \beta = \alpha \cdot f_*\beta$. Further, $(f^{-1})^* = (f^*)^{-1}$. And since $\lim_{n\to\infty} ||f_*^n||^{1/n} = \lim_{n\to\infty} ||f^{*n}||^{1/n}$, we have $\lambda(f) = \lambda(f^{-1})$.

Theorem 2.8. *Let $f \in Aut(X)$ be an automorphism of a Kähler manifold with $\lambda(f) > 1$. Then λ is an eigenvalue of f^* with multiplicity 1, and it is the unique eigenvalue with modulus > 1.*

Proof. Let $\omega_1, \dots, \omega_k$ denote the eigenvectors for f^* for which the associated eigenvalues μ_j has modulus > 1. For $1 \le j \le k$ we have

$$(\omega_j, \omega_k) = (f^*\omega_j, f^*\omega_k) = \mu_j \bar{\mu}_k (\omega_j, \omega_k),$$

so $(\omega_j, \omega_k) = 0$. Letting L denote the linear span of $\omega_1, \dots, \omega_k$, we see that each element $\omega = \sum c_j \omega_j \in L$ satisfies $(\omega, \omega) = 0$. By the Signature Theorem, it follows that the dimension of L is ≤ 1. On the other hand, since $\lambda(f) > 1$, L is spanned by a unique nontrivial eigenvector. If ω has eigenvalue μ, then $\bar{\omega}$ has eigenvalue $\bar{\mu}$, so we must have $\mu = \bar{\mu} = \lambda$.

Now we claim that λ has multiplicity one. Otherwise there exists θ such that $f^*\theta = \lambda\theta + c\omega$. In this case,

$$(\theta, \omega) = (f^*\theta, f^*\omega) = (\lambda\theta + c\omega, \lambda\omega) = \lambda^2(\theta, \omega),$$

so $(\theta, \omega) = 0$. Similarly, we have $(\theta, \theta) = 0$, so by the Signature Theorem again, the space spanned by θ and ω must have dimension 1, so λ is a simple eigenvalue. □

Corollary 2.9. *If η is an eigenvalue of f^*, then either $\eta = \lambda, \lambda^{-1}$, or $|\eta| = 1$.*

Proof. We have seen that $\lambda(f)$ is the only eigenvalue of modulus > 1. Now we know that $(f^*)^{-1} = (f^{-1})^*$, so if η is an eigenvalue of f^*, then η^{-1} is an eigenvalue of $(f^{-1})^*$. Applying the Theorem to f^{-1}, we conclude that λ is the only eigenvalue for $(f^{-1})^*$ which is > 1. □

Let χ_f denote the characteristic polynomial of f^*. It follows that χ_f is monic, and the constant term (the determinant of f^*, an invertible matrix) is ± 1. Let ψ_f denote the minimal polynomial of λ. By the Theorem, we see that except for λ and $1/\lambda$, all zeros of χ_f (and thus all zeros of ψ_f) lie on the unit circle. Such a polynomial ψ_f is called a *Salem polynomial*, and λ is a *Salem number*. We may factor $\chi_f = C \cdot \psi_f$, where C is a polynomial whose coefficients belong to $\mathbf{Z}$, and the roots of C lie in the unit circle. It follows by elementary number theory that the zeros of C are roots of unity.

The following is a heuristic argument for the existence of a positive, closed invariant 1,1-current. Suppose that X is a Kähler surface and that $f \in Aut(X)$ has $\lambda(f) > 1$. Let ω^+ be a positive, smooth cohomology class which is an eigenvector of λ. By this we mean that there is a smooth form ω^+ such that the cohomology class is expanded

$$\{f^*\omega^+\} = f^*\{\omega^+\} = \lambda\{\omega^+\}.$$

Thus $f^*\omega^+ - \lambda\omega^+$ is cohomologous to zero, and thus there is a smooth function γ^+ such that

$$\frac{1}{\lambda} f^*\omega^+ - \omega^+ = dd^c\gamma^+.$$

If we can take ω^+ to be poisitive, then $f^*\omega^+$ will be positive, and we see that γ^+ is essentially pluri-subharmonic, i.e., $dd^c\gamma^+ + \omega^+ \geq 0$. Applying $\lambda^{-1} f^*$ repeatedly, we find

$$\begin{aligned} \frac{1}{\lambda^2} f^{*2}\omega^+ - \frac{1}{\lambda} f^*\omega^+ &= \frac{1}{\lambda} dd^c f^*\gamma^+ = \frac{1}{\lambda} dd^c\gamma^+ \circ f \\ \frac{1}{\lambda^n} f^{*n}\omega^+ - \frac{1}{\lambda^{n-1}} f^{*(n-1)}\omega^+ &= \frac{1}{\lambda^{n-1}} dd^c\gamma^+ \circ f^{n-1}. \end{aligned}$$

If we define

$$g^+ = \sum_{n=0}^{\infty} \frac{\gamma^+ \circ f^n}{\lambda^n}, \quad T^+ = \omega^+ + dd^c g^+,$$

then we see that g^+ is continuous, since the defining series converges uniformly. Further, T^+ is a positive, closed current with the invariance property $f^*T^+ = \lambda T^+$. We may apply the same argument with f^{-1} and the eigenvalue λ^{-1} obtain a positive, closed current T^- with the property that $(f^{-1})^*T^- = \lambda T^-$. We may obtain an invariant measure $\mu := T^+ \wedge T^-$.

Notes The currents $T^{\pm}$ and measure μ have been shown to have many of the same properties that were found for the Hénon family (see [C2, Du]). Much of this theory may be carried over to more general meromorphic surface mappings; one recent work in this direction is [DDG1–3].

2.4 Three Involutions

Let us apply the preceding discussion to three birational maps that are involutions. These are quadratic maps, presented in the order of increasing degeneracy. All three can be turned into regular involutions after blowing up. The degree of degeneracy will correspond to the depth of blowup that is necessary to remove the exceptional curves and indeterminate points. The linear fractional recurrences, discussed in the following section, are of the form $A \circ J$, where A is linear, and J is the Cremona inversion, our first involution. We note, too, that the general quadratic Hénon map is also given by composing the third involution L with an affine map:

$$(y, y^2 + c - \delta x) = L \circ A, \quad A(x,y) = (y, -\delta x + -c).$$

These involutions play a basic role in the classification of the Cremona group of birational maps of the plane. Namely, if f is a quadratic, birational map of the plane, then f is linearly conjugate to a map of the form $A \circ \phi$, where ϕ is one of the involutions studied in this section (see [CD]). The analogous classification of the cubic Cremona transformations, which is more complicated, is also given in [CD].

The first of these involutions is the Cremona inversion J of $\mathbf{P}^2$ which is defined by

$$J[x_0 : x_1 : x_2] = [J_0 : J_1 : J_2] = [x_0^{-1} : x_1^{-1} : x_2^{-1}] = [x_1 x_2 : x_0 x_2 : x_0 x_1].$$

It is evident that $J = J^{-1}$. We set $p_0 = [1:0:0]$, $p_1 = [0:1:0]$, and $p_2 = [0:0:1]$, and let L_j, $0 \le j \le 2$, denote the side of the triangle $p_0 p_1 p_2$ which is opposite p_j. We let X denote the manifold obtained by blowing up $\mathbf{P}^2$ at the points p_j, $0 \le j \le 2$. Figure 7 shows that we may visualize this as an inversion in a triangle sending $p_j \leftrightarrow \Sigma_j$ for $j = 0, 1, 2$.

Exercise: Let $J_X : X \to X$. Show that J_X is holomorphic and maps $L_j \leftrightarrow P_j$ for $0 \le j \le 2$.

In $H^2(X)$, we have $L_0 = H - P_1 - P_2$, etc. Specifically, if we pull back a line $\sum a_j x_j = 0$, then we have $\sum a_j J_j = 0$, which is a quadric which contains all three points p_j, $0 \le j \le 2$. Thus we see that $J^* H = 2H$ on $H^2(\mathbf{P}^2)$, and

$$J_X^* H_X = 2H_X - P_0 - P_1 - P_2 \in H^2(X).$$

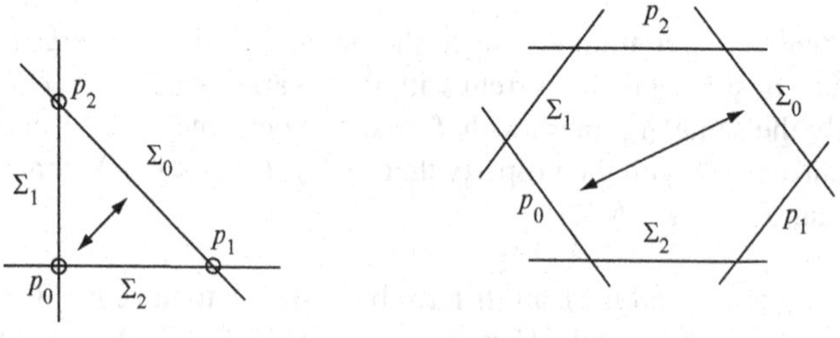

Fig. 7 Inversion in a triangle

By the exercise above, we have $J_X^* P_j = L_j = H_X - \sum_{i \neq j} P_i$. So with respect to the ordered basis $\langle H_X, P_0, P_1, P_2 \rangle$, we find that J_X^* is represented by the matrix

$$\begin{pmatrix} 2 & 1 & 1 & 1 \\ -1 & 0 & -1 & -1 \\ -1 & -1 & 0 & -1 \\ -1 & -1 & -1 & 0 \end{pmatrix} \tag{6}$$

The second involution. The next involution is $K(x,y) = (y^2/x, y)$, which in homogeneous coordinates is given by

$$K[x_0 : x_1 : x_2] = [x_0 x_1 : x_2^2 : x_1 x_2].$$

We observe that K is quadratic, and its jacobian determinant is $-2x_1x_2^2$. Thus the exceptional locus consists of the y-axis $A_Y = \{x_2 = 0\}$ and the x-axis $A_X = \{x_1 = 0\}$, which has multiplicity 2. They map according to $K:\ A_Y \to p_0 := [1:0:0]$, and $A_X \to p_1 := [0:1:0]$. Now let us construct the space $\pi_1 : \mathscr{X}_1 \to \mathbf{P}^2$ which is the blow up the points p_1 and p_2. The induced map is then $K_{\mathscr{X}_1} = \pi_1^{-1} \circ K \circ \pi_1$.

Exercise: Show that neither A_Y nor P_1 is exceptional for $K_{\mathscr{X}_1}$. In fact they map: $A_Y \leftrightarrow P_1$. One choice for local coordinates at P_1 is $\pi_1(t,\eta) = [t : 1 : t\eta] = [x_0 : 1 : x_2]$, so $\pi_1^{-1}(x_0, x_2) = (t = x_0, \eta = x_2/x_0)$. Let us look at the behavior of $K_{\mathscr{X}_1}$ on P_2 and A_X. We use the coordinate chart (ξ, s), with projection $\pi_1(\xi, s) = [1 : \xi s : s] \in \mathbf{P}^2$. Thus $\pi_1^{-1}[x_0 : x_1 : x_2] = (\xi = x_1/x_2, s = x_2/x_0)$. In this chart, the blowup fiber is $P_2 = \{s = 0\}$, and $A_X = \{\xi = 0\}$. In this chart we have $K_{\mathscr{X}_1} : (\xi, s) \to [1 : \xi s : s] \to [\xi s : s^2 : \xi s^2] = [1 : s/\xi : s] = [x_0 : x_1 : x_2]$. Thus $\{s = 0\}$ maps to $p_2 = [1:0:0]$. If we follow this map by π_1^{-1}, then we have $K_{\mathscr{X}_1} : (\xi, s) \to (\xi^{-1}, s)$. In other words the mapping $K_{\mathscr{X}_1}|_{P_2}$ of P_2 to itself is given by $\xi \mapsto \xi^{-1}$.

Now we look at the behavior of $K_{\mathscr{X}_1}$ on the set A_X. Since A_X maps to p_2, we look at the map $\pi_1^{-1} \circ K$. We see that this is given by

$$[1 : x : y] \to [x_0 = x : x_1 = y^2 : x_2 = xy] \to (\xi = x_1/x_2 = y/x, s = x_2/x_0 = y)$$

Thus $A_X \to p_3 := (\xi = 0) \in P_2$. Now we create the space $\pi_2 : \mathscr{X}_2 \to \mathscr{X}_1$ by blowing up the point p_3. It is an exercise like the one above to show that A_X is not exceptional for the induced map $K_{\mathscr{X}_2} = \pi_2^{-1} \circ \pi_1^{-1} K \circ \pi_1 \circ \pi_2$. Further, $K_{\mathscr{X}_2}$ has no exceptional curves and no points of indeterminacy. Thus $K_{\mathscr{X}_2} \in Aut(\mathscr{X}_2)$.

Now we find what is happening with the action on the Picard group. We will use $H = H_{\mathscr{X}_2}, P_1, P_2, P_3$ as an ordered basis for $Pic(\mathscr{X}_2)$. Let us write the elements $A_X, A_Y \in Pic(\mathscr{X}_2)$ in terms of this basis. In $\mathbf{P}^2$, we have $A_Y = H$. Pulling this back by π_1^* to $\mathscr{X}_1$, we have $A_Y + P_2 = H \in Pic(\mathscr{X}_1)$. Now we pull this back by π_2^* to $\mathscr{X}_2$ to have $A_Y + P_2 + 2P_3 = H \in Pic(\mathscr{X}_2)$. (Note that p_3 belongs to P_2 and A_Y both, so we have 2 occurrences of P_3.) Reasoning in a similar way, we have $A_X = H \in Pic(\mathbf{P}^2)$, $A_X + P_1 + P_2 = H \in Pic(\mathscr{X}_1)$, and $A_X + P_1 + P_2 + P_3 = H \in Pic(\mathscr{X}_2)$.

From Figure 8, we see that

$$K^* : P_1 \to A_Y = H - P_2 - 2P_3, \quad P_2 \to P_2, \quad P_3 \to A_X = H - P_1 - P_2 - P_3.$$

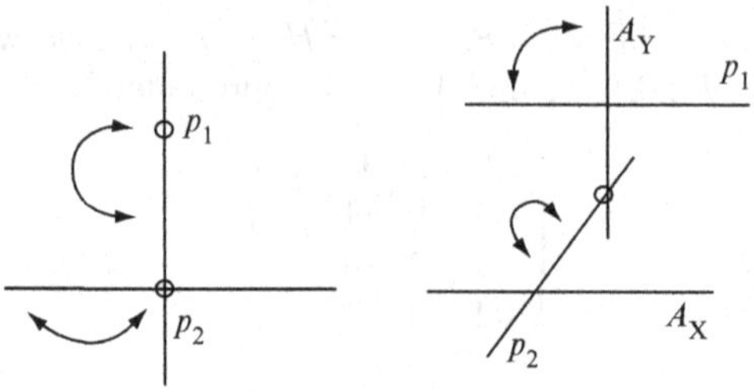

Fig. 8 Involution K and its lifts to $\mathscr{X}_1$ and $\mathscr{X}_2$

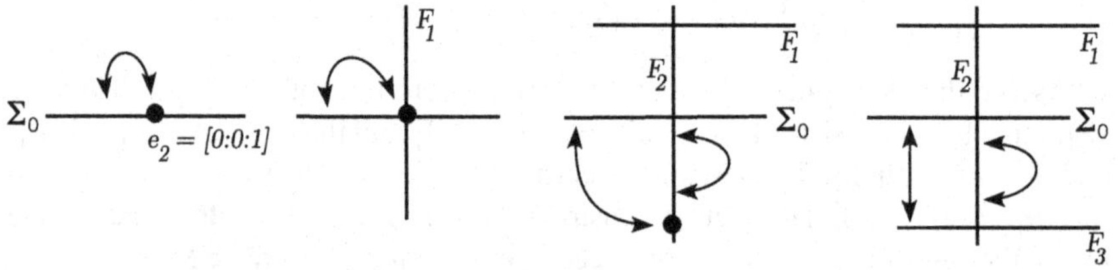

Fig. 9 Involution L and its lifts to $\mathscr{X}_1$, $\mathscr{X}_2$ and $\mathscr{X}_3$

Finally, if H is a line in $\mathbf{P}^2$, then H will intersect both A_X and A_Y. The pullback of H will intersect both points of indeterminacy p_1 and p_2. Looking at Figure 8, we see that in $\mathscr{X}_1$, p_3 blows up to A_X, so the pullback of H will pass through p_3. It follows that we will have $K^*H = 2H - P_1 - P_2 - 2P_3$. Thus we have

$$K^* = \begin{pmatrix} 2 & 1 & 0 & 1 \\ -1 & 0 & 0 & -1 \\ -1 & -1 & 1 & -1 \\ -2 & -2 & 0 & -1 \end{pmatrix} \tag{7}$$

on $Pic(\mathscr{X}_2)$.

Exercise: Define a new ordered basis $\langle H, E_1, E_2, E_3 \rangle$ for $Pic(\mathscr{X}_2)$ by setting $E_1 = P_1$, $E_2 = P_2 + P_3$, $E_3 = P_3$. Show that with respect to this basis, K^* is represented by the matrix (6).

Exercise: Give an analysis of the involutions $K_j(x,y) = (y^j/x, y)$ for $j = 1,2,3,\ldots$ along similar lines. That is, can you construct suitable spaces $\mathscr{X}_j$ such that K_j becomes an automorphism? How does K_j^* act on $Pic(\mathscr{X}_j)$?

The third involution. Finally, we consider $L(x,y) = (x, -y + x^2)$. In homogeneous coordinates, this is written as $L : [x_0 : x_1 : x_2] \to [x_0^2 : x_0x_1 : -x_0x_2 + x_1^2]$. Thus $\Sigma_0 = \{x_0 = 0\}$ is the unique exceptional curve, and $e_2 = [0 : 0 : 1]$ is the unique point of indeterminacy.

We will regularize this map L by performing the sequence of three blowups which is shown in Figure 9. Let $\pi_1 : \mathscr{X}_1 \to \mathbf{P}^2$ be the blowup of e_2, and denote the blowup fiber by F_1. We will use the local coordinate chart $\pi_1(\xi, s) = [\xi s : s : 1] = [x_0 : x_1 : x_2]$, so the fiber is $F_1 = \{s = 0\}$, and $\{\xi = 0\} = \Sigma_0$. The inverse is $\pi_1^{-1}[x_0 : x_1 : x_2] = (\xi = x_0/x_1, s = x_1/x_2)$. In this coordinate chart, we have $L_{\mathscr{X}_1} = \pi_1^{-1} \circ L \circ \pi_1(\xi, s) = (\xi, \xi s/(s-\xi))$. The restriction to $F_1 = \{s = 0\}$ is given by $(\xi, 0) \mapsto (\xi, 0)$. Thus each point of F_1 is fixed under $L_{\mathscr{X}_1}$, and F_1 is not exceptional. For the behavior near Σ_0, we map

$$[x_0 : x_1 : x_2] \to \pi_1^{-1}[x_0^2 : x_0 x_1 : -x_0 x_2 + x_1^2] = (\xi' = x_0/x_1, s' = x_0 x_1/(-x_0 x_2 + x_1^2)).$$

We see that $\Sigma_0 \to \{\xi = 0\} = F_1 \cap \Sigma_0 \in F_1$.

The next step is to construct $\pi_2 : \mathscr{X}_2 \to \mathscr{X}_1$ by blowing up the point $F_1 \cap \Sigma_0$. We let F_2 denote the exceptional blowup fiber, and we use local coordinates (u, η) with coordinate projection $\pi_2(u, \eta) = (\xi = u, s = u\eta)$. The inverse is $\pi_2^{-1}(\xi, s) = (u = \xi, \eta = s/\xi)$. Thus we use $L_{\mathscr{X}_2} = \pi_2^{-1} \circ \pi_1^{-1} \circ L \circ \pi_1 \circ \pi_2$. We find that $L_{\mathscr{X}_2}(u, \eta) = (u, \eta/(\eta - 1))$, which means that $F_2 = \{u = 0\}$ is mapped to itself in an invertible way. For the image of Σ_0, we see that

$$\pi_2^{-1} \circ \pi_1^{-1} \circ L[x_0 : x_1 : x_2] = (u = x_0/x_1, \eta = x_1^2/(x_1^2 - x_0 x_2))$$

Setting $x_0 = 0$ we find that $L_{\mathscr{X}_2} : \Sigma_0 \to \eta = 1 \in F^2$.

Finally, we construct the blowup $\pi_3 : \mathscr{X}_3 \to \mathscr{X}_2$ centered at $\eta = 1 \in F^2$. For this we use local coordinates (t, μ) with coordinate projection $\pi_3(t, \mu) = (t = u, t\mu + 1 = \eta)$. Thus the exceptional fiber F_3 is $\{t = 0\}$. We find that

$$L \circ \pi_3 \circ \pi_2 \circ \pi_1(t, \mu) = [x_0 = t(t\mu + 1) : x_1 = t\mu + 1 : x_2 = \mu].$$

Thus, setting $t = 0$, we see that $L_{\mathscr{X}_3}$ maps $F_3 \ni \mu \to [0 : 1 : \mu]$. We conclude that Σ_0 is no longer exceptional. Since L is an involution, it follows that $F_3 \to \Sigma_0$ is also not exceptional.

Now we discuss the pullback L^* on $Pic(\mathscr{X}_3)$. We use the ordered basis $\langle H, F_1, F_2, F_3 \rangle$. We start by observing that $\Sigma_0 = H$ in $Pic(\mathbf{P}^2)$. Pulling back to $\mathscr{X}_1$, we have $\Sigma_0 + F_1 = H$ in $Pic(\mathscr{X}_1)$. The next center of blowup is $\Sigma_0 \cap F_1 \in \mathscr{X}_1$, so when we pull back, we get 2 copies of F_2, which gives $\Sigma_0 + F_1 + 2F_2 = H$ in $Pic(\mathscr{X}_2)$. Finally, we pull back the point $\eta = 1 \in F_2 - (\Sigma_0 \cup F_1)$, so $\Sigma_0 + F_1 + 2F_2 + 2F_3 = H$ in $Pic(\mathscr{X}_3)$. Now we pull back $H = \{\sum a_j x_j = 0\}$ and have $\{a_0 x_0^2 + a_1 x_0 x_1 + a_2(-x_0 x_2 + x_1^2) = 0\}$, which is a quadric, so $L^* H = 2H + \sum m_j P_j$. We need to determine the a_j. For m_1, we pull back $\ell = \sum a_j x_j$ by $L \circ \pi_1$ and find $a_0 \xi^2 s^2 + a_1 \xi s^2 + a_2(-\xi s + s^2)$. This vanishes to order 1 on $P_1 = \{s = 0\}$, so $m_1 = 1$. For m_2, we look at $\ell \circ L \circ \pi_1 \circ \pi_2$ to obtain the function $u^2 \eta (a_0 u^2 \eta + a_1 u \eta + a_2(-1 + \eta))$, which vanishes to order 2 on $P_2 = \{u = 0\}$. Thus $m_2 = 2$. For m_3, we pull back $\sum a_j x_j$ by $L \circ \pi_1 \circ \pi_2 \circ \pi_3$ to obtain $t^3(1 + t\mu)(a_0 t(1 + t\mu) + a_1(1 + t\mu) + a_2 \mu)$ which vanishes to order 3 on $F_3 = \{t = 0\}$, so $m_3 = 3$.

In conclusion, we see that L^* is given by

$$H \to 2H - F_1 - 2F_2 - 3F_3, \;\; P_1 \to P_1, \;\; P_2 \to P_2, \;\; P_3 \to \Sigma_0 = H - F_1 - 2F_2 - 2F_3$$

which corresponds to the matrix

$$\begin{pmatrix} 2 & 0 & 0 & 1 \\ -1 & 1 & 0 & -1 \\ -2 & 0 & 1 & -2 \\ -3 & 0 & 0 & -2 \end{pmatrix} \tag{8}$$

As expected, the square of this matrix is the identity.

Exercise: Define a new ordered basis $\langle H, E_1, E_2, E_3 \rangle$ for $Pic(\mathscr{X}_3)$ by setting $E_1 = F_1 + F_2 + F_3$, $E_2 = F_2 + F_3$, $E_3 = F_3$. Show that with respect to this basis, L^* is represented by the matrix (6).

Exercise: Perform an analysis on the maps $L_j(x,y) = (x, -y + x^j)$ for $j = 1, 2, 3, \ldots$

2.5 Linear Fractional Recurrences

Here we consider the possibility of producing a space X and an automorphism $f \in Aut(X)$ by the following procedure. That is, we consider a birational map f of projective space $\mathbf{P}^2$, and we would like to perform some blowups $\pi : X \to \mathbf{P}^2$ such that the induced map f_X will be an automorphism. Now if the original map f fails to be an automorphism, then there will be an exceptional curve C which is blown down to a point p_1. We can try to fix this by blowing up the point p_1 to produce a space $\pi_1 : X_1 \to \mathbf{P}^2$. If we are lucky, then the induced map f_{X_1} will not map C to a point but will map it to the whole curve P_1. However, this is only a temporary success if p_1 is not a point of indeterminacy, because f_{X_1} will map the fiber P_1 to the point $p_2 := fp_1$. Thus we see that the blowing-up procedure cannot work unless we have the property that any exceptional curve C in $\mathbf{P}^2$ is eventually mapped to a point of indeterminacy.

We spend this section looking at the example of linear fractional recurrences:

$$f(x,y) = \left(y, \frac{y+a}{x+b}\right). \tag{9}$$

See [BK3] for further discussion of this family. In projective coordinates this map takes the form

$$f[x_0 : x_1 : x_2] = [x_0(bx_0 + x_1) : x_2(bx_0 + x_1) : x_0(ax_0 + x_2)].$$

We explain that to obtain the homogeneous representation, we write

$$[x_0 : x_1 : x_2] = [1 : x_1/x_0 : x_2/x_0] = [1 : x : y]$$

and so

$$f = \left[1 : y : \frac{y+a}{x+b}\right] = \left[1 : \frac{x_2}{x_0} : \frac{\frac{x_2}{x_0} + a}{\frac{x_1}{x_0} + b}\right] = \left[1 : \frac{x_2}{x_0} : \frac{x_2 + ax_0}{x_1 + bx_0}\right].$$

Inspection shows that the exceptional curves are given by

$$\begin{aligned}
\Sigma_0 &:= \{x_0 = 0\} \mapsto p_1 := [0:1:0] \\
\Sigma_\beta &:= \{x+b=0\} = \{x_1 + bx_0 = 0\} \mapsto p_2 := [0:0:1] \\
\Sigma_\gamma &:= \{y+a=0\} = \{ax_0 + x_2 = 0\} \mapsto q := [1:-a:0].
\end{aligned}$$

This means that Σ_0 minus the indeterminacy locus maps to p_1, etc. In order to find the exceptional curves in a more systematic way, we may take the jacobian of the homogeneous form of f, which gives $2x_0(bx_0+x_1)(ax_0+x_2)$ and which shows us the exceptional curves directly.

If we set $p_* = (-b,-a) = [1:-b:-a]$, then the points of indeterminacy are given by

$$\begin{aligned}
p_2 &= \Sigma_0 \cap \Sigma_\beta \mapsto \overline{p_1 p_2} = \Sigma_0 \\
p_1 &= \Sigma_0 \cap \Sigma_\gamma \mapsto \overline{p_1 q} =: \Sigma_B \\
p_* &= \Sigma_\beta \cap \Sigma_\gamma \mapsto \overline{p_2 q} =: \Sigma_C
\end{aligned}$$

We note that the map f is not actually defined at the points of indeterminacy, and we interpret these formulas to mean $f^{-1} : \Sigma_0 \mapsto p_2$, etc.

Let $\pi : Y \to \mathbf{P}^2$ be the space obtained by blowing up p_1 and p_2. Let us show that Σ_0 is not exceptional for the induced map f_Y. We may use local coordinates $(t,x) \mapsto [t:x:1]$ at $\Sigma_0 = \{t=0\}$. Local coordinate chart U' at $p_1 = [0:1:0]$ is given by $(s_1,\eta_1) \mapsto [s_1 : 1 : s_1\eta_1]$. In these coordinates the map becomes:

$$\begin{aligned}
(t,x) \mapsto [t:x:1] \mapsto f[t:x:1] &= [f_0 : f_1 : f_2] \\
&= [f_0[t:x:1]/f_1[t:x:1] : 1 : f_2[t:x:1]/f_1[t:x:1]] \\
&= [t/x : 1 : (1+at)/(x(x+bt))] = [s_1 : 1 : s_1\eta_1]
\end{aligned}$$

So we have

$$(t,x) \mapsto (s_1,\eta_1) = (t/x, (1+at)(bt+x)),$$

which means that we have $\Sigma_0 = \{t=0\} \ni [0:x:1] \mapsto (0,1/x) \in P_1$, so Σ_0 is not exceptional. Similarly, we have

$$\Sigma_\beta \to P_2 \to \Sigma_0 \to P_1 \to \Sigma_B = \{y=0\}. \tag{10}$$

Exercise: Carry out the details of the remark concerning (10). In particular, show that the only exceptional curve for the rational map $f_Y : Y \dashrightarrow Y$ is Σ_γ, and the only point of indeterminacy is p_*.

Exercise: Compare Figures 7 and 10, and show that a map f corresponding to Figure 10 must be linearly conjugate to a map of the form $L \circ J$, where L is linear, and J is the involution from the previous section.

Now we can look at $H^2(Y) = \langle H, P_1, P_2 \rangle$. We observe that Σ_B is a line which contains exactly one center of blowup, namely p_1. Thus we have $\Sigma_B = H - P_1 \in H^2(Y)$. Similarly, since Σ_0 contains both p_1 and p_2, we have $\Sigma_0 = H - P_1 - P_2$.

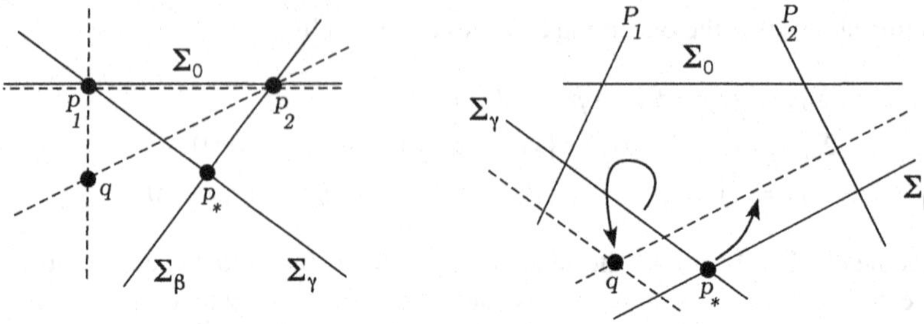

Fig. 10 Linear fractional recurrence on $\mathbf{P}^2$ and on $\mathscr{Y}$

Now, if we want to know what f_{Y*} does on $H^2(Y)$, we must determine the action on H. A generic line intersects Σ_0, Σ_β and Σ_γ. Thus fH will be a quadric passing through the images of these lines, namely p_1, p_2 and q. Thus $f_{Y*}H = 2H - P_1 - P_2$. Using (10), we have that f_{Y*} is given with respect to the ordered basis H, P_1, P_2 by the matrix

$$\begin{pmatrix} 2 & 1 & 1 \\ -1 & -1 & -1 \\ -1 & 0 & -1 \end{pmatrix}$$

The characteristic polynomial of this matrix is $t^3 - t - 1$.

The following is how we may use f_{Y*} to get information about f_Y. Let L be a line, and let $\{L\}$ denote its class in $Pic(Y)$. If L does not intersect P_1 or P_2, then $\{L\} = H$. We have just seen that $f_{Y*}H = 2H - P_1 - P_1$. The $2H$ means that $f_Y L$ has degree 2, and the $-P_1 - P_2$ means that it intersects P_1 and P_2, each with multiplicity 1. If L contains p_*, then $f_Y L$ is the union of two lines, one of which is Σ_C. In addition, if L is disjoint from P_1 and p_*, it must intersect Σ_γ, which means that $f_Y L$ must contain q.

Now suppose that $p_* \notin L \cup f_Y L$. Then $\{f_Y^2 L\} = (2, 0, -1) = 2H - P_2$ is obtained by multiplying the square of the matrix above by $H = (1, 0, 0)$. Thus $f_Y^2 L$ is a quadric intersecting P_2 but not P_1. Similarly, if $p_* \notin L \cup f_Y L \cup f_Y^2 L$, then we multiply $H = (1, 0, 0)$ by the cube of this matrix to find that $\{f_Y^3 L\} = 3H - P_1 - P_2$, so $f_Y^3 L$ is a cubic intersecting P_1 and P_2 with multiplicity 1. If $p_* \notin L \cup \cdots \cup f_Y^{n-1} L$, then the iterates of f_Y are holomorphic in a neighborhood of L, so we have $(f_Y^*)^n H = \{f_Y^n L\}$. We will say that the parameters a and b are *generic* if $p_* \notin \bigcup_{j=0}^{\infty} f_Y^j L$.

Theorem 2.10. *For generic a and b, we have $(f_Y^*)^n = (f_Y^n)^*$, and $\lambda(f) \sim 1.324\ldots$ is the largest root of the polynomial $t^3 - t - 1$.*

2.6 Linear Fractional Automorphisms

Let $C_{a,b}^2$ denote the parameter space of $a, b \in C$, and for $(a, b) \in C_{a,b}^2$ let $f_{a,b}$ be as in (9). We construct the space Y as in the previous section, and let $f_Y : Y \dashrightarrow Y$

be the associated birational map. The point $p_* = (-b, -a)$ is the unique point of indeterminacy for f_Y, and Σ_γ is the exceptional locus. Setting $q = (-a, 0)$, let us define the following subset of parameter space:

$$\{\mathscr{V}_n = \{(a,b) \in C^2_{a,b} : f^j_Y q \neq p_*, 0 \leq j < n,\ f^n_Y q = p_*\}.$$

Theorem 2.11. *f_Y is a rational surface automorphism if and only if $(a,b) \in \mathscr{V}_n$ for some n.*

Proof. If $(a,b) \notin \mathscr{V}_n$ for any n, then by Theorem 2.10 we have that $\lambda(f_Y)$ is the largest root of $t^3 - t - 1$. This is not a Salem number, so $f_{a,b}$ is not an automorphism. Conversely, let us suppose that $(a,b) \in \mathscr{V}_n$ for some n. Let Z denote the manifold obtained by blowing up the $n+1$ points in the orbit q, $f_Y q, \ldots, f^n_Y q = p_*$. It follows that the induced map f_Z is an automorphism. □

The action of f_{Z*} on cohomology is given by:

$$P_2 \to \Sigma_0 = H - P_1 - P_2 \to P_1 \to \Sigma_B = H - P_1 - Q \tag{11}$$

which is like what we have seen already from the action of f_Y, except that now the point $q \in \Sigma_B$ has been blown up, so we must subtract Q to obtain the representation of $\Sigma_B = \{y = 0\}$ as an element of $Pic(Z)$. The behavior of the new blowup fibers

$$Q \to fQ \to \cdots \to f^n Q = P_* \to \Sigma_C = \overline{p_2 q} = H - P_2 - Q \tag{12}$$

Finally, since a generic line L intersects all three lines Σ_0, Σ_β, and Σ_γ with multiplicity one, the image fL will be a quadric passing through e_2, e_1, and q. Thus we have

$$H \to 2H - P_1 - P_2 - Q. \tag{13}$$

Theorem 2.12. *If $(a,b) \in \mathscr{V}_n$, then the characteristic polynomial of f_{Z*} is*

$$\chi_n = x^{n+1}(x^3 - x - 1) + x^3 + x^2 - 1.$$

Thus $\delta(f) = \lambda_n$, which is the largest root of χ_n, and $\lambda_n > 1$ if $n \geq 7$.

We note that λ_n increases to the number $\lambda \sim 1.324\ldots$ from the previous section. An interesting consideration is to ask whether $f_{a,b}$ has an invariant curve. The maps which posses invariant curves have a number of interesting properties; we describe one of them below.

There are rational functions $\varphi_j : \mathbf{C} \to \mathbf{C}^2_{a,b}$ such that if $(a,b) = \varphi_j(t)$ for some $t \in \mathbf{C}$, then $f_{a,b}$ has an invariant curve S with j irreducible components. The curve S is a singular cubic. For instance, the first of these functions is

$$\varphi_1(t) = \left(\frac{t - t^3 - t^4}{1 + 2t + t^2}, \frac{1 - t^5}{t^2 + t^3}\right).$$

Theorem 2.13. *Suppose that $(a,b) = \varphi_j(t)$, and j divides n. Then $(a,b) \in \mathscr{V}_n$ if and only if $\chi_n(t) = 0$.*

In every case, the map $f_{a,b}$ has two fixed points. If $(a,b) = \varphi_1(t)$, then one of the fixed points is (x_s, y_s) with $x_s = y_s = t^3/(1+t)$ and is the singular point of S. (There are similar formulas for $j = 2$ and 3.) Thus multipliers of $df_{a,b}$ at (x_s, y_s) are t^2 and t^3.

Now we consider the case where t is a root of χ_n. We may factor $\chi_n = \psi_n \cdot C$ where ψ_n is the minimal polynomial for λ_n, and C is cyclotomic. As we saw earlier, λ_n and λ_n^{-1} are roots of ψ_n, and all the other roots of ψ_n have modulus one and are not roots of unity. In fact, the degree of the cyclotomic may be shown to be bounded by 26. Thus almost all of the roots of χ_n will have modulus 1 and not be roots of unity.

Theorem 2.14. *Suppose that $n \geq 7$, t is a root of χ_n with $|t| = 1$, and t is not a root of unity. If $(a,b) = \varphi_1(t)$, then $f_{a,b}$ has a rank 1 rotation domain about the fixed point (x_s, y_x).*

Notes The linear fractional automorphisms are discussed in [BK1, 2] and [M]. Other automorphisms in the same general spirit have been found by J. Diller [Di].

2.7 Cremona Representation

Let us discuss the representation $\rho : Aut(X) \to GL(H^2(X;\mathbf{Z}))$ which is given by $\rho(f) = f_*$. It is evident that the image of ρ consists of isometries with respect to the intersection product, as well as some other relevant properties. In the most familiar case $X = \mathbf{P}^2$, we see that $\rho(Aut(\mathbf{P}^2))$ consists only of the identity element acting on $H^2(\mathbf{P}^2;\mathbf{Z}) \cong \mathbf{Z}$. Similarly, $\rho(Aut(\mathbf{P}^1 \times \mathbf{P}^1))$ consists of two elements: the identity, and the interchange of coordinates on $\mathbf{Z} \times \mathbf{Z}$.

Exercise: Let $\tilde{\mathbf{P}}^2_j$ denote $\mathbf{P}^2$ blown up at j points in general position, for $1 \leq j \leq 4$. Determine $Aut(\tilde{\mathbf{P}}^2_j)$ and its image under ρ. What happens if the points are not in general position?

The previous exercise shows that if we blow up a small number of points p_j, $1 \leq j \leq 4$ in $\mathbf{P}^2$, then the possible automorphisms are relatively limited. What happens if we also blow up a point of the exceptional fiber P_j over p_j?

Exercise: Let Z_1 be $\mathbf{P}^2$ blown up at a point $p_1 \in \mathbf{P}^2$, and let $p_2, p_3 \in P_1$ denote two points of the blowup fiber P_1 over p_1. Let Z denote the space Z_1 blown up at the points p_2 and p_3. What is the automorphism group of Z? (One way of visualizing Z_1 is as follows. Let $[x:y:z]$ be coordinates on $\mathbf{P}^2$, and let $p_1 = [1:0:0]$ be the point where the x-axis intersects $\{z = 0\}$, which we take to be the line at infinity. The points of the blowup fiber P_1 are then identified with the points where the horizontal lines $L_c = \{[x:y:1] : y = c\}$ intersect $\{z = 0\}$.)

The automorphism group is limited in the exercises above is because we have created a finite number of rational curves with self intersection -1, and these curves must be mapped around among themselves. That is, for each pair p_1, p_2 of centers of blowup, the strict transform of the line $\overline{p_1 p_2}$ will have self intersection ≤ -1 in X (or more precisely, the self intersection is $1-k$, where k is the number of indices j such that $p_j \in \overline{p_1 p_2}$). If Q is a quadric containing exactly 5 centers of blowup, then the strict transform $\tilde{Q}$ will have self-intersection -1. And as we have seen in the previous section, there can be infinitely many curves with self-intersection -1 if we blow up ≥ 9 "correctly chosen" points. Let us recall a result of Nagata.

Theorem 2.15. *(Nagata) Let X be a rational surface, and let $f \in Aut(X)$ be an automorphism for which f_* has infinite order. Then there is a sequence of holomorphic maps $\pi_{j+1} : X_{j+1} \to X_j$ such that $X_1 = \mathbf{P}^2$, $X_{N+1} = X$, and π_{j+1} is the blowup of a point $p_j \in X_j$.*

This gives a very useful starting point if we are looking for rational surface automorphisms. It says that every one must be given by a "model" birational map of $\mathbf{P}^2$. That is, if $F \in Aut(X)$, then the (birational) projection $\pi = \pi_1 \circ \cdots \circ \pi_{N+1}$ conjugates f to a birational map f of $\mathbf{P}^2$. Or conversely, we can start with birational maps of $\mathbf{P}^2$ which are promising candidates and see whether there might be a blowup X which turns them into automorphisms.

There are further limitations on the image $\rho(Aut(X))$ in $GL(H^2(X;\mathbf{Z}))$. For this, we need to start with a good basis for H^2. Specifically, let us consider a basis $\{e_0, \dots, e_N\}$ for $H^2(X;\mathbf{Z})$ such that the intersection form with respect to this basis is given by the diagonal matrix with eigenvalues $1, -1, \dots, -1$. Such a basis is called a *geometric basis*, and has the properties: $e_0 \cdot e_0 = 1$, $e_j \cdot e_j = -1$ if $1 \le j \le N$, and $e_i \cdot e_j = 0$ if $i \ne j$. If X is obtained by blowing up N distinct points of $\mathbf{P}^2$ as in section §B1, then the basis $\langle H_X, P_1, \dots, P_N \rangle$ as in the Theorem B.1.1 is geometric. Now let us consider a space $X = X_{N+1} \to X_N \to \cdots \to X_1 = \mathbf{P}^2$ as in Nagata's theorem. Let $P_j = \pi_{j+1}^{-1} p_j \subset X_{j+1}$ denote the exceptional fiber, and let $e_j := \pi_{N+1}^* \cdots \pi_{j+1}^* P_j$.

Exercise: Show that if X is as above, and if we set $e_0 = H_X$, then $e_0, \dots, e_N$ is a geometric basis. Try this first on the space Z constructed in the exercise above. Show that in Z we have $e_1 = P_1 + P_2 + 2P_3$, $e_2 = P_2 + P_3$, $e_3 = P_3$ and that $P_1 \cdot P_1 = -3$, $P_2 \cdot P_2 = -2$. From this, conclude that $\{e_j, 0 \le j \le 3\}$ is a geometric basis.

Exercise: Let X be as above. Show that $\langle H_X, P_1, \dots, P_N \rangle$ and $\langle e_0, e_1, \dots, e_N \rangle$ are dual bases. That is, $H_X \cdot e_0 = 1$, $P_j \cdot e_k = 0$ if $j \ne k$, and $P_j \cdot e_j = -1$ for $1 \le j \le N$.

If α is an element of H^2 such that $\alpha \cdot \alpha = -2$, then $R_\alpha(x) = x - (x \cdot \alpha)\alpha$ is a reflection in the direction α: this means that R_α sends $\alpha \to -\alpha$, and R_α fixes all elements of $\alpha^\perp$.

Let us define the vectors

$$\alpha_0 = e_0 - e_1 - e_2 - e_3, \quad \alpha_j = e_{j+1} - e_j, \quad 1 \le j \le N-1.$$

It follows that $\alpha_j \cdot \alpha_j = -2$ for $0 \le j \le N-1$. For $1 \le j \le N-1$, the reflection R_{α_j} interchanges $e_j \leftrightarrow e_{j+1}$. Thus the subgroup generated by R_{α_j}, $1 \le j \le N-1$, is exactly the set of permutations on the elements $\{e_1, \dots, e_N\}$. The reflection R_{α_0}

corresponds to the Cremona inversion. That is, if we write the action of R_{e_0} on the subspace with ordered basis $\langle e_0, e_1, e_2, e_3 \rangle$, then the restriction of R_{e_0} is represented by the matrix:

$$\begin{pmatrix} 2 & 1 & 1 & 1 \\ -1 & 0 & -1 & -1 \\ -1 & -1 & 0 & -1 \\ -1 & -1 & -1 & 0 \end{pmatrix}$$

In §2.3, we saw that after the involutions J, K, and L have been regularized to become automorphisms, the actions of J^*, K^* and L^* can all be written in the form of this matrix.

Let W_N denote the group generated by the reflections R_{α_j}, $0 \le j \le N-1$. Thus W_N is generated by the permutations of the e_j's, $1 \le j \le N$, together with the Cremona inversion. The following result is classical (see [Do, Theorem 5.2]):

Theorem 2.16. *Let X be a rational surface, and let $e_0, \dots, e_N$ be a geometric basis for $H^2(X)$. If $f \in Aut(X)$, then $f_* \in W_N$.*

Problem: It remains unknown which elements of W_N can arise from rational surface automorphisms.

Let us return to the linear fractional automorphisms and see how f^* is related to W_N. Since the space $\mathscr{Z}$ was constructed by simple blowups, we have a geometric basis for $Pic(Z)$ by letting e_0 be the class of a general line and then setting $e_1 = P$, $e_2 = E_2$, $e_3 = E_1$, $e_4 = Q$, ..., $e_j = f^{j-4}Q$, $4 \le j \le N-4$. We see that cyclic permutation $\sigma = (123\dots N)$ is equal to the composition $R_{\alpha_1} \circ R_{\alpha_2} \circ \cdots \circ R_{\alpha_{N-1}}$, and that f_* itself (cf. equation (11-13)) is equal to $R_{\alpha_0} \circ \sigma = R_{\alpha_0} \circ \cdots R_{\alpha_{N-1}}$ is a product of the reflections that generate W_N.

2.8 More Automorphisms

We give some more examples of rational surface automorphisms of positive entropy. Our goal here is to give families of maps which illustrate that such automorphisms can occur in continuous families of arbitrarily high dimension. Namely, for each even k we give a family $\{f_a : a \in \mathbf{C}^{\frac{k}{2}-1}\}$ of birational maps of $\mathbf{P}^2$, and for each map f_a there is a rational surface $\pi : \mathscr{X}_a \to \mathbf{P}^2$, and f_a lifts to an automorphism of $\mathscr{X}_a$. We note that the complex structures of the surfaces $\mathscr{X}_a$ are allowed to vary with a, but the smooth structures are locally constant. The smooth dynamical systems $(f_a, \mathscr{X}_a)$, however, may be shown to vary nontrivially with a. The maps we discuss are

$$f(x,y) = \left(y, -x + cy + \sum_{\substack{\ell=2 \\ \ell \text{ even}}}^{k-2} \frac{a_\ell}{y^\ell} + \frac{1}{y^k} \right) \tag{14}$$

where the sum is taken only over even values of ℓ.

We will describe a number of properties of the maps f_a; the details are given in [BK4], and we do not repeat them here. Each f_a maps the line $\{y=0\}$ to the point $[0:1:0]\in\mathbf{P}^2$, and f_a^{-1} maps $\{x=0\}$ to $[1:0:0]$. The line at infinity $\{z=0\}$ is invariant and is mapped according to $f[1:w:0]=[1:c-1/w:0]$. For $1\le s$ we set $f^s[0:1:0]=[1:w_s:0]$. We note that if $(j,n)=1$ and if we set $c=\pm 2\cos(j\pi/n)$, then $f^{n-1}[0:1:0]=[1:0:0]$, and the restriction of f to $\{z=0\}$ has period n. We define $C_n=\{\pm 2\cos(j\pi/n):(j,n)=1\}$, where we choose "+", "−", or both, according to the condition that $w_1\cdots w_{n-2}=1$. With $c\in C_n$, we obtain the surface $\mathscr{X}_a$ by performing $2k+1$ iterated blowups over each point $[1:w_s:0]$, $0\le s\le n-1$. The fibers over $[1:w_s:0]$ are denoted $\mathscr{F}_s^j$, $1\le j\le 2k+1$. From this construction, it follows that the fibers map as follows:

$$\mathscr{F}_0^1\to\cdots\to\mathscr{F}_s^1\to\mathscr{F}_{s+1}^1\to\cdots\to\mathscr{F}_{n-1}^1\to\mathscr{F}_0^1$$

$$\mathscr{F}_0^j\to\cdots\to\mathscr{F}_s^j\to\mathscr{F}_{s+1}^j\to\cdots\to\mathscr{F}_{n-1}^j\to\mathscr{F}_0^{2k+2-j}\to\cdots\to\mathscr{F}_{n-1}^{2k+2-j}\to\mathscr{F}_0^{2k+2-j}$$

$$\{y=0\}\to\mathscr{F}_0^{2k+1}\to\cdots\to\mathscr{F}_{n-1}^{2k+1}\to\{x=0\}$$

A further observation is that these maps give rational surface automorphisms:

Theorem 2.17. *Let $1\le j<n$ satisfy $(j,n)=1$. There is a nonempty set $C_n\subset\mathbf{R}$ such that for even $k\ge 2$ and for all choices of $c\in C_n$ and $a_\ell\in\mathbf{C}$, the map f in (14) is an automorphism.*

Now we let S denote the subgroup of $Pic(\mathscr{X}_a)$ spanned by $\{z=0\}$ and the fibers $\mathscr{F}_s^j$, $0\le s\le n-1$, $1\le j\le 2k+1$. From [BK4] we also have:

Proposition 2.18. *The intersection form of $\mathscr{X}_a$, when restricted to S, is negative definite.*

We let $T:=S^\perp$ denote the vectors of $Pic(\mathscr{X}_a)$ which are orthogonal to S. By the Proposition, we see that $S\cap T=0$, so we have $Pic(\mathscr{X}_a)=S\oplus T$. Since S is invariant, T is also invariant. For each $0\le s\le n-1$, we let γ_s denote the projection of the class $\{\mathscr{F}_s^{2k+1}\}\in Pic(\mathscr{X})$ to T. Thus the γ_s, $0\le s\le n-1$ give a basis of T. Following $\{x=0\}$ through the various blowups in the construction of $\mathscr{X}$, we may show that $\{x=0\}=-\gamma_0+k\gamma_1+\cdots k\gamma_{n-1}$. And from the mapping of the fibers, we see that f_* maps according to

$$\gamma_0\to\gamma_1\to\cdots\to\gamma_{n-1}\to\{x=0\}=-\gamma_0+k\gamma_1+\cdots+k\gamma_{n-1}. \tag{15}$$

Computing the characteristic polynomial for the transformation (14), we obtain:

Theorem 2.19. *The dynamical degree $\delta(f_a)$ is the largest root of the polynomial*

$$\chi_{n,k}(x)=1-k\sum_{\ell=1}^{n-1}x^\ell+x^n. \tag{16}$$

We have seen that Σ_0 is invariant under f_a, and from the mapping of the fibers we see that certain unions of the fibers are invariant. One consequence of the Proposition is the following: If C is an irreducible curve in $\mathscr{X}$, and if the class of its divisor $\{C\}$ belongs to S, then C must be one of the curves Σ_0 or $\mathscr{F}_s^j$, $0 \le s \le n-1$, $1 \le j \le 2k$. A consequence of this is:

Proposition 2.20. *If C is a curve which is invariant under f_a, then C is a union of components from the collection Σ_0 and $\mathscr{F}_s^j$, $0 \le s \le n-1$, $1 \le j \le 2k$.*

Proof. Let t denote the projection of the class of C to T. Since C is invariant, we have $f_* C = C$, and thus $f_* t = t$. But since $\chi_{n,k}(1) = 2 - k(n-1) < 0$, 1 is not an eigenvalue of f_*. Thus $t = 0$, which means that $C \in S$. And we saw above that each element of S is represented uniquely as a union of the generating curves. □

At this point, it may be evident that there is a certain amount of arbitrariness in our choice of blowups. Specifically, if $f \in Aut(\mathscr{X})$, and $a, fa, \ldots, f^k a = a$ is an orbit of period k, then we may construct a new space $\pi : \mathscr{Y} \to \mathscr{X}$ by blowing up this orbit. The induced map $f_{\mathscr{Y}}$ is an automorphism of $\mathscr{Y}$. If $\mathscr{Z}$ is a smooth surface, and $f \in Aut(\mathscr{Z})$, we say that the dynamical system $(f, \mathscr{Z})$ is *minimal* if whenever $\mathscr{W}$ is a smooth surface and $g \in Aut(\mathscr{W})$, and whenever $\pi : \mathscr{Z} \to \mathscr{W}$ is a regular, birational map such that $g \circ \pi = \pi \circ f$, then π is a biregular conjugacy. It is evident that if we blow up a periodic orbit, then the resulting dynamical system is not minimal. On the other hand, we do not obtain uniqueness simply by requiring that our model be minimal. For instance, let $J(x,y) = (1/x, 1/y)$ be the Cremona involution. We have seen that the space $\mathscr{X}$ obtained by blowing up three points gives $J_{\mathscr{X}} \in Aut(\mathscr{X})$. In addition, it is clear that if $\mathscr{Y} = \mathbf{P}^1 \times \mathbf{P}^1$, then $J_{\mathscr{Y}} \in Aut(\mathscr{Y})$.

Here is another example. Let $\mathscr{X}$ denote the space obtained by blowing up $\mathbf{P}^2$ at $[1:1:1]$, and let L denote the involution $[x:y:z] \to [-x:y:z]$, acting on $\mathscr{X}$. Thus the fiber over $[1:1:1]$ is exceptional and is blown down to $[-1:1:1]$, which is indeterminate. We can turn L into an automorphism by blowing up $\mathscr{X}$ at the point $[-1:1:1]$. The result, of course, is not minimal, since we can now blow both exceptional fibers back down, and L will be a linear automorphism of $\mathbf{P}^2$.

Theorem 2.21. *$(f, \mathscr{X}_f)$ is minimal if $n > 2$. If $n = 2$, then it becomes minimal after we blow down Σ_0.*

Proof. Suppose that $\varphi : \mathscr{X}_f \to \mathscr{Y}$ is a morphism. Consider the curve $\mathscr{C}$ consisting of all the varieties in $\mathscr{X}_f$ which are blown down to points under φ. It follows that $\mathscr{C}$ is invariant under f, so by Theorem 3.5, $\mathscr{C}$ must be a union of components coming from Σ_0 and $\mathscr{F}_s^j$. If $n > 2$, then the self-intersection of each of the components Σ_0 and $\mathscr{F}_s^j$ is ≤ -2, so it is not possible to blow any of them down. On the other hand, if $n = 2$, then the self-intersection of Σ_0 is -1, so we can blow it down. This leaves the self-intersection of all the other fibers unchanged, except for $\mathscr{F}_s^1$, which increases to $-k$. This is strictly less than -1, so nothing further can be blown down. □

References

[BHPV] Barth, W.P., Hulek, K., Peters, C.A.M., Van de Ven.: Antonius Compact Complex Surfaces. Second edition. Ergebnisse der Mathematik und ihrer Grenzgebiete. Springer-Verlag, Berlin (2004)

[BK1] Bedford, E., Kim, K.H.: On the degree growth of birational mappings in higher dimension. J. Geom. Anal. **14**, 567–596 (2004). arXiv:math.DS/0406621

[BK2] Bedford, E., Kim, K.H.: Periodicities in linear fractional recurrences: degree growth of birational surface maps. Mich. Math. J. bf 54, 647–670 (2006). arxiv:math.DS/0509645

[BK3] Bedford, E., Kim, K.H.: Dynamics of rational surface automorphisms: linear fractional recurrences. J. Geomet. Anal. **19**, 553–583 (2009). arXiv:math/0611297

[BK4] Bedford, E., Kim, K.H.: Continuous families of rational surface automorphisms with positive entropy, Math. Ann., arXiv:0804.2078

[BLS] Bedford, E., Lyubich, M., Smillie, J.: Polynomial diffeomorphisms of C^2. IV: the measure of maximal entropy and laminar currents. Invent. Math. **112**, 77–125 (1993)

[BS1] Bedford, E., Smillie, J.: Polynomial diffeomorphisms of C^2: currents, equilibrium measure, and hyperbolicity. Invent. Math. **87**, 69–99 (1990)

[BS2] Bedford, E., Smillie, J.: Polynomial diffeomorphisms of C^2. II: Stable manifolds and recurrence. J. Am. Math. Soc. **4**(4), 657–679 (1991)

[BS7] Bedford, E., Smillie, J.: Polynomial diffeomorphisms of C^2. VII: Hyperbolicity and external rays. Ann. Sci. Ecole Norm. Sup. **32**, 455–497 (1999)

[BSn] Bedford, E., Smillie, J.: External rays in the dynamics of polynomial automorphisms of C^2. Complex geometric analysis in Pohang (1997), 41–79, Contemp. Math., **222**, Am. Math. Soc., Providence, RI (1999)

[C1] Cantat, S.: Dynamique des automorphismes des surfaces projectives complexes. C. R. Acad. Sci. Paris Sér. I Math. **328**(10), 901–906 (1999)

[C2] Cantat, S.: Dynamique des automorphismes des surfaces $K3$. Acta Math. **187**(1), 1–57 (2001)

[C3] Cantat, S.: Quelques aspects des systèmes dynamiques polynomiaux: Existence, exemples, rigidité. Panorama et Synthèse, vol. 30, arXiv:0811.2325

[CD] Cerveau, D., Déserti, J.: Transformations Birationelles de Petit Degré.

[D] Diller, J.: Cremona transformations, surface automorphisms, and the group law, arXiv:0811.3038

[DDG1] Diller, J., Dujardin, R., Guedj, V.: Dynamics of meromorphic maps with small topological degree I: from cohomology to currents. Indiana U. Math. J., arXiv:0803.0955

[DDG2] Diller, J., Dujardin, R., Guedj, V.: Dynamics of meromorphic maps with small topological degree II: Energy and invariant measure. Comment. Math. Helvet., arXiv:0805.3842

[DDG3] Diller, J., Dujardin, R., Guedj, V.: Dynamics of meromorphic maps with small topological degree III: geometric currents and ergodic theory. Annales Scientifiques de l'ENS **43**, 235–278 (2010). arXiv:0806.0146

[DF] Diller, J., Favre, C.: Dynamics of bimeromorphic maps of surfaces. Am. J. Math. **123**, 1135–1169 (2001)

[Do] Dolgachev, I.: Reflection groups in algebraic geometry. Bull. AMS **45**, 1–60 (2008)

[Du] Dujardin, R.: Laminar currents and birational dynamics. Duke Math. J. **131**(2), 219–247 (2006)

[FS1] Fornæss, J.E., Sibony, N.: Complex Hénon mappings in C^2 and Fatou-Bieberbach domains. Duke Math. J. **65**, 345–380 (1992)

[FS2] Fornæss, J.E., Sibony, N.: Classification of recurrent domains for some holomorphic maps. Math. Ann. **301**, 813–820 (1995)

[FS3] Fornæss, J.E., Sibony, N.: Complex dynamics in higher dimensions. Notes partially written by Estela A. Gavosto. NATO Adv. Sci. Inst. Ser. C Math. Phys. Sci. **439**, Complex potential theory (Montreal, PQ, 1993), 131–186, Kluwer Acad. Publ., Dordrecht, 1994.

[FM] Friedland, S., Milnor, J.: Dynamical properties of plane polynomial automorphisms. Ergod. Theor. Dyn. Syst. **9**(1), 67–99 (1989)

[GF] Grauert, H., Fritzsche, K.: From holomorphic functions to complex manifolds. Graduate Texts in Mathematics, 213. Springer-Verlag, New York (2002)

[GH] Griffiths, P., Harris, J.: Principles of Algebraic Geometry. John Wiley, New York (1978)

[Gu] Guedj, V.: Propriétés ergodiques des applications rationnelles, arXiv:math/0611302

[He] Herman, M.-R.: Recent results and some open questions on Siegel's linearization theorem of germs of complex analytic diffeomorphisms of C^n near a fixed point. VIIIth international congress on mathematical physics (Marseille, 1986), 138–184, World Sci. Publishing, Singapore (1987)

[Hi] Hirschowitz, A.: Sumétries des surfaces rationnelles génériques. Math. Ann. **281**(2), 255–261 (1988)

[HO1] Hubbard, J.H., Oberste-Vorth, R.: Hénon mappings in the complex domain. I. The global topology of dynamical space. Inst. Hautes Études Sci. Publ. Math. **79**, 5–46 (1994)

[HO2] Hubbard, J.H., Oberste-Vorth, R.: Hénon mappings in the complex domain. II. Projective and inductive limits of polynomials. Real and Complex Dynamical Systems (Hillerød), 1993 ED B. Branner and P. Hjorth, 1995 pp. 89–132

[I] Ishii, Y.: Hyperbolic polynomial diffeomorphisms of $\mathbf{C}^2$. II. Hubbard trees. Adv. Math. **220**(4), 985–1022 (2009)

[K] Koitabashi, M.: Automorphism groups of generic rational surfaces. J. Algebra **116**(1), 130–142 (1988)

[M1] McMullen, C.: Dynamics on $K3$ surfaces: Salem numbers and Siegel disks. J. Reine Angew. Math. **545**, 201–233 (2002)

[M2] McMullen, C.: Dynamics on blowups of the projective plane. Pub. Sci. IHES, **105**, 49–89 (2007)

[MNTU] Morosawa, S., Nishimura, Y., Taniguchi, M., Ueda, T.: Holomorphic Dynamics. Cambridge University Press, Cambridge (2000)

[N] Narasimhan, R.: Several complex variables. Reprint of the 1971 original. Chicago Lectures in Mathematics. University of Chicago Press, Chicago, IL (1995)

[O] Oliva, R.: On the combinatorics of external rays in the dynamics of the complex Hénon map. Ph. D. Thesis, Cornell University (1997)

[S] Sibony, N.: Dynamique des applications rationnelles de $\mathbf{P}^k$, Dynamique et géométrie complexes (Lyon, 1997). Panor. Synthèses, **8**, Soc. Math. France, Paris, 9–185 (1999)

[U1] Ueda, T.: Local structure of analytic transformations of two complex variables. I. J. Math. Kyoto Univ. **26**(2), 233–261 (1986)

[U2] Ueda, T.: Local structure of analytic transformations of two complex variables. II. J. Math. Kyoto Univ. **31**, 695–711 (1991)

Uniformisation of Foliations by Curves

Marco Brunella

Abstract These lecture notes provide a full discussion of certain analytic aspects of the uniformisation theory of foliations by curves on compact Kähler manifolds, with emphasis on convexity properties and their consequences on positivity properties of the corresponding canonical bundles.

1 Foliations by Curves and their Uniformisation

Let X be a complex manifold. A **foliation by curves** $\mathscr{F}$ on X is defined by a holomorphic line bundle $T_{\mathscr{F}}$ on X and a holomorphic linear morphism

$$\tau_{\mathscr{F}} : T_{\mathscr{F}} \to TX$$

which is injective outside an analytic subset $Sing(\mathscr{F}) \subset X$ of codimension at least 2, called the **singular set** of the foliation. Equivalently, we have an open covering $\{U_j\}$ of X and a collection of holomorphic vector fields $v_j \in \Theta(U_j)$, with zero set of codimension at least 2, such that

$$v_j = g_{jk} v_k \qquad \text{on} \quad U_j \cap U_k,$$

where $g_{jk} \in \mathscr{O}^*(U_j \cap U_k)$ is a multiplicative cocycle defining the dual bundle $T^*_{\mathscr{F}} = K_{\mathscr{F}}$, called the **canonical bundle** of $\mathscr{F}$.

These vector fields can be locally integrated, and by the relations above these local integral curves can be glued together (without respecting the time parametrization), giving rise to the **leaves** of the foliation $\mathscr{F}$.

M. Brunella
IMB - CNRS UMR 5584, 9 Avenue Savary, 21078 Dijon, France
e-mail: Marco.Brunella@u-bourgogne.fr

G. Gentili et al. (eds.), *Holomorphic Dynamical Systems*,
Lecture Notes in Mathematics 1998, DOI 10.1007/978-3-642-13171-4_3,

By the classical Uniformisation Theorem, the universal covering of each leaf is either the unit disc $\mathbb{D}$ (hyperbolic leaf) or the affine line $\mathbb{C}$ (parabolic leaf) or the projective line $\mathbb{P}$ (rational leaf).

In these notes we shall assume that the ambient manifold X is a compact connected Kähler manifold, and we will be concerned with the following problem: how the universal covering $\widetilde{L_p}$ of the leaf L_p through the point p depends on p ? For instance, we may first of all ask about the structure of the subset of X formed by those points through which the leaf is hyperbolic, resp. parabolic, resp. rational: is the set of hyperbolic leaves open in X? Is the set of parabolic leaves analytic? But, even if all the leaves are, say, hyperbolic, there are further basic questions: the uniformising map of every leaf is almost unique (unique modulo automorphisms of the disc), and after some normalization (to get uniqueness) we may ask about the way in which the uniformising map of L_p depends on the point p. Equivalently, we may put on every leaf its Poincaré metric, and we may ask about the way in which this leafwise metric varies in the directions transverse to the foliation.

Our main result will be that these universal coverings of leaves can be glued together in a vaguely "holomorphically convex" way. That is, the *leafwise* universal covering of the foliated manifold $(X, \mathscr{F})$ can be defined and it has a sort of "holomorphically convex" structure [Br2, Br3]. This was inspired by a seminal work of Il'yashenko [Il1, Il2], who proved a similar result when X is a Stein manifold instead of a compact Kähler one. Related ideas can also be found in Suzuki's paper [Suz], still in the Stein case. Another source of inspiration was Shafarevich conjecture on the holomorphic convexity of universal coverings of projective (or compact Kähler) manifolds [Nap].

This main result will allow us to apply results by Nishino [Nis] and Yamaguchi [Ya1, Ya2, Ya3, Kiz] concerning the transverse variation of the leafwise Poincaré metric and other analytic invariants. As a consequence of this, for instance, we shall obtain that if the foliation has at least one hyperbolic leaf, then: (1) there are no rational leaves; (2) parabolic leaves fill a subset of X which is *complete pluripolar*, i.e. locally given by the poles of a plurisubharmonic function. In other words, the set of hyperbolic leaves of $\mathscr{F}$ is either empty or potential-theoretically full in X.

These results are related also to positivity properties of the canonical bundle $K_{\mathscr{F}}$, along a tradition opened by Arakelov [Ara, BPV] in the case of algebraic fibrations by curves and developed by Miyaoka [Miy, ShB] and then McQuillan and Bogomolov [MQ1, MQ2, BMQ, Br1] in the case of foliations on projective manifolds. From this point of view, our final result is the following ruledness criterion for foliations:

Theorem 1.1. [Br3, Br5] *Let X be a compact connected Kähler manifold and let $\mathscr{F}$ be a foliation by curves on X. Suppose that the canonical bundle $K_{\mathscr{F}}$ is* not *pseudoeffective. Then through every point $p \in X$ there exists a rational curve tangent to $\mathscr{F}$.*

Recall that a line bundle on a compact connected manifold is *pseudoeffective* if it admits a (singular) hermitian metric with positive curvature in the sense of currents

[Dem]. When X is projective the above theorem follows also from results of [BMQ] and [BDP], but with a totally different proof, untranslatable in our Kähler context.

Let us now describe in more detail the content of these notes.

In Section 2 we shall recall the results by Nishino and Yamaguchi on Stein fibrations that we shall use later, and also some of Il'yashenko's results. In Section 3 and 4 we construct the leafwise universal covering of $(X,\mathscr{F})$: we give an appropriate definition of leaf L_p of $\mathscr{F}$ through a point $p \in X \setminus Sing(\mathscr{F})$ (this requires some care, because some leaves are allowed to pass through some singular points), and we show that the universal coverings $\widetilde{L_p}$ can be glued together to get a complex manifold. In Section 5 we prove that the complex manifold so constructed enjoys some "holomorphic convexity" property. This is used in Section 6 and 8, together with Nishino and Yamaguchi results, to prove (among other things) Theorem 1.1 above. The parabolic case requires also an extension theorem for certain meromorphic maps into compact Kähler manifolds, which is proved in Section 7.

All this work has been developed in our previous papers [Br2,Br3,Br4,Br5] (with few imprecisions which will be corrected here). Further results and application can be found in [Br6] and [Br7].

2 Some Results on Stein Fibrations

2.1 Hyperbolic Fibrations

In a series of papers, Nishino [Nis] and then Yamaguchi [Ya1, Ya2, Ya3] studied the following situation. It is given a Stein manifold U, of dimension $n+1$, equipped with a holomorphic submersion $P: U \to \mathbb{D}^n$ with connected fibers. Each fiber $P^{-1}(z)$ is thus a smooth connected curve, and as such it has several potential theoretic invariants (Green functions, Bergman Kernels, harmonic moduli...). One is interested in knowing how these invariants vary with z, and then in using this knowledge to obtain some information on the structure of U.

For our purposes, the last step in this program has been carried out by Kizuka [Kiz], in the following form.

Theorem 2.1. [Ya1, Ya3, Kiz] *If U is Stein, then the fiberwise Poincaré metric on $U \stackrel{P}{\to} \mathbb{D}^n$ has a plurisubharmonic variation.*

This means the following. On each fiber $P^{-1}(z)$, $z \in \mathbb{D}^n$, we put its Poincaré metric, i.e. the (unique) complete hermitian metric of curvature -1 if $P^{-1}(z)$ is uniformised by $\mathbb{D}$, or the identically zero "metric" if $P^{-1}(z)$ is uniformised by $\mathbb{C}$ (U being Stein, there are no other possibilities). If v is a holomorphic nonvanishing vector field, defined in some open subset $V \subset U$ and tangent to the fibers of P, then we can take the function on V

$$F = \log \|v\|_{Poin}$$

where, for every $q \in V$, $\|v(q)\|_{Poin}$ is the norm of $v(q)$ evaluated with the Poincaré metric on $P^{-1}(P(q))$. The statement above means that, whatever v is, the function F is *plurisubharmonic*, or identically $-\infty$ if all the fibers are parabolic. Note that if we replace v by $v' = g \cdot v$, with g a holomorphic nonvanishing function on V, then F is replaced by $F' = F + G$, where $G = \log|g|$ is pluriharmonic. A more intrinsic way to state this property is: the fiberwise Poincaré metric (if not identically zero) defines on the relative canonical bundle of $U \xrightarrow{P} \mathbb{D}^n$ a hermitian metric (possibly singular) whose curvature is a *positive current* [Dem]. Note also that the plurisubharmonicity of F along the fibers is just a restatement of the negativity of the curvature of the Poincaré metric. The important fact here is the plurisubharmonicity along the directions transverse to the fibers, whence the *variation* terminology.

Remark that the poles of F correspond exactly to parabolic fibers of U. We therefore obtain the following dichotomy: either all the fibers are parabolic ($F \equiv 0$), or the parabolic fibers correspond to a complete pluripolar subset of $\mathbb{D}^n$ ($F \not\equiv 0$).

The theorem above is a generalization of, and was motivated by, a classical result of Hartogs [Ran, II.5], asserting (in modern language) that for a domain U in $\mathbb{D}^n \times \mathbb{C}$ of the form (Hartogs tube)

$$U = \{\ (z,w) \mid \ \ |w| < e^{-f(z)}\},$$

where $f : \mathbb{D}^n \to [-\infty, +\infty)$ is an upper semicontinuous function, the Steinness of U implies that f is plurisubharmonic. Indeed, in this special case the Poincaré metric is easily computed, and one checks that the plurisubharmonicity of f is equivalent to the plurisubharmonic variation of the fiberwise Poincaré metric. By Oka, also the converse holds: if f is plurisubharmonic, then U is Stein. This special case suggests that some converse statement to Theorem 2.1 could be true.

We give the proof of Theorem 2.1 only in a particular case, which is anyway the only case that we shall actually use.

We start with a fibration $P : U \to \mathbb{D}^n$ as above, but without assuming U Stein. We consider an open subset $U_0 \subset U$ such that:

(i) for every $z \in \mathbb{D}^n$, the intersection $U_0 \cap P^{-1}(z)$ is a disc, relatively compact in the fiber $P^{-1}(z)$;
(ii) the boundary ∂U_0 is real analytic and transverse to the fibers of P;
(iii) the boundary ∂U_0 is pseudoconvex in U.

Then we restrict our attention to the fibration by discs $P_0 = P|_{U_0} : U_0 \to \mathbb{D}^n$. It is not difficult to see that U_0 is Stein, but this fact will not really be used below.

Proposition 2.2. [Ya1, Ya3] *The fiberwise Poincaré metric on* $U_0 \xrightarrow{P_0} \mathbb{D}^n$ *has a plurisubharmonic variation.*

Proof. It is sufficient to consider the case $n = 1$. The statement is local on the base, and for every $z_0 \in \mathbb{D}$ we can embed a neighbourhood of $\overline{P_0^{-1}(z_0)}$ in U into $\mathbb{C}^2$ in such a way that P becomes the projection to the first coordinate (see, e.g., [Suz, §3]). Thus we may assume that $U_0 \subset \mathbb{D} \times \mathbb{C}$, $P_0(z,w) = z$, and $P^{-1}(z) = D_z$ is a disc in $\{z\} \times \mathbb{C} = \mathbb{C}$, with real analytic boundary, depending on z in a real analytic and pseudoconvex way.

Take a holomorphic section $\alpha : \mathbb{D} \to U_0$ and a holomorphic vertical vector field v along α, i.e. for every $z \in \mathbb{D}$, $v(z)$ is a vector in $T_{\alpha(z)}U_0$ tangent to the fiber over z (and nonvanishing). We need to prove that $\log \|v(z)\|_{Poin(D_z)}$ is a subharmonic function on $\mathbb{D}$. By another change of coordinates, we may assume that $\alpha(z) = (z,0)$ and $v(z) = \frac{\partial}{\partial w}|_{(z,0)}$.

For every z, let

$$g(z,\cdot) : \overline{D_z} \to [0,+\infty]$$

be the Green function of D_z with pole at 0. That is, $g(z,\cdot)$ is harmonic on $D_z \setminus \{0\}$, zero on ∂D_z, and around $w=0$ it has the development

$$g(z,w) = \log \frac{1}{|w|} + \lambda(z) + O(|w|).$$

The constant $\lambda(z)$ (Robin constant) is related to the Poincaré metric of D_z: more precisely, we have

$$\lambda(z) = -\log \left\| \frac{\partial}{\partial w}|_{(z,0)} \right\|_{Poin(D_z)}$$

(indeed, recall that the Green function gives the radial part of a uniformisation of D_z). Hence, we are reduced to show that $z \mapsto \lambda(z)$ is *super*harmonic.

Fix $z_0 \in \mathbb{D}$. By real analyticity of ∂U_0, the function g is (outside the poles) also real analytic, and thus extensible (in a real analytic way) beyond ∂U_0. This means that if z is sufficiently close to z_0, then $g(z,\cdot)$ is actually defined on $\overline{D_{z_0}}$, and harmonic on $D_{z_0} \setminus \{0\}$. Of course, $g(z,\cdot)$ does not need to vanish on ∂D_{z_0}. The difference $g(z,\cdot) - g(z_0,\cdot)$ is harmonic on D_{z_0} (the poles annihilate), equal to $\lambda(z) - \lambda(z_0)$ at 0, and equal to $g(z,\cdot)$ on ∂D_{z_0}. By Green formula:

$$\lambda(z) - \lambda(z_0) = -\frac{1}{2\pi} \int_{\partial D_{z_0}} g(z,w) \frac{\partial g}{\partial n}(z_0,w) ds$$

and consequently:

$$\frac{\partial^2 \lambda}{\partial z \partial \bar{z}}(z_0) = -\frac{1}{2\pi} \int_{\partial D_{z_0}} \frac{\partial^2 g}{\partial z \partial \bar{z}}(z_0,w) \frac{\partial g}{\partial n}(z_0,w) ds.$$

We now compute the z-laplacian of $g(\cdot, w_0)$ when w_0 is a point of the boundary ∂D_{z_0}.

The function $-g$ is, around (z_0,w_0), a defining function for U_0. By pseudoconvexity, the Levi form of g at (z_0,w_0) is therefore nonpositive on the complex tangent space $T^{\mathbb{C}}_{(z_0,w_0)}(\partial U_0)$, i.e. on the Kernel of ∂g at (z_0,w_0) [Ran, II.2]. By developing, and using also the fact that g is w-harmonic, we obtain

$$\frac{\partial^2 g}{\partial z \partial \bar{z}}(z_0,w_0) \leq 2Re \left\{ \frac{\frac{\partial^2 g}{\partial w \partial \bar{z}}(z_0,w_0) \cdot \frac{\partial g}{\partial z}(z_0,w_0)}{\frac{\partial g}{\partial w}(z_0,w_0)} \right\}.$$

We put this inequality into the expression of $\frac{\partial^2\lambda}{\partial z\partial\bar{z}}(z_0)$ derived above from Green formula, and then we apply Stokes theorem. We find

$$\frac{\partial^2\lambda}{\partial z\partial\bar{z}}(z_0) \leq -\frac{2}{\pi}\int_{D_{z_0}}\left|\frac{\partial^2 g}{\partial w\partial\bar{z}}(z_0,w)\right|^2 idw\wedge d\bar{w} \leq 0$$

from which we see that λ is superharmonic. □

A similar result can be proved, by the same proof, even when we drop the simply connectedness hypothesis on the fibers, for instance when the fibers of U_0 are annuli instead of discs; however, the result is that the Bergman fiberwise metric, and not the Poincaré one, has a plurisubharmonic variation. This is because on a multiply connected curve the Green function is more directly related to the Bergman metric [Ya3]. The case of the Poincaré metric is done in [Kiz], by a covering argument. The general case of Theorem 2.1 requires also to understand what happens when ∂U_0 is still pseudoconvex but no more transverse to the fibers, so that U_0 is no more a differentiably trivial family of curves. This is rather delicate, and it is done in [Ya1]. Then Theorem 2.1 is proved by an exhaustion argument.

2.2 Parabolic Fibrations

Theorem 2.1, as stated, is rather empty when all the fibers are isomorphic to $\mathbb{C}$. However, in that case Nishino proved that if U is Stein then it is isomorphic to $\mathbb{D}^n \times \mathbb{C}$ [Nis, II]. A refinement of this was found in [Ya2].

As before, we consider a fibration $P : U \to \mathbb{D}^n$ and we do not assume that U is Stein. We suppose that there exists an embedding $j : \mathbb{D}^n \times \mathbb{D} \to U$ such that $P \circ j$ coincides with the projection from $\mathbb{D}^n \times \mathbb{D}$ to $\mathbb{D}^n$ (this can always be done, up to restricting the base). For every $\varepsilon \in [0,1)$, we set

$$U_\varepsilon = U \setminus j(\mathbb{D}^n \times \overline{\mathbb{D}(\varepsilon)})$$

with $\overline{\mathbb{D}(\varepsilon)} = \{z \in \mathbb{C} | \quad |z| \leq \varepsilon\}$, and we denote by

$$P_\varepsilon : U_\varepsilon \to \mathbb{D}^n$$

the restriction of P. Thus, the fibers of P_ε are obtained from those of P by removing a closed disc (if $\varepsilon > 0$) or a point (if $\varepsilon = 0$).

Theorem 2.3. [Nis, Ya2, II] *Suppose that:*

(i) for every $z \in \mathbb{D}^n$, the fiber $P^{-1}(z)$ is isomorphic to $\mathbb{C}$;

(ii) for every $\varepsilon > 0$ the fiberwise Poincaré metric on $U_\varepsilon \xrightarrow{P_\varepsilon} \mathbb{D}^n$ has a plurisubharmonic variation.

Then U is isomorphic to a product:

$$U \simeq \mathbb{D}^n \times \mathbb{C}.$$

Proof. For every $z \in \mathbb{D}^n$ we have a unique isomorphism

$$f(z,\cdot) : P^{-1}(z) \to \mathbb{C}$$

such that, using the coordinates given by j,

$$f(z,0) = 0 \qquad \text{and} \qquad f'(z,0) = 1.$$

We want to prove that f is holomorphic in z.

Set $R_\varepsilon(z) = f(z, P_\varepsilon^{-1}(z)) \subset \mathbb{C}$. By Koebe's Theorem, the distorsion of $f(z,\cdot)$ on compact subsets of $\mathbb{D}$ is uniformly bounded, and so $\mathbb{D}(\frac{1}{k}\varepsilon) \subset f(z, \mathbb{D}(\varepsilon)) \subset \mathbb{D}(k\varepsilon)$ for every $\varepsilon \in (0, \frac{1}{2})$ and for some constant k, independent on z. Therefore, for every ε and z,

$$\mathbb{C} \setminus \mathbb{D}\left(\frac{1}{k}\varepsilon\right) \subset R_\varepsilon(z) \subset \mathbb{C} \setminus \mathbb{D}(k\varepsilon).$$

In a similar way [Nis, II], Koebe's Theorem gives also the continuity of the above map f.

On the fibers of P_0, which are all isomorphic to $\mathbb{C}^*$, we put the unique complete hermitian metric of zero curvature and period (=length of closed simple geodesics) equal to $\sqrt{2}\pi$. On the fibers of P_ε, $\varepsilon > 0$, which are all hyperbolic, we put the Poincaré metric multiplied by $\log \varepsilon$, whose (constant) curvature is therefore equal to $-\frac{1}{(\log \varepsilon)^2}$. By a simple and explicit computation, the Poincaré metric on $\mathbb{C} \setminus \mathbb{D}(c\varepsilon)$ multiplied by $\log \varepsilon$ converges uniformly to the flat metric of period $\sqrt{2}\pi$ on $\mathbb{C}^*$, as $\varepsilon \to 0$. Using this and the above bounds on $R_\varepsilon(z)$, we obtain that our fiberwise metric on $U_\varepsilon \xrightarrow{P_\varepsilon} \mathbb{D}^n$ converges uniformly, as $\varepsilon \to 0$, to our fiberwise metric on $U_0 \xrightarrow{P_0} \mathbb{D}^n$ (see [Br4] for more explicit computations). Hence, from the plurisubharmonic variation of the former we deduce the plurisubharmonic variation of the latter.

Our flat metric on $P_0^{-1}(z)$ is the pull-back by $f(z,\cdot)$ of the metric $\frac{idx \wedge d\bar{x}}{4|x|^2}$ on $R_0(z) = \mathbb{C}^*$. In the coordinates given by j, we have

$$f(z,w) = w \cdot e^{g(z,w)},$$

with g holomorphic in w and $g(z,0) = 0$ for every z, by the choice of the normalization. Hence, in these coordinates our metric takes the form

$$\left| 1 + w \frac{\partial g}{\partial w}(z,w) \right|^2 \cdot \frac{idw \wedge d\bar{w}}{4|w|^2}.$$

Set $F = \log |1 + w \frac{\partial g}{\partial w}|^2$. We know, by the previous arguments, that F is plurisubharmonic. Moreover, $\frac{\partial^2 F}{\partial w \partial \bar{w}} \equiv 0$, by flatness of the metric. By semipositivity of the Levi form we then obtain $\frac{\partial^2 F}{\partial w \partial \bar{z}_k} \equiv 0$ for every k. Hence the function $\frac{\partial F}{\partial w}$ is holomorphic, that is the function $(\frac{\partial g}{\partial w} + w \frac{\partial^2 g}{\partial w^2})(1 + w \frac{\partial g}{\partial w})^{-1}$ is holomorphic. Taking into account that $g(z,0) \equiv 0$, we obtain from this that g also is fully holomorphic. Thus f is fully holomorphic in the chart given by j, and hence everywhere. It follows that U is isomorphic to a product. □

Remark that if U is Stein then the hypothesis on the plurisubharmonic variation is automatically satisfied, by Theorem 2.1, and because if U is Stein then also U_ε are Stein, for every ε. That was the situation originally considered by Nishino and Yamaguchi.

A standard illustration of Theorem 2.3 is the following one. Take a continuous function $h : \mathbb{D} \to \mathbb{P}$, let $\Gamma \subset \mathbb{D} \times \mathbb{P}$ be its graph, and set $U = (\mathbb{D} \times \mathbb{P}) \setminus \Gamma$. Then U fibers over $\mathbb{D}$ and all the fibers are isomorphic to $\mathbb{C}$. Clearly U is isomorphic to a product $\mathbb{D} \times \mathbb{C}$ if and only if h is holomorphic, which in turn is equivalent, by a classical result (due, once a time, to Hartogs), to the Steinness of U.

2.3 Foliations on Stein Manifolds

Even if we shall not need Il'yashenko's results [Il1, Il2], let us briefly explain them, as a warm-up for some basic ideas.

Let X be a Stein manifold, of dimension n, and let $\mathscr{F}$ be a foliation by curves on X. In order to avoid some technicalities (to which we will address later), let us assume that $\mathscr{F}$ is *nonsingular*, i.e. $Sing(\mathscr{F}) = \emptyset$.

Take an embedded $(n-1)$-disc $T \subset X$ transverse to $\mathscr{F}$. For every $t \in T$, let L_t be the leaf of $\mathscr{F}$ through t, and let $\widetilde{L}_t$ be its universal covering with basepoint t. Remark that, because X is Stein, every $\widetilde{L}_t$ is isomorphic either to $\mathbb{D}$ or to $\mathbb{C}$. In [Il1] Il'yashenko proves that these universal coverings $\{\widetilde{L}_t\}_{t\in T}$ can be glued together to get a complex manifold of dimension n, a sort of "long flow box". More precisely, there exists a complex n-manifold U_T with the following properties:

(i) U_T admits a submersion $P_T : U_T \to T$ and a section $p_T : T \to U_T$ such that, for every $t \in T$, the pointed fiber $(P_T^{-1}(t), p_T(t))$ is identified (in a natural way) with $(\widetilde{L}_t, t)$;
(ii) U_T admits an immersion (i.e., local biholomorphism) $\Pi_T : U_T \to X$ which sends each fiber $(\widetilde{L}_t, t)$ to the corresponding leaf (L_t, t), as universal covering.

We shall not prove here these facts, because we shall prove later (Section 4) some closely related facts in the context of (singular) foliations on compact Kähler manifolds.

Theorem 2.4. [Il1, Il2] *The manifold U_T is Stein.*

Proof. Following Suzuki [Suz], it is useful to factorize the immersion $U_T \to X$ through another manifold V_T, which is constructed in a similar way as U_T except that the universal coverings $\widetilde{L}_t$ are replaced by the holonomy coverings $\widehat{L}_t$.

Here is Suzuki's construction. Fix a foliated chart $\Omega \subset X$ around T, i.e. $\Omega \simeq \mathbb{D}^{n-1} \times \mathbb{D}$, $T \simeq \mathbb{D}^{n-1} \times \{0\}$, $\mathscr{F}|_\Omega =$ vertical foliation, with leaves $\{*\} \times \mathbb{D}$. Let $\mathscr{O}_\mathscr{F}(\Omega)$ be the set of holomorphic functions on Ω which are constant on the leaves of $\mathscr{F}|_\Omega$, i.e. which depend only on the first $(n-1)$ coordinates. Let $\overline{V}_T$ be the *existence domain* of $\mathscr{O}_\mathscr{F}(\Omega)$ over X: by definition, this is the maximal holomorphically separable Riemann domain

$$\overline{V}_T \to X$$

which contains Ω and such that every $f \in \mathcal{O}_{\mathcal{F}}(\Omega)$ extends to some $\tilde{f} \in \mathcal{O}(\overline{V}_T)$. The classical Cartan-Thullen-Oka theory [GuR] says that $\overline{V}_T$ is a Stein manifold.

The projection $\Omega \to T$ extends to a map

$$\overline{Q}_T : \overline{V}_T \to T$$

thanks to $\mathcal{O}_{\mathcal{F}}(\Omega) \hookrightarrow \mathcal{O}(\overline{V}_T)$. Consider a fiber $\overline{Q}_T^{-1}(t)$. It is not difficult to see that the connected component of $\overline{Q}_T^{-1}(t)$ which cuts Ω ($\subset \overline{V}_T$) is exactly the holonomy covering $\widehat{L}_t$ of L_t, with basepoint t. The reason is the following one. Firstly, if $\gamma : [0,1] \to L_t$ is a path contained in a leaf, with $\gamma(0) = t$, then any function $f \in \mathcal{O}_{\mathcal{F}}(\Omega)$ can be analytically prolonged along γ, by preserving the constancy on the leaves. Secondly, if γ_1 and γ_2 are two such paths with the same endpoint $s \in L_t$, then the germs at s obtained by the two continuations of f along γ_1 and γ_2 may be different. If the foliation has trivial holonomy along $\gamma_1 * \gamma_2^{-1}$, then the two germs are certainly equal; conversely, if the holonomy is not trivial, then we can find f such that the two final germs are different. This argument shows that $\widehat{L}_t$ is naturally contained into $\overline{Q}_T^{-1}(t)$. The fact that it is a connected component is just a "maximality" argument (note that $\overline{V}_T$ is foliated by the pull-back of $\mathcal{F}$, and fibers of $\overline{Q}_T$ are closed subvarieties invariant by this foliation).

We denote by $V_T \subset \overline{V}_T$ (open subset) the union of these holonomy coverings, and by Q_T the restriction of $\overline{Q}_T$ to V_T.

Let us return to U_T. We have a natural map (local biholomorphism)

$$F_T : U_T \to V_T$$

which acts as a covering between fibers (but not globally: see Examples 4.7 and 4.8 below). In particular, U_T is a Riemann domain over the Stein manifold $\overline{V}_T$.

Lemma 2.5. *U_T is holomorphically separable.*

Proof. Given $p,q \in U_T$, $p \neq q$, we want to construct $f \in \mathcal{O}(U_T)$ such that $f(p) \neq f(q)$. The only nontrivial case (V_T being holomorphically separable) is the case where $F_T(p) = F_T(q)$, in particular p and q belong to the same fiber $\widetilde{L}_t$.

We use the following procedure. Take a path γ in $\widetilde{L}_t$ from p to q. It projects by F_T to a closed path γ_0 in $\widehat{L}_t$. Suppose that $[\gamma_0] \neq 0$ in $H_1(\widehat{L}_t, \mathbb{R})$. Then we may find a holomorphic 1-form $\omega \in \Omega^1(\widehat{L}_t)$ such that $\int_{\gamma_0} \omega = 1$. This 1-form can be holomorphically extended from $\widehat{L}_t$ to $V_T \subset \overline{V}_T$, because $\overline{V}_T$ is Stein and $\widehat{L}_t$ is a closed submanifold of it. Call $\widehat{\omega}$ such an extension, and $\widetilde{\omega} = F_T^*(\widehat{\omega})$ its lift to U_T. On every (simply connected!) fiber $\widetilde{L}_t$ of U_T the 1-form $\widetilde{\omega}$ is exact, and can be integrated giving a holomorphic function $f_t(z) = \int_t^z \widetilde{\omega}|_{\widetilde{L}_t}$. We thus obtain a holomorphic function f on U_T, which separates p and q: $f(p) - f(q) = \int_\gamma \widetilde{\omega} = \int_{\gamma_0} \widehat{\omega} = 1$.

This procedure does not work if $[\gamma_0] = 0$: in that case, every $\omega \in \Omega^1(\widehat{L}_t)$ has period equal to zero on γ_0. But, in that case, we may find two 1-forms $\omega_1, \omega_2 \in \Omega^1(\widehat{L}_t)$ such that the iterated integral of (ω_1, ω_2) along γ_0 is not zero (this iterated integral [Che] is just the integral along γ of $\phi_1 d\phi_2$, where ϕ_j is a primitive of ω_j

lifted to $\widetilde{L_t}$). Then we can repeat the argument above: the fiberwise iterated integral of $(\widetilde{\omega}_1, \widetilde{\omega}_2)$ is a holomorphic function on U_T which separates p and q. □

Having established that U_T is a holomorphically separable Riemann domain over $\overline{V}_T$, it is again a fundamental result of Cartan-Thullen-Oka theory [GuR] that there exists a Stein Riemann domain

$$\overline{F}_T : \overline{U}_T \to \overline{V}_T$$

which contains U_T and such that $\mathscr{O}(\overline{U}_T) = \mathscr{O}(U_T)$. The map $P_T : U_T \to T$ extends to

$$\overline{P}_T : \overline{U}_T \to T,$$

and U_T can be identified with the open subset of $\overline{U}_T$ composed by the connected components of fibers of $\overline{P}_T$ which cut $\Omega \subset \overline{U}_T$. But, in fact, much better is true:

Lemma 2.6. *Every fiber of $\overline{P}_T$ is connected, that is:*

$$\overline{U}_T = U_T.$$

Proof. If not, then, by a connectivity argument, we may find $a_0, b_0 \in \overline{P}_T^{-1}(t_0)$, $a_k, b_k \in \overline{P}_T^{-1}(t_k)$, with $a_k \to a_0$ and $b_k \to b_0$, such that:

(i) $a_0 \in \widetilde{L_{t_0}}$, $b_0 \in \overline{P}_T^{-1}(t_0) \setminus \widetilde{L_{t_0}}$;
(ii) $a_k, b_k \in \widetilde{L_{t_k}}$.

Denote by $\mathscr{M}_{t_0}$ the maximal ideal of $\mathscr{O}_{t_0}$ (on T), and for every $p \in \overline{P}_T^{-1}(t_0)$ denote by $\mathscr{I}_p \subset \mathscr{O}_p$ the ideal generated by $(\overline{P}_T)^*(\mathscr{M}_{t_0})$. At points of $\widetilde{L_{t_0}}$, this is just the ideal of functions vanishing along $\widetilde{L_{t_0}}$; whereas at points of $\overline{P}_T^{-1}(t_0) \setminus \widetilde{L_{t_0}}$, at which $\overline{P}_T$ may fail to be a submersion, this ideal may correspond to a "higher order" vanishing. Because $\overline{U}_T$ is Stein and $\overline{P}_T^{-1}(t_0)$ is a closed subvariety, we may find a holomorphic function $f \in \mathscr{O}(\overline{U}_T)$ such that:

(iii) $f \equiv 0$ on $\widetilde{L_{t_0}}$, $f \equiv 1$ on $\overline{P}_T^{-1}(t_0) \setminus \widetilde{L_{t_0}}$;
(iv) for every $p \in \overline{P}_T^{-1}(t_0)$, the differential df_p of f at p belongs to the ideal $\mathscr{I}_p \Omega^1_p$.

Let $\{z_1, \ldots, z_{n-1}\}$ denote the coordinates on T lifted to $\overline{U}_T$. Then, by property (iv), we can factorize

$$df = \sum_{j=1}^{n-1} (z_j - z_j(t_0)) \cdot \beta_j$$

where β_j are holomorphic 1-forms on $\overline{U}_T$.

As in Lemma 2.5, each β_j can be integrated along the simply connected fibers of U_T (with starting point on T), giving a function $g_j \in \mathscr{O}(U_T)$. This function can be holomorphically extended to the envelope $\overline{U}_T$. By the factorization above, and (ii), we have

$$f(b_k) - f(a_k) = \sum_{j=1}^{n-1} (z_j(t_k) - z_j(t_0)) \cdot (g_j(b_k) - g_j(a_k))$$

and this expression tends to 0 as $k \to +\infty$. Therefore $f(b_0) = f(a_0)$, in contradiction with (i) and (iii). □

It follows from this Lemma that $U_T = \overline{U}_T$ is Stein. □

Remark 2.7. We do not know if V_T also is Stein, i.e. if $V_T = \overline{V}_T$.

This Theorem allows to apply the results of Nishino and Yamaguchi discussed above to holomorphic foliations on Stein manifolds. For instance: the set of parabolic leaves of such a foliation is either full or complete pluripolar. A similar point of view is pursued in [Suz].

3 The Unparametrized Hartogs Extension Lemma

In order to construct the leafwise universal covering of a foliation, we shall need an extension lemma of Hartogs type. This is done in this Section.

Let X be a compact Kähler manifold. Denote by A_r, $r \in (0,1)$, the semiclosed annulus $\{r < |w| \leq 1\}$, with boundary $\partial A_r = \{|w| = 1\}$. Given a holomorphic immersion

$$f : A_r \to X$$

we shall say that $f(A_r)$ **extends to a disc** if there exists a holomorphic map

$$g : \overline{\mathbb{D}} \to X,$$

not necessarily immersive, such that f factorizes as $g \circ j$ for some embedding $j : A_r \to \overline{\mathbb{D}}$, sending ∂A_r to $\partial\mathbb{D}$. That is, f itself does not need to extend to the full disc $\{|w| \leq 1\}$, but it extends "after a reparametrization", given by j.

Remark that if f is an embedding, and $f(A_r)$ extends to a disc, then we can find g as above which is moreover injective outside a finite subset. The image $g(\overline{\mathbb{D}})$ is a (possibly singular) disc in X with boundary $f(\partial A_r)$. Such an extension g or $g(\overline{\mathbb{D}})$ will be called **simple** extension of f or $f(A_r)$. Note that such a g is uniquely defined up to a Moëbius reparametrization of $\overline{\mathbb{D}}$.

Given a holomorphic immersion

$$f : \mathbb{D}^k \times A_r \to X$$

we shall say that $f(\mathbb{D}^k \times A_r)$ **extends to a meromorphic family of discs** if there exists a meromorphic map

$$g : W \dashrightarrow X$$

such that:

(i) W is a complex manifold of dimension $k+1$ with boundary, equipped with a holomorphic submersion $W \to \mathbb{D}^k$ all of whose fibers W_z, $z \in \mathbb{D}^k$, are isomorphic to $\overline{\mathbb{D}}$;

(ii) f factorizes as $g \circ j$ for some embedding $j : \mathbb{D}^k \times A_r \to W$, sending $\mathbb{D}^k \times \partial A_r$ to ∂W and $\{z\} \times A_r$ into W_z, for every $z \in \mathbb{D}^k$.

In particular, the restriction of g to the fiber W_z gives, after removal of indeterminacies, a disc which extends $f(z, A_r)$, and these discs depend on z in a meromorphic way. The manifold W is differentiably a product of $\mathbb{D}^k$ with $\overline{\mathbb{D}}$, but in general this does not hold holomorphically. However, note that by definition W is around its boundary ∂W isomorphic to a product $\mathbb{D}^k \times A_r$.

We shall say that an immersion $f : \mathbb{D}^k \times A_r \to X$ is an **almost embedding** if there exists a proper analytic subset $I \subset \mathbb{D}^k$ such that the restriction of f to $(\mathbb{D}^k \setminus I) \times A_r$ is an embedding. In particular, for every $z \in \mathbb{D}^k \setminus I$, $f(z, A_r)$ is an embedded annulus in X, and $f(z, A_r)$, $f(z', A_r)$ are disjoint if $z, z' \in \mathbb{D}^k \setminus I$ are different.

The following result is a sort of "unparametrized" Hartogs extension lemma [Siu, Iv1], in which the extension of maps is replaced by the extension of their images. Its proof is inspired by [Iv1] and [Iv2]. The new difficulty is that we need to construct not only a map but also the space where it is defined. The necessity of this unparametrized Hartogs lemma for our future constructions, instead of the usual parametrized one, has been observed in [ChI].

Theorem 3.1. *Let X be a compact Kähler manifold and let $f : \mathbb{D}^k \times A_r \to X$ be an almost embedding. Suppose that there exists an open nonempty subset $\Omega \subset \mathbb{D}^k$ such that $f(z, A_r)$ extends to a disc for every $z \in \Omega$. Then $f(\mathbb{D}^k \times A_r)$ extends to a meromorphic family of discs.*

Proof. Consider the subset

$$Z = \{\, z \in \mathbb{D}^k \setminus I \mid f(z, A_r) \text{ extends to a disc } \}.$$

Our first aim is to give to Z a complex analytic structure with countable base. This is a rather standard fact, see [Iv2] for related ideas and [CaP] for a larger perspective.

For every $z \in Z$, fix a simple extension

$$g_z : \overline{\mathbb{D}} \to X$$

of $f(z, A_r)$. We firstly put on Z the following metrizable topology: we define the distance between $z_1, z_2 \in Z$ as the Hausdorff distance in X between the discs $g_{z_1}(\overline{\mathbb{D}})$ and $g_{z_2}(\overline{\mathbb{D}})$. Note that this topology may be *finer* than the topology induced by the inclusion $Z \subset \mathbb{D}^k$: if $z_1, z_2 \in Z$ are close each other in $\mathbb{D}^k$ then $g_{z_1}(\overline{\mathbb{D}})$, $g_{z_2}(\overline{\mathbb{D}})$ are close each other near their boundaries, but their interiors may be far each other (think to blow-up).

Take $z \in Z$ and take a Stein neighbourhood $U \subset X$ of $g_z(\overline{\mathbb{D}})$. Consider the subset $A \subset \mathbb{D}^k \setminus I$ of those points z' such that the circle $f(z', \partial A_r)$ is the boundary of a compact complex curve $C_{z'}$ contained in U. Note that, by the maximum principle, such a curve is Hausdorff-close to $g_z(\overline{\mathbb{D}})$, if z' is close to z. According to a theorem of Wermer or Harvey-Lawson [AWe, Ch.19], this condition is equivalent to say that $\int_{f(z', \partial A_r)} \beta = 0$ for every holomorphic 1-form β on U (moment condition). These integrals depend holomorphically on z', for every β. We deduce (by noetherianity) that A is an analytic subset of $\mathbb{D}^k \setminus I$, on a neighbourhood of z. For every $z' \in A$, however, the curve $C_{z'}$ is not necessarily the image of a disc: recall that $g_z(\overline{\mathbb{D}})$ may be singular and may have selfintersections, and so a curve close to it may have positive genus, arising from smoothing the singularities.

Set $\mathscr{A} = \{ (z',x) \in A \times U \mid x \in C_{z'} \}$. By inspection of the proof of Wermer-Harvey-Lawson theorem [AWe, Ch.19], we see that $\mathscr{A}$ is an analytic subset of $A \times U$ (just by the holomorphic dependence on parameters of the Cauchy transform used in that proof to construct $C_{z'}$). We have a tautological fibration $\pi : \mathscr{A} \to A$ and a tautological map $\tau : \mathscr{A} \to U$ defined by the two projections. Let $B \subset A$ be the subset of those points z' such that the fiber $\pi^{-1}(z') = C_{z'}$ has geometric genus zero. This is an analytic subset of A (the function $z' \mapsto \{$ geometric genus of $\pi^{-1}(z')\}$ is Zariski lower semicontinuous). By restriction, we have a tautological fibration $\pi : \mathscr{B} \to B$ and a tautological map $\tau : \mathscr{B} \to U \subset X$. Each fiber of π over B is a disc, sent by τ to a disc in U with boundary $f(z', \partial A_r)$. In particular, B is contained in Z.

Now, a neighbourhood of z in B can be identified with a neighbourhood of z in Z (in the Z-topology above): if $z' \in Z$ is Z-close to z then $g_{z'}(\overline{\mathbb{D}})$ is contained in U and then $z' \in B$. In this way, the analytic structure of B is transferred to Z. Note that, with this complex analytic structure, the inclusion $Z \hookrightarrow \mathbb{D}^k$ is holomorphic. More precisely, each irreducible component of Z is a *locally analytic subset* of $\mathbb{D}^k \setminus I$ (where, as usual, "locally analytic" means "analytic in a neighbourhood of it"; of course, a component does not need to be closed in $\mathbb{D}^k \setminus I$).

Let us now prove that the complex analytic space Z has a *countable* number of irreducible components.

To see this, we use the area function $\mathbf{a} : Z \to \mathbb{R}^+$, defined by

$$\mathbf{a}(z) = \text{ area of } g_z(\overline{\mathbb{D}}) = \int_{\overline{\mathbb{D}}} g_z^*(\omega)$$

($\omega =$ Kähler form of X). This function is continuous on Z. Let $c > 0$ be the minimal area of rational curves in X. Set, for every $m \in \mathbb{N}$,

$$Z_m = \left\{ z \in Z \,\middle|\, \mathbf{a}(z) \in \left(m\frac{c}{2}, (m+2)\frac{c}{2}\right) \right\},$$

so that Z is covered by $\cup_{m=0}^{+\infty} Z_m$. Each Z_m is open in Z, and we claim that on each Z_m the Z-topology coincides with the $\mathbb{D}^k$-topology. Indeed, take a sequence $\{z_n\} \subset Z_m$ which $\mathbb{D}^k$-converges to $z_\infty \in Z_m$. We thus have, in X, a sequence of discs $g_{z_n}(\overline{\mathbb{D}})$ with boundaries $f(z_n, \partial A_r)$ and areas in the interval $(m\frac{c}{2}, (m+2)\frac{c}{2})$. By Bishop's compactness theorem [Bis] [Iv1, Prop.3.1], up to subsequencing, $g_{z_n}(\overline{\mathbb{D}})$ converges, in the Hausdorff topology, to a compact complex curve of the form $D \cup Rat$, where D is a disc with boundary $f(z_\infty, \partial A_r)$ and Rat is a finite union of rational curves (the bubbles). Necessarily, $D = g_{z_\infty}(\overline{\mathbb{D}})$. Moreover,

$$\lim_{n \to +\infty} \text{area}(g_{z_n}(\overline{\mathbb{D}})) = \text{area}(g_{z_\infty}(\overline{\mathbb{D}})) + \text{area}(Rat).$$

From $\mathbf{a}(z_\infty), \mathbf{a}(z_n) \in (m\frac{c}{2}, (m+2)\frac{c}{2})$ it follows that $\text{area}(Rat) < c$, hence, by definition of c, $Rat = \emptyset$. Hence $g_{z_n}(\overline{\mathbb{D}})$ converges, in the Hausdorff topology, to $g_{z_\infty}(\overline{\mathbb{D}})$, i.e. z_n converges to z_∞ in the Z-topology.

Therefore, if $L_m \subset Z_m$ is a countable $\mathbb{D}^k$-dense subset then L_m is also Z-dense in Z_m, and $\cup_{m=0}^{+\infty} L_m$ is countable and Z-dense in Z. It follows that Z has countably many irreducible components.

After these preliminaries, we can really start the proof of the theorem.

The hypotheses imply that the space Z has (at least) one irreducible component V which is open in $\mathbb{D}^k \setminus I$. Let us consider again the area function $\mathbf{a}$ on V. The following lemma is classical, and it is at the base of every extension theorem for maps into Kähler manifolds [Siu, Iv1].

Lemma 3.2. *For every compact $K \subset \mathbb{D}^k$, the function $\mathbf{a}$ is bounded on $V \cap K$.*

Proof. If $z_0, z_1 \in V$, then we can join them by a continuous path $\{z_t\}_{t\in[0,1]} \subset V$, so that we have in X a continuous family of discs $g_{z_t}(\overline{\mathbb{D}})$, with boundaries $f(z_t, \partial A_r)$. By Stokes formula, the difference between the area of $g_{z_1}(\overline{\mathbb{D}})$ and $g_{z_0}(\overline{\mathbb{D}})$ is equal to the integral of the Kähler form ω on the "'tube" $\cup_{t\in[0,1]} g_{z_t}(\partial\mathbb{D}) = f(\cup_{t\in[0,1]}\{z_t\} \times \partial A_r)$. Now, for topological reasons, $f^*(\omega)$ admits a primitive λ on $\mathbb{D}^k \times A_r$. Therefore

$$\mathbf{a}(z_1) - \mathbf{a}(z_0) = \int_{\{z_1\}\times\partial A_r} \lambda - \int_{\{z_0\}\times\partial A_r} \lambda .$$

Remark that the function $z \mapsto \int_{\{z\}\times\partial A_r} \lambda$ is defined (and smooth) on the full $\mathbb{D}^k$, not only on V, and so it is bounded on every compact $K \subset \mathbb{D}^k$. The conclusion follows immediately. □

We use this lemma to study the boundary of V, and to show that the complement of V is small.

Take a point $z_\infty \in (\mathbb{D}^k \setminus I) \cap \partial V$ and a sequence $z_n \in V$ converging to z_∞. By the boundedness of $\mathbf{a}(z_n)$ and Bishop compactness theorem, we obtain a disc in X with boundary $f(z_\infty, \partial A_r)$ (plus, perhaps, some rational bubbles, but we may forget them). In particular, the point z_∞ belongs to Z. Obviously, the irreducible component of Z which contains z_∞ is not open in $\mathbb{D}^k \setminus I$, because $z_\infty \in \partial V$, and so that component is a locally analytic subset of $\mathbb{D}^k \setminus I$ of *positive* codimension. It follows that the boundary ∂V is a *thin subset* of $\mathbb{D}^k$, i.e. it is contained in a countable union of locally analytic subsets of positive codimension (certain components of Z, plus the analytic subset I). Disconnectedness properties of thin subsets show that also *the complement* $\mathbb{D}^k \setminus V$ *(*$= \partial V$*) is thin in* $\mathbb{D}^k$.

Recall now that over V we have the (normalized) tautological fibration $\pi : \mathscr{V} \to V$, equipped with the tautological map $\tau : \mathscr{V} \to X$. Basically, this provides the desired extension of f over the large open subset V. As in [Iv1], we shall get the extension over the full $\mathbb{D}^k$ by reducing to the Thullen type theorem of Siu [Siu].

By construction, $\partial\mathscr{V}$ has a neigbourhood isomorphic to $V \times A_r$, the isomorphism being realized by f. Hence we can glue to $\mathscr{V}$ the space $\mathbb{D}^k \times A_r$, using the same f. We obtain a new space $\mathscr{W}$ equipped with a fibration $\pi : \mathscr{W} \to \mathbb{D}^k$ and a map $\tau : \mathscr{W} \to X$ such that:

(i) $\pi^{-1}(z) \simeq \overline{\mathbb{D}}$ for $z \in V$, $\pi^{-1}(z) \simeq A_r$ for $z \in \mathbb{D}^k \setminus V$;
(ii) f factorizes through τ.

In other words, and recalling how $\mathscr{V}$ was defined, up to normalization $\mathscr{W}$ is simply the analytic subset of $\mathbb{D}^k \times X$ given by the union of all the discs $\{z\} \times g_z(\overline{\mathbb{D}})$, $z \in V$, and all the annuli $\{z\} \times f(z, A_r)$, $z \in \mathbb{D}^k \setminus V$.

Lemma 3.3. *There exists an embedding* $\mathscr{W} \to \mathbb{D}^k \times \mathbb{P}$, *which respects the fibrations over* $\mathbb{D}^k$.

Proof. Set $B_r = \{\, w \in \mathbb{P} \mid |w| > r \,\}$. By construction, $\partial\mathscr{W}$ has a neighbourhood isomorphic to $\mathbb{D}^k \times A_r$. We can glue $\mathbb{D}^k \times B_r$ to $\mathscr{W}$ by identification of that neighbourhood with $\mathbb{D}^k \times A_r \subset \mathbb{D}^k \times B_r$, i.e. by prolonging each annulus A_r to a disc B_r. The result is a new space $\widehat{\mathscr{W}}$ with a fibration $\widehat{\pi} : \widehat{\mathscr{W}} \to \mathbb{D}^k$ such that:

(i) $\widehat{\pi}^{-1}(z) \simeq \mathbb{P}$ for every $z \in V$;
(ii) $\widehat{\pi}^{-1}(z) \simeq B_r$ for every $z \in \mathbb{D}^k \setminus V$.

We shall prove that $\widehat{\mathscr{W}}$ (and hence $\mathscr{W}$) embeds into $\mathbb{D}^k \times \mathbb{P}$ (incidentally, note the common features with the proof of Theorem 2.3).

For every $z \in V$, there exists a unique isomorphism

$$\varphi_z : \widehat{\pi}^{-1}(z) \to \mathbb{P}$$

such that

$$\varphi_z(\infty) = 0\,, \quad \varphi_z'(\infty) = 1\,, \quad \varphi_z(r) = \infty$$

where $\infty, r \in \overline{B_r} \subset \widehat{\pi}^{-1}(z)$ and the derivative at ∞ is computed using the coordinate $\frac{1}{w}$. Every $\mathbb{P}$-fibration is locally trivial, and so this isomorphism φ_z depends holomorphically on z. Thus we obtain a biholomorphism

$$\Phi : \widehat{\pi}^{-1}(V) \to V \times \mathbb{P}$$

and we want to prove that Φ extends to the full $\widehat{\mathscr{W}}$.

By Koebe's Theorem, the distorsion of φ_z on any compact $K \subset B_r$ is uniformly bounded (note that $\varphi_z(B_r) \subset \mathbb{C}$). Hence, for every $w_0 \in B_r$ the holomorphic function $z \mapsto \varphi_z(w_0)$ is bounded on V. Because the complement of V in $\mathbb{D}^k$ is thin, by Riemann's extension theorem this function extends holomorphically to $\mathbb{D}^k$. This permits to extends the above Φ also to fibers over $\mathbb{D}^k \setminus V$. Still by bounded distorsion, this extension is an embedding of $\widehat{\mathscr{W}}$ into $\mathbb{D}^k \times \mathbb{P}$. □

Now we can finish the proof of the theorem. Thanks to the previous embedding, we may "fill in" the holes of $\mathscr{W}$ and obtain a $\overline{\mathbb{D}}$-fibration W over $\mathbb{D}^k$. Then, by the Thullen type theorem of Siu [Siu] (and transfinite induction) the map $\tau : \mathscr{W} \to X$ can be meromorphically extended to W. This is the meromorphic family of discs which extends $f(\mathbb{D}^k \times A_r)$.

By comparison with the usual "parametrized" Hartogs extension lemma [Iv1], one could ask if the almost embedding hypothesis in Theorem 3.1 is really indispensable. In some sense, the answer is yes. Indeed, we may easily construct a fibered immersion $f : \mathbb{D} \times A_r \to \mathbb{D} \times \mathbb{P} \subset \mathbb{P} \times \mathbb{P}$, $f(z,w) = (z, f_0(z,w))$, such that: (i) for some $z_0 \in \mathbb{D}$, $f_0(z_0, \partial A_r)$ is an embedded circle in $\mathbb{P}$; (ii) for some other $z_1 \in \mathbb{D}$, $f_0(z_1, \partial A_r)$ is an immersed but not embedded circle in $\mathbb{P}$. Then, for some neighbourhhod $U \subset \mathbb{D}$ of z_0, $f(U \times A_r)$ can be obviously extended to a meromorphic (even holomorphic) family of discs, but such a U cannot be enlarged to contain z_1, because $f_0(z_1, \partial A_r)$ bounds no disc in $\mathbb{P}$. Note, however, that $f_0(z_1, \partial A_r)$ bounds a so called holomorphic chain in $\mathbb{P}$ [AWe, Ch.19]: if $\Omega_1, \ldots, \Omega_m$ are the connected components of $\mathbb{P} \setminus f_0(z_1, \partial A_r)$, then $f_0(z_1, \partial A_r)$ is the "algebraic" boundary of $\sum_{j=1}^{m} n_j \Omega_j$, for suitable integers n_j. It is conceivable that Theorem 3.1 holds under the sole assumption that f is an immersion, provided that the manifold W is replaced by a (suitably defined) "meromorphic family of 1-dimensional chains".

4 Holonomy Tubes and Covering Tubes

Here we shall define leaves, holonomy tubes and covering tubes, following [Br3].

Let X be a compact Kähler manifold, of dimension n, and let $\mathscr{F}$ be a foliation by curves on X. Set $X^0 = X \setminus Sing(\mathscr{F})$ and $\mathscr{F}^0 = \mathscr{F}|_{X^0}$. We could define the "leaves" of the singular foliation $\mathscr{F}$ simply as the usual leaves of the nonsingular foliation $\mathscr{F}^0$. However, for our purposes we shall need that the universal coverings of the leaves glue together in a nice way, producing what we shall call covering tubes. This shall require a sort of semicontinuity of the fundamental groups of the leaves. With the naïve definition "leaves of $\mathscr{F}$ = leaves of $\mathscr{F}^0$", such a semicontinuity can fail, in the sense that a leaf (of $\mathscr{F}^0$) can have a larger fundamental group than nearby leaves (of $\mathscr{F}^0$). To remedy to this, we give now a less naïve definition of leaf of $\mathscr{F}$, which has the effect of killing certain homotopy classes of cycles, and the problem will be settled almost by definition (but we will require also the unparametrized Hartogs extension lemma of the previous Section).

4.1 Vanishing Ends

Take a point $p \in X^0$, and let L_p^0 be the leaf of $\mathscr{F}^0$ through p. It is a smooth complex connected curve, equipped with an injective immersion

$$i_p^0 : L_p^0 \to X^0,$$

and sometimes we will tacitly identify L_p^0 with its image in X^0 or X. Recall that, given a local transversal $\mathbb{D}^{n-1} \hookrightarrow X^0$ to $\mathscr{F}^0$ at p, we have a holonomy representation [CLN]

$$hol_p : \pi_1(L_p^0, p) \to Diff(\mathbb{D}^{n-1}, 0)$$

of the fundamental group of L_p^0 with basepoint p into the group of germs of holomorphic diffeomorphisms of $(\mathbb{D}^{n-1}, 0)$.

Let $E \subset L_p^0$ be a *parabolic end* of L_p^0, that is a closed subset isomorphic to the punctured closed disc $\overline{\mathbb{D}}^* = \{ 0 < |w| \leq 1 \}$, and suppose that the holonomy of $\mathcal{F}^0$ along the cycle ∂E is trivial. Then, for some $r \in (0,1)$, the inclusion $A_r \subset \overline{\mathbb{D}}^* = E$ can be extended to an embedding $\mathbb{D}^{n-1} \times A_r \to X^0$ which sends each $\{z\} \times A_r$ into a leaf of $\mathcal{F}^0$, and $\{0\} \times A_r$ to $A_r \subset E$ (this is because A_r is Stein, see for instance [Suz, §3]). More generally, if the holonomy of $\mathcal{F}^0$ along ∂E is finite, of order k, then we can find an immersion $\mathbb{D}^{n-1} \times A_r \to X^0$ which sends each $\{z\} \times A_r$ into a leaf of $\mathcal{F}^0$ and $\{0\} \times A_r$ to $A_{r'} \subset E$, in such a way that $\{0\} \times A_r \to A_{r'}$ is a regular covering of order k. Such an immersion is (or can be chosen as) an almost embedding: the exceptional subset $I \subset \mathbb{D}^{n-1}$, outside of which the map is an embedding, corresponds to leaves which intersect the transversal, over which the holonomy is computed, at points whose holonomy orbit has cardinality strictly less than k. This is an analytic subset of the transversal. Such an almost embedding will be called *adapted to* E.

We shall use the following terminology: a meromorphic map is a *meromorphic immersion* if it is an immersion outside its indeterminacy set.

Definition 4.1. Let $E \subset L_p^0$ be a parabolic end with finite holonomy, of order $k \geq 1$. Then E is a **vanishing end**, of order k, if there exists an almost embedding $f : \mathbb{D}^{n-1} \times A_r \to X^0 \subset X$ adapted to E such that:

(i) $f(\mathbb{D}^{n-1} \times A_r)$ extends to a meromorphic family of discs $g : W \dashrightarrow X$;
(ii) g is a meromorphic immersion.

In other words, E is a vanishing end if, firstly, it can be compactified in X to a disc, by adding a singular point of $\mathcal{F}$, and, secondly, this disc-compactification can be meromorphically and immersively deformed to nearby leaves, up to a ramification given by the holonomy. This definition mimics, in our context, the classical definition of vanishing cycle for real codimension one foliations [CLN].

Remark 4.2. If $g : W \dashrightarrow X$ is as in Definition 4.1, then the indeterminacy set $F = Indet(g)$ cuts each fiber W_z, $z \in \mathbb{D}^{n-1}$, along a finite subset $F_z \subset W_z$. The restricted map $g_z : W_z \to X$ sends $W_z \setminus F_z$ into a leaf of $\mathcal{F}^0$, in an immersive way, and F_z into $Sing(\mathcal{F})$. Each point of F_z corresponds to a parabolic end of $W_z \setminus F_z$, which is sent by g_z to a parabolic end of a leaf; clearly, this parabolic end is a vanishing one (whose order, however, may be smaller than k), and the corresponding meromorphic family of discs is obtained by restricting g. Remark also that, F being of codimension at least 2, we have $F_z = \emptyset$ for every z outside an analytic subset of $\mathbb{D}^{n-1}$ of positive codimension. This means (as we shall see better below) that "most" leaves have no vanishing end.

If $E \subset L_p^0$ is a vanishing end of order k, then we compactify it by adding one point, i.e. by prolonging $\overline{\mathbb{D}}^*$ to $\overline{\mathbb{D}}$. But we do such a compactification in an *orbifold sense*: the added point has, by definition, a *multiplicity* equal to k. By doing such a end-compactification for every vanishing end of L_p^0, we finally obtain a connected curve (with orbifold structure) L_p, which is by definition the **leaf** of $\mathcal{F}$ through p. The initial inclusion $i_p^0 : L_p^0 \to X^0$ can be extended to a holomorphic map

$$i_p : L_p \to X$$

which sends the discrete subset $L_p \setminus L_p^0$ into $Sing(\mathscr{F})$. Note that i_p may fail to be immersive at those points. Moreover, it may happen that two different points of $L_p \setminus L_p^0$ are sent by i_p to the same singular point of $\mathscr{F}$ (see Example 4.4 below). In spite of this, we shall sometimes identify L_p with its image in X. For instance, to say that a map $f : Z \to X$ "has values into L_p" shall mean that f factorizes through i_p.

Remark that we have not defined, and shall not define, leaves L_p through $p \in Sing(\mathscr{F})$: a leaf may pass through $Sing(\mathscr{F})$, but its basepoint must be chosen outside $Sing(\mathscr{F})$.

Let us see two examples.

Example 4.3. Take a compact Kähler surface S foliated by an elliptic fibration $\pi : S \to C$, and let $c_0 \in C$ be such that the fiber $F_0 = \pi^{-1}(c_0)$ is of Kodaira type II [BPV, V.7], i.e. a rational curve with a cusp q. If $p \in F_0$, $p \neq q$, then the leaf L_p^0 is equal to $F_0 \setminus \{q\} \simeq \mathbb{C}$. This leaf has a parabolic end with trivial holonomy, which is *not* a vanishing end. Indeed, this end can be compactified to a cuspidal disc, which however cannot be meromorphically deformed *as a disc* to nearby leaves, because nearby leaves have positive genus close to q. Hence $L_p = L_p^0$.

Let now $\widetilde{S} \to S$ be the composition of three blow-ups which transforms F_0 into a tree of four smooth rational curves $\widetilde{F_0} = G_1 + G_2 + G_3 + G_6$ of respective self-intersections $-1, -2, -3, -6$ [BPV, V.10]. Let $\widetilde{\pi} : \widetilde{S} \to C$ be the new elliptic fibration/foliation. Set $p_j = G_1 \cap G_j$, $j = 2, 3, 6$. If $p \in G_1$ is different from those three points, then $L_p^0 = G_1 \setminus \{p_2, p_3, p_6\}$. The parabolic end of L_p^0 corresponding to p_2 (resp. p_3, p_6) has holonomy of order 2 (resp. 3, 6). This time, this is a vanishing end: a disc D in G_1 through p_2 (resp. p_3, p_6) ramified at order 2 (resp. 3, 6) can be deformed to nearby leaves as discs close to $2D + G_2$ (resp. $3D + G_3$, $6D + G_6$), and also the "meromorphic immersion" condition can be easily respected. Thus L_p is isomorphic to the orbifold "$\mathbb{P}$ with three points of multiplicity 2, 3, 6". Note that the universal covering (in orbifold sense) of L_p is isomorphic to $\mathbb{C}$, and the holonomy covering (defined below) is a smooth elliptic curve.

Finally, if $p \in G_j$, $p \neq p_j$, $j = 2, 3, 6$, then L_p^0 has a parabolic end with trivial holonomy, which is not a vanishing end, and so $L_p = L_p^0 \simeq \mathbb{C}$.

A more systematic analysis of the surface case, from a slightly different point of view, can be found in [Br1].

Example 4.4. Take a projective threefold M containing a smooth rational curve C with normal bundle $N_C = \mathscr{O}(-1) \oplus \mathscr{O}(-1)$. Take a foliation $\mathscr{F}$ on M, nonsingular around C, such that: (i) for every $p \in C$, $T_p\mathscr{F}$ is different from T_pC; (ii) $T_{\mathscr{F}}$ has degree -1 on C. It is easy to see that there are a lot of foliations on M satisfying these two requirements. Note that, on a neighbourhood of C, we can glue together the local leaves (discs) of $\mathscr{F}$ through C, and obtain a smooth surface S containing C; condition (ii) means that the selfintersection of C in S is equal to -1.

We now perform a *flop* of M along C. That is, we firstly blow-up M along C, obtaining a threefold $\widetilde{M}$ containing an exceptional divisor D naturally $\mathbb{P}$-fibered over C. Because $N_C = \mathscr{O}(-1) \oplus \mathscr{O}(-1)$, this divisor D is in fact isomorphic to $\mathbb{P} \times \mathbb{P}$, hence it admits a second $\mathbb{P}$-fibration, transverse to the first one. Each fibre of this second fibration can be blow-down to a point (Moishezon's criterion [Moi]), and the result is a smooth threefold M', containing a smooth rational curve C' with normal bundle $N_{C'} = \mathscr{O}(-1) \oplus \mathscr{O}(-1)$, over which D fibers. (At this point, M' could be no more projective, nor Kähler, but this is not an important fact in this example). The strict transform $\overline{S}$ of S in $\widetilde{M}$ cuts the divisor D along one of the fibers of the second fibration $D \to C'$, by condition (ii) above, therefore its image S' in M' is a bidimensional disc which cuts C' transversely at some point p.

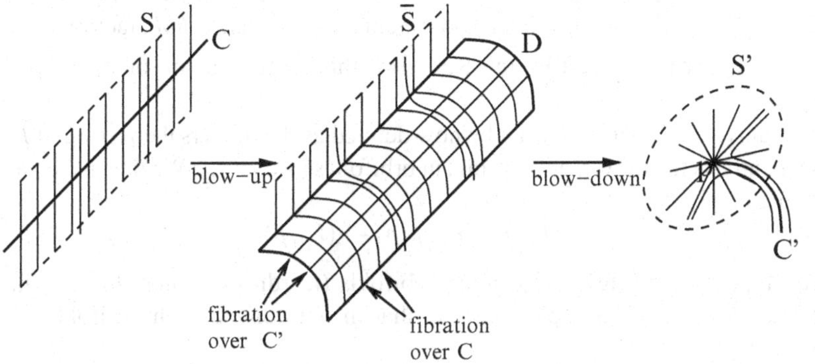

Let us look at the transformed foliation $\mathscr{F}'$ on M'. The point p is a singular point of $\mathscr{F}'$, the only one on a neighbourhood of C'. The curve C' is invariant by $\mathscr{F}'$. The surface S' is tangent to $\mathscr{F}'$, and over it the foliation has a radial type singularity. In fact, in appropriate coordinates around p the foliation is generated by the vector field $x\frac{\partial}{\partial x} + y\frac{\partial}{\partial y} - z\frac{\partial}{\partial z}$, with $S' = \{z = 0\}$ and $C' = \{x = y = 0\}$.

If L^0 is a leaf of $(\mathscr{F}')^0$, then each component D^0 of $L^0 \cap S'$ is a parabolic end converging to p. It is a vanishing end, of order 1: the meromorphic family of discs of Definition 4.1 is obviously constructed from a flow box of $\mathscr{F}$, around a suitably chosen point of C. Generic fibers of this family are sent to discs in M' close to $D^0 \cup C'$; other fibers are sent to discs in S' passing through p, and close to $D^0 \cup \{p\}$. Remark that it can happen that $L^0 \cap S'$ has several, or even infinitely many, components; in that case the map $i : L \to M'$ sends several, or even infinitely many, points to the same $p \in M'$.

Having defined the leaf L_p through $p \in X^0$, we can now define its **holonomy covering** $\widehat{L_p}$ and its **universal covering** $\widetilde{L_p}$. The first one is the covering defined by the Kernel of the holonomy. More precisely, we start with the usual holonomy covering $(\widehat{L_p^0}, p) \to (L_p^0, p)$ with basepoint p (it is useful to think to $\widehat{L_p^0}$ as equivalence classes of paths in L_p^0 starting at p, so that the basepoint $p \in \widehat{L_p^0}$ is the class of the constant path). If $E \subset L_p^0$ is a vanishing end of order k, then its preimage in $\widehat{L_p^0}$ is a (finite or infinite) collection of parabolic ends $\{\widehat{E}_j\}$, each one regularly covering E with degree k. Each such map $\overline{\mathbb{D}}^* \simeq \widehat{E}_j \to \overline{\mathbb{D}}^* \simeq E$ can be extended to a map $\overline{\mathbb{D}} \to \overline{\mathbb{D}}$, with a ramification at 0 of order k. By definition, $\widehat{L_p}$ is obtained by compactifying all these parabolic ends of $\widehat{L_p^0}$, over all the vanishing ends of L_p^0. Therefore, we have a covering map

$$(\widehat{L_p}, p) \to (L_p, p)$$

which ramifies over $L_p \setminus L_p^0$. However, from the orbifold point of view such a map is a *regular* covering: $w = z^k$ is a regular covering if $z = 0$ has multiplicity 1 and $w = 0$ has multiplicity k. Note that we do not need anymore a orbifold structure on $\widehat{L_p}$, in the sense that all its points have multiplicity 1.

In a more algebraic way, the orbifold fundamental group $\pi_1(L_p, p)$ is a quotient of $\pi_1(L_p^0, p)$, through which the holonomy representation hol_p factorizes. Then $\widehat{L_p}$ is the covering defined by the Kernel of this representation of $\pi_1(L_p, p)$ into $Diff(\mathbb{D}^{n-1}, 0)$.

The universal covering $\widetilde{L_p}$ can be now defined as the universal covering of $\widehat{L_p}$, or equivalently as the universal covering, in orbifold sense, of L_p. We then have natural covering maps

$$(\widetilde{L_p}, p) \to (\widehat{L_p}, p) \to (L_p, p).$$

Recall that there are few exceptional orbifolds (teardrops) which do not admit a universal covering. It is a pleasant fact that in our context such orbifolds do not appear.

4.2 Holonomy Tubes

We now analyze how the maps $p \mapsto \widehat{L_p}$ and then $p \mapsto \widetilde{L_p}$ depend on p. Propositions 4.5 and 4.6 below say that, in some sense, the dependence on p is holomorphic: holonomy coverings and universal coverings can be holomorphically glued together, producing fibered complex manifolds.

Let $T \subset X^0$ be a (local) transversal to $\mathcal{F}^0$.

Proposition 4.5. *There exists a complex manifold V_T of dimension n, a holomorphic submersion*

$$Q_T : V_T \to T,$$

a holomorphic section

$$q_T : T \to V_T,$$

and a meromorphic immersion

$$\pi_T : V_T \dashrightarrow X$$

such that:

(i) for every $t \in T$, the pointed fiber $(Q_T^{-1}(t), q_T(t))$ is isomorphic to $(\widehat{L_t}, t)$;
(ii) the indeterminacy set $Indet(\pi_T)$ of π_T cuts each fiber $Q_T^{-1}(t) = \widehat{L_t}$ along the discrete subset $\widehat{L_t} \setminus \widehat{L_t^0}$;

(iii) for every $t \in T$, the restriction of π_T to $Q_T^{-1}(t) = \widehat{L_t}$ coincides, after removal of indeterminacies, with the holonomy covering $\widehat{L_t} \to L_t \xrightarrow{i_t} i_t(L_t) \subset X$.

Proof. We firstly prove a similar statement for the regular foliation $\mathscr{F}^0$ on X^0. We use Il'yashenko's methodology [Il1]; an alternative but equivalent one can be found in [Suz], we have already seen it at the beginning of the proof of Theorem 2.4. In fact, in the case of a regular foliation the construction of V_T^0 below is a rather classical fact in foliation theory, which holds in the much more general context of smooth foliations with real analytic holonomy.

Consider the space $\Omega_T^{\mathscr{F}^0}$ composed by continuous paths $\gamma : [0,1] \to X^0$ tangent to $\mathscr{F}^0$ and such that $\gamma(0) \in T$, equipped with the uniform topology. On $\Omega_T^{\mathscr{F}^0}$ we put the following equivalence relation: $\gamma_1 \sim \gamma_2$ if $\gamma_1(0) = \gamma_2(0)$, $\gamma_1(1) = \gamma_2(1)$, and the loop $\gamma_1 * \gamma_2^{-1}$, obtained by juxtaposing γ_1 and γ_2^{-1}, has trivial holonomy.

Set

$$V_T^0 = \Omega_T^{\mathscr{F}^0} / \sim$$

with the quotient topology. Note that we have natural continuous maps

$$Q_T^0 : V_T^0 \to T$$

and

$$\pi_T^0 : V_T^0 \to X^0$$

defined respectively by $[\gamma] \mapsto \gamma(0) \in T$ and $[\gamma] \mapsto \gamma(1) \in X^0$. We also have a natural section

$$q_T^0 : T \to V_T^0$$

which associates to $t \in T$ the equivalence class of the constant path at t. Clearly, for every $t \in T$ the pointed fiber $((Q_T^0)^{-1}(t), q_T^0(t))$ is the same as $(\widehat{L_t^0}, t)$, by the very definition of holonomy covering, and π_T^0 restricted to that fiber is the holonomy covering map. Therefore, we just have to find a complex structure on V_T^0 such that all these maps become holomorphic.

We claim that V_T^0 is a Hausdorff space. Indeed, if $[\gamma_1], [\gamma_2] \in V_T^0$ are two nonseparated points, then $\gamma_1(0) = \gamma_2(0) = t$, $\gamma_1(1) = \gamma_2(1)$, and the loop $\gamma_1 * \gamma_2^{-1}$ in the leaf L_t^0 can be uniformly approximated by loops $\gamma_{1,n} * \gamma_{2,n}^{-1}$ in the leaves $L_{t_n}^0$ ($t_n \to t$) with trivial holonomy (so that $[\gamma_{1,n}] = [\gamma_{2,n}]$ is a sequence of points of V_T^0 converging to both $[\gamma_1]$ and $[\gamma_2]$). But this implies that also the loop $\gamma_1 * \gamma_2^{-1}$ has trivial holonomy, by the identity principle: if $h \in Diff(\mathbb{D}^{n-1}, 0)$ is the identity on a sequence of open sets accumulating to 0, then h is the identity everywhere. Thus $[\gamma_1] = [\gamma_2]$, and V_T^0 is Hausdorff.

Now, note that $\pi_T^0 : V_T \to X^0$ is a local homeomorphism. Hence we can pull back to V_T^0 the complex structure of X^0, and in this way V_T^0 becomes a complex manifold of dimension n with all the desired properties. Remark that, at this point, π_T^0 has not yet indeterminacy points, and so V_T^0 is a so-called Riemann Domain over X^0.

In order to pass from V_T^0 to V_T, we need to add to each fiber $\widehat{L_t^0}$ of V_T^0 the discrete set $\widehat{L_t} \setminus \widehat{L_t^0}$.

Take a vanishing end $E \subset L_t^0$, of order k, let $f : \mathbb{D}^{n-1} \times A_r \to X^0$ be an almost embedding adapted to E, and let $g : W \dashrightarrow X$ be a meromorphic family of discs extending f, immersive outside $F = Indet(g)$. Take also a parabolic end $\widehat{E} \subset \widehat{L_t^0}$ projecting to E, with degree k. By an easy holonomic argument, the immersion $g|_{W \setminus F} : W \setminus F \to X^0$ can be lifted to V_T^0, as a proper embedding

$$\widetilde{g} : W \setminus F \to V_T^0$$

which sends the central fiber $W_0 \setminus F_0$ to $\widehat{E}$. Each fiber $W_z \setminus F_z$ is sent by $\widetilde{g}$ to a closed subset of a fiber $\widehat{L^0_{t(z)}}$, and each point of F_z corresponds to a parabolic end of $\widehat{L^0_{t(z)}}$ projecting to a vanishing end of $L^0_{t(z)}$.

Now we can glue W to V_T^0 using $\widetilde{g}$: this corresponds to compactify all parabolic ends of fibers of V_T^0 which project to vanishing ends and which are close to $\widehat{E}$. By doing this operation for every E and $\widehat{E}$, we finally construct our manifold V_T, fibered over T with fibers $\widehat{L_t}$. The map π_T extending (meromorphically) π_T^0 is then deduced from the maps g above. □

The manifold V_T will be called **holonomy tube** over T. The meromorphic immersion π_T is, of course, *very complicated*: it contains all the dynamics of the foliation, so that it is, generally speaking, very far from being, say, finite-to-one. Note, however, that most fibers do not cut the indeterminacy set of π_T, so that π_T sends that fibers to leaves of $\mathscr{F}^0$; moreover, most leaves have trivial holonomy (it is a general fact [CLN] that leaves with non trivial holonomy cut any transversal along a thin subset), and so on most fibers π_T is even an isomorphism between the fiber and the corresponding leaf of $\mathscr{F}^0$. But be careful: a leaf may cut a transversal T infinitely many times, and so V_T will contain infinitely many fibers sent by π_T to the same leaf, as holonomy coverings (possibly trivial) with different basepoints.

4.3 Covering Tubes

The following proposition is similar, in spirit, to Proposition 4.5, but, as we shall see, its proof is much more delicate. Here the Kähler assumption becomes really indispensable, via the unparametrized Hartogs extension lemma. Without the Kähler hypothesis it is easy to find counterexamples (say, for foliations on Hopf surfaces).

Proposition 4.6. *There exists a complex manifold U_T of dimension n, a holomorphic submersion*

$$P_T : U_T \to T,$$

a holomorphic section

$$p_T : T \to U_T,$$

and a surjective holomorphic immersion

$$F_T : U_T \to V_T$$

such that:

(i) for every $t \in T$, the pointed fiber $(P_T^{-1}(t), p_T(t))$ is isomorphic to $(\widetilde{L_t}, t)$;
(ii) for every $t \in T$, F_T sends the fiber $(\widetilde{L_t}, t)$ to the fiber $(\widehat{L_t}, t)$, as universal covering.

Proof. We use the same methodology as in the first part of the previous proof, with $\mathscr{F}^0$ replaced by the fibration V_T and $\Omega_T^{\mathscr{F}^0}$ replaced by $\Omega_T^{V_T}$ = space of continuous paths $\gamma : [0,1] \to V_T$ tangent to the fibers and starting from $q_T(T) \subset V_T$. But now the equivalence relation $\sim$ is given by homotopy, not holonomy: $\gamma_1 \sim \gamma_2$ if they have the same extremities and the loop $\gamma_1 * \gamma_2^{-1}$ is homotopic to zero in the fiber containing it. The only thing that we need to prove is that the quotient space

$$U_T = \Omega_T^{V_T} / \sim$$

is Hausdorff; then everything is completed as in the previous proof, with F_T associating to a homotopy class of paths its holonomy class. The Hausdorff property can be spelled as follows ("nonexistence of vanishing cycles"):

(*) if $\gamma : [0,1] \to \widehat{L_t} \subset V_T$ is a loop (based at $q_T(t)$) uniformly approximated by loops $\gamma_n : [0,1] \to \widehat{L_{t_n}} \subset V_T$ (based at $q_T(t_n)$) homotopic to zero in $\widehat{L_{t_n}}$, then γ is homotopic to zero in $\widehat{L_t}$.

Let us firstly consider the case in which γ is a simple loop. We may assume that $\Gamma = \gamma([0,1])$ is a real analytic curve in $\widehat{L_t}$, and we may find an embedding

$$f : \mathbb{D}^{n-1} \times A_r \to V_T$$

sending fibers to fibers and such that $\Gamma = f(0, \partial A_r)$. Thus $\Gamma_n = f(z_n, \partial A_r)$ is homotopic to zero in its fiber, for some sequence $z_n \to 0$. For evident reasons, if z_n' is sufficiently close to z_n, then also $f(z_n', \partial A_r)$ is homotopic to zero in its fiber. Thus, we have an open nonempty subset $U \subset \mathbb{D}^{n-1}$ such that, for every $z \in U$, $f(z, \partial A_r)$ is homotopic to zero in its fiber. Denote by D_z the disc in the fiber bounded by such $f(z, \partial A_r)$.

We may also assume that Γ is disjoint from the discrete subset $\widehat{L_t} \setminus \widehat{L_t^0}$, so that, after perhaps restricting $\mathbb{D}^{n-1}$, the composite map

$$f' : \pi_T \circ f : \mathbb{D}^{n-1} \times A_r \to X$$

is holomorphic, and therefore it is an almost embedding. We already know that, for every $z \in U$, $f'(z, A_r)$ extends to a disc, image by π_T of D_z. Therefore, by Theorem 3.1, $f'(\mathbb{D}^{n-1} \times A_r)$ extends to a meromorphic family of discs

$$g : W \dashrightarrow X.$$

It may be useful to observe that such a g is a meromorphic immersion. Indeed, setting $F = Indet(g)$, the set of points of $W \setminus F$ where g is not an immersion is (if not empty) an hypersurface. Such a hypersurface cannot cut a neighbourhood of the boundary ∂W, where g is a reparametrization of the immersion f'. Also, such a hypersurface cannot cut the fiber W_z when $z \in U$ is generic (i.e. $W_z \cap F = \emptyset$), because on a neighbourhood of such a W_z the map g is a reparametrization of the immersion π_T on a neighbourhood of D_z. It follows that such a hypersurface is empty.

As in the proof of Proposition 4.5, $g|_{W \setminus F}$ can be lifted, holomorphically, to V_T^0, and then g can be lifted to V_T, giving an embedding $\widetilde{g} : W \to V_T$. Then $\widetilde{g}(W_0)$ is a disc in $\widehat{L_t}$ with boundary Γ, and consequently γ is homotopic to zero in the fiber $\widehat{L_t}$.

Consider now the case in which γ is possibly not simple. We may assume that γ is a smooth immersion with some points of transverse selfintersection, and idem for γ_n. We reduce to the previous simple case, by a purely topological argument.

Take the immersed circles $\Gamma = \gamma([0,1])$ and $\Gamma_n = \gamma_n([0,1])$. Let $R_n \subset \widehat{L_{t_n}}$ be the open bounded subset obtained as the union of a small tubular neighbourhood of Γ_n and all the bounded components of $\widehat{L_{t_n}} \setminus \Gamma_n$ isomorphic to the disc. Thus, each connected component of $R_n \setminus \Gamma_n$ is either a disc with boundary in Γ_n (union of arcs between selfintersection points), or an annulus with one boundary component in Γ_n and another one in ∂R_n; this last one is not the boundary of a disc in $\widehat{L_{t_n}} \setminus R_n$. We have the following elementary topological fact: if Γ_n is homotopic to zero in $\widehat{L_{t_n}}$, then it is homotopic to zero also in R_n.

Let $R \subset \widehat{L_t}$ be defined in a similar way, starting from Γ. By the first part of the proof, if $D_n \subset R_n \setminus \Gamma_n$ is a disc with boundary in Γ_n, then for $t_n \to t$ such a disc converges to a disc $D \subset \widehat{L_t}$ with boundary in Γ, i.e. to a disc $D \subset R \setminus \Gamma$. Conversely, but by elementary reasons, any disc $D \subset R \setminus \Gamma$ with boundary in Γ can be deformed to discs $D_n \subset R_n \setminus \Gamma_n$ with boundaries in Γ_n. We deduce that R is diffeomorphic to R_n, or more precisely that the pair (R,Γ) is diffeomorphic to the pair (R_n,Γ_n), for n large. Hence from Γ_n homotopic to zero in R_n we infer Γ homotopic to zero in R, and a fortiori in $\widehat{L_t}$. This completes the proof of the Hausdorff property (*). □

The manifold U_T will be called **covering tube** over T. We have a meromorphic immersion

$$\Pi_T = \pi_T \circ F_T : U_T \dashrightarrow X$$

whose indeterminacy set $Indet(\Pi_T)$ cuts each fiber $P_T^{-1}(t) = \widetilde{L_t}$ along the discrete subset which is the preimage of $\widehat{L_t} \setminus \widehat{L_t^0}$ under the covering map $\widetilde{L_t} \to \widehat{L_t}$. For every $t \in T$, the restriction of Π_T to $P^{-1}(t)$ coincides, after removal of indeterminacies, with the universal covering $\widetilde{L_t} \to L_t \xrightarrow{i_t} i_t(L_t) \subset X$.

The local biholomorphism $F_T : U_T \to V_T$ is a fiberwise covering, but globally it may have a quite wild structure. Let us see two examples.

Example 4.7. We take again the elliptic fibration $S \xrightarrow{\pi} C$ of Example 4.3. Let $T \subset S$ be a small transverse disc centered at $t_0 \in F_0 \setminus \{q\}$. Then, because the holonomy is trivial, $\widehat{L_t} = L_t$ for every t. We have already seen that $L_{t_0} = F_0 \setminus \{q\}$, and obviously for $t \neq t_0$, L_t is the smooth elliptic curve through t. The covering tube V_T is simply $\pi^{-1}(\pi(T)) \setminus \{q\}$. Remark that its central fiber is simply connected, but the other fibers are not. All the fibers of U_T are isomorphic to $\mathbb{C}$ (in fact, one can see that $U_T \simeq T \times \mathbb{C}$). The map $F_T : U_T \to V_T$, therefore, is injective on the central fiber, but not on the other ones.

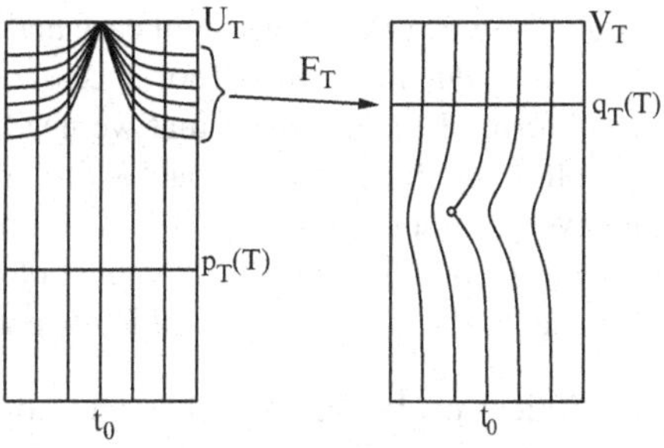

To see better what is happening, take the basepoints $q_T(T) \subset V_T$ and consider the preimage $F_T^{-1}(q_T(T)) \subset U_T$. This preimage has infinitely many components: one of them is $p_T(T) \subset U_T$, and each other one is the graph over $T \setminus \{t_0\}$ of a 6-valued section of U_T. This follows from the fact that the monodromy of the elliptic fibration around a fiber of type II has order 6 [BPV, V.10]. The map F_T sends this 6-valued graph to $q_T(T \setminus \{t_0\})$, as a regular 6-fold covering. There is a "virtual" ramification of order 6 over $q_T(t_0)$, which is however pushed-off U_T, to the point at infinity of the central fiber.

Example 4.8. We take again an elliptic fibration $S \xrightarrow{\pi} C$, but now with a fiber $\pi^{-1}(c_0) = F_0$ of Kodaira type I_1, i.e. a rational curve with a node q. As before, V_T coincides with $\pi^{-1}(\pi(T)) \setminus \{q\}$, but now the central fiber is isomorphic to $\mathbb{C}^*$. Again $U_T \simeq T \times \mathbb{C}$. The map $F_T : U_T \to V_T$ is a $\mathbb{Z}$-covering over the central fiber, a $\mathbb{Z}^2$-covering over the other fibers. The preimage of $q_T(T)$ by F_T has still infinitely many components. One of them is $p_T(T)$. Some of them are graphs of (1-valued) sections over T, passing through the (infinitely many) points of $F_T^{-1}(q_T(t_0))$. But most of them are graphs of ∞-valued sections over $T \setminus \{t_0\}$ (like the graph of the logarithm). Indeed, the monodromy of the elliptic fibration around a fiber of type I_1 has infinite order [BPV, V.10]. If $t \neq t_0$, then $F_T^{-1}(q_T(t))$ is a lattice in $\widetilde{L_t} \simeq \mathbb{C}$, with generators 1 and $\lambda(t) \in \mathbb{H}$. For $t \to t_0$, this second generator diverges to $+i\infty$, and the lattices reduces to $\mathbb{Z} = F_T^{-1}(q_T(t_0))$. The monodromy acts as $(n, m\lambda(t)) \mapsto (n+m, m\lambda(t))$. Then each connected component of $F_T^{-1}(q_T(T))$ intersects $\widetilde{L_t}$ either at a single point $(n,0)$, fixed by the monodromy, or along an orbit $(n+m\mathbb{Z}, m\lambda(t))$, $m \neq 0$.

More examples concerning elliptic fibrations can be found in [Br4].

Remark 4.9. As we recalled in Section 2, similar constructions of U_T and V_T have been done, respectively, by Il'yashenko [Il1] and Suzuki [Suz], in the case where the ambient manifold X is a Stein manifold. However, the Stein case is much simpler than the compact Kähler one. Indeed, the meromorphic maps $g : W \dashrightarrow X$ with which we work are automatically *holomorphic* if X is Stein. Thus, in the Stein case there are no vanishing ends, i.e. $L_p = L_p^0$ for every p and leaves of $\mathscr{F}$ = leaves of $\mathscr{F}^0$. Then the maps π_T and Π_T are *holomorphic* immersions of V_T and U_T into X^0 (and so V_T and U_T are Riemann Domains over X^0). Also, our unparametrized Hartogs extension lemma still holds in the Stein case, but with a much simpler proof, because we do not need to worry about "rational bubbles" arising in Bishop's Theorem.

In fact, there is a common framework for the Stein case and the compact Kähler case: the framework of *holomorphically convex* (not necessarily compact) Kähler manifolds. Indeed, the only form of compactness that we need, in this Section and also in the next one, is the following: for every compact $K \subset X$, there exists a (larger) compact $\hat{K} \subset X$ such that every holomorphic disc in X with boundary in K is fully contained in $\hat{K}$. This property is obviously satisfied by any holomorphically convex Kähler manifold, with $\hat{K}$ equal to the usual holomorphically convex hull of K.

A more global point of view on holonomy tubes and covering tubes will be developed in the last Section, on parabolic foliations.

4.4 Rational Quasi-Fibrations

We conclude this Section with a result which can be considered as an analog, in our context, of the classical Reeb Stability Theorem for real codimension one foliations [CLN].

Proposition 4.10. *Let X be a compact connected Kähler manifold and let $\mathcal{F}$ be a foliation by curves on X. Suppose that there exists a rational leaf L_p (i.e., $\widetilde{L_p} = \mathbb{P}$). Then all the leaves are rational. Moreover, there exists a compact connected Kähler manifold Y, $\dim Y = \dim X - 1$, a meromorphic map $B : X \dashrightarrow Y$, and Zariski open and dense subsets $X_0 \subset X$, $Y_0 \subset Y$, such that:*

(i) B is holomorphic on X_0 and $B(X_0) = Y_0$;
(ii) $B : X_0 \to Y_0$ is a proper submersive map, all of whose fibers are smooth rational curves, leaves of $\mathcal{F}$.

Proof. It is sufficient to verify that all the leaves are rational; then the second part follows by standard arguments of complex analytic geometry, see e.g. [CaP].

By connectivity, it is sufficient to prove that, given a covering tube U_T, if some fiber is rational then all the fibers are rational. We can work, equivalently, with the holonomy tube V_T. Now, such a property was actually already verified in the proof of Proposition 4.6, in the form of "nonexistence of vanishing cycles". Indeed, the set of rational fibers of V_T is obviously open. To see that it is also closed, take a fiber $\widehat{L_t}$ approximated by fibers $\widehat{L_{t_n}} \simeq \mathbb{P}$. Take an embedded cycle $\Gamma \subset \widehat{L_t}$, approximated by cycles $\Gamma_n \subset \widehat{L_{t_n}}$. Each Γ_n bounds in $\widehat{L_{t_n}}$ *two* discs, one on each side. As in the proof of Proposition 4.6, we obtain that Γ also bounds in $\widehat{L_t}$ *two* discs, one on each side. Hence $\widehat{L_t}$ is rational. □

Such a foliation will be called **rational quasi-fibration**. A meromorphic map B as in Proposition 4.10 is sometimes called *almost holomorphic*, because the image of its indeterminacy set is a proper subset of Y, of positive codimension, contained in $Y \setminus Y_0$. If $\dim X = 2$ then B is necessarily holomorphic, and the foliation is a rational fibration (with possibly some singular fibers). In higher dimensions one may think that the foliation is obtained from a rational fibration by a meromorphic transformation which does not touch generic fibers (like flipping along a codimension two subset).

Note that, as the proof shows, for a rational quasi-fibration every holonomy tube and every covering tube is isomorphic to $T \times \mathbb{P}$, provided that the transversal T is sufficiently small (every $\mathbb{P}$-fibration is locally trivial).

There are certainly many interesting issues concerning rational quasi-fibrations, but basically this is a chapter of Algebraic Geometry. In the following, we shall forget about them, and we will concentrate on foliations with parabolic and hyperbolic leaves.

5 A Convexity Property of Covering Tubes

Let X be a compact Kähler manifold, of dimension n, and let $\mathscr{F}$ be a foliation by curves on X, different from a rational quasi-fibration. Fix a transversal $T \subset X^0$ to $\mathscr{F}^0$, and consider the covering tube U_T over T, with projection $P_T : U_T \to T$, section $p_T : T \to U_T$, and meromorphic immersion $\Pi_T : U_T \dashrightarrow X$. Each fiber of U_T is either $\mathbb{D}$ or $\mathbb{C}$.

We shall establish in this Section, following [Br2] and [Br3], a certain convexity property of U_T, which later will allow us to apply to U_T the results of Section 2 of Nishino and Yamaguchi.

We fix also an embedded closed disc $S \subset T$ ($S \simeq \overline{\mathbb{D}}$, and the embedding in T is holomorphic up to the boundary), and we denote by U_S, P_S, p_S, Π_S the corresponding restrictions. Set $\partial U_S = P_S^{-1}(\partial S)$. We shall assume that S satisfies the following properties:

(a) U_S, as a subset of U_T, intersects $Indet(\Pi_T)$ along a discrete subset, necessarily equal to $Indet(\Pi_S)$, and ∂U_S does not intersect $Indet(\Pi_T)$;
(b) for every $z \in \partial S$, the area of the fiber $P_S^{-1}(z)$ is infinite.

In (b), the area is computed with respect to the pull-back by Π_S of the Kähler form ω of X. Without loss of generality, we take ω real analytic. We will see later that these assumptions (a) and (b) are "generic", in a suitable sense.

Theorem 5.1. *For every compact subset $K \subset \partial U_S$ there exists a real analytic bidimensional torus $\Gamma \subset \partial U_S$ such that:*

(i) Γ is transverse to the fibers of $\partial U_S \xrightarrow{P_S} \partial S$, and cuts each fiber $P_S^{-1}(z)$, $z \in \partial S$, along a circle $\Gamma(z)$ which bounds a disc $D(z)$ which contains $K \cap P_S^{-1}(z)$ and $p_S(z)$;
(ii) Γ is the boundary of a real analytic Levi-flat hypersurface $M \subset U_S$, filled by a real analytic family of holomorphic discs D^θ, $\theta \in \mathbb{S}^1$; each D^θ is the image of a section $s^\theta : S \to U_S$, holomorphic up to the boundary, with $s^\theta(\partial S) \subset \Gamma$;
(iii) M bounds in U_S a domain Ω, which cuts each fiber $P_S^{-1}(z)$, $z \in S$, along a disc $\Omega(z)$ which contains $p_S(z)$ ($\Omega(z) = D(z)$ when $z \in \partial S$).

This statement should be understood as expressing a variant of Hartogs-convexity [Ran, II.2], in which the standard Hartogs figure is replaced by $p_S(S)\cup(\cup_{z\in\partial S}D(z))$, and its envelope is replaced by Ω. By choosing a large compact K, condition (i) says that $\cup_{z\in\partial S}D(z)$ almost fills the lateral boundary ∂U_S; conditions (ii) and (iii) say that the family of discs $D(z)$, $z\in\partial S$, can be pushed inside S, getting a family of discs $\Omega(z)$, $z\in S$, in such a way that the boundaries $\partial\Omega(z)$, $z\in S$, vary with z in a "holomorphic" manner ("variation analytique" in the terminology of [Ya3]). It is a sort of "geodesic" convexity of U_S, in which the extremal points of the geodesic are replaced by Γ and the geodesic is replaced by M.

Theorem 5.1 will be proved by solving a nonlinear Riemann-Hilbert problem, see [For] and [AWe, Ch. 20] and reference therein for some literature on this subject. An important difference with this classical literature, however, is that the torus Γ is not fixed a priori: we want just to prove that *some* torus Γ, enclosing the compact K as in (i), is the boundary of a Levi-flat hypersurface M as in (ii); we do not pretend that *every* torus Γ has such a property. Even if, as we shall see below, we have a great freedom in the choice of Γ.

We shall use the continuity method. The starting point is the following special (but not so much) family of tori.

Lemma 5.2. *Given $K\subset\partial U_S$ compact, there exists a real analytic embedding*

$$F:\partial S\times\overline{\mathbb{D}}\to\partial U_S,$$

sending fibers to fibers, such that:

(i) $\partial S\times\{0\}$ is sent to $p_S(\partial S)\subset\partial U_S$;
(ii) $\partial S\times\{|w|=t\}$, $t\in(0,1]$, is sent to a real analytic torus $\Gamma_t\subset\partial U_S$ transverse to the fibers of P_S, so that for every $z\in\partial S$, $\Gamma_t(z)=\Gamma_t\cap P_S^{-1}(z)$ is a circle bounding a disc $D_t(z)$ containing $p_S(z)$;
(iii) $D_1(z)$ contains $K\cap P_S^{-1}(z)$, for every $z\in\partial S$;
(iv) for every $t\in(0,1]$ the function

$$\mathbf{a}_t:\partial S\to\mathbb{R}^+,\qquad \mathbf{a}_t(z)=\mathrm{area}(D_t(z))$$

is constant (the constant depending on t, of course).

Proof. Recall that the area in the fibers is computed with respect to the pull-back of the Kähler form ω. Because the fibers over ∂S have infinite area, we can certainly find a smooth torus $\Gamma'\subset\partial U_S$ which encloses K and $p_S(\partial S)$, and such that all the discs $D'(z)$, bounded by $\Gamma'(z)$, have the same area, say equal to k. We may approximate Γ' with a real analytic torus Γ''; the corresponding discs $D''(z)$ have now variable area, but close to k, say between k and $k+\varepsilon$.

For every $z\in\partial S$ we have in $\overline{D''(z)}\setminus p_S(z)$ a canonical foliation by circles, the standard circles under the uniformisation $(\overline{D''(z)},p_S(z))\simeq(\overline{\mathbb{D}},0)$. For every $t\in(0,1]$, let $\Gamma_t(z)$ be the circle of that foliation which bounds a disc of area equal to kt. Then, because all the data (Γ'', ω,...) are real analytic, the union $\Gamma_t=\cup_{z\in\partial S}\Gamma_t(z)$ is a real analytic torus, and these tori glue together in a real analytic way, producing the map F. If the initial perturbation is sufficiently small, Γ_1 encloses K. And the function $\mathbf{a}_t$ is constantly equal to kt. □

Given F as in Lemma 5.2, we shall say that a real analytic embedding

$$G : S \times \overline{\mathbb{D}} \to U_S$$

is a **Levi-flat extension** of F if G sends fibers to fibers and:

(i) $G(S \times \{0\}) = p_S(S)$;
(ii) $G(S \times \{|w| = t\}), t \in (0,1]$, is a real analytic Levi-flat hypersurface $M_t \subset U_S$ with boundary Γ_t, filled by images of holomorphic sections over S with boundary values in Γ_t.

Our aim is to construct such a G. Then $\Gamma = \Gamma_1$ and $M = M_1$ gives Theorem 5.1.

The continuity method consists in analyzing the set of those $t_0 \in (0,1]$ such that a similar G can be constructed over $S \times \overline{\mathbb{D}(t_0)}$. We need to show that this set is nonempty, open and closed.

Nonemptyness is a consequence of classical results [For]. Just note that a neighbourhood of $p_S(S)$ can be embedded in $\mathbb{C}^2$, in such a way that P_S becomes the projection to the first coordinate, and $p_S(S)$ becomes the closed unit disc in the first axis. Hence Γ_t, t small, becomes a torus in $\partial\mathbb{D} \times \mathbb{C}$ enclosing $\partial\mathbb{D} \times \{0\}$. Classical results on the Riemann-Hilbert problem in $\mathbb{C}^2$ imply that, for $t_0 > 0$ sufficiently small, there exists a Levi-flat extension on $S \times \overline{\mathbb{D}(t_0)}$.

Openness is a tautology. By definition, a real analytic embedding defined on $S \times \overline{\mathbb{D}(t_0)}$ is in fact defined on $S \times \mathbb{D}(t_0 + \varepsilon)$, for some $\varepsilon > 0$, and obviously if G is a Levi-flat extension on $S \times \overline{\mathbb{D}(t_0)}$, then it is a Levi-flat extension also on $S \times \overline{\mathbb{D}(t_0 + \varepsilon')}$, for every $\varepsilon' < \varepsilon$.

The heart of the matter is closedness. In other words, we need to prove that:

if a Levi-flat extension exists on $S \times \mathbb{D}(t_0)$, then it exists also on $S \times \overline{\mathbb{D}(t_0)}$.

The rest of this Section is devoted to the proof of this statement.

5.1 Boundedness of Areas

We shall denote by D_t^θ, $\theta \in \mathbb{S}^1$, the closed holomorphic discs filling M_t, $0 < t < t_0$. Each D_t^θ is the image of a section $s_t^\theta : S \to U_S$, holomorphic up to the boundary, with boundary values in Γ_t.

Consider the areas of these discs. These areas are computed with respect to $\Pi_S^*(\omega) = \omega_0$, which is a real analytic Kähler form on $U_S \setminus Indet(\Pi_S)$. Because $H^2(U_S \setminus Indet(\Pi_S), \mathbb{R}) = 0$ (for U_S is a contractible complex surface and $Indet(\Pi_S)$ is a discrete subset), this Kähler form is exact:

$$\omega_0 = d\lambda$$

for some real analytic 1-form λ on $U_S \setminus Indet(\Pi_S)$. If D_t^θ is disjoint from $Indet(\Pi_S)$, then its area $\int_{D_t^\theta} \omega_0$ is simply equal, by Stokes formula, to $\int_{\partial D_t^\theta} \lambda$. If D_t^θ intersects $Indet(\Pi_S)$, this is no more true, but still we have the inequality

$$\text{area}(D_t^\theta) = \int_{D_t^\theta} \omega_0 \leq \int_{\partial D_t^\theta} \lambda.$$

The reason is the following: by the meromorphic map Π_S the disc D_t^θ is mapped not really to a disc in X, but rather to a disc *plus* some rational bubbles coming from indeterminacy points of Π_S; then $\int_{\partial D_t^\theta} \lambda$ is equal to the area (in X) of the disc *plus* the areas of these rational bubbles, whence the inequality above. Remark that, by our standing assumptions, the boundary of D_t^θ is contained in ∂U_S and hence it is disjoint from $Indet(\Pi_S)$.

Now, the important fact is that, thanks to the crucial condition (iv) of Lemma 5.2, we may get a *uniform* bound of these areas.

Lemma 5.3. *There exists a constant $C > 0$ such that for every $t \in (0, t_0)$ and every $\theta \in \mathbb{S}^1$:*

$$\text{area}(D_t^\theta) \leq C.$$

Proof. By the previous remarks, we just have to bound the integrals $\int_{\partial D_t^\theta} \lambda$. The idea is the following one. For t fixed the statement is trivial, and we need just to understand what happens for $t \to t_0$. Look at the curves $\partial D_t^\theta \subset \Gamma_t$. They are graphs of sections over ∂S. For $t \to t_0$ these graphs could oscillate more and more. But, using condition (iv) of Lemma 5.2, we will see that these oscillations do not affect the integral of λ. This would be evident if the tori Γ_t were lagrangian (i.e. $\omega_0|_{\Gamma_t} \equiv 0$, i.e. $\lambda|_{\Gamma_t}$ closed), so that the integrals of λ would have a cohomological meaning, not affected by the oscillations. Our condition (iv) of Lemma 5.2 expresses a sort of half-lagrangianity in the direction along which oscillations take place, and this is sufficient to bound the integrals.

Fix real analytic coordinates $(\varphi, \psi, r) \in \mathbb{S}^1 \times \mathbb{S}^1 \times (-\varepsilon, \varepsilon)$ around Γ_{t_0} in ∂U_S such that:

(i) $P_S : \partial U_S \to \partial S$ is given by $(\varphi, \psi, r) \mapsto \varphi$;
(ii) $\Gamma_t = \{r = t - t_0\}$ for every t close to t_0.

Each curve ∂D_t^θ, $t < t_0$ close to t_0, is therefore expressed by

$$\partial D_t^\theta = \{\psi = h_t^\theta(\varphi),\ r = t - t_0\}$$

for some real analytic function $h_t^\theta : \mathbb{S}^1 \to \mathbb{S}^1$. Because the discs D_t^θ form a continuous family, all these functions h_t^θ have the same degree, and we may suppose that it is zero up to changing ψ to $\psi + \ell\varphi$.

The 1-form λ, restricted to ∂U_S, in these coordinates is expressed by

$$\lambda = a(\varphi, \psi, r)d\varphi + b(\varphi, \psi, r)d\psi + c(\varphi, \psi, r)dr$$

for suitable real analytic functions a, b, c on $\mathbb{S}^1 \times \mathbb{S}^1 \times (-\varepsilon, \varepsilon)$. Setting $b_0(\varphi, r) = \int_{\mathbb{S}^1} b(\varphi, \psi, r) d\psi$, we can write $b(\varphi, \psi, r) = b_0(\varphi, r) + \frac{\partial b_1}{\partial \psi}(\varphi, \psi, r)$, for some real analytic function b_1 (the indefinite integral of $b - b_0$ along ψ), and therefore

$$\lambda = a_0(\varphi, \psi, r) d\varphi + b_0(\varphi, r) d\psi + c_0(\varphi, \psi, r) dr + db_1$$

with $a_0 = a - \frac{\partial b_1}{\partial \varphi}$ and $c_0 = c - \frac{\partial b_1}{\partial r}$.

Remark now that $b_0(\varphi, r)$ is just equal to $\int_{\partial D_t(z)} \lambda$, for $r = t - t_0$ and $\varphi =$ the coordinate of $z \in \partial S$. By Stokes formula, this is equal to the area of the disc $D_t(z)$, and by condition (iv) of Lemma 5.2 this does not depend on φ. That is, the function b_0 depends only on r, and not on φ:

$$b_0(\varphi, r) = b_0(r).$$

In particular, if we restrict λ to a torus Γ_t we obtain, up to an exact term, a 1-form

$$a_0(\varphi, \psi, t - t_0) d\varphi + b_0(t - t_0) d\psi$$

which is perhaps not closed (this would be the lagrangianity of Γ_t), but its component along ψ is closed. And note that the oscillations of the curves ∂D_t^θ are directed along ψ.

If we now integrate λ along ∂D_t^θ we obtain

$$\int_{\partial D_t^\theta} \lambda = \int_{\mathbb{S}^1} a_0(\varphi, h_t^\theta(\varphi), t - t_0) d\varphi + b_0(t - t_0) \cdot \int_{\mathbb{S}^1} \frac{\partial h_t^\theta}{\partial \varphi}(\varphi) d\varphi.$$

The first integral is bounded by $C = \sup |a_0|$, and the second integral is equal to zero because the degree of h_t^θ is zero. □

Take now any sequence of discs

$$D_n = D_{t_n}^{\theta_n}, \; n \in \mathbb{N},$$

with $t_n \to t_0$. Our next aim is to prove that $\{D_n\}$ converges (up to subsequencing) to some disc $D_\infty \subset U_S$, with boundary in Γ_{t_0}. The limit discs so obtained will be then glued together to produce the Levi-flat hypersurface M_{t_0}.

5.2 Convergence Around the Boundary

We firstly prove that everything is good around the boundary. Recall that every disc D_n is the image of a section $s_n = s_{t_n}^{\theta_n} : S \to U_S$ with boundary values in $\Gamma_n = \Gamma_{t_n}$.

Lemma 5.4. *There exists a neighbourhood $V \subset S$ of ∂S and a section*

$$s_\infty : V \to U_S$$

such that $s_n|_V$ converges uniformly to s_∞ (up to subsequencing).

Proof. We want to apply Bishop compactness theorem [Bis, Chi] to the sequence of analytic subsets of bounded area $D_n \subset U_S$. This requires some care due to the presence of the boundary.

Let us work on some slightly larger open disc $S' \subset T$ containing the closed disc S. Every torus $\Gamma_t \subset U_{S'}$ has a neighbourhood $W_t \subset U_{S'}$ over which we have a well defined Schwarz reflection with respect to Γ_t (which is totally real and of half dimension in $U_{S'}$). Thus, the complex curve $D_t^\theta \cap W_t$ with boundary in Γ_t can be doubled to a complex curve without boundary A_t^θ, properly embedded in W_t. Moreover, using the fact that the tori Γ_t form a real analytic family up to t_0, we see that the size of the neighbourhoods W_t is uniformly bounded from below. That is, there exists a neighbourhood W of Γ_{t_0} in $U_{S'}$ which is contained in every W_t, for t sufficiently close to t_0, and therefore every A_t^θ restricts to a properly embedded complex curve in W, still denoted by A_t^θ. Set

$$\widehat{D}_t^\theta = D_t^\theta \cup A_t^\theta.$$

Because the Schwarz reflection respects the fibration of $U_{S'}$, it is clear that $\widehat{D}_t^\theta$ is still the image of a section $\widehat{s}_t^\theta$, defined over some open subset $R_t^\theta \subset S'$ which contains S. The area of A_t^θ is roughly the double of the area of $D_t^\theta \cap W$, and therefore the properly embedded analytic subsets $\widehat{D}_t^\theta \subset U_S \cup W$ also have uniformly bounded areas.

Having in mind this uniform extension of the discs D_t^θ into the neighbourhood W of Γ_{t_0}, we now apply Bishop Theorem to the sequence $\{D_n\}$. Remark that $\partial D_n \subset \Gamma_n$ cannot exit from W, as $n \to +\infty$, because Γ_n converges to Γ_{t_0}. Up to subsequencing, we obtain that $\{D_n\}$ Hausdorff-converges to a complex curve $D_\infty \subset U_S$ with boundary in Γ_{t_0}. Moreover, and taking into account that D_n are graphs over S, we see that D_∞ has a graph-type irreducible component plus, possibly, some vertical components. More precisely (compare with [Iv1, Prop. 3.1]):

(i) $D_\infty = D_\infty^0 \cup E_1 \cup \ldots \cup E_m \cup F_1 \cup \ldots \cup F_\ell$;
(ii) D_∞^0 is the image of a section $s_\infty : V \to U_S$, over some open subset $V \subset S$ which contains ∂S;
(iii) each E_j is equal to $P_S^{-1}(p_j)$, for some $p_j \in S \setminus \partial S$ (interior bubble);
(iv) each F_j is equal to the closure of a connected component of $P_S^{-1}(q_j) \setminus \Gamma_{t_0}(q_j)$, for some $q_j \in \partial S$ (boundary bubble);
(v) for every compact $K \subset V \setminus \{p_1, \ldots, p_m, q_1, \ldots, q_\ell\}$, $s_n|_K$ converges uniformly to $s_\infty|_K$, as $n \to +\infty$.

We have just to prove that there are no boundary bubbles, i.e. that the set $\{q_1, \ldots, q_\ell\}$ is in fact empty. Then the conclusion follows by taking a smaller V, which avoids interior bubbles.

Consider the family of Levi-flat hypersurfaces $M_t \subset U_S$ with boundary Γ_t, for $t < t_0$. Each M_t is a "lower barrier", which prevents the approaching of D_n to the bounded component of $P_S^{-1}(q) \setminus \Gamma_{t_0}(q)$, for every $q \in \partial S$. More precisely, for any compact R in that bounded component we may select $t_1 < t_0$ such that $\cup_{0 \le t < t_1} \Gamma_t$ contains R, and so $\cup_{0 \le t < t_1} M_t$ is a neighbourhood of R in U_S. For n sufficiently large (so that $t_n > t_1$), $D_n \subset M_{t_n}$ is disjoint from that neighbourhood of R. Hence the sequence D_n cannot accumulate to the bounded component of $P_S^{-1}(q) \setminus \Gamma_{t_0}(q)$.

But neither D_n can accumulate to the unbounded component of $P_S^{-1}(q) \setminus \Gamma_{t_0}(q)$, because that component has infinite area, by our standing assumptions. Therefore, as desired,

$$\{q_1, \ldots, q_\ell\} = \emptyset.$$

□

Remark that by the same barrier argument we have also

$$\{p_1, \ldots, p_m\} = \emptyset$$

but this fact will not be used below. The proof above shows in fact the following: there is a maximal V over which s_∞ is defined, and the image $s_\infty(V)$ in U_S is a properly embedded complex curve. For every $z \in V$ the sequence $s_n(z)$ is convergent to $s_\infty(z)$, whereas for every $z \notin V$ the sequence $s_n(z)$ is divergent in the fiber $P_S^{-1}(z)$.

5.3 Convergence on the Interior

In order to extend the convergence above from V to the full S, we need to use the map Π_S into X. Consider the discs

$$f_n = \Pi_S \circ s_n : S \to X$$

in the compact Kähler manifold X. They have bounded area, and, once a time, we apply to them Bishop compactness theorem [Bis] [Iv1, Prop.3.1]. We obtain a holomorphic map

$$f_\infty : S \cup B \to X$$

which obviously coincides with $\Pi_S \circ s_\infty$ on the neighbourhood V of ∂S of Lemma 5.4. The set B is a union of trees of rational curves, each one attached to some point of S outside V. We will prove that $f_\infty|_S$ can be lifted to U_S, providing the extension of the section s_∞ to the full S.

The map f_∞ is an immersion around ∂S. Let us even suppose that it is an embedding (anyway, this is true up to moving a little ∂S inside S, and this does not affect the following reasoning). In some sufficiently smooth tubular neighbourhood $X_0 \subset X$ of $f_\infty(\partial S)$, we have a properly embedded complex surface with boundary Y, given by the image by Π_S of a neighbourhood of $s_\infty(\partial S)$ in U_S. The boundary ∂Y of Y in X_0 is filled by the images by Π_S of part of the tori Γ_t, t close to t_0; denote them by Γ_t' (with a good choice of X_0, each Γ_t' is a real annulus). Thus f_n, n large, sends S to a disc in X whose (embedded) boundary is contained in $\Gamma_n' = \Gamma_{t_n}'$, and f_∞ sends $S \cup B$ to a disc with rational bubbles in X whose (embedded) boundary is contained in $\Gamma_\infty' = \Gamma_{t_0}'$. Inspired by [IvS], but avoiding any infinite dimensional tool due to our special context, we now prove that f_∞ and f_n, for some large n, can be holomorphically interpolated by discs with boundaries in ∂Y.

Lemma 5.5. *There exists a complex surface with boundary W, a proper map $\pi : W \to \mathbb{D}$, a holomorphic map $g : W \to X$, such that:*

(i) for every $w \neq 0$, the fiber $W_w = \pi^{-1}(w)$ is isomorphic to S, and g sends that fiber to a disc in X with boundary in ∂Y;
(ii) for some $e \neq 0$, g coincides on $W_e = \pi^{-1}(e)$ with f_n, for some n (large);
(iii) $W_0 = \pi^{-1}(0)$ is isomorphic to $S \cup B$, and g on that fiber coincides with f_∞.

Proof. Let us work on the complex manifold $\widehat{X} = X \times \mathbb{D}(t_0, \varepsilon)$, where the second factor is a small disc in $\mathbb{C}$ centered at t_0. The real surfaces Γ_t' in X_0 can be seen as a single real analytic submanifold of dimension three Γ' in $\widehat{X_0} = X_0 \times \mathbb{D}(t_0, \varepsilon)$, by considering Γ_t' as a subset of $X_0 \times \{t\}$. Remark that Γ' is totally real. Similarly, the discs $f_n(S)$ can be seen as discs in $X \times \{t_n\} \subset \widehat{X}$, and the disc with bubbles $f_\infty(S \cup B)$ can be seen as a disc with bubbles in $X \times \{t_0\} \subset \widehat{X}$; all these discs have boundaries in Γ'. In $\widehat{X_0}$ we also have a complex submanifold of dimension three with boundary $\widehat{Y} = Y \times \mathbb{D}(t_0, \varepsilon)$, which is "half" of the complexification of Γ'.

Around the circle $f_\infty(\partial S) \subset \widehat{X}$, we may find holomorphic coordinates $z_1, \ldots, z_{n+1}$, with $|z_j| < \delta$ for $j \leq n$, $1 - \delta < |z_{n+1}| < 1 + \delta$, such that:

(i) $f_\infty(\partial S) = \{z_1 = \ldots = z_n = 0,\ |z_{n+1}| = 1\}$;
(ii) $\Gamma' = \{z_1 = \ldots = z_{n-2} = 0,\ Im\, z_{n-1} = Im\, z_n = 0,\ |z_{n+1}| = 1\}$;
(iii) $\widehat{Y} = \{z_1 = \ldots = z_{n-2} = 0,\ |z_{n+1}| \leq 1\}$.

We consider, in these cordinates, the Schwarz reflection $(z_1, \ldots, z_n, z_{n+1}) \mapsto (\bar{z}_1, \ldots \bar{z}_n, \frac{1}{\bar{z}_{n+1}})$. It is a antiholomorphic involution, which fixes in particular every point of Γ'. Using it, we may double a neighbourhood Z_0 of $f_\infty(S \cup B)$ in $\widehat{X}$: we take Z_0 and $\overline{Z}_0$ (i.e., Z_0 with the opposite complex structure), and we glue them together using the Schwarz reflection. Call Z this double of Z_0. Then Z naturally contains a tree of rational curves R_∞ which comes from doubling $f_\infty(S \cup B)$, because this last has boundary in the fixed point set of the Schwarz reflection. Similarly, each $f_n(S)$ doubles to a rational curve $R_n \subset Z$, close to R_∞ for n large. Moreover, in some neighbourhood $N \subset Z$ of the median circle of R_∞ (arising from $f_\infty(\partial S)$), we have a complex threefold $\widetilde{Y} \subset N$, arising by doubling $\widehat{Y}$ or by complexifying Γ', which contains $R_\infty \cap N$ and every $R_n \cap N$.

Now, the space of trees of rational curves in Z close to R_∞ has a natural structure of complex analytic space $\mathscr{R}$, see e.g. [CaP] or [IvS]. Those trees which, in N, are contained in $\widetilde{Y}$ form a complex analytic subspace $\mathscr{R}_0 \subset \mathscr{R}$. The curve R_n above correspond to points of $\mathscr{R}_0$ converging to a point corresponding to R_∞. Therefore we can find a disc in $\mathscr{R}_0$ centered at R_∞ and passing through some R_n. This gives a holomorphic family of trees of rational curves in Z interpolating R_∞ and R_n. Restricting to $Z_0 \subset Z$ and projecting to X, we obtain the desired family of discs g. □

Note the the doubling trick used in the previous lemma is not so far from the similar trick used in the proof of Lemma 3.3. In both cases, a problem concerning discs is reduced to a more tractable problem concerning rational curves.

The map g can be lifted to U_S around ∂W and around W_e, thanks to properties (i) and (ii). In this way, and up to a reparametrization, we obtain an embedding

$$h : H \to U_S,$$

where $H = \{(z,w) \in S \times \mathbb{D} \mid z \in V \text{ or } |w-e| < \varepsilon\}$ (for some $\varepsilon > 0$ small), such that:

(i) $h(\cdot, w)$ is a section of U_S over S (if $|w-e| < \varepsilon$) or over V (if $|w-e| \geq \varepsilon$);
(ii) $h(z,e) = s_n(z)$ for every $z \in S$;
(iii) $h(z,0) = s_\infty(z)$ for every $z \in V$.

(note, however, that generally speaking the section $h(\cdot, w)$ has not boundary values in some torus Γ_t, when $w \neq 0, e$). In some sense, we are in a situation similar to the one already encountered in the construction of covering tubes in Section 4, but rotated by 90 degrees.

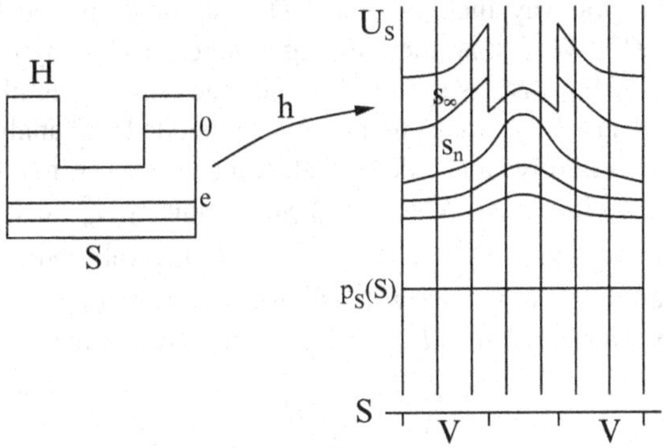

Consider now the meromorphic immersion $\Pi_S \circ h : H \dashrightarrow X$. By [Iv1], this map can be meromorphically extended to the envelope $S \times \mathbb{D}$, and clearly this extension is still a meromorphic immersion. Each vertical fiber $\{z\} \times \mathbb{D}$ is sent to a disc tangent to the foliation $\mathscr{F}$, and possibly passing through $Sing(\mathscr{F})$. But for every $z \in V$ we already have, by construction, that such a disc can be lifted to U_S. By our definition and construction of U_S, it then follows that the same holds for every $z \in S$: every intersection point with $Sing(\mathscr{F})$ is a vanishing end. Hence the full family $S \times \mathbb{D}$ can be lifted to U_S, or in other words the embedding $h : H \to U_S$ can be extended to $\widehat{h} : S \times \mathbb{D} \to U_S$.

Take now $\widehat{h}(\cdot, 0)$: it is a section over S which extends s_∞. Thus, the section s_∞ from Lemma 5.4 can be extended from V to S, and the sequence of discs $D_n \subset U_S$ uniformly converges to $D_\infty = s_\infty(S)$.

5.4 Construction of the Limit Levi-flat Hypersurface

Let us resume. We are assuming that our Levi-flat extension exists over $S \times \mathbb{D}(t_0)$, providing an embedded real analytic family of Levi-flat hypersurfaces $M_t \subset U_S$ with boundaries Γ_t, $t < t_0$. Given any sequence of holomorphic discs $D_{t_n}^{\theta_n} \subset M_{t_n}$, $t_n \to t_0$, we have proved that (up to subsequencing) $D_{t_n}^{\theta_n}$ converges uniformly to some disc D_∞ with $\partial D_\infty \subset \Gamma_{t_0}$. Given any point $p \in \Gamma_{t_0}$, we may choose the sequence $D_{t_n}^{\theta_n}$ so that ∂D_∞ will contain p. It remains to check that all the discs so constructed glue together in a real analytic way, giving M_{t_0}, and that this M_{t_0} glues to M_t, $t < t_0$, in a real analytic way, giving the Levi-flat extension over $S \times \overline{\mathbb{D}(t_0)}$.

This can be seen using a Lemma from [BeG, §5]. It says that if D is an embedded disc in a complex surface Y with boundary in a real analytic totally real surface $\Gamma \subset Y$, and if the winding number (Maslov index) of Γ along ∂D is zero, then D belongs to a unique embedded real analytic family of discs D^ε, $\varepsilon \in (-\varepsilon_0, \varepsilon_0)$, $D^0 = D$, with boundaries in Γ (incidentally, in our real analytic context this can be easily proved by the doubling argument used in Lemma 5.5, which reduces the statement to the well known fact that a smooth rational curve of zero selfintersection belongs to a unique local fibration by smooth rational curves). Moreover, if Γ is moved in a real analytic way, then the family D^ε also moves in a real analytic way.

For our discs $D_t^\theta \subset M_t$, $t < t_0$, the winding number of Γ_t along ∂D_t^θ is zero. By continuity of this index, if D_∞ is a limit disc then the winding number of Γ_{t_0} along ∂D_∞ is also zero. Thus, D_∞ belongs to a unique embedded real analytic family D_∞^ε, with $\partial D_\infty^\varepsilon \subset \Gamma_{t_0}$. This family can be deformed, real analytically, to a family D_t^ε with $\partial D_t^\varepsilon \subset \Gamma_t$, for every t close to t_0. When $t = t_n$, such a family $D_{t_n}^\varepsilon$ necessarily contains $D_{t_n}^{\theta_n}$, and thus coincides with $D_{t_n}^\theta$ for θ in a suitable interval around θ_n. Hence, for every $t < t_0$ the family D_t^ε coincides with D_t^θ, for θ in a suitable interval.

In this way, for every limit disc D_∞ we have constructed a piece

$$\bigcup_{\varepsilon \in (-\varepsilon_0, \varepsilon_0)} D_\infty^\varepsilon$$

of our limit M_{t_0}, this piece is real analytic and glues to M_t, $t < t_0$, in a real analytic way.

Because each $p \in \Gamma_{t_0}$ belongs to some limit disc D_∞, we have completed in this way our construction of the Levi-flat hypersurface M_{t_0}, and the proof of Theorem 5.1 is finished.

6 Hyperbolic Foliations

We can now draw the first consequences of the convexity of covering tubes given by Theorem 5.1, still following [Br2] and [Br3].

As in the previous Section, let X be a compact Kähler manifold of dimension n, equipped with a foliation by curves $\mathscr{F}$ which is not a rational quasi-fibration. Let $T \subset X^0$ be local transversal to $\mathscr{F}^0$. We firstly need to discuss the pertinence of hypotheses (a) and (b) that we made at the beginning of Section 5.

Concerning (a), let us simply observe that $Indet(\Pi_T)$ is an analytic subset of codimension at least two in U_T, and therefore its projection to T by P_T is a countable union of locally analytic subsets of positive codimension in T (a thin subset of T).

Hypothesis (a) means that the closed disc $S \subset T$ is chosen so that it is not contained in that projection, and its boundary ∂S is disjoint from that projection.

Concerning (b), let us set

$$R = \{z \in T \mid \text{area}(P_T^{-1}(z)) < +\infty\}.$$

Lemma 6.1. *Either R is a countable union of analytic subsets of T of positive codimension, or $R = T$. In this second case, U_T is isomorphic to $T \times \mathbb{C}$.*

Proof. If $z \in R$, then $\widetilde{L_z}$ has finite area and, a fortiori, L_z^0 has finite area. In particular, L_z^0 is properly embedded in X^0: otherwise, L_z^0 should cut some foliated chart, where $\mathscr{F}^0$ is trivialized, along infinitely many plaques, and so L_z^0 would have infinite area. Because $X \setminus X^0$ is an analytic subset of X, the fact that $L_z^0 \subset X^0$ is properly embedded and with finite area implies that its closure $\overline{L_z^0}$ in X is a complex compact curve, by Bishop extension theorem [Siu, Chi]. This closure coincides with $\overline{L_z}$, the closure of L_z.

The finiteness of the area of $\widetilde{L_z}$ implies also that the covering $\widetilde{L_z} \to L_z$ has finite order, i.e. the orbifold fundamental group of L_z is finite. By the previous paragraph, L_z can be compactified (as a complex curve) by adding a finite set. This excludes the case $\widetilde{L_z} = \mathbb{D}$: a finite quotient of the disc does not enjoy such a property. Also, the case $\widetilde{L_z} = \mathbb{P}$ is excluded by our standing assumptions. Therefore $\widetilde{L_z} = \mathbb{C}$. Moreover, again the finiteness of the orbifold fundamental group implies that L_z is equal to $\mathbb{C}$ with at most one multiple point. The closure $\overline{L_z}$ is a rational curve in X.

Now, by general principles of analytic geometry [CaP], rational curves in X (Kähler) constitute an analytic space with countable base, each irreducible component of which can be compactified by adding points corresponding to trees of rational curves. It follows easily from this fact that the subset

$$R' = \{z \in T \mid \overline{L_z} \text{ is rational}\}$$

is either a countable union of analytic subsets of T of positive codimension, or it is equal to the full T. Moreover, if A' is a component of R' then we can find a meromorphic map $A' \times \mathbb{P} \dashrightarrow X$ sending $\{z\} \times \mathbb{P}$ to $\overline{L_z}$, for every $z \in A'$ (compare with the arguments used at the beginning of the proof of Theorem 3.1).

Not every $z \in A'$, however, belongs to R: a point $z \in A'$ belongs to R if and only if among the points of $\{z\} \times \mathbb{P}$ sent to $Sing(\mathscr{F})$ only one does not correspond to a vanishing end of L_z^0, and at most one corresponds to a vanishing end of order $m \geq 2$. By a simple semicontinuity argument, $A' \cap R = A$ is an analytic subset of A'. Hence R also satisfies the above dichotomy.

Finally, if $R = T$ then we have a map $T \times \mathbb{P} \dashrightarrow X$ sending each fiber $\{z\} \times \mathbb{P}$ to $\overline{L_z}$ and (z, ∞) to the unique nonvanishing end of L_z^0. It follows that $U_T = T \times \mathbb{C}$. □

Let now $U \subset X$ be an open connected subset where $\mathscr{F}$ is generated by a holomorphic vector field $v \in \Theta(U)$, vanishing precisely on $Sing(\mathscr{F}) \cap U$. Set $U^0 = U \setminus (Sing(\mathscr{F}) \cap U)$, and consider the real function

$$F : U^0 \to [-\infty, +\infty)$$

$$F(q) = \log \|v(q)\|_{Poin}$$

where, as usual, $\|v(q)\|_{Poin}$ is the norm of $v(q)$ measured with the Poincaré metric on L_q. Recall that this "metric" is identically zero when L_q is parabolic, so that F is equal to $-\infty$ on the intersection of U^0 with parabolic leaves.

Proposition 6.2. *The function F above is either plurisubharmonic or identically $-\infty$.*

Proof. Let $T \subset U^0$ be a transversal to $\mathscr{F}^0$, and let U_T be the corresponding covering tube. Put on the fibers of U_T their Poincaré metric. The vector field v induces a nonsingular vertical vector field on U_T along $p_T(T)$, which we denote again by v. Due to the arbitrariness of T, and by a connectivity argument, we need just to verify that the function on T defined by

$$F(z) = \log \|v(p_T(z)\|_{Poin}$$

is either plurisubharmonic or identically $-\infty$. That is, the fiberwise Poincaré metric on U_T has a plurisubharmonic variation.

The upper semicontinuity of F being evident (see e.g. [Suz, §3] or [Kiz]), let us consider the submean inequality over discs in T.

Take a closed disc $S \subset T$ as in Theorem 5.1, i.e. satisfying hypotheses (a) and (b) of Section 5. By that Theorem, and by choosing an increasing sequence of compact subsets K_j in ∂U_S, we can find a sequence of relatively compact domains $\Omega_j \subset U_S$, $j \in \mathbb{N}$, such that:

(i) the relative boundary of Ω_j in U_S is a real analytic Levi-flat hypersurface $M_j \subset U_S$, with boundary $\Gamma_j \subset \partial U_S$, filled by a $\mathbb{S}^1$-family of graphs of holomorphic sections of U_S with boundary values in Γ_j;
(ii) for every $z \in S$, the fiber $\Omega_j(z) = \Omega_j \cap P_S^{-1}(z)$ is a disc, centered at $p_S(z)$; moreover, for $z \in \partial S$ we have $\cup_{j=1}^{+\infty} \Omega_j(z) = P_S^{-1}(z)$.

Note that one cannot hope that the exhaustive property in (ii) holds also for z in the interior of S.

We may apply to Ω_j, whose boundary is Levi-flat and hence pseudoconvex, the result of Yamaguchi discussed in Section 2, more precisely Proposition 2.2. It says that the function on S

$$F_j(z) = \log \|v(p_S(z)\|_{Poin(j)},$$

where $\|v(p_S(z)\|_{Poin(j)}$ is the norm with respect to the Poincaré metric on the disc $\Omega_j(z)$, is plurisubharmonic. Hence we have at the center 0 of $S \simeq \overline{\mathbb{D}}$ the submean inequality:

$$F_j(0) \leq \frac{1}{2\pi} \int_0^{2\pi} F_j(e^{i\theta}) d\theta.$$

We now pass to the limit $j \to +\infty$. For every $z \in \partial S$ we have $F_j(z) \to F(z)$, by the exhaustive property in (ii) above. Moreover, we may assume that $\Omega_j(z)$ is an increasing sequence for every $z \in \partial S$ (and in fact for every $z \in S$, but this is not important), so that $F_j(z)$ converges to $F(z)$ in a decreasing way, by the monotonicity property of the Poincaré metric. It follows that the boundary integral in the submean inequality above converges, as $j \to +\infty$, to $\frac{1}{2\pi}\int_0^{2\pi} F(e^{i\theta})d\theta$ (which may be $-\infty$, of course).

Concerning $F_j(0)$, it is sufficient to observe that, obviously, $F(0) \leq F_j(0)$, because $\Omega_j(0) \subset P_S^{-1}(0)$, and so $F(0) \leq \liminf_{j\to+\infty} F_j(0)$. In fact, and because $\Omega_j(0)$

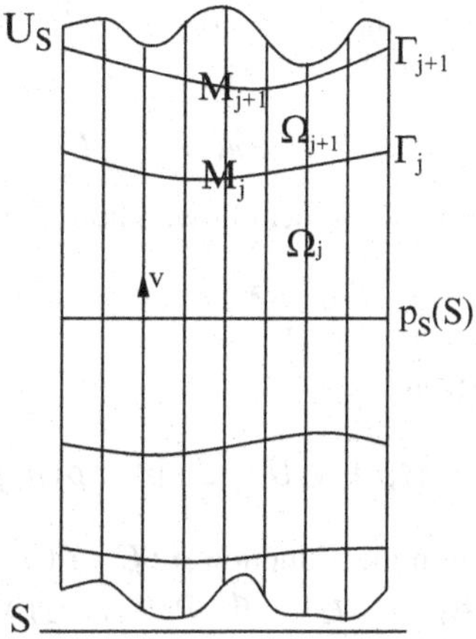

is increasing, $F_j(0)$ converges to some value c in $[-\infty, +\infty)$, but we may have the strict inequality $F(0) < c$ if $\Omega_j(0)$ do not exhaust $P_S^{-1}(0)$. Therefore the above submean inequality gives, at the limit,

$$F(0) \leq \frac{1}{2\pi}\int_0^{2\pi} F(e^{i\theta})d\theta$$

that is, the submean inequality for F on S.

Take now an arbitrary closed disc $S \subset T$, centered at some point $p \in T$. By Lemma 6.1 and the remarks before it, we may approximate S by a sequence of closed discs S_j with the same center p and satisfying moreover hypotheses (a) and (b) before Theorem 5.1 (unless $R = T$, but in that case $U_T = T \times \mathbb{C}$ and $F \equiv -\infty$). More precisely, if $\varphi : \overline{\mathbb{D}} \to T$ is a parametrization of S, $\varphi(0) = p$, then we may uniformly approximate φ by a sequence of embeddings $\varphi_j : \overline{\mathbb{D}} \to T$, $\varphi_j(0) = p$, such that $S_j = \varphi_j(\overline{\mathbb{D}})$ satisfies the assumptions of Theorem 5.1. Hence we have, by the previous arguments and for every j,

$$F(p) \leq \frac{1}{2\pi}\int_0^{2\pi} F(\varphi_j(e^{i\theta}))d\theta$$

and passing to the limit, using Fatou Lemma, and taking into account the upper semicontinuity of F, we finally obtain

$$F(p) \leq \limsup_{j\to+\infty} \frac{1}{2\pi}\int_0^{2\pi} F(\varphi_j(e^{i\theta}))d\theta \leq \frac{1}{2\pi}\int_0^{2\pi} \limsup_{j\to+\infty} F(\varphi_j(e^{i\theta}))d\theta \leq$$
$$\leq \frac{1}{2\pi}\int_0^{2\pi} F(\varphi(e^{i\theta}))d\theta.$$

This is the submean inequality on an arbitrary disc $S \subset T$, and so F is, if not identically $-\infty$, plurisubharmonic. □

Because $U \setminus U^0$ is an analytic subset of codimension at least two, the above plurisubharmonic function F on U^0 admits a (unique) plurisubharmonic extension to the full U, given explicitely by

$$F(q) = \limsup_{p \in U^0,\ p \to q} F(p)\,, \quad q \in U \setminus U^0.$$

Proposition 6.3. *We have $F(q) = -\infty$ for every $q \in U \setminus U^0$.*

Proof. The vector field v on U has a local flow: a holomorphic map

$$\Phi : \mathscr{D} \to U$$

defined on a domain of the form

$$\mathscr{D} = \{(p,t) \in U \times \mathbb{C} \mid |t| < \rho(p)\}$$

for a suitable lower semicontinuous function $\rho : U \to (0, +\infty]$, such that $\Phi(p,0) = p$, $\frac{\partial \Phi}{\partial t}(p,0) = v(p)$, and $\Phi(p, t_1 + t_2) = \Phi(\Phi(p,t_1), t_2)$ whenever it makes sense. Standard results on ordinary differential equations show that we may choose the function ρ so that $\rho \equiv +\infty$ on $U \setminus U^0 =$ the zero set of v.

Take $q \in U \setminus U^0$ and $p \in U^0$ close to it. Then $\Phi(p, \cdot)$ sends the large disc $\mathbb{D}(\rho(p))$ into $L_p^0 \cap U^0$, and consequently into L_p, with derivative at 0 equal to $v(p)$. It follows, by monotonicity of the Poincaré metric, that the Poincaré norm of $v(p)$ is bounded from above by something like $\frac{1}{\rho(p)}$, which tends to 0 as $p \to q$. We therefore obtain that $\log \|v(p)\|_{Poin}$ tends to $-\infty$ as $p \to q$. □

The functions $F : U \to [-\infty, +\infty)$ so constructed can be seen [Dem] as local weights of a (singular) hermitian metric on the tangent bundle $T_{\mathscr{F}}$ of $\mathscr{F}$, and by duality on the canonical bundle $K_{\mathscr{F}} = T_{\mathscr{F}}^*$. Indeed, if $v_j \in \Theta(U_j)$ are local generators of $\mathscr{F}$, for some covering $\{U_j\}$ of X, with $v_j = g_{jk} v_k$ for a multiplicative cocycle g_{jk} generating $K_{\mathscr{F}}$, then the functions $F_j = \log \|v_j\|_{Poin}$ are related by $F_j - F_k = \log |g_{jk}|$. The curvature of this metric on $K_{\mathscr{F}}$ is the current on X, of bidegree $(1,1)$, locally defined by $\frac{i}{\pi} \partial \bar{\partial} F_j$. Hence Propositions 6.2 and 6.3 can be restated in the following more intrinsic form, where we set

$$Parab(\mathscr{F}) = \{p \in X^0 \mid \widetilde{L_p} = \mathbb{C}\}.$$

Theorem 6.4. *Let X be a compact connected Kähler manifold and let $\mathscr{F}$ be a foliation by curves on X. Suppose that $\mathscr{F}$ has at least one hyperbolic leaf. Then the Poincaré metric on the leaves of $\mathscr{F}$ induces a hermitian metric on the canonical bundle $K_{\mathscr{F}}$ whose curvature is positive, in the sense of currents. Moreover, the polar set of this metric coincides with $Sing(\mathscr{F}) \cup Parab(\mathscr{F})$.*

A foliation with at least one hyperbolic leaf will be called **hyperbolic foliation**. The existence of a hyperbolic leaf (and the connectedness of X) implies that $\mathscr{F}$

is not a rational quasi-fibration, and all the local weights F introduced above are plurisubharmonic, and not identically $-\infty$.

Let us state two evident but important Corollaries.

Corollary 6.5. *The canonical bundle $K_{\mathscr{F}}$ of a hyperbolic foliation $\mathscr{F}$ is pseudoeffective.*

Corollary 6.6. *Given a hyperbolic foliation $\mathscr{F}$, the subset*

$$Sing(\mathscr{F}) \cup Parab(\mathscr{F})$$

is complete pluripolar in X.

We think that the conclusion of this last Corollary could be strengthened. The most optimistic conjecture is that $Sing(\mathscr{F}) \cup Parab(\mathscr{F})$ is even an *analytic subset* of X. At the moment, however, we are very far from proving such a fact (except when $\dim X = 2$, where special techniques are available, see [MQ1] and [Br1]). Even the *closedness* of $Sing(\mathscr{F}) \cup Parab(\mathscr{F})$ seems an open problem! This is related to the more general problem of the continuity of the leafwise Poincaré metric (which would give, in particular, the closedness of its polar set). Let us prove a partial result in this direction, following a rather standard hyperbolic argument [Ghy, Br2]. Recall that a complex compact analytic space Z is *hyperbolic* if every holomorphic map of $\mathbb{C}$ into Z is constant [Lan].

Theorem 6.7. *Let $\mathscr{F}$ be a foliation by curves on a compact connected Kähler manifold M. Suppose that:*

(i) every leaf is hyperbolic;
(ii) $Sing(\mathscr{F})$ is hyperbolic.

Then the leafwise Poincaré metric is continuous.

Proof. Let us consider the function

$$F : U^0 \to \mathbb{R}, \quad F(q) = \log \|v(q)\|_{Poin}$$

introduced just before Proposition 6.2. We have to prove that F is continuous (the continuity on the full U is then a consequence of Proposition 6.3). We have already observed, during the proof of Proposition 6.2, that F is upper semicontinuous, hence let us consider its lower semicontinuity.

Take $q_\infty \in U^0$ and take a sequence $\{q_n\} \subset U^0$ converging to q_∞. For every n, let $\varphi_n : \mathbb{D} \to X$ be a holomorphic map into $L_{q_n} \subset X$, sending $0 \in \mathbb{D}$ to $q_n \in L_{q_n}$. For every compact subset $K \subset \mathbb{D}$, consider

$$I_K = \{\|\varphi_n'(t)\| \mid t \in K, n \in \mathbb{N}\} \subset \mathbb{R}$$

(the norm of φ_n' is here computed with the Kähler metric on X).

Claim: I_K is a bounded subset of $\mathbb{R}$.

Indeed, in the opposite case we may find a subsequence $\{n_j\} \subset \mathbb{N}$ and a sequence $\{t_j\} \subset K$ such that $\|\varphi'_{n_j}(t_j)\| \to +\infty$ as $j \to +\infty$. By Brody's Reparametrization Lemma [Lan, Ch. III], we may reparametrize these discs so that they converge to an entire curve: there exists maps $h_j : \mathbb{D}(r_j) \to \mathbb{D}$, with $r_j \to +\infty$, such that the maps

$$\psi_j = \varphi_j \circ h_j : \mathbb{D}(r_j) \to X$$

converge, uniformly on compact subsets, to a nonconstant map

$$\psi : \mathbb{C} \to X.$$

It is clear that ψ is tangent to $\mathscr{F}$, more precisely $\psi'(t) \in T_{\psi(t)}\mathscr{F}$ whenever $\psi(t) \notin Sing(\mathscr{F})$, because each ψ_j has the same property. Moreover, by hypothesis (ii) we have that the image of ψ is not contained in $Sing(\mathscr{F})$. Therefore, $S = \psi^{-1}(Sing(\mathscr{F}))$ is a discrete subset of $\mathbb{C}$, and $\psi(\mathbb{C} \setminus S)$ is contained in some leaf L^0 of $\mathscr{F}^0$.

Take now $t_0 \in S$. It corresponds to a parabolic end of L^0. On a small compact disc B centered at t_0, $\psi|_B$ is uniform limit of $\psi_j|_B : B \to X$, which are maps into leaves of $\mathscr{F}$. If U_T is a covering tube associated to some transversal T cutting L^0, then the maps $\psi_j|_B$ can be lifted to U_T, in such a way that they converge on ∂B to some map which lifts $\psi|_{\partial B}$. The structure of U_T (absence of vanishing cycles) implies that, in fact, we have convergence on the full B, to a map which lifts $\psi|_B$. By doing so at every $t_0 \in S$, we see that $\psi : \mathbb{C} \to X$ can be fully lifted to U_T, i.e. $\psi(\mathbb{C})$ is contained in the leaf L of $\mathscr{F}$ obtained by completion of L^0. But this contradicts hypothesis (i), and proves the Claim.

The Claim implies now that, up to subsequencing, the maps $\varphi_n : \mathbb{D} \to X$ converge, uniformly on compact subsets, to some $\varphi_\infty : \mathbb{D} \to X$, with $\varphi_\infty(0) = q_\infty$. As before, we obtain $\varphi_\infty(\mathbb{D}) \subset L_{q_\infty}$.

Recall now the extremal propery of the Poincaré metric: if we write $\varphi'_n(0) = \lambda_n \cdot v(q_n)$, then $\|v(q_n)\|_{Poin} \leq \frac{1}{|\lambda_n|}$, and equality is atteined if φ_n is a uniformization of L_{q_n}. Hence, with this choice of $\{\varphi_n\}$, we see that

$$\|v(q_\infty)\|_{Poin} \leq \frac{1}{|\lambda_\infty|} = \lim_{n \to +\infty} \frac{1}{|\lambda_n|} = \lim_{n \to +\infty} \|v(q_n)\|_{Poin}$$

i.e. $F(q_\infty) \leq \lim_{n \to +\infty} F(q_n)$. Due to the arbitrariness of the initial sequence $\{q_n\}$, this gives the lower semicontinuity of F. □

Of course, due to hypothesis (i) such a result says nothing about the possible closedness of $Sing(\mathscr{F}) \cup Parab(\mathscr{F})$, when $Parab(\mathscr{F})$ is not empty, but at least it leaves some hope. The above proof breaks down when there are parabolic leaves, because Brody's lemma does not allow to control where the limit entire curve ψ is located: even if each ψ_j passes through q_{n_j}, it is still possible that ψ does not pass through q_∞, because the points in $\psi_j^{-1}(q_{n_j})$ could exit from every compact subset of $\mathbb{C}$. Hence, the only hypothesis "L_{q_∞} is hyperbolic" (instead of "all the leaves are hyperbolic") is not sufficient to get a contradiction and prove the Claim. In other words, the (parabolic) leaf L appearing in the Claim above could be "far" from q_∞, but still could have some influence on the possible discontinuity of the leafwise Poincaré metric at q_∞.

The subset $Sing(\mathscr{F}) \cup Parab(\mathscr{F})$ being complete pluripolar, a natural question concerns the computation of its Lelong numbers. For instance, if these Lelong numbers were positive, then, by Siu Theorem [Dem], we should get that $Sing(\mathscr{F}) \cup Parab(\mathscr{F})$ is a countable union of analytic subsets, a substantial step toward the conjecture above. However, we generally expect that these Lelong numbers are zero, even when $Sing(\mathscr{F}) \cup Parab(\mathscr{F})$ is analytic.

Example 6.8. Let E be an elliptic curve and let $X = \mathbb{P} \times E$. Let $\alpha = f(z)dz$ be a meromorphic 1-form on $\mathbb{P}$, with poles $P = \{z_1, \dots, z_k\}$ of orders $\{\nu_1, \dots, \nu_k\}$. Consider the (nonsingular) foliation $\mathscr{F}$ on X defined by the (saturated) Kernel of the meromorphic 1-form $\beta = f(z)dz - dw$, i.e. by the differential equation $\frac{dw}{dz} = f(z)$. Then each fiber $\{z_j\} \times E$, $z_j \in P$, is a leaf of $\mathscr{F}$, whereas each other fiber $\{z\} \times E$, $z \notin P$, is everywhere transverse to $\mathscr{F}$. In [Br1] such a foliation is called *turbulent.* Outside the elliptic leaves $P \times E$, every leaf is a regular covering of $\mathbb{P} \setminus P$, by the projection $X \to \mathbb{P}$. Hence, if $k \geq 3$ then these leaves are hyperbolic, and their Poincaré metric coincides with the pull-back of the Poincaré metric on $\mathbb{P} \setminus P$.

Take a point $(z_j, w) \in P \times E = Parab(\mathscr{F})$. Around it, the foliation is generated by the holomorphic and nonvanishing vector field $v = f(z)^{-1}\frac{\partial}{\partial z} + \frac{\partial}{\partial w}$, whose z-component has at $z = z_j$ a zero of order ν_j. The weight $F = \log \|v\|_{Poin}$ is nothing but than the pull-back of $\log \|f(z)^{-1}\frac{\partial}{\partial z}\|_{Poin}$, where the norm is measured in the Poincaré metric of $\mathbb{P} \setminus P$. Recalling that the Poincaré metric of the punctured disc $\mathbb{D}^*$ is $\frac{idz \wedge d\bar{z}}{|z|^2(\log|z|^2)^2}$, we see that F is something like

$$\log |z - z_j|^{\nu_j - 1} - \log \big| \log |z - z_j|^2 \big|.$$

Hence the Lelong number along $\{z_j\} \times E$ is positive if and only if $\nu_j \geq 2$, which can be considered as an "exceptional" case; in the "generic" case $\nu_j = 1$ the pole of F along $\{z_j\} \times E$ is a weak one, with vanishing Lelong number.

Remark 6.9. We used the convexity property stated by Theorem 5.1 as a substitute of the Stein property required by the results of Nishino, Yamaguchi, Kizuka discussed in Section 2. One could ask if, after all, such a convexity property can be used to prove the Steinness of U_T, when T is Stein. If the ambient manifold X is Stein, instead of Kähler compact, Il'yashenko proved in [Il1] and [Il2] (see Section 2) that indeed U_T is Stein, using Cartan-Thullen-Oka convexity theory over Stein manifolds. See also [Suz] for a similar approach to V_T, [Br6] for some result in the case of projective manifolds, close in spirit to [Il2], and [Nap] and [Ohs] for related results in the case of proper fibrations by curves.

For instance, suppose that all the fibers of U_T are hyperbolic, and that the fiberwise Poincaré metric is of class C^2. Then we can take the function $\psi : U_T \to \mathbb{R}$ defined by $\psi = \psi_0 + \varphi \circ P_T$, where $\psi_0(q)$ is the squared hyperbolic distance (in the fiber) between q and the basepoint $p_T(P_T(q))$, and $\varphi : T \to \mathbb{R}$ is a strictly plurisubharmonic exhaustion of T. A computation shows that ψ is strictly plurisubharmonic (thanks to the plurisubharmonic variation of the fiberwise Poincaré metric on U_T), and being also exhaustive we deduce that U_T is Stein. Probably, this can be done also if the fiberwise Poincaré metric is less regular, say C^0. But when there are parabolic fibers such a simple argument cannot work, because ψ is no more exhaustive (one can try perhaps to use a renormalization argument like the one used in the proof of Theorem 2.3). However, if *all* the fibers are parabolic then we shall see later that U_T is a product $T \times \mathbb{C}$ (if T is small), and hence it is Stein.

A related problem concerns the existence on U_T of holomorphic functions which are not constant on the fibers. By Corollary 6.5, $K_{\mathscr{F}}$ is pseudoeffective, if $\mathscr{F}$ is hyperbolic. Let us assume a little more, namely that it is effective. Then any nontrivial section of $K_{\mathscr{F}}$ over X can be lifted to U_T, giving a holomorphic section of the relative canonical bundle of the fibration. As in Lemmata 2.5 and 2.6, this section can be integrated along the (simply connected and pointed) fibers, giving a holomorphic function on U_T not constant on generic fibers.

7 Extension of Meromorphic Maps from Line Bundles

In order to generalize Corollary 6.5 to cover (most) parabolic foliations, we need an extension theorem for certain meromorphic maps. This is done in the present Section, following [Br5].

7.1 Volume Estimates

Let us firstly recall some results of Dingoyan [Din], in a slightly simplified form due to our future use.

Let V be a connected complex manifold, of dimension n, and let ω be a smooth closed semipositive $(1,1)$-form on V (e.g., the pull-back of a Kähler form by some holomorphic map from V). Let $U \subset V$ be an open subset, with boundary ∂U compact in V. Suppose that the mass of ω^n on U is finite: $\int_U \omega^n < +\infty$. We look for some condition ensuring that also the mass on V is finite: $\int_V \omega^n < +\infty$. In other words, we look for the boundedness of the ω^n-volume of the ends $V \setminus U$.

Set

$$P_\omega(V,U) = \{\varphi : V \to [-\infty,+\infty) \text{ u.s.c.} \mid dd^c\varphi + \omega \geq 0,\ \varphi|_U \leq 0\}$$

where u.s.c. means upper semicontinuous, and the first inequality is in the sense of currents. This first inequality defines the so-called *ω-plurisubharmonic functions*. Note that locally the space of ω-plurisubharmonic functions can be identified with a translation of the space of the usual plurisubharmonic functions: locally the form ω admits a smooth potential ϕ ($\omega = dd^c\phi$), and so φ is ω-plurisubharmonic is and only if $\varphi + \phi$ is plurisubharmonic. In this way, most local problems on ω-plurisubharmonic functions can be reduced to more familiar problems on plurisubharmonic functions.

Remark that the space $P_\omega(V,U)$ is not empty, for it contains at least all the constant nonpositive functions on V.

Suppose that $P_\omega(V,U)$ satisfies the following condition:

(A) the functions in $P_\omega(V,U)$ are locally uniformly bounded from above: for every $z \in V$ there exists a neighbourhood $V_z \subset V$ of z and a constant c_z such that $\varphi|_{V_z} \leq c_z$ for every $\varphi \in P_\omega(V,U)$.

Then we can introduce the upper envelope

$$\Phi(z) = \sup_{\varphi \in P_\omega(V,U)} \varphi(z) \qquad \forall z \in V$$

and its upper semicontinuous regularization

$$\Phi^*(z) = \limsup_{w \to z} \Phi(w) \qquad \forall z \in V.$$

The function

$$\Phi^* : V \to [0, +\infty)$$

is identically zero on U, upper semicontinuous, and ω-plurisubharmonic (Brelot-Cartan [Kli]), hence it belongs to the space $P_\omega(V,U)$. Moreover, by results of Bedford and Taylor [BeT, Kli] the wedge product $(dd^c\Phi^* + \omega)^n$ is well defined, as a locally finite measure on V, and it is identically zero outside $\overline{U}$:

$$(dd^c\Phi^* + \omega)^n \equiv 0 \qquad \text{on } V \setminus \overline{U}.$$

Indeed, let $B \subset V \setminus \overline{U}$ be a ball around which ω has a potential. Let $P_\omega(B, \Phi^*)$ be the space of ω-plurisubharmonic functions ψ on B such that $\limsup_{z\to w} \psi(z) \leq \Phi^*(w)$ for every $w \in \partial B$. Let Ψ^* be the regularized upper envelope of the family $P_\omega(B, \Phi^*)$ (which is bounded from above by the maximum principle). Remark that $\Phi^*|_B$ belongs to $P_\omega(B, \Phi^*)$, and so $\Psi^* \geq \Phi^*$ on B. By [BeT], Ψ^* satisfies the homogeneous Monge-Ampère equation $(dd^c\Psi^* + \omega)^n = 0$ on B, with Dirichlet boundary condition $\limsup_{z\to w} \Psi^*(z) = \Phi^*(w)$, $w \in \partial B$ ("balayage"). Then the function $\widetilde{\Phi}^*$ on V, which is equal to Ψ^* on B and equal to Φ^* on $V \setminus \overline{B}$, still belongs to $P_\omega(V,U)$, and it is everywhere not smaller than Φ^*. Hence, by definition of Φ^*, we must have $\widetilde{\Phi}^* = \Phi^*$, i.e. $\Phi^* = \Psi^*$ on B and so Φ^* satisfies the homogeneous Monge-Ampère equation on B.

Suppose now that the following condition is also satisfied:

(B) $\Phi^* : V \to [0, +\infty)$ is exhaustive on $V \setminus U$: for every $c > 0$, the subset $\{\Phi^* < c\} \setminus U$ is relatively compact in $V \setminus U$.

Roughly speaking, this means that the function Φ^* solves on $V \setminus \overline{U}$ the homogeneous Monge-Ampère equation, with boundary conditions 0 on ∂U and $+\infty$ on the "boundary at infinity" of $V \setminus U$.

Theorem 7.1. [Din] *Under assumptions (A) and (B), the ω^n-volume of V is finite:*

$$\int_V \omega^n < +\infty.$$

Proof. The idea is that, using Φ^*, we can push all the mass of ω^n on $V \setminus U$ to the compact set ∂U. Note that we certainly have

$$\int_V (dd^c\Phi^* + \omega)^n < +\infty$$

because, after decomposing $V = U \cup (V \setminus U) \cup \partial U$, we have:

(i) on U, $\Phi^* \equiv 0$ and $\int_U \omega^n$ is finite by standing assumptions;
(ii) on $V \setminus U$, $(dd^c\Phi^* + \omega)^n \equiv 0$;
(iii) ∂U is compact (but, generally speaking, ∂U is charged by the measure $(dd^c\Phi^* + \omega)^n$).

Hence the theorem follows from the next inequality.

Lemma 7.2. [Din, Lemma 4]

$$\int_V \omega^n \leq \int_V (dd^c\Phi^* + \omega)^n.$$

Proof. More generally, we shall prove that for every $k = 0, \ldots, n-1$:

$$\int_V (dd^c\Phi^* + \omega)^{k+1} \wedge \omega^{n-k-1} \geq \int_V (dd^c\Phi^* + \omega)^k \wedge \omega^{n-k},$$

so that the desired inequality follows by concatenation. We can decompose the integral on the left hand side as

$$\int_V (dd^c\Phi^* + \omega)^k \wedge \omega^{n-k} + \int_V dd^c\Phi^* \wedge (dd^c\Phi^* + \omega)^k \wedge \omega^{n-k-1}$$

and so we need to prove that, setting $\eta = (dd^c\Phi^* + \omega)^k \wedge \omega^{n-k-1}$, we have

$$I = \int_V dd^c\Phi^* \wedge \eta \geq 0.$$

Here all the wedge products are well defined, because Φ^* is locally bounded, and moreover η is a closed positive current of bidegree $(n-1, n-1)$ [Kli].

Take a sequence of smooth functions $\chi_n : \mathbb{R} \to [0,1]$, $n \in \mathbb{N}$, such that $\chi_n(t) = 1$ for $t \leq n$, $\chi_n(t) = 0$ for $t \geq n+1$, and $\chi_n'(t) \leq 0$ for every t. Thus, for every $z \in V$ we have $(\chi_n \circ \Phi^*)(z) = 0$ for $n \leq \Phi^*(z) - 1$ and $(\chi_n \circ \Phi^*)(z) = 1$ for $n \geq \Phi^*(z)$. Hence it is sufficient to prove that

$$I_n = \int_V (\chi_n \circ \Phi^*) \cdot dd^c\Phi^* \wedge \eta \geq 0$$

for every n. By assumption (B), the support of $\chi_n \circ \Phi^*$ intersects $V \setminus U$ along a compact subset. Moreover, Φ^* is identically zero on U. Thus, the integrand above has compact support in V, as well as $(\chi_n \circ \Phi^*) \cdot d^c\Phi^* \wedge \eta$. Hence, by Stokes formula,

$$I_n = -\int_V d(\chi_n \circ \Phi^*) \wedge d^c\Phi^* \wedge \eta = -\int_V (\chi_n' \circ \Phi^*) \cdot d\Phi^* \wedge d^c\Phi^* \wedge \eta.$$

Now, $d\Phi^* \wedge d^c\Phi^*$ is a positive current, and its product with η is a positive measure. From $\chi_n' \leq 0$ we obtain $I_n \geq 0$, for every n. □

This inequality completes the proof of the theorem. □

7.2 Extension of Meromorphic Maps

As in [Din, §6], we shall use the volume estimate of Theorem 7.1 to get an extension theorem for certain meromorphic maps into Kähler manifolds.

Consider the following situation. It is given a compact connected Kähler manifold X, of dimension n, and a line bundle L on X. Denote by E the total space of L, and by $\Sigma \subset E$ the graph of the null section of L. Let $U_\Sigma \subset E$ be a connected (tubular) neighbourhood of Σ, and let Y be another compact Kähler manifold, of dimension m.

Theorem 7.3. [Br5] *Suppose that L is not pseudoeffective. Then any meromorphic map*

$$f : U_\Sigma \setminus \Sigma \dashrightarrow Y$$

extends to a meromorphic map

$$\bar{f} : U_\Sigma \dashrightarrow Y.$$

Before the proof, let us make a link with [BDP]. In the special case where X is projective, and not only Kähler, the non pseudoeffectivity of L translates into the existence of a covering family of curves $\{C_t\}_{t\in B}$ on X such that $L|_{C_t}$ has negative degree for every $t \in B$ [BDP]. This means that the normal bundle of Σ in E has negative degree on every $C_t \subset \Sigma \simeq X$. Hence the restriction of E over C_t is a surface E_t which contains a compact curve Σ_t whose selfintersection is negative, and thus contractible to a normal singularity. By known results [Siu] [Iv1], every meromorphic map from $U_t \setminus \Sigma_t$ (U_t being a neighbourhood of Σ_t in E_t) into a compact Kähler manifold can be meromorphically extended to U_t. Because the curves C_t cover the full X, this is sufficient to extends from $U_\Sigma \setminus \Sigma$ to U_Σ.

Of course, if X is only Kähler then such a covering family of curves could not exist, and we need a more global approach, which avoids any restriction to curves. Even in the projective case, this seems a more natural approach than evoking [BDP].

Proof. We begin with a simple criterion for pseudoeffectivity, analogous to the well known fact that a line bundle is ample if and only if its dual bundle has strongly pseudoconvex neighbourhoods of the null section. Recall that an open subset W of a complex manifold E is *locally pseudoconvex in* E if for every $w \in \partial W$ there exists a neighbourhood $U_w \subset E$ of w such that $W \cap U_w$ is Stein.

Lemma 7.4. *Let X be a compact connected complex manifold and let L be a line bundle on X. The following two properties are equivalent:*

(i) L is pseudoeffective;
(ii) in the total space E^* of the dual line bundle L^* there exists a neighbourhood $W \neq E^*$ of the null section Σ^* which is locally pseudoconvex in E^*.

Proof. The implication (i) $\Rightarrow$ (ii) is quite evident. If h is a (singular) hermitian metric on L with positive curvature [Dem], then in a local trivialization $E|_{U_j} \simeq U_j \times \mathbb{C}$ the unit ball is expressed by $\{(z,t) \mid |t| < e^{h_j(z)}\}$, where $h_j : U_j \to [-\infty, +\infty)$ is the plurisubharmonic weight of h. In the dual local trivialization $E^*|_{U_j} \simeq U_j \times \mathbb{C}$, the unit ball of the dual metric is expressed by $\{(z,s) \mid |s| < e^{-h_j(z)}\}$. The plurisubharmonicity of h_j gives (and is equivalent to) the Steinness of such an open subset of $U_j \times \mathbb{C}$ (recall Hartogs Theorem on Hartogs Tubes mentioned in Section 2). Hence we get (ii), with W equal to the unit ball in E^*.

The implication (ii) $\Rightarrow$ (i) is not more difficult. Let $W \subset E^*$ be as in (ii). On E^* we have a natural $\mathbb{S}^1$-action, which fixes Σ^* and rotates each fiber. For every $\vartheta \in \mathbb{S}^1$, let W_ϑ be the image of W by the action of ϑ. Then

$$W' = \cap_{\vartheta \in \mathbb{S}^1} W_\vartheta$$

is still a nontrivial locally pseudoconvex neighbourhood of Σ^*, for local pseudoconvexity is stable by intersections. For every $z \in X$, W' intersects the fiber E^*_z along an open subset which is $\mathbb{S}^1$-invariant, a connected component of which is therefore a disc W^0_z centered at the origin (possibly $W^0_z = E^*_z$ for certain z, but not for all); the other components are annuli around the origin. Using the local pseudoconvexity of W', i.e. its Steinness in local trivializations $E^*|_{U_j} \simeq U_j \times \mathbb{C}$, it is easy to see that these annuli and discs cannot merge when z moves in X. In other words,

$$W'' = \cup_{z \in X} W^0_z$$

is a connected component of W', and of course it is still a nontrivial pseudoconvex neighbourhood of Σ^*. We may use W'' as unit ball for a metric on L^*. As in the first part of the proof, the corresponding dual metric on L has positive curvature, in the sense of currents. □

Consider now, in the space $U_\Sigma \times Y$, the graph Γ_f of the meromorphic map $f : U^0_\Sigma = U_\Sigma \setminus \Sigma \dashrightarrow Y$. By definition of meromorphicity, Γ_f is an irreducible analytic subset of $U^0_\Sigma \times Y \subset U_\Sigma \times Y$, whose projection to U^0_Σ is proper and generically bijective. It may be singular, and in that case we replace it by a resolution of its singularities, still denoted by Γ_f. The (new) projection

$$\pi : \Gamma_f \to U^0_\Sigma$$

is a proper map, and it realizes an isomorphism between $\Gamma_f \setminus Z$ and $U_\Sigma^0 \setminus B$, for suitable analytic subsets $Z \subset \Gamma_f$ and $B \subset U_\Sigma^0$, with B of codimension at least two.

The manifold $U_\Sigma \times Y$ is Kähler. The Kähler form restricted to the graph of f and pulled-back to its resolution gives a smooth, semipositive, closed $(1,1)$-form ω on Γ_f. Fix a smaller (tubular) neighbourhood U'_Σ of Σ, and set $U_0 = U_\Sigma \setminus \overline{U'_\Sigma}$, $U = \pi^{-1}(U_0) \subset \Gamma_f$. Up to restricting a little the initial U_Σ, we may assume that the $\omega^{n'}$-volume of the shell U is finite ($n' = n+1 = \dim \Gamma_f$). Our aim is to prove that

$$\int_{\Gamma_f} \omega^{n'} < +\infty.$$

Indeed, this is the volume of the graph of f. Its finiteness, together with the analyticity of the graph in $U_\Sigma^0 \times Y$, imply that the closure of that graph in $U_\Sigma \times Y$ is still an analytic subset of dimension n', by Bishop's extension theorem [Siu, Chi]. This closure, then, is the graph of the desired meromorphic extension $\bar{f} : U_\Sigma \dashrightarrow Y$.

We shall apply Theorem 7.1. Hence, consider the space $P_\omega(\Gamma_f, U)$ of ω-plurisubharmonic functions on Γ_f, nonpositive on U, and let us check conditions (A) and (B) above, at the beginning of this Section.

Consider the open subset $\Omega \subset \Gamma_f$ where the functions of $P_\omega(\Gamma_f, U)$ are locally uniformly bounded from above. It contains U, and it is a general fact that it is locally pseudoconvex in Γ_f [Din, §3]. Therefore $\Omega' = \Omega \cap (\Gamma_f \setminus Z)$ is locally pseudoconvex in $\Gamma_f \setminus Z$. Its isomorphic projection $\pi(\Omega')$ is therefore locally pseudoconvex in $U_\Sigma^0 \setminus B$. Classical characterizations of pseudoconvexity [Ran, II.5] show that $\Omega_0 = \text{interior}\{\pi(\Omega') \cup B\}$ is locally pseudoconvex in U_Σ^0. From $\Omega \supset U$ we also have $\Omega_0 \supset U_0$.

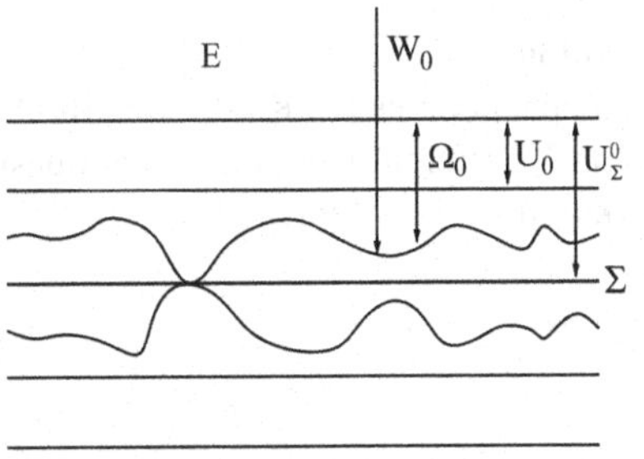

Take now in E the neighbourhood of infinity $W_0 = \Omega_0 \cup (E \setminus U_\Sigma)$. Because $E \setminus \Sigma$ is naturally isomorphic to $E^* \setminus \Sigma^*$, the isomorphism exchanging null sections and sections at infinity, we can see W_0 as an open subset of E^*, so that $W = W_0 \cup \Sigma^*$ is a neighbourhood of Σ^*, locally pseudoconvex in E^*. Because L is not pseudoeffective by assumption, Lemma 7.4 says that $W = E^*$. That is, $\Omega_0 = U_\Sigma^0$.

This implies that the original $\Omega \subset \Gamma_f$ contains, at least, $\Gamma_f \setminus Z$. But, by the maximum principle, a family of ω-plurisubharmonic functions locally bounded outside an analytic subset is automatically bounded also on the same analytic subset. Therefore $\Omega = \Gamma_f$, and condition (A) of Theorem 7.1 is fulfilled.

Condition (B) is simpler [Din, §4]. We just have to exhibit a ω-plurisubharmonic function on Γ_f which is nonpositive on U and exhaustive on $\Gamma_f \setminus U$. On U_Σ^0 we take the function

$$\psi(z) = -\log dist(z, \Sigma)$$

where $dist(\cdot, \Sigma)$ is the distance function from Σ, with respect to the Kähler metric ω_0 on U_Σ. Classical estimates (Takeuchi) give $dd^c\psi \geq -C \cdot \omega_0$, for some positive constant C. Thus

$$dd^c(\psi \circ \pi) \geq -C \cdot \pi^*(\omega_0) \geq -C \cdot \omega$$

because $\omega \geq \pi^*(\omega_0)$. Hence $\frac{1}{C}(\psi \circ \pi)$ is ω-plurisubharmonic on Γ_f. For a sufficiently large $C' > 0$, $\frac{1}{C}(\psi \circ \pi) - C'$ is moreover negative on U, and it is exhaustive on $\Gamma_f \setminus U$. Thus condition (B) is fulfilled.

Finally we can apply Theorem 7.1, obtain the finiteness of the volume of the graph of f, and conclude the proof of the theorem. □

Remark 7.5. We think that Theorem 7.3 should be generalizable to the following "nonlinear" statement: if U_Σ is any Kähler manifold and $\Sigma \subset U_\Sigma$ is a compact hypersurface whose normal bundle is not pseudoeffective, then any meromorphic map $f : U_\Sigma \setminus \Sigma \dashrightarrow Y$ (Y Kähler compact) extends to $\bar{f} : U_\Sigma \dashrightarrow Y$. The difficulty is to show that a locally pseudoconvex subset of $U_\Sigma^0 = U_\Sigma \setminus \Sigma$ like Ω_0 in the proof above can be "lifted" in the total space of the normal bundle of Σ, preserving the local pseudoconvexity.

8 Parabolic Foliations

We can now return to foliations.

As usual, let X be a compact connected Kähler manifold, $\dim X = n$, and let $\mathcal{F}$ be a foliation by curves on X different from a rational quasi-fibration. Let us start with some general remarks, still following [Br5].

8.1 Global Tubes

The construction of holonomy tubes and covering tubes given in Section 4 can be easily modified by replacing the transversal $T \subset X^0$ with the full X^0. That is, all the holonomy coverings $\widehat{L_p}$ and universal coverings $\widetilde{L_p}$, $p \in X^0$, can be glued together, without the restriction $p \in T$. The results are complex manifolds $V_\mathcal{F}$ and $U_\mathcal{F}$, of dimension $n+1$, equipped with submersions

$$Q_\mathcal{F} : V_\mathcal{F} \to X^0, \qquad P_\mathcal{F} : U_\mathcal{F} \to X^0$$

sections

$$q_\mathcal{F} : X^0 \to V_\mathcal{F}, \qquad p_\mathcal{F} : X^0 \to U_\mathcal{F}$$

and meromorphic maps

$$\pi_{\mathscr{F}} : V_{\mathscr{F}} \dashrightarrow X, \qquad \Pi_{\mathscr{F}} : U_{\mathscr{F}} \dashrightarrow X$$

such that, for any transversal $T \subset X^0$, we have $Q_{\mathscr{F}}^{-1}(T) = V_T$, $q_{\mathscr{F}}|_T = q_T$, $\pi_{\mathscr{F}}|_{Q_{\mathscr{F}}^{-1}(T)} = \pi_T$, etc.

Remark that if $D \subset X^0$ is a small disc contained in some leaf L_p of $\mathscr{F}$, $p \in D$, then $Q_{\mathscr{F}}^{-1}(D)$ is naturally isomorphic to the product $\widehat{L_p} \times D$: for every $q \in D$, $\widehat{L_q}$ is the same as $\widehat{L_p}$, but with a different basepoint. More precisely, thinking to points of $\widehat{L_q}$ as equivalence classes of paths starting at q, we see that for every $q \in D$ the isomorphism between $\widehat{L_q}$ and $\widehat{L_p}$ is completely canonical, once D is fixed and because D is contractible. This means that D can be lifted, in a canonical way, to a foliation by discs in $Q_{\mathscr{F}}^{-1}(D)$, transverse to the fibers. In this way, by varying D in X^0, we get in the full space $V_{\mathscr{F}}$ a nonsingular foliation by curves $\widehat{\mathscr{F}}$, which projects by $Q_{\mathscr{F}}$ to $\mathscr{F}^0$.

If $\gamma : [0,1] \to X^0$ is a loop in a leaf, $\gamma(0) = \gamma(1) = p$, then this foliation $\widehat{\mathscr{F}}$ permits to define a *monodromy map* of the fiber $\widehat{L_p}$ into itself. This monodromy map is just the covering transformation of $\widehat{L_p}$ corresponding to γ (which may be trivial, if the holonomy of the foliation along γ is trivial).

In a similar way, in the space $U_{\mathscr{F}}$ we get a canonically defined nonsingular foliation by curves $\widetilde{\mathscr{F}}$, which projects by $P_{\mathscr{F}}$ to $\mathscr{F}^0$. And we have a fiberwise covering

$$F_{\mathscr{F}} : U_{\mathscr{F}} \to V_{\mathscr{F}}$$

which is a local diffeomorphism, sending $\widetilde{\mathscr{F}}$ to $\widehat{\mathscr{F}}$.

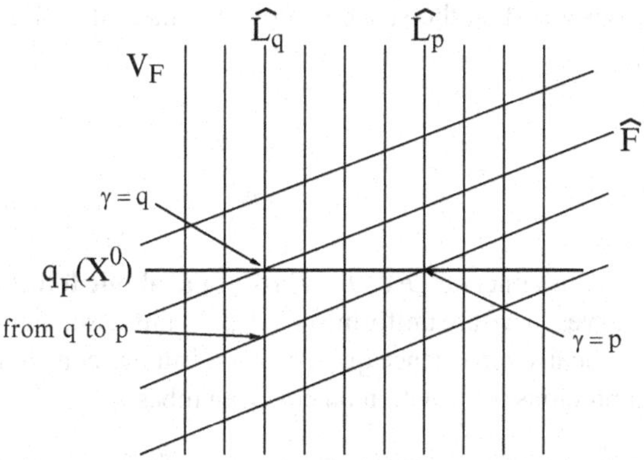

(over a leaf)

In the spaces $U_{\mathscr{F}}$ and $V_{\mathscr{F}}$ we also have the graphs of the sections $p_{\mathscr{F}}$ and $q_{\mathscr{F}}$. They are *not* invariant by the foliations $\widetilde{\mathscr{F}}$ and $\widehat{\mathscr{F}}$: in the notation above, with D in a leaf and $p, q \in D$, the basepoint $q_{\mathscr{F}}(q) \in \widehat{L_q}$ corresponds to the constant path $\gamma(t) \equiv q$, whereas the point of $\widehat{L_q}$ in the same leaf (of $\widehat{\mathscr{F}}$) of $q_{\mathscr{F}}(p) \in \widehat{L_p}$ corresponds to the class of a path in D from q to p. In fact, the graphs $p_{\mathscr{F}}(X^0) \subset U_{\mathscr{F}}$ and $q_{\mathscr{F}}(X^0) \subset V_{\mathscr{F}}$ are hypersurfaces everywhere transverse to $\widetilde{\mathscr{F}}$ and $\widehat{\mathscr{F}}$.

A moment of reflection shows also the following fact: the normal bundle of the hypersurface $p_{\mathscr{F}}(X^0)$ in $U_{\mathscr{F}}$ (or $q_{\mathscr{F}}(X^0)$ in $V_{\mathscr{F}}$) is naturally isomorphic to $T_{\mathscr{F}}|_{X^0}$, the tangent bundle of the foliation restricted to X^0. That is, the manifold $U_{\mathscr{F}}$ (resp. $V_{\mathscr{F}}$) can be thought as an "integrated form" of the (total space of the) tangent bundle of the foliation, in which tangent lines to the foliation are replaced by universal coverings (resp. holonomy coverings) of the corresponding leaves. From this perspective, which will be useful below, the map $\Pi_{\mathscr{F}} : U_{\mathscr{F}} \dashrightarrow X$ is a sort of "skew flow" associated to $\mathscr{F}$, in which the "time" varies not in $\mathbb{C}$ but in the universal covering of the leaf.

Let us conclude this discussion with a trivial but illustrative example.

Example 8.1. Suppose $n = 1$, i.e. X is a compact connected curve and $\mathscr{F}$ is the foliation with only one leaf, X itself. The manifold $V_{\mathscr{F}}$ is composed by equivalence classes of paths in X, where two paths are equivalent if they have the same starting point and the same ending point (here holonomy is trivial!). Clearly, $V_{\mathscr{F}}$ is the product $X \times X$, $Q_{\mathscr{F}}$ is the projection to the first factor, $q_{\mathscr{F}}$ is the diagonal embedding of X into $X \times X$, and $\pi_{\mathscr{F}}$ is the projection to the second factor. Note that the normal bundle of the diagonal $\Delta \subset X \times X$ is naturally isomorphic to TX. The foliation $\widehat{\mathscr{F}}$ is the horizontal foliation, and note that its monodromy is trivial, corresponding to the fact that the holonomy of the foliation is trivial. The manifold $U_{\mathscr{F}}$ is the fiberwise universal covering of $V_{\mathscr{F}}$, with basepoints on the diagonal. It is *not* the product of X with the universal covering $\widetilde{X}$ (unless $X = \mathbb{P}$, of course). It is only a locally trivial $\widetilde{X}$-bundle over X. The foliation $\widetilde{\mathscr{F}}$ has nontrivial monodromy: if $\gamma : [0,1] \to X$ is a loop based at p, then the monodromy of $\widetilde{\mathscr{F}}$ along γ is the covering transformation of the fiber over p (i.e. the universal covering of X with basepoint p) associated to γ. The foliation $\widetilde{\mathscr{F}}$ can be described as the suspension of the natural representation $\pi_1(X) \to Aut(\widetilde{X})$ [CLN].

8.2 Parabolic Foliations

After these preliminaries, let us concentrate on the class of **parabolic foliations**, i.e. let us assume that all the leaves of $\mathscr{F}$ are uniformised by $\mathbb{C}$. In this case, the Poincaré metric on the leaves is identically zero, hence quite useless. But our convexity result Theorem 5.1 still gives a precious information on covering tubes.

Theorem 8.2. *Let X be a compact connected Kähler manifold and let $\mathscr{F}$ be a parabolic foliation on X. Then the global covering tube $U_{\mathscr{F}}$ is a locally trivial $\mathbb{C}$-fibration over X^0, isomorphic to the total space of $T_{\mathscr{F}}$ over X^0, by an isomorphism sending $p_{\mathscr{F}}(X^0)$ to the null section.*

Proof. By the discussion above (local triviality of $U_{\mathscr{F}}$ along the leaves), the first statement is equivalent to say that, if $T \subset X^0$ is a small transversal (say, isomorphic to $\mathbb{D}^{n-1}$), then $U_T \simeq T \times \mathbb{C}$.

We use for this Theorem 2.3 of Section 2. We may assume that there exists an embedding $T \times \mathbb{D} \stackrel{j}{\to} U_T$ sending fibers to fibers and $T \times \{0\}$ to $p_T(T)$. Then we set

$$U_T^{\varepsilon} = U_T \setminus \{j(T \times \overline{\mathbb{D}(\varepsilon)})\}.$$

In order to apply Theorem 2.3, we need to prove that the fiberwise Poincaré metric on U_T^{ε} has a plurisubharmonic variation, for every $\varepsilon > 0$ small.

But this follows from Theorem 5.1 in exactly the same way as we did in Proposition 6.2 of Section 6. We just replace, in that proof, the open subsets $\Omega_j \subset U_S$ (for $S \subset T$ a generic disc) with

$$\Omega_j^{\varepsilon} = \Omega_j \setminus \{j(S \times \overline{\mathbb{D}(\varepsilon)})\}.$$

Then the fibration $\Omega_j^{\varepsilon} \to S$ is, for j large, a fibration by annuli, and its boundary in U_S has two components: one is the Levi flat M_j, and the other one is the Levi-flat $j(S \times \partial\mathbb{D}(\varepsilon))$. Then Theorem 2.1 of Section 2, or more simply the annular generalization of Proposition 2.2, gives the desired plurisubharmonic variation on Ω_j^{ε}, and then on U_S^{ε} by passing to the limit, and finally on U_T^{ε} by varying S.

Hence $U_T \simeq T \times \mathbb{C}$ and $U_{\mathscr{F}}$ is a locally trivial $\mathbb{C}$-fibration over X^0.

Let us now define explicitely the isomorphism between $U_{\mathscr{F}}$ and the total space $E_{\mathscr{F}}$ of $T_{\mathscr{F}}$ over X^0.

Take $p \in X^0$ and let $v_p \in E_{\mathscr{F}}$ be a point over p. Then v_p is a tangent vector to L_p at p, and it can be lifted to $\widetilde{L_p}$ as a tangent vector $\widetilde{v_p}$ at p. Suppose $\widetilde{v_p} \neq 0$. Then, because $\widetilde{L_p} \simeq \mathbb{C}$, $\widetilde{v_p}$ can be extended, in a uniquely defined way, to a complete holomorphic and nowhere vanishing vector field $\widetilde{v}$ on $\widetilde{L_p}$. If, instead, we have $\widetilde{v_p} = 0$, then we set $\widetilde{v} \equiv 0$. Take $q \in \widetilde{L_p}$ equal to the image of p by the time-one flow of $\widetilde{v}$. We have in this way defined a map $(E_{\mathscr{F}})_p \to \widetilde{L_p}$, $v_p \mapsto q$, which obviously is an isomorphism, sending the origin of $(E_{\mathscr{F}})_p$ to the basepoint of $\widetilde{L_p}$. In other words: because L_p is parabolic, we have a canonically defined isomorphism between $(T_pL_p, 0)$ and $(\widetilde{L_p}, p)$.

By varying p in X^0, we thus have a bijective map

$$E_{\mathscr{F}}|_{X^0} \to U_{\mathscr{F}}$$

sending the null section to $p_{\mathscr{F}}(X^0)$, and we need to verify that this map is *holomorphic*. This follows from the fact that $U_{\mathscr{F}}$ (and $E_{\mathscr{F}}$ also, of course) is a locally trivial fibration. In terms of the previous construction, we take a local transversal $T \subset X^0$ and a nowhere vanishing holomorphic section v_p, $p \in T$, of $E_{\mathscr{F}}$ over T. The previous construction gives a vertical vector field $\widetilde{v}$ on U_T, which is, on every fiber, complete holomorphic and nowhere vanishing, and moreover it is holomorphic along $p_T(T) \subset U_T$. After a trivialization $U_T \simeq T \times \mathbb{C}$, sending $p_T(T)$ to $\{w = 0\}$, this vertical vector field $\widetilde{v}$ becomes something like $F(z,w)\frac{\partial}{\partial w}$, with F nowhere vanishing, $F(z,\cdot)$ holomorphic for every fixed z, and $F(\cdot,0)$ also holomorphic. The completeness on fibers gives that F is in fact *constant* on every fiber, i.e. $F = F(z)$, and so F is in fact fully holomorphic. Thus $\widetilde{v}$ is fully holomorphic on the tube. This means precisely that the above map $E_{\mathscr{F}}|_{X^0} \to U_{\mathscr{F}}$ is holomorphic. □

Example 8.3. Consider a foliation $\mathscr{F}$ generated by a global holomorphic vector field $v \in \Theta(X)$, vanishing precisely on $Sing(\mathscr{F})$. This means that $T_{\mathscr{F}}$ is the trivial line bundle, and $E_{\mathscr{F}} = X \times \mathbb{C}$. The compactness of X permits to define the *flow* of v

$$\Phi : X \times \mathbb{C} \to X$$

which sends $\{p\} \times \mathbb{C}$ to the orbit of v through p, that is to L_p^0 if $p \in X^0$ or to $\{p\}$ if $p \in Sing(\mathscr{F})$. Recalling that $L_p = L_p^0$ for a generic leaf, and observing that every L_p^0 is obviously parabolic, we see that $\mathscr{F}$ is a parabolic foliation. It is also not difficult to see that, in fact, $L_p = L_p^0$ for every leaf, i.e. there are no vanishing ends, and so the map

$$\Pi_{\mathscr{F}} : U_{\mathscr{F}} \to X^0$$

is everywhere holomorphic, with values in X^0. We have $U_{\mathscr{F}} = X^0 \times \mathbb{C}$ (by Theorem 8.2, which is however quite trivial in this special case), and the map $\Pi_{\mathscr{F}} : X^0 \times \mathbb{C} \to X^0$ can be identified with the restricted flow $\Phi : X^0 \times \mathbb{C} \to X^0$.

Remark 8.4. It is a general fact [Br3] that vanishing ends of a foliation $\mathscr{F}$ produce rational curves in X over which the canonical bundle $K_{\mathscr{F}}$ has negative degree. In particular, if $K_{\mathscr{F}}$ is algebraically nef (i.e. $K_{\mathscr{F}} \cdot C \geq 0$ for every compact curve $C \subset X$) then $\mathscr{F}$ has no vanishing end.

8.3 Foliations by Rational Curves

We shall say that a foliation by curves $\mathscr{F}$ is a **foliation by rational curves** if for every $p \in X^0$ there exists a rational curve $R_p \subset X$ passing through p and tangent to $\mathscr{F}$. This class of foliations should not be confused with the smaller class of rational quasi-fibrations: certainly a rational quasi-fibration is a foliation by rational curves, but the converse is in general false, because the above rational curves R_p can pass through $Sing(\mathscr{F})$ and so L_p (which is equal to R_p minus those points of $R_p \cap Sing(\mathscr{F})$ not corresponding to vanishing ends) can be parabolic or even hyperbolic. Thus the class of foliations by rational curves is transversal to our fundamental trichotomy rational quasi-fibrations / parabolic foliations / hyperbolic foliations.

A typical example is the radial foliation in the projective space $\mathbb{C}P^n$, i.e. the foliation generated (in an affine chart) by the radial vector field $\sum z_j \frac{\partial}{\partial z_j}$: it is a foliation by rational curves, but it is parabolic. By applying a birational map of the projective space, we can get also a hyperbolic foliation by rational curves.On the other hand, it is a standard exercise in bimeromorphic geometry to see that any foliation by rational curves can be transformed, by a bimeromorphic map, into a rational quasi-fibration. For instance, the radial foliation above can be transformed into a rational quasi-fibration, and even into a $\mathbb{P}$-bundle, by blowing-up the origin.

We have seen in Section 6 that the canonical bundle $K_{\mathscr{F}}$ of a hyperbolic foliation is always pseudoeffective. At the opposite side, for a rational quasi-fibration $K_{\mathscr{F}}$ is never pseudoeffective: its degree on a generic leaf (a smooth rational curve disjoint from $Sing(\mathscr{F})$) is equal to -2, and this prevents pseudoeffectivity. For parabolic foliations, the situation is mixed: the radial foliation in $\mathbb{C}P^n$ has canonical bundle equal to $\mathscr{O}(-1)$, which is not pseudoeffective; a foliation like in Example 8.3 has trivial canonical bundle, which is pseudoeffective. One can also easily find examples of parabolic foliations with ample canonical bundle, for instance most foliations arising from complete polynomial vector fields in $\mathbb{C}^n$ [Br4].

The following result, which combines Theorem 8.2 and Theorem 7.3, shows that most parabolic foliations have pseudoeffective canonical bundle.

Theorem 8.5. *Let $\mathscr{F}$ be a parabolic foliation on a compact connected Kähler manifold X. Suppose that its canonical bundle $K_{\mathscr{F}}$ is not pseudoeffective. Then $\mathscr{F}$ is a foliation by rational curves.*

Proof. Consider the meromorphic map

$$\Pi_{\mathscr{F}} : E_{\mathscr{F}}|_{X^0} \simeq U_{\mathscr{F}} \dashrightarrow X$$

given by Theorem 8.2. Because $Sing(\mathscr{F}) = X \setminus X^0$ has codimension at least two, such a map meromorphically extends [Siu] to the full space $E_{\mathscr{F}}$:

$$\Pi_{\mathscr{F}} : E_{\mathscr{F}} \dashrightarrow X.$$

The section at infinity of $E_{\mathscr{F}}$ is the same as the null section of $E^*_{\mathscr{F}}$, the total space of $K_{\mathscr{F}}$. If $K_{\mathscr{F}}$ is not pseudoeffective, then by Theorem 7.3 $\Pi_{\mathscr{F}}$ extends to the full $\overline{E_{\mathscr{F}}} = E_{\mathscr{F}} \cup \{$ section at $\infty\}$, as a meromorphic map

$$\overline{\Pi_{\mathscr{F}}} : \overline{E_{\mathscr{F}}} \dashrightarrow X.$$

By construction, $\overline{\Pi_{\mathscr{F}}}$ sends the rational fibers of $\overline{E_{\mathscr{F}}}$ to rational curves in X tangent to $\mathscr{F}$, which is therefore a foliation by rational curves. □

Note that the converse to this theorem is not always true: for instance, a parabolic foliation like in Example 8.3 has trivial (pseudoeffective) canonical bundle, yet it can be a foliation by rational curves, for some special v. A parabolic foliation is a foliation by rational curves if and only if the meromorphic map $\Pi_{\mathscr{F}} : E_{\mathscr{F}} \dashrightarrow X$ introduced in the proof above extends to the section at infinity, and this can possibly occur even if $K_{\mathscr{F}}$ is pseudoeffective, or even ample.

We have now completed our analysis of positivity properties of the canonical bundle of a foliation, and their relation to uniformisation. We may resume the various inclusions of the various classes of foliations in the diagram below.

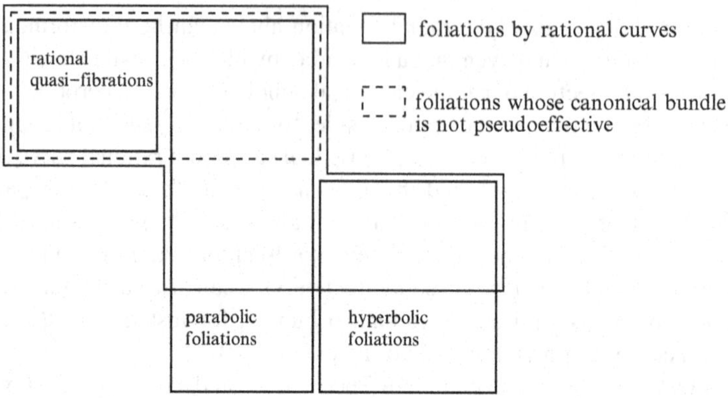

Let us discuss the classical case of fibrations.

Example 8.6. Suppose that $\mathcal{F}$ is a fibration over some base B, i.e. there exists a holomorphic map $f : X \to B$ whose generic fiber is a leaf of $\mathcal{F}$ (but there may be singular fibers, and even some higher dimensional fibers). Let g be the genus of a generic fiber, and suppose that $g \geq 1$. The relative canonical bundle of f is defined as

$$K_f = K_X \otimes f^*(K_B^{-1}).$$

It is related to the canonical bundle $K_{\mathcal{F}}$ of $\mathcal{F}$ by the relation

$$K_f = K_{\mathcal{F}} \otimes \mathcal{O}_X(D)$$

where D is an *effective* divisor which takes into account the possible ramifications of f along nongeneric fibers. Indeed, by adjunction along the leaves, we have $K_X = K_{\mathcal{F}} \otimes N^*_{\mathcal{F}}$, where $N^*_{\mathcal{F}}$ denotes the determinant conormal bundle of $\mathcal{F}$. If ω is a local generator of K_B, then $f^*(\omega)$ is a local section of $N^*_{\mathcal{F}}$ which vanishes along the ramification divisor D of f, hence $f^*(K_B) = N^*_{\mathcal{F}} \otimes \mathcal{O}_X(-D)$, whence the relation above.

Because $\mathcal{F}$ is not a foliation by rational curves, we have, by the Theorems above, that $K_{\mathcal{F}}$ is pseudoeffective, and therefore also K_f is pseudoeffective. In particular, $f_*(K_{\mathcal{F}})$ and $f_*(K_f)$ are "pseudoeffective sheaves" on B, in the sense that their degrees with respect to Kähler metrics on B are nonnegative. This must be compared with Arakelov's positivity theorem [Ara] [BPV, Ch. III]. But, as in Arakelov's results, something more can be said. Suppose that B is a curve (or restrict the fibration f over some curve in B) and let us distinguish between the hyperbolic and the parabolic case.

- $g \geq 2$. Then the pseudoeffectivity of $K_{\mathcal{F}}$ is realized by the leafwise Poincaré metric (Theorem 6.4). A subtle computation [Br2, Br1] shows that this leafwise (or fiberwise) Poincaré metric has a *strictly* plurisubharmonic variation, unless the fibration is isotrivial. This means that if f is *not* isotrivial then the degree of $f_*(K_{\mathcal{F}})$ (and, a fortiori, the degree of $f_*(K_f)$) is *strictly positive*.

- $g = 1$. We put on every smooth elliptic leaf of $\mathcal{F}$ the (unique) flat metric with total area 1. It is shown in [Br4] (using Theorem 8.2 above) that this leafwise metric extends to a metric on $K_{\mathcal{F}}$ with positive curvature. In other words, the pseudoeffectivity of $K_{\mathcal{F}}$ is realized by a leafwise flat metric. Moreover, still in [Br4] it is observed that if the fibration is not isotrivial then the curvature of such a metric on $K_{\mathcal{F}}$ is strictly positive on directions transverse to the fibration. We thus get the same conclusion as in the hyperbolic case: if f is *not* isotrivial then the degree of $f_*(K_{\mathcal{F}})$ (and, a fortiori, the degree of $f_*(K_f)$) is *strictly positive*.

Let us conclude with several remarks around the pseudoeffectivity of $K_{\mathcal{F}}$.

Remark 8.7. In the case of hyperbolic foliations, Theorem 6.4 is very efficient: not only $K_{\mathcal{F}}$ is pseudoeffective, but even this pseudoeffectivity is realized by an explicit metric, induced by the leafwise Poincaré metric. This gives further useful properties. For instance, we have seen that the polar set of the metric is filled by singularities and parabolic leaves. Hence, for example, if all the leaves are hyperbolic and the singularities are isolated, then $K_{\mathcal{F}}$ is not only pseudoeffective but even nef (numerically eventually free [Dem]). This efficiency is unfortunately lost in the case of parabolic foliations, because in Theorem 8.5 the pseudoeffectivity of $K_{\mathcal{F}}$ is obtained via a more abstract argument. In particular, we do not know how to control the polar set of the metric. See, however, [Br4] for some special cases in which a distinguished metric on $K_{\mathcal{F}}$ can be constructed even in the parabolic case, besides the case of elliptic fibrations discussed in Example 8.6 above.

Remark 8.8. According to general principles [BDP], once we know that $K_{\mathcal{F}}$ is pseudoeffective we should try to understand its discrepancy from being nef. There is on X a unique maximal countable collection of compact analytic subsets $\{Y_j\}$ such that $K_{\mathcal{F}}|_{Y_j}$ is *not* pseudoeffective. It seems reasonable to try to develop the above theory in a "relative" context, by replacing X with Y_j, and then to prove something like this: every Y_j is $\mathcal{F}$-invariant, and the restriction of $\mathcal{F}$ to Y_j is a foliation by rational curves. Note, however, that the restriction of a foliation to an invariant analytic subspace Y is a dangerous operation. Usually, we like to work with "saturated" foliations, i.e. with a singular set of codimension at least two (see, e.g., the beginning of the proof of Theorem 8.5 for the usefulness of this condition). If $Z = Sing(\mathcal{F}) \cap Y$ has codimension one in Y, this means that our "restriction of $\mathcal{F}$ to Y" is not really $\mathcal{F}|_Y$, but rather its saturation. Consequently, the canonical bundle of that restriction is not really $K_{\mathcal{F}}|_Y$, but rather $K_{\mathcal{F}}|_Y \otimes \mathcal{O}_Y(-\mathcal{Z})$, where $\mathcal{Z}$ is an effective divisor supported in Z. If $Z = Sing(\mathcal{F}) \cap Y$ has codimension zero in Y (i.e., $Y \subset Sing(\mathcal{F})$), the situation is even worst, because then there is not a really well defined notion of restriction to Y.

Remark 8.9. The previous remark is evidently related to the problem of constructing minimal models of foliations by curves, i.e. birational models (on possibly singular varieties) for which the canonical bundle is nef. In the projective context, results in this direction have been obtained by McQuillan and Bogomolov [BMQ, MQ2]. From this birational point of view, however, we rapidly meet another open and difficult problem: the resolution of the singularities of $\mathcal{F}$. A related problem is the construction of birational models for which there are no vanishing ends in the leaves, compare with Remark 8.4 above.

Remark 8.10. Finally, the pseudoeffectivity of $K_{\mathscr{F}}$ may be measured by finer invariants, like Kodaira dimension or numerical Kodaira dimension. When $\dim X = 2$ then the picture is rather clear and complete [MQ1, Br1]. When $\dim X > 2$ then almost everything seems open (see, however, the case of fibrations discussed above). Note that already in dimension two the so-called "abundance" does not hold: there are foliations (Hilbert Modular Foliations [MQ1, Br1]) whose canonical bundle is pseudoeffective, yet its Kodaira dimension is $-\infty$. The classification of these exceptional foliations was the first motivation for the plurisubharmonicity result of [Br2].

References

[AWe] Alexander, H., Wermer, J.: Several complex variables and Banach algebras. Graduate Texts in Mathematics 35. Springer Verlag, Berlin (1998)

[Ara] Ju. Arakelov, S.: Families of algebraic curves with fixed degeneracies. Izv. Akad. Nauk SSSR **35**, 1269–1293 (1971)

[BPV] Barth, W., Peters, C., Van de Ven, A.: Compact complex surfaces. Ergebnisse der Mathematik (3) 4. Springer Verlag, Berlin (1984)

[BeG] Bedford, E., Gaveau, B.: Envelopes of holomorphy of certain 2-spheres in $\mathbb{C}^2$. Am. J. Math. **105**, 975–1009 (1983)

[BeT] Bedford, E., Taylor, B.A.: The Dirichlet problem for a complex Monge-Ampère equation. Invent. Math. **37**, 1–44 (1976)

[Bis] Bishop, E.: Conditions for the analyticity of certain sets. Michigan Math. J. **11**, 289–304 (1964)

[BMQ] Bogomolov, F.A., McQuillan, M.: Rational curves on foliated varieties. Preprint IHES (2001)

[BDP] Boucksom, S., Demailly, J.-P., Paun, M., Peternell, Th.: The pseudo-effective cone of a compact Kähler manifold and varieties of negative Kodaira dimension. Preprint (2004)

[Br1] Brunella, M.: Foliations on complex projective surfaces, in Dynamical systems. Pubbl. Cent. Ric. Mat. Ennio De Giorgi, SNS Pisa, 49–77 (2003)

[Br2] Brunella, M.: Subharmonic variation of the leafwise Poincaré metric. Invent. Math. **152**, 119–148 (2003)

[Br3] Brunella, M.: Plurisubharmonic variation of the leafwise Poincaré metric. Internat. J. Math. **14**, 139–151 (2003)

[Br4] Brunella, M.: Some remarks on parabolic foliations, in Geometry and dynamics. Contemp. Math. **389**, 91–102 (2005)

[Br5] Brunella, M.: A positivity property for foliations on compact Kähler manifolds. Internat. J. Math. **17**, 35–43 (2006)

[Br6] Brunella, M.: On the plurisubharmonicity of the leafwise Poincaré metric on projective manifolds. J. Math. Kyoto Univ. **45**, 381–390 (2005)

[Br7] Brunella, M.: On entire curves tangent to a foliation. J. Math. Kyoto Univ. **47**, 717–734 (2007)

[CLN] Camacho, C., Lins Neto, A.: Geometric theory of foliations. Birkhäuser (1985)

[CaP] Campana, F., Peternell, Th.: Cycle spaces, in Several complex variables VII. Encyclopaedia Math. Sci. 74. Springer Verlag, Berlin, 319–349 (1994)

[Che] Chen, K.T.: Iterated path integrals. Bull. Amer. Math. Soc. **83**, 831–879 (1977)

[Chi] Chirka, E.M.: Complex analytic sets. Mathematics and its Applications 46, Kluwer, Dordecht (1989)

[ChI] Chirka, E.M., Ivashkovich, S.: On nonimbeddability of Hartogs figures into complex manifolds. Bull. Soc. Math. France **134**, 261–267 (2006)

[Dem] Demailly, J.-P.: L^2 vanishing theorems for positive line bundles and adjunction theory, in Transcendental methods in algebraic geometry (Cetraro, 1994), Lecture Notes in Math. 1646, 1–97 (1996)

[Din] Dingoyan, P.: Monge-Ampère currents over pseudoconcave spaces. Math. Ann. **320**, 211–238 (2001)

[For] Forstneric, F.: Polynomial hulls of sets fibered over the circle. Indiana Univ. Math. J. **37**, 869–889 (1988)

[Ghy] Ghys, E.: Laminations par surfaces de Riemann, in Dynamique et géométrie complexes. Panor. Synthèses **8**, 49–95 (1999)

[GuR] Gunning, R.C., Rossi, H.: Analytic Functions of Several Complex Variables. Prentice-Hall, Englewood Cliffs, NJ, USA (1965)

[Il1] Il'yashenko, Ju. S.: Foliations by analytic curves. Mat. Sb. (N.S.) **88**(130), 558–577 (1972)

[Il2] Il'yashenko, Ju. S.: Covering manifolds for analytic families of leaves of foliations by analytic curves. Topol. Methods Nonlinear Anal. **11**, 361–373 (1998) (addendum: 23 (2004), 377–381)

[Iv1] Ivashkovich, S.: The Hartogs-type extension theorem for meromorphic maps into compact Kähler manifolds. Invent. Math. **109**, 47–54 (1992)

[Iv2] Ivashkovich, S.: Extension properties of meromorphic mappings with values in non-Kähler complex manifolds. Ann. Math. **160**, 795–837 (2004)

[IvS] Ivashkovich, S., Shevchishin, V.: Structure of the moduli space in a neighborhood of a cusp-curve and meromorphic hulls. Invent. Math. **136**, 571–602 (1999)

[Kiz] Kizuka, T.: On the movement of the Poincaré metric with the pseudoconvex deformation of open Riemann surfaces. Ann. Acad. Sci. Fenn. **20**, 327–331 (1995)

[Kli] Klimek, M.: Pluripotential theory. London Mathematical Society Monographs 6, Oxford University Press, Oxford (1991)

[Lan] Lang, S.: Introduction to complex hyperbolic spaces. Springer Verlag, Berlin (1987)

[MQ1] McQuillan, M.: Canonical models of foliations. Pure Appl. Math. Q. **4**, 877–1012 (2008)

[MQ2] McQuillan, M.: Semi-stable reduction of foliations. Preprint IHES (2005)

[Miy] Miyaoka, Y.: Deformations of a morphism along a foliation and applications, in Algebraic geometry (Bowdoin 1985). Proc. Sympos. Pure Math. **46**, 245–268 (1987)

[Moi] Moishezon, B.G.: On n-dimensional compact varieties with n algebraically independent meromorphic functions. Am. Math. Soc. Transl. **63**, 51–177 (1967)

[Nap] Napier, T.: Convexity properties of coverings of smooth projective varieties. Math. Ann. **286**, 433–479 (1990)

[Nis] Nishino, T.: Nouvelles recherches sur les fonctions entières de plusieurs variables complexes I-V. J. Math. Kyoto Univ. **8, 9, 10, 13, 15** (1968–1975)

[Ohs] Ohsawa, T.: A note on the variation of Riemann surfaces. Nagoya Math. J. **142**, 1–4 (1996)

[Ran] Range, R.M.: Holomorphic functions and integral representations in several complex variables. Graduate Texts in Mathematics 108, Springer Verlag, Berlin (1986)

[ShB] Shepherd-Barron, N.: Miyaoka's theorems on the generic seminegativity of T_X and on the Kodaira dimension of minimal regular threefolds, in Flips and abundance for algebraic threefolds. Astérisque **211**, 103–114 (1992)

[Siu] Siu, Y.T.: Techniques of extension of analytic objects, Lecture Notes in Pure and Applied Mathematics 8. Marcel Dekker, New York (1974)

[Suz] Suzuki, M.: Sur les intégrales premières de certains feuilletages analytiques complexes, in Fonctions de plusieurs variables complexes III (Sém. Norguet 1975-77), Lecture Notes in Math. **670**, 53–79 (1978)

[Ya1] Yamaguchi, H.: Sur le mouvement des constantes de Robin. J. Math. Kyoto Univ. **15**, 53–71 (1975)

[Ya2] Yamaguchi, H.: Parabolicité d'une fonction entière. J. Math. Kyoto Univ. **16**, 71–92 (1976)

[Ya3] Yamaguchi, H.: Calcul des variations analytiques. Jpn. J. Math. (N.S.) **7**, 319–377 (1981)

Dynamics in Several Complex Variables: Endomorphisms of Projective Spaces and Polynomial-like Mappings

Tien-Cuong Dinh and Nessim Sibony

Abstract The emphasis of this introductory course is on pluripotential methods in complex dynamics in higher dimension. They are based on the compactness properties of plurisubharmonic (p.s.h.) functions and on the theory of positive closed currents. Applications of these methods are not limited to the dynamical systems that we consider here. Nervertheless, we choose to show their effectiveness and to describe the theory for two large families of maps: the endomorphisms of projective spaces and the polynomial-like mappings. The first section deals with holomorphic endomorphisms of the projective space $\mathbb{P}^k$. We establish the first properties and give several constructions for the Green currents T^p and the equilibrium measure $\mu = T^k$. The emphasis is on quantitative properties and speed of convergence. We then treat equidistribution problems. We show the existence of a proper algebraic set $\mathscr{E}$, totally invariant, i.e. $f^{-1}(\mathscr{E}) = f(\mathscr{E}) = \mathscr{E}$, such that when $a \notin \mathscr{E}$, the probability measures, equidistributed on the fibers $f^{-n}(a)$, converge towards the equilibrium measure μ, as n goes to infinity. A similar result holds for the restriction of f to invariant subvarieties. We survey the equidistribution problem when points are replaced with varieties of arbitrary dimension, and discuss the equidistribution of periodic points. We then establish ergodic properties of μ: K-mixing, exponential decay of correlations for various classes of observables, central limit theorem and large deviations theorem. We heavily use the compactness of the space $\mathrm{DSH}(\mathbb{P}^k)$ of differences of quasi-p.s.h. functions. In particular, we show that the measure μ is moderate, i.e. $\langle \mu, e^{\alpha|\varphi|}\rangle \leq c$, on bounded sets of φ in $\mathrm{DSH}(\mathbb{P}^k)$, for suitable positive constants α, c. Finally, we study the entropy, the Lyapounov exponents and the dimension of μ. The second section develops the theory of polynomial-like maps, i.e. proper

T.-C. Dinh
UPMC Univ Paris 06, UMR 7586, Institut de Mathématiques de Jussieu, F-75005 Paris, France
e-mail: dinh@math.jussieu.fr

N. Sibony
Université Paris-Sud 11, Mathématique - Bâtiment 425, 91405 Orsay, France
e-mail: nessim.sibony@math.u-psud.fr

G. Gentili et al. (eds.), *Holomorphic Dynamical Systems*,
Lecture Notes in Mathematics 1998, DOI 10.1007/978-3-642-13171-4_4,

holomorphic maps $f : U \to V$ where U, V are open subsets of $\mathbb{C}^k$ with V convex and $U \Subset V$. We introduce the dynamical degrees for such maps and construct the equilibrium measure μ of maximal entropy. Then, under a natural assumption on the dynamical degrees, we prove equidistribution properties of points and various statistical properties of the measure μ. The assumption is stable under small pertubations on the map. We also study the dimension of μ, the Lyapounov exponents and their variation. Our aim is to get a self-contained text that requires only a minimal background. In order to help the reader, an appendix gives the basics on p.s.h. functions, positive closed currents and super-potentials on projective spaces. Some exercises are proposed and an extensive bibliography is given.

Introduction

These notes are based on a series of lectures given by the authors at IHP in 2003, Luminy in 2007, Cetraro in 2008 and Bedlewo 2008. The purpose is to provide an introduction to some developments in dynamics of several complex variables. We have chosen to treat here only two sections of the theory: the dynamics of endomorphisms of the projective space $\mathbb{P}^k$ and the dynamics of polynomial-like mappings in higher dimension. Besides the basic notions and results, we describe the recent developments and the new tools introduced in the theory. These tools are useful in other settings. We tried to give a complete picture of the theory for the above families of dynamical systems. Meromorphic maps on compact Kähler manifolds, in particular polynomial automorphisms of $\mathbb{C}^k$, will be studied in a forthcoming survey.

Let us comment on how complex dynamics fits in the general theory of dynamical systems. The abstract ergodic theory is well-developed with remarkable achievements like the Oseledec-Pesin theory. It is however difficult to show in concrete examples that an invariant measure satisfies exponential decay of correlations for smooth observables or is hyperbolic, i.e. has only non-zero Lyapounov exponents, see e.g. Benedicks-Carleson [BC], Viana [V], L.S. Young [Y, Y1]. One of our goals is to show that holomorphic dynamics in several variables provides remarkable examples of non-uniformly hyperbolic systems where the abstract theory can be applied. Powerful tools from the theory of several complex variables permit to avoid delicate combinatorial estimates. Complex dynamics also require a development of new tools like the calculus on currents and the introduction of new spaces of observables, which are of independent interest.

Complex dynamics in dimension one, i.e. dynamics of rational maps on $\mathbb{P}^1$, is well-developed and has in some sense reached maturity. The main tools there are Montel's theorem on normal families, the Riemann measurable mapping theorem and the theory of quasi-conformal maps, see e.g. Beardon, Carleson-Gamelin [BE, CG]. When dealing with maps in several variables such tools are not available: the Kobayashi hyperbolicity of a manifold and the possibility to apply normal family arguments, are more difficult to check. Holomorphic maps in several variables are not conformal and there is no Riemann measurable mapping theorem.

The theory in higher dimension is developed using mostly pluripotential theory, i.e. the theory of plurisubharmonic (p.s.h. for short) functions and positive closed currents. The Montel's compactness property is replaced with the compactness properties of p.s.h. or quasi-p.s.h. functions. Another crucial tool is the use of good estimates for the dd^c-equation in various settings. One of the main ideas is: in order to study the statistical behavior of orbits of a holomorphic map, we consider its action on some appropriate functional spaces. We then decompose the action into the "harmonic" part and the "non-harmonic" one. This is done solving a dd^c-equation with estimates. The non-harmonic part of the dynamical action may be controled thanks to good estimates for the solutions of a dd^c-equation. The harmonic part can be treated using either Harnack's inequality in the local setting or the linear action of maps on cohomology groups in the case of dynamics on compact Kähler manifolds. This approach has permitted to give a satisfactory theory of the ergodic properties of holomorphic and meromorphic dynamical systems: construction of the measure of maximal entropy, decay of correlations, central limit theorem, large deviations theorem, etc. with respect to that measure.

In order to use the pluripotential methods, we are led to develop the calculus on positive closed currents. Readers not familiar with these theories may start with the appendix at the end of these notes where we have gathered some notions and results on currents and pluripotential theory. A large part in the appendix is classical but there are also some recent results, mostly on new spaces of currents and on the notion of super-potential associated to positive closed currents in higher bidegree. Since we only deal here with projective spaces and open sets in $\mathbb{C}^k$, this is easier and the background is limited.

The main problem in the dynamical study of a map is to understand the behavior of the orbits of points under the action of the map. Simple examples show that in general there is a set (Julia set) where the dynamics is unstable: the orbits may diverge exponentially. Moreover, the geometry of the Julia set is in general very wild. In order to study complex dynamical systems, we follow the classical concepts. We introduce and establish basic properties of some invariants associated to the system, like the topological entropy and the dynamical degrees which are the analogues of volume growth indicators in the real dynamical setting. These invariants give a rough classification of the system. The remarkable fact in complex dynamics is that they can be computed or estimated in many non-trivial situations.

A central question in dynamics is to construct interesting invariant measures, in particular, measures with positive entropy. Metric entropy is an indicator of the complexity of the system with respect to an invariant measure. We focus our study on the measure of maximal entropy. Its support is in some sense the most chaotic part of the system. For the maps we consider here, measures of maximal entropy are constructed using pluripotential methods. For endomorphisms in $\mathbb{P}^k$, they can be obtained as self-intersections of some invariant positive closed $(1,1)$-currents (Green currents). We give estimates on the Hausdorff dimension and on Lyapounov exponents of these measures. The results give the behavior on the most chaotic part. Lyapounov exponents are shown to be strictly positive. This means in some sense that the system is expansive in all directions, despite of the existence of a critical set.

Once, the measure of maximal entropy is constructed, we study its fine dynamical properties. Typical orbits can be observed using test functions. Under the action of the map, each observable provides a sequence of functions that can be seen as dependent random variables. The aim is to show that the dependence is weak and then to establish stochastic properties which are known for independent random variables in probability theory. Mixing, decay of correlations, central limit theorem, large deviations theorems, etc. are proved for the measure of maximal entropy. It is crucial here that the Green currents and the measures of maximal entropy are obtained using an iterative process with estimates; we can then bound the speed of convergence.

Another problem, we consider in these notes, is the equidistribution of periodic points or of preimages of points with respect to the measure of maximal entropy. For endomorphisms of $\mathbb{P}^k$, we also study the equidistribution of varieties with respect to the Green currents. Results in this direction give some informations on the rigidity of the system and also some strong ergodic properties that the Green currents or the measure of maximal entropy satisfy. The results we obtain are in spirit similar to a second main theorem in value distribution theory and should be useful in order to study the arithmetic analogues. We give complete proofs for most results, but we only survey the equidistribution of hypersurfaces and results using super-potentials, in particular, the equidistribution of subvarieties of higher codimension.

The text is organized as follows. In the first section, we study holomorphic endomorphisms of $\mathbb{P}^k$. We introduce several methods in order to construct and to study the Green currents and the Green measure, i.e. equilibrium measure or measure of maximal entropy. These methods were not originally introduced in this setting but here they are simple and very effective. The reader will find a description and the references of the earlier approach in the ten years old survey by the second author [S3]. The second section deals with a very large family of maps: polynomial-like maps. In this case, $f : U \to V$ is proper and defined on an open set U, strictly contained in a convex domain V of $\mathbb{C}^k$. Holomorphic endomorphisms of $\mathbb{P}^k$ can be lifted to a polynomial-like maps on some open set in $\mathbb{C}^{k+1}$. So, we can consider polynomial-like maps as a semi-local version of the endomorphisms studied in the first section. They can appear in the study of meromorphic maps or in the dynamics of transcendental maps. The reader will find in the end of these notes an appendix on the theory of currents and an extensive bibliography. We have given exercises, basically in each section, some of them are not straightforward.

1 Endomorphisms of Projective Spaces

In this section, we give the main results on the dynamics of holomorphic maps on the projective space $\mathbb{P}^k$. Several results are recent and some of them are new even in dimension 1. The reader will find here an introduction to methods that can be developed in other situations, in particular, in the study of meromorphic maps on arbitrary compact Kähler manifolds. The main references for this section are [BD1, BD2, DNS, DS9, DS10, FS1, S3].

1.1 Basic Properties and Examples

Let $f : \mathbb{P}^k \to \mathbb{P}^k$ be a holomorphic endomorphism. Such a map is always induced by a polynomial self-map $F = (F_0, \ldots, F_k)$ on $\mathbb{C}^{k+1}$ such that $F^{-1}(0) = \{0\}$ and the components F_i are homogeneous polynomials of the same degree $d \geq 1$. Given an endomorphism f, the associated map F is unique up to a multiplicative constant and is called *a lift* of f to $\mathbb{C}^{k+1}$. From now on, assume that f is non-invertible, i.e. the *algebraic degree* d is at least 2. Dynamics of an invertible map is simple to study. If $\pi : \mathbb{C}^{k+1} \setminus \{0\} \to \mathbb{P}^k$ is the natural projection, we have $f \circ \pi = \pi \circ F$. Therefore, dynamics of holomorphic maps on $\mathbb{P}^k$ can be deduced from the polynomial case in $\mathbb{C}^{k+1}$. We will count preimages of points, periodic points, introduce Fatou and Julia sets and give some examples.

It is easy to construct examples of holomorphic maps in $\mathbb{P}^k$. The family of homogeneous polynomial maps F of a given degree d is parametrized by a complex vector space of dimension $N_{k,d} := (k+1)(d+k)!/(d!k!)$. The maps satisfying $F^{-1}(0) = \{0\}$ define a Zariski dense open set. Therefore, the parameter space $\mathscr{H}_d(\mathbb{P}^k)$, of holomorphic endomorphisms of algebraic degree d, is a Zariski dense open set in $\mathbb{P}^{N_{k,d}-1}$, in particular, it is connected.

If $f : \mathbb{C}^k \to \mathbb{C}^k$ is a polynomial map, we can extend f to $\mathbb{P}^k$ but the extension is not always holomorphic. The extension is holomorphic when the dominant homogeneous part f^+ of f, satisfies $(f^+)^{-1}(0) = \{0\}$. Here, if d is the maximal degree in the polynomial expression of f, then f^+ is composed by the monomials of degree d in the components of f. So, it is easy to construct examples using products of one dimensional polynomials or their pertubations.

A general meromorphic map $f : \mathbb{P}^k \to \mathbb{P}^k$ of algebraic degree d is given in homogeneous coordinates by

$$f[z_0 : \cdots : z_k] = [F_0 : \cdots : F_k],$$

where the components F_i are homogeneous polynomials of degree d without common factor, except constants. The map $F := (F_0, \ldots, F_k)$ on $\mathbb{C}^{k+1}$ is still called *a lift* of f. In general, f is not defined on the analytic set $I = \{[z] \in \mathbb{P}^k, F(z) = 0\}$ which is of codimension ≥ 2 since the F_i's have no common factor. This is the *indeterminacy set of* f which is empty when f is holomorphic.

It is easy to check that if f is in $\mathscr{H}_d(\mathbb{P}^k)$ and g is in $\mathscr{H}_{d'}(\mathbb{P}^k)$, the composition $f \circ g$ belongs to $\mathscr{H}_{dd'}(\mathbb{P}^k)$. This is in general false for meromorphic maps: the algebraic degree of the composition is not necessarily equal to the product of the algebraic degrees. It is enough to consider the meromorphic involution of algebraic degree k

$$f[z_0 : \cdots : z_k] := \left[\frac{1}{z_0} : \cdots : \frac{1}{z_k}\right] = \left[\frac{z_0 \ldots z_k}{z_0} : \cdots : \frac{z_0 \ldots z_k}{z_k}\right].$$

The composition $f \circ f$ is the identity map.

We say that f is *dominant* if $f(\mathbb{P}^k \setminus I)$ contains a non-empty open set. The space of dominant meromorphic maps of algebraic degree d, is denoted by $\mathscr{M}_d(\mathbb{P}^k)$. It is also a Zariski dense open set in $\mathbb{P}^{N_{k,d}-1}$. A result by Guelfand, Kapranov and

Zelevinsky shows that $\mathscr{M}_d(\mathbb{P}^k) \setminus \mathscr{H}_d(\mathbb{P}^k)$ is an irreducible algebraic variety [GK]. We will be concerned in this section mostly with holomorphic maps. We show that they are open and their topological degree, i.e. the number of points in a generic fiber, is equal to d^k. We recall here the classical Bézout's theorem which is a central tool for the dynamics in $\mathbb{P}^k$.

Theorem 1.1 (Bézout). *Let $P_1, \ldots, P_k$ be homogeneous polynomials in $\mathbb{C}^{k+1}$ of degrees $d_1, \ldots, d_k$ respectively. Let Z denote the set of common zeros of P_i, in $\mathbb{P}^k$, i.e. the set of points $[z]$ such that $P_i(z) = 0$ for $1 \leq i \leq k$. If Z is discrete, then the number of points in Z, counted with multiplicity, is $d_1 \ldots d_k$.*

The multiplicity of a point a in Z can be defined in several ways. For instance, if U is a small neighbourhood of a and if P_i' are generic homogeneous polynomials of degrees d_i close enough to P_i, then the hypersurfaces $\{P_i' = 0\}$ in $\mathbb{P}^k$ intersect transversally. The number of points of the intersection in U does not depend on the choice of P_i' and is *the multiplicity* of a in Z.

Proposition 1.2. *Let f be an endomorphism of algebraic degree d of $\mathbb{P}^k$. Then for every a in $\mathbb{P}^k$, the fiber $f^{-1}(a)$ contains exactly d^k points, counted with multiplicity. In particular, f is open and defines a ramified covering of degree d^k.*

Proof. For the multiplicity of f and the notion of ramified covering, we refer to Appendix A.1. Let $f = [F_0 : \cdots : F_k]$ be an expression of f in homogeneous coordinates. Consider a point $a = [a_0 : \cdots : a_k]$ in $\mathbb{P}^k$. Without loss of generality, we can assume $a_0 = 1$, hence $a = [1 : a_1 : \cdots : a_k]$. The points in $f^{-1}(a)$ are the common zeros, in $\mathbb{P}^k$, of the polynomials $F_i - a_i F_0$ for $i = 1, \ldots, k$.

We have to check that the common zero set is discrete, then Bézout's theorem asserts that the cardinality of this set is equal to the product of the degrees of $F_i - a_i F_0$, i.e. to d^k. If the set were not discrete, then the common zero set of $F_i - a_i F_0$ in $\mathbb{C}^{k+1}$ is analytic of dimension ≥ 2. This implies that the set of common zeros of the F_i's, $0 \leq i \leq k$, in $\mathbb{C}^{k+1}$ is of positive dimension. This is impossible when f is holomorphic. So, f is a ramified covering of degree d^k. In particular, it is open.

Note that when f is a map in $\mathscr{M}_d(\mathbb{P}^k) \setminus \mathscr{H}_d(\mathbb{P}^k)$ with indeterminacy set I, we can prove that the generic fibers of $f : \mathbb{P}^k \setminus I \to \mathbb{P}^k$ contains at most $d^k - 1$ points. Indeed, for every a, the hypersurfaces $\{F_i - a_i F_0 = 0\}$ in $\mathbb{P}^k$ contain I. □

Periodic points of order n, i.e. points which satisfy $f^n(z) = z$, play an important role in dynamics. Here, $f^n := f \circ \cdots \circ f$, n times, is *the iterate of order n* of f. Periodic points of order n of f are fixed points of f^n which is an endomorphism of algebraic degree d^n. In the present case, their counting is simple. We have the following result.

Proposition 1.3. *Let f be an endomorphism of algebraic degree $d \geq 2$ in $\mathbb{P}^k$. Then the number of fixed points of f, counted with multiplicity, is equal to $(d^{k+1} - 1)/(d - 1)$. In particular, the number of periodic points of order n of f is $d^{kn} + o(d^{kn})$.*

Proof. There are several methods to count the periodic points. In $\mathbb{P}^{k+1}$, with homogeneous coordinates $[z:t] = [z_0 : \cdots : z_k : t]$, we consider the system of equations $F_i(z) - t^{d-1}z_i = 0$. The set is discrete since it is analytic and does not intersect the hyperplane $\{t = 0\}$. So, we can count the solutions of the above system using Bézout's theorem and we find d^{k+1} points counting with multiplicity. Each point $[z:t]$ in this set, except $[0 : \cdots : 0 : 1]$, corresponds to a fixed point $[z]$ of f. The correspondence is $d-1$ to 1. Indeed, if we multiply t by a $(d-1)$-th root of unity, we get the same fixed point. Hence, the number of fixed points of f counted with multiplicity is $(d^{k+1}-1)/(d-1)$.

The number of fixed points of f is also the number of points in the intersection of the graph of f with the diagonal of $\mathbb{P}^k \times \mathbb{P}^k$. So, we can count these points using the cohomology classes associated to the above analytic sets, i.e. using the Lefschetz fixed point formula, see [GH]. We can also observe that this number depends continuously on f. So, it is constant for f in $\mathscr{H}_d(\mathbb{P}^k)$ which is connected. We obtain the result by counting the fixed points of an example, e.g. for $f[z] = [z_0^d : \cdots : z_k^d]$. □

Note that the periodic points of period n are isolated. If p is such a point, a theorem due to Shub-Sullivan [KH, p.323] implies that the multiplicity at p of the equation $f^{mn}(p) = p$ is bounded independently on m. The result holds for $\mathscr{C}^1$ maps. We deduce from the above result that f admits infinitely many distinct periodic points.

The set of fixed points of a meromorphic map could be empty or infinite. One checks easily that the map $(z_1, z_2) \mapsto (z_1^2, z_2)$ in $\mathbb{C}^2$ admits $\{z_1 = 0\}$ as a curve of fixed points.

Example 1.4. Consider the following map:

$$f(z_1, z_2) := (z_1 + 1, P(z_1, z_2)),$$

where P is a homogeneous polynomial of degree $d \geq 2$ such that $P(0,1) = 0$. It is clear that f has no periodic point in $\mathbb{C}^2$. The meromorphic extension of f is given in homogeneous coordinates $[z_0 : z_1 : z_2]$ by

$$f[z] = [z_0^d : z_0^{d-1}z_1 + z_0^d : P(z_1, z_2)].$$

Here, $\mathbb{C}^2$ is identified to the open set $\{z_0 = 1\}$ of $\mathbb{P}^2$. The indeterminacy set I of f is defined by $z_0 = P(z_1, z_2) = 0$ and is contained in the line at infinity $L_\infty := \{z_0 = 0\}$. We have $f(L_\infty \setminus I) = [0:0:1]$ which is an indeterminacy point. So, $f : \mathbb{P}^2 \setminus I \to \mathbb{P}^2$ has no periodic point.

Example 1.5. Consider the holomorphic map f on $\mathbb{P}^2$ given by

$$f[z] := [z_0^d + P(z_1, z_2) : z_2^d + \lambda z_0^{d-1} z_1 : z_1^d],$$

with P homogeneous of degree $d \geq 2$. Let $p := [1:0:0]$, then $f^{-1}(p) = p$. Such a point is called *totally invariant*. In general, p is not necessarily an attractive point. Indeed, the eigenvalues of the differential of f at p are 0 and λ. When $|\lambda| > 1$, there is an expansive direction for f in a neighbourhood of p. In dimension one, totally invariant points are always attractive.

For a holomorphic map f on $\mathbb{P}^k$, a point a in $\mathbb{P}^k$ is *critical* if f is not injective in a neighbourhood of a or equivalently the multiplicity of f at a in the fiber $f^{-1}(f(a))$ is strictly larger than 1, see Theorem A.3. We say that a is a critical point of *multiplicity* m if the multiplicity of f at a in the fiber $f^{-1}(f(a))$ is equal to $m+1$.

Proposition 1.6. *Let f be a holomorphic endomorphism of algebraic degree $d\geq 2$ of $\mathbb{P}^k$. Then, the critical set of f is an algebraic hypersurface of degree $(k+1)(d-1)$ counted with multiplicity.*

Proof. If F is a lift of f to $\mathbb{C}^{k+1}$, the Jacobian $\mathrm{Jac}(F)$ is a homogeneous polynomial of degree $(k+1)(d-1)$. The zero set of $\mathrm{Jac}(F)$ in $\mathbb{P}^k$ is exactly the critical set of f. The result follows. □

Let $\mathscr{C}$ denote the critical set of f. The orbit $\mathscr{C}, f(\mathscr{C}), f^2(\mathscr{C}),\ldots$ is either a hypersurface or a countable union of hypersurfaces. We say that f is *postcritically finite* if this orbit is a hypersurface, i.e. has only finitely many irreducible components. Besides very simple examples, postcritically finite maps are difficult to construct, because the image of a variety is often a variety of larger degree; so we have to increase the multiplicity in order to get only finitely many irreducible components. We give few examples of postcritically finite maps, see [FS, FS7].

Example 1.7. We can check that for $d\geq 2$ and $(1-2\lambda)^d=1$

$$f[z_0:\cdots:z_k] := [z_0^d:\lambda(z_0-2z_1)^d:\cdots:\lambda(z_0-2z_k)^d]$$

is postcritically finite. For some parameters $\alpha\in\mathbb{C}$ and $0\leq l\leq d$, the map

$$f_\alpha[z] := [z_0^d:z_1^d:z_2^d+\alpha z_1^{d-l}z_2^l]$$

is also postcritically finite. In particular, for $f_0[z]=[z_0^d:z_1^d:z_2^d]$, the associated critical set is equal to $\{z_0z_1z_2=0\}$ which is invariant under f_0. So, f_0 is postcritically finite.

Arguing as above, using Bézout's theorem, we can prove that if Y is an analytic set of pure codimension p in $\mathbb{P}^k$ then $f^{-1}(Y)$ is an analytic set of pure codimension p. Its degree, counting with multiplicity, is equal to $d^p\deg(Y)$. Recall that the degree $\deg(Y)$ of Y is the number of points in the intersection of Y with a generic projective subspace of dimension p. We deduce that the pull-back operator f^* on the Hodge cohomology group $H^{p,p}(\mathbb{P}^k,\mathbb{C})$ is simply a multiplication by d^p. Since f is a ramified covering of degree d^k, $f_*\circ f^*$ is the multiplication by d^k. Therefore, the push-forward operator f_* acting on $H^{p,p}(\mathbb{P}^k,\mathbb{C})$ is the multiplication by d^{k-p}. In particular, the image $f(Y)$ of Y by f is an analytic set of pure codimension p and of degree $d^{k-p}\deg(Y)$, counted with multiplicity.

We now introduce the Fatou and Julia sets associated to an endomorphism. The following definition is analogous to the one variable case.

Definition 1.8. *The Fatou set* of f is the largest open set $\mathscr{F}$ in $\mathbb{P}^k$ where the sequence of iterates $(f^n)_{n\geq 1}$ is locally equicontinuous. The complement $\mathscr{J}$ of $\mathscr{F}$ is called *the Julia set* of f.

Fatou and Julia sets are totally invariant by $\mathscr{F}$, that is, $f^{-1}(\mathscr{F}) = f(\mathscr{F}) = \mathscr{F}$ and the same property holds for $\mathscr{J}$. Julia and Fatou sets associated to f^n are also equal to $\mathscr{J}$ and $\mathscr{F}$. We see here that the space $\mathbb{P}^k$ is divided into two parts: on $\mathscr{F}$ the dynamics is stable and tame while the dynamics on $\mathscr{J}$ is a priori chaotic. If x is a point in $\mathscr{F}$ and y is close enough to x, the orbit of y is close to the orbit of x when the time n goes to infinity. On the Julia set, this property is not true. Attractive fixed points and their basins are examples of points in the Fatou set. Siegel domains, i.e. invariant domains on which f is conjugated to a rotation, are also in the Fatou set. Repelling periodic points are always in the Julia set. Another important notion in dynamics is the non-wandering set.

Definition 1.9. A point a in $\mathbb{P}^k$ is *non-wandering* with respect to f if for every neighbourhood U of a, there is an $n \geq 1$ such that $f^n(U) \cap U \neq \varnothing$.

The study of the Julia and Fatou sets is a fundamental problem in dynamics. It is quite well-understood in the one variable case where the Riemann measurable theorem is a basic tool. The help of computers is also important there. In higher dimension, Riemann measurable theorem is not valid and the use of computers is more delicate. The most important tool in higher dimension is pluripotential theory.

For instance, Fatou and Julia sets for a general map are far from being understood. Many fundamental questions are still open. We do not know if wandering Fatou components exist in higher dimension. In dimension one, a theorem due to Sullivan [SU] says that such a domain does not exist. The classification of Fatou components is not known, see [FS8] for a partial answer in dimension 2 and [FS, S3, U] for the case of postcritically finite maps. The reader will find in the survey [S3] some results on local dynamics near a fixed point, related to the Fatou-Julia theory. We now give few examples.

The following construction is due to Ueda [U]. It is useful in order to obtain interesting examples, in particular, to show that some properties hold for generic maps. The strategy is to check that the set of maps satisfying these properties is a Zariski open set in the space of parameters and then to produce an example using Ueda's construction.

Example 1.10. Let $h : \mathbb{P}^1 \to \mathbb{P}^1$ be a rational map of degree $d \geq 2$. Consider the multi-projective space $\mathbb{P}^1 \times \cdots \times \mathbb{P}^1$, k times. The permutations of coordinates define a finite group Γ acting on this space and the quotient of $\mathbb{P}^1 \times \cdots \times \mathbb{P}^1$ by Γ is equal to $\mathbb{P}^k$. Let $\Pi : \mathbb{P}^1 \times \cdots \times \mathbb{P}^1 \to \mathbb{P}^k$ denote the canonical projection. Let $\widetilde{f}$ be the endomorphism of $\mathbb{P}^1 \times \cdots \times \mathbb{P}^1$ defined by $\widetilde{f}(z_1, \ldots, z_k) := (h(z_1), \ldots, h(z_k))$. If σ is a permutation of coordinates $(z_1, \ldots, z_k)$, then $\sigma \circ \widetilde{f} = \widetilde{f} \circ \sigma$. It is not difficult to deduce that there is an endomorphism f on $\mathbb{P}^k$ of algebraic degree d semi-conjugated to $\widetilde{f}$, that is, $f \circ \Pi = \Pi \circ \widetilde{f}$. One can deduce dynamical properties of f from properties of h. For example, if h is chaotic, i.e. has a dense orbit, then f is also chaotic. The first chaotic maps on $\mathbb{P}^1$ were constructed by Lattès. Ueda's construction gives Lattès maps in higher dimension. A *Lattès map* f on $\mathbb{P}^k$ is a map semi-conjugated to an affine map on a torus. More precisely, there is an open holomorphic map $\Psi : \mathbb{T} \to \mathbb{P}^k$ from a k-dimensional torus $\mathbb{T}$ onto $\mathbb{P}^k$ and an affine map $A : \mathbb{T} \to \mathbb{T}$ such that $f \circ \Psi = \Psi \circ A$. We refer to [BR, BL, D4, DS0, MIL] for a discussion of Lattès maps.

The following map is the simplest in our context. Its iterates can be explicitely computed. The reader may use this map and its pertubations as basic examples in order to get a picture on the objects we will introduce latter.

Example 1.11. Let $f : \mathbb{C}^k \to \mathbb{C}^k$ be the polynomial map defined by

$$f(z_1,\ldots,z_k) := (z_1^d,\ldots,z_k^d), \quad d \geq 2.$$

We can extend f holomorphically to $\mathbb{P}^k$. Let $[z_0 : \cdots : z_k]$ denote the homogeneous coordinates on $\mathbb{P}^k$ such that $\mathbb{C}^k$ is identified to the chart $\{z_0 \neq 0\}$. Then, the extension of f to $\mathbb{P}^k$ is

$$f[z_0 : \cdots : z_k] = [z_0^d : \cdots : z_k^d].$$

The Fatou set is the union of the basins of the $k+1$ attractive fixed points $[0 : \cdots : 0 : 1 : 0 : \cdots : 0]$. These components are defined by

$$\mathscr{F}_i := \{z \in \mathbb{P}^k, \quad |z_j| < |z_i| \quad \text{for every } j \neq i\}.$$

The Julia set of f is the union of the following sets $\mathscr{J}_{ij}$ with $0 \leq i < j \leq k$, where

$$\mathscr{J}_{ij} := \{z \in \mathbb{P}^k, \quad |z_i| = |z_j| \quad \text{and} \quad |z_l| \leq |z_i| \quad \text{for every } l\}.$$

We have $f^n(z) = (z_1^{d^n},\ldots,z_k^{d^n})$ for $n \geq 1$.

Exercise 1.12. Let $h : \mathbb{P}^1 \to \mathbb{P}^1$ be a rational map. Discuss Fatou components for the associated map f defined in Example 1.10. Prove in particular that there exist Fatou components which are bi-holomorphic to a disc cross an annulus. Describe the set of non-wandering points of f.

Exercise 1.13. Let a be a fixed point of f. Show that the eigenvalues of the differential Df of f at a do not depend on the local coordinates. Assume that a is in the Fatou set. Show that these eigenvalues are of modulus ≤ 1. If all the eigenvalues are of modulus 1, show that $Df(a)$ is diagonalizable.

Exercise 1.14. Let f be a Lattès map associated to an affine map A as in Example 1.10. Show that f is postcritically finite. Show that $d^{-1/2}DA$ is an unitary matrix where DA is the differential of A. Deduce that the orbit of a is dense in $\mathbb{P}^k$ for almost every a in $\mathbb{P}^k$. Show that the periodic points of f are dense in $\mathbb{P}^k$.

Exercise 1.15. Let $f : \mathbb{P}^k \to \mathbb{P}^k$ be a dominant meromorphic map. Let I be the indeterminacy set of f, defined as above. Show that f cannot be extended to a holomorphic map on any open set which intersects I.

1.2 Green Currents and Julia Sets

Let f be an endomorphism of algebraic degree $d \geq 2$ as above. In this paragraph, we give the first construction of canonical invariant currents T^p associated to f (Green currents). The construction is now classical and is used in most of the references, see [FS1, FS3, HJ, S3]. We will show that the support of the Green $(1,1)$-current is exactly the Julia set of f [FS1]. In some examples, Green currents describe the distribution of stable varieties but in general their geometric structure is not yet well-understood. We will see later that $\mu := T^k$ is the invariant measure of maximal entropy.

Theorem 1.16. *Let S be a positive closed $(1,1)$-current of mass 1 on $\mathbb{P}^k$. Assume that S has bounded local potentials. Then $d^{-n}(f^n)^*(S)$ converge weakly to a positive closed $(1,1)$-current T of mass 1. This current has continuous local potentials and does not depend on S. Moreover, it is totally invariant: $f^*(T) = dT$ and $f_*(T) = d^{k-1}T$. We also have for a smooth $(k-1,k-1)$-form Φ*

$$\left|\langle d^{-n}(f^n)^*(S) - T, \Phi\rangle\right| \leq cd^{-n}\|\Phi\|_{\text{DSH}},$$

where $c > 0$ is a constant independent of Φ and of n.

Proof. We refer to Appendix for the basic properties of quasi-p.s.h. functions, positive closed currents and DSH currents. Since S has mass 1, it is cohomologous to ω_{FS}. Therefore, we can write $S = \omega_{\text{FS}} + dd^c u$ where u is a quasi-p.s.h. function. By hypothesis, this function is bounded. The current $d^{-1}f^*(\omega_{\text{FS}})$ is smooth and of mass 1 since $f^* : H^{1,1}(\mathbb{P}^k, \mathbb{C}) \to H^{1,1}(\mathbb{P}^k, \mathbb{C})$ is the multiplication by d and the mass of a positive closed current can be computed cohomologically. So, we can also write $d^{-1}f^*(\omega_{\text{FS}}) = \omega_{\text{FS}} + dd^c v$ where v is a quasi-p.s.h. function. Here, v is smooth since ω_{FS} and $f^*(\omega_{\text{FS}})$ are smooth. We have

$$\begin{aligned} d^{-1}f^*(S) &= d^{-1}f^*(\omega_{\text{FS}}) + dd^c(d^{-1}u \circ f) \\ &= \omega_{\text{FS}} + dd^c v + dd^c(d^{-1}u \circ f). \end{aligned}$$

By induction, we obtain

$$d^{-n}(f^n)^*(S) = \omega_{\text{FS}} + dd^c\big(v + \cdots + d^{-n+1}v \circ f^{n-1}\big) + dd^c\big(d^{-n}u \circ f^n\big).$$

Observe that, since v is smooth, the sequence of smooth functions $v + \cdots + d^{-n+1}v \circ f^{n-1}$ converges uniformly to a continuous function g. Since u is bounded, the functions $d^{-n}u \circ f^n$ tend to 0. It follows that $d^{-n}(f^n)^*(S)$ converge weakly to a current T which satisfies

$$T = \omega_{\text{FS}} + dd^c \mathrm{g}.$$

From the definition, we also have

$$d^{-1}\mathrm{g} \circ f + v = \mathrm{g}.$$

Clearly, this current does not depend on S since g does not depend on S. Moreover, the currents $d^{-n}(f^n)^*(S)$ are positive closed of mass 1. So, T is also a positive closed current of mass 1. We deduce that g is quasi-p.s.h. since it is continuous and satisfies $dd^c\mathrm{g} \geq -\omega_{\text{FS}}$.

Applying the above computation to T instead of S, we obtain that

$$d^{-1}f^*(T) = \omega_{\rm FS} + dd^c v + dd^c(d^{-1}g\circ f) = \omega_{\rm FS} + dd^c g.$$

Hence, $f^*(T) = dT$. On smooth forms $f_*\circ f^*$ is equal to d^k times the identity; this holds by continuity for positive closed currents. Therefore,

$$f_*(T) = f_*(f^*(d^{-1}T)) = d^{k-1}T.$$

It remains to prove the estimate in the theorem. Recall that we can write $dd^c\Phi = R^+ - R^-$ where $R^\pm$ are positive measures such that $\|R^\pm\| \leq \|\Phi\|_{\rm DSH}$. We have

$$\begin{aligned}\big|\langle d^{-n}(f^n)^*(S) - T, \Phi\rangle\big| &= \big|\langle dd^c(v + \cdots + d^{-n+1}v\circ f^{n-1} + d^{-n}u\circ f^n - g), \Phi\rangle\big| \\ &= \big|\langle v + \cdots + d^{-n+1}v\circ f^{n-1} + d^{-n}u\circ f^n - g, dd^c\Phi\rangle\big| \\ &= \Big|\Big\langle d^{-n}u\circ f^n - \sum_{i\geq n} d^{-i}v\circ f^i, R^+ - R^-\Big\rangle\Big|.\end{aligned}$$

Since u and v are bounded, the mass estimate for $R^\pm$ implies that the last integral is $\lesssim d^{-n}\|\Phi\|_{\rm DSH}$. The result follows. □

Theorem 1.16 gives a convergence result for S quite diffuse (with bounded potentials). It is like the first main theorem in value distribution theory. The question that we will address is the convergence for singular S, e.g. hypersurfaces.

Definition 1.17. We call T *the Green* $(1,1)$*-current* and g *the Green function* of f. The power $T^p := T\wedge \ldots \wedge T$, p factors, is *the Green* (p,p)*-current* of f, and its support $\mathscr{J}_p$ is called *the Julia set of order* p.

Note that the Green function is defined up to an additive constant and since T has a continuous quasi-potential, T^p is well-defined. Green currents are totally invariant: we have $f^*(T^p) = d^pT^p$ and $f_*(T^p) = d^{k-p}T^p$. The Green (k,k)-current $\mu := T^k$ is also called *the Green measure, the equilibrium measure or the measure of maximal entropy*. We will give in the next paragraphs results which justify the terminologies. The iterates f^n, $n\geq 1$, have the same Green currents and Green function. We have the following result.

Proposition 1.18. *The local potentials of the Green current T are γ-Hölder continuous for every γ such that $0<\gamma<\min(1,\log d/\log d_\infty)$, where $d_\infty := \lim \|Df^n\|_\infty^{1/n}$. In particular, the Hausdorff dimension of T^p is strictly larger than $2(k-p)$ and T^p has no mass on pluripolar sets and on proper analytic sets of $\mathbb{P}^k$.*

Since $Df^{n+m}(x) = Df^m(f^n(x))\circ Df^n(x)$, it is not difficult to check that the sequence $\|Df^n\|_\infty^{1/n}$ is decreasing. So, $d_\infty = \inf \|Df^n\|_\infty^{1/n}$. The last assertion of the proposition is deduced from Corollary A.32 and Proposition A.33 in Appendix. The first assertion is equivalent to the Hölder continuity of the Green function g, it was obtained by Sibony [SI] for one variable polynomials and by Briend [BJ] and Kosek [KO] in higher dimension.

The following lemma, due to Dinh-Sibony [DS4, DS10], implies the above proposition and can be applied in a more general setting. Here, we apply it to $\Lambda := f^m$ with m large enough and to the above smooth function v. We choose $\alpha := 1, A := \|Df^m\|_\infty$ and d is replaced with d^m.

Lemma 1.19. *Let K be a metric space with finite diameter and $\Lambda : K \to K$ be a Lipschitz map: $\|\Lambda(a) - \Lambda(b)\| \leq A\|a-b\|$ with $A > 0$. Here, $\|a-b\|$ denotes the distance between two points a, b in K. Let v be an α-Hölder continuous function on K with $0 < \alpha \leq 1$. Then, $\sum_{n\geq 0} d^{-n} v \circ \Lambda^n$ converges pointwise to a function which is β-Hölder continuous on K for every β such that $0 < \beta < \alpha$ and $\beta \leq \log d/\log A$.*

Proof. By hypothesis, there is a constant $A' > 0$ such that $|v(a) - v(b)| \leq A'\|a-b\|^\alpha$. Define $A'' := \|v\|_\infty$. Since K has finite diameter, A'' is finite and we only have to consider the case where $\|a-b\| \ll 1$. If N is an integer, we have

$$\begin{aligned}
&\Big|\sum_{n\geq 0} d^{-n} v\circ\Lambda^n(a) - \sum_{n\geq 0} d^{-n} v\circ\Lambda^n(b)\Big| \\
&\quad\leq \sum_{0\leq n\leq N} d^{-n}|v\circ\Lambda^n(a) - v\circ\Lambda^n(b)| + \sum_{n>N} d^{-n}|v\circ\Lambda^n(a) - v\circ\Lambda^n(b)| \\
&\quad\leq A' \sum_{0\leq n\leq N} d^{-n}\|\Lambda^n(a) - \Lambda^n(b)\|^\alpha + 2A'' \sum_{n>N} d^{-n} \\
&\quad\lesssim \|a-b\|^\alpha \sum_{0\leq n\leq N} d^{-n}A^{n\alpha} + d^{-N}.
\end{aligned}$$

If $A^\alpha \leq d$, the last sum is of order at most equal to $N\|a-b\|^\alpha + d^{-N}$. For a given $0 < \beta < \alpha$, choose $N \simeq -\beta \log\|a-b\|/\log d$. So, the last expression is $\lesssim \|a-b\|^\beta$. In this case, the function is β-Hölder continuous for every $0 < \beta < \alpha$. When $A^\alpha > d$, the sum is $\lesssim d^{-N}A^{N\alpha}\|a-b\|^\alpha + d^{-N}$. For $N \simeq -\log\|a-b\|/\log A$, the last expression is $\lesssim \|a-b\|^\beta$ with $\beta := \log d/\log A$. Therefore, the function is β-Hölder continuous. □

Remark 1.20. Lemma 1.19 still holds for K with infinite diameter if v is Hölder continuous and bounded. We can also replace the distance on K with any positive symmetric function on $K \times K$ which vanishes on the diagonal. Consider a family (f_s) of endomorphisms of $\mathbb{P}^k$ depending holomorphically on s in a space of parameters Σ. In the above construction of the Green current, we can locally on Σ, choose $v_s(z)$ smooth such that $dd^c_{s,z} v_s(z) \geq -\omega_{\rm FS}(z)$. Lemma 1.19 implies that the Green function $g_s(z)$ of f_s is locally Hölder continuous on (s,z) in $\Sigma \times \mathbb{P}^k$. Then, $\omega_{\rm FS}(z) + dd^c_{s,z} g_s(z)$ is a positive closed $(1,1)$-current on $\Sigma \times \mathbb{P}^k$. Its slices by $\{s\} \times \mathbb{P}^k$ are the Green currents T_s of f_s.

We want to use the properties of the Green currents in order to establish some properties of the Fatou and Julia sets. We will show that the Julia set coincides with the Julia set of order 1. We recall the notion of Kobayashi hyperbolicity on a complex manifold M. Let p be a point in M and ξ a tangent vector of M at p. Consider the holomorphic maps $\tau : \Delta \to M$ on the unit disc Δ in $\mathbb{C}$ such that

$\tau(0) = p$ and $D\tau(0) = c\xi$ where $D\tau$ is the differential of τ and c is a constant. *The Kobayashi-Royden pseudo-metric* is defined by

$$K_M(p,\xi) := \inf_\tau |c|^{-1}.$$

It measures the size of a disc that one can send in M. In particular, if M contains an image of $\mathbb{C}$ passing through p in the direction ξ, we have $K_M(p,\xi) = 0$.

Kobayashi-Royden pseudo-metric is contracting for holomorphic maps: if $\Psi : N \to M$ is a holomorphic map between complex manifolds, we have

$$K_M(\Psi(p), D\Psi(p)\cdot\xi) \le K_N(p,\xi).$$

The Kobayashi-Royden pseudo-metric on Δ coincides with the Poincaré metric. A complex manifold M is *Kobayashi hyperbolic* if K_M is a metric [K]. In which case, holomorphic self-maps of M, form a locally equicontinuous family of maps. We have the following result where the norm of ξ is with respect to a smooth metric on X.

Proposition 1.21. *Let M be a relatively compact open set of a compact complex manifold X. Assume that there is a bounded function ρ on M which is strictly p.s.h., i.e. $dd^c\rho \ge \omega$ on M for some positive Hermitian form ω on X. Then M is Kobayashi hyperbolic and hyperbolically embedded in X. More precisely, there is a constant $\lambda > 0$ such that $K_M(p,\xi) \ge \lambda\|\xi\|$ for every $p \in M$ and every tangent vector ξ of M at p.*

Proof. If not, we can find holomorphic discs $\tau_n : \Delta \to M$ such that $\|D\tau_n(0)\| \ge n$ for $n \ge 1$. So, this family is not equicontinuous. A lemma due to Brody [K] says that, after reparametrization, there is a subsequence converging to an image of $\mathbb{C}$ in $\overline{M}$. More precisely, up to extracting a subsequence, there are holomorphic maps $\Psi_n : \Delta_n \to \Delta$ on discs Δ_n centered at 0, of radius n, such that $\tau_n \circ \Psi_n$ converge locally uniformly to a non-constant map $\tau_\infty : \mathbb{C} \to \overline{M}$. Since ρ is bounded, up to extracting a subsequence, the subharmonic functions $\rho_n := \rho \circ \tau_n \circ \Psi_n$ converge in $L^1_{loc}(\mathbb{C})$ to some subharmonic function ρ_∞. Since the function ρ_∞ is bounded, it should be constant.

For simplicity, we use here the metric on X induced by ω. Let L, K be arbitrary compact subsets of $\mathbb{C}$ such that $L \Subset K$. For n large enough, the area of $\tau_n(\Psi_n(L))$ counted with multiplicity, satisfies

$$\text{area}(\tau_n(\Psi_n(L))) = \int_L (\tau_n \circ \Psi_n)^*(\omega) \le \int_L dd^c\rho_n.$$

We deduce that

$$\text{area}(\tau_\infty(L)) = \lim_{n\to\infty} \text{area}(\tau_n(\Psi_n(L))) \le \int_K dd^c\rho_\infty = 0.$$

This is a contradiction. □

The following result was obtained by Fornæss-Sibony in [FS3, FS2] and by Ueda for the assertion on the Kobayashi hyperbolicity of the Fatou set [U2].

Theorem 1.22. *Let f be an endomorphism of algebraic degree $d \geq 2$ of $\mathbb{P}^k$. Then, the Julia set of order* 1 *of f, i.e. the support $\mathscr{J}_1$ of the Green* $(1,1)$*-current T, coincides with the Julia set $\mathscr{J}$. The Fatou set $\mathscr{F}$ is Kobayashi hyperbolic and hyperbolically embedded in $\mathbb{P}^k$. Moreover, for $p \leq k/2$, the Julia set of order p of f is connected.*

Proof. The sequence (f^n) is equicontinuous on the Fatou set $\mathscr{F}$ and f^n are holomorphic, hence the differential Df^n are locally uniformly bounded on $\mathscr{F}$. Therefore, $(f^n)^*(\omega_{\rm FS})$ are locally uniformly bounded on $\mathscr{F}$. We deduce that $d^{-n}(f^n)^*(\omega_{\rm FS})$ converge to 0 on $\mathscr{F}$. Hence, T is supported on the Julia set $\mathscr{J}$.

Let $\mathscr{F}'$ denote the complement of the support of T in $\mathbb{P}^k$. Observe that $\mathscr{F}'$ is invariant under f^n and that $-g$ is a smooth function which is strictly p.s.h. on $\mathscr{F}'$. Therefore, by Proposition 1.21, $\mathscr{F}'$ is Kobayashi hyperbolic and hyperbolically embedded in $\mathbb{P}^k$. Therefore, the maps f^n, which are self-maps of $\mathscr{F}'$, are equicontinuous with respect to the Kobayashi-Royden metric. On the other hand, the fact that $\mathscr{F}'$ is hyperbolically embedded implies that the Kobayashi-Royden metric is bounded from below by a constant times the Fubini-Study metric. It follows that (f^n) is locally equicontinuous on $\mathscr{F}'$ with respect to the Fubini-Study metric. We conclude that $\mathscr{F}' \subset \mathscr{F}$, hence $\mathscr{F} = \mathscr{F}'$ and $\mathscr{J} = \text{supp}(T) = \mathscr{J}_1$.

In order to show that $\mathscr{J}_p$ are connected, it is enough to prove that if S is a positive closed current of bidegree (p,p) with $p \leq k/2$ then the support of S is connected. Assume that the support of S is not connected, then we can write $S = S_1 + S_2$ with S_1 and S_2 non-zero, positive closed with disjoint supports. Using a convolution on the automorphism group of $\mathbb{P}^k$, we can construct smooth positive closed (p,p)-forms S_1', S_2' with disjoint supports. So, we have $S_1' \wedge S_2' = 0$. This contradicts that the cup-product of the classes $[S_1']$ and $[S_2']$ is non-zero in $H^{2p,2p}(\mathbb{P}^k, \mathbb{R}) \simeq \mathbb{R}$: we have $[S_1'] = \|S_1'\|[\omega_{\rm FS}^p]$, $[S_2'] = \|S_2'\|[\omega_{\rm FS}^p]$ and $[S_1'] \smile [S_2'] = \|S_1'\|\|S_2'\|[\omega_{\rm FS}^{2p}]$, a contradiction. Therefore, the support of S is connected. □

Example 1.23. Let f be a polynomial map of algebraic degree $d \geq 2$ on $\mathbb{C}^k$ which extends holomorphically to $\mathbb{P}^k$. If B is a ball large enough centered at 0, then $f^{-1}(B) \Subset B$. Define $G_n := d^{-n}\log^+ \|f^n\|$, where $\log^+ := \max(\log, 0)$. As in Theorem 1.16, we can show that G_n converge uniformly to a continuous p.s.h. function G such that $G \circ f = dG$. On $\mathbb{C}^k$, the Green current T of f is equal to dd^cG and $T^p = (dd^cG)^p$. The Green measure is equal to $(dd^cG)^k$. If $\mathscr{K}$ denotes the set of points in $\mathbb{C}^k$ with bounded orbit, then μ is supported on $\mathscr{K}$. Indeed, outside $\mathscr{K}$ we have $G = \lim d^{-n}\log\|f^n\|$ and the convergence is locally uniform. It follows that $(dd^cG)^k = \lim d^{-kn}(dd^c\log\|f^n\|)^k$ on $\mathbb{C}^k \setminus \mathscr{K}$. One easily check that $(dd^c\log\|f^n\|)^k = 0$ out of $f^{-n}(0)$. Therefore, $(dd^cG)^k = 0$ on $\mathbb{C}^k \setminus \mathscr{K}$. The set $\mathscr{K}$ is called *the filled Julia set*. We can show that $\mathscr{K}$ is the zero set of G. In particular, if $f(z) = (z_1^d, \ldots, z_k^d)$, then $G(z) = \sup_i \log^+|z_i|$. One can check that the support of T^p is foliated (except for a set of zero measure with respect to the trace of T^p) by stable manifolds of dimension $k-p$ and that $\mu = T^k$ is the Lebesgue measure on the torus $\{|z_i| = 1,\ i = 1, \ldots, k\}$.

Example 1.24. We consider Example 1.10. Let ν be the Green measure of h on $\mathbb{P}^1$, i.e. $\nu = \lim d^{-n}(h^n)^*(\omega_{\rm FS})$. Here, $\omega_{\rm FS}$ denotes also the Fubini-Study form on $\mathbb{P}^1$.

Let π_i denote the projections of $\mathbb{P}^1 \times \cdots \times \mathbb{P}^1$ on the factors. Then, the Green current of f is equal to

$$T = \frac{1}{k!}\pi_*\big(\pi_1^*(v) + \cdots + \pi_k^*(v)\big),$$

as can be easily checked.

Example 1.25. The following family of maps on $\mathbb{P}^2$ was studied in [FS6]:

$$f[z_0 : z_1 : z_2] := [z_0^d + \lambda z_0 z_1^{d-1}, v(z_1 - 2z_2)^d + cz_0^d : z_1^d + cz_0^d].$$

For appropriate choices of the parameters c and λ, one can show that $\mathrm{supp}(T)$ and $\mathrm{supp}(\mu)$ coincide. Moreover, f has an attracting fixed point, so the Fatou set is not empty. Observe that the restriction of f to the projective line $\{z_0 = 0\}$, for appropriate v, is chaotic, i.e. has dense orbits. One shows that $\{z_0 = 0\}$ is in the support of μ and that for appropriate m, $\mathbb{P}^2 \setminus \cup_{i=0}^m f^{-i}\{z_0 = 0\}$ is Kobayashi hyperbolic. Hence using the total invariance of $\mathrm{supp}(\mu)$, we get that the complement of $\mathrm{supp}(\mu)$ is in the Fatou set. It is possible to choose the parameters so that $\mathbb{P}^2 \setminus \mathrm{supp}(\mu)$ contains an attractive fixed point. Several other examples are discussed in [FS6]. For example it is possible that $\mathrm{supp}(T)$ has non-empty interior and the Fatou set has also non-empty interior. The situation is then quite different from the one variable case, where either the Julia set is equal to $\mathbb{P}^1$ or it has empty interior.

We now give a characterization of the Julia sets in term of volume growth. There is an interesting gap in the possible volume growth.

Proposition 1.26. *Let f be a holomorphic endomorphism of algebraic degree $d \geq 2$ of $\mathbb{P}^k$. Let T be its Green $(1,1)$-current. Then the following properties are equivalent:*

1. *x is a point in the Julia set of order p, i.e. $x \in \mathcal{J}_p := \mathrm{supp}(T^p)$;*
2. *For every neighbourhood U of x, we have*

$$\liminf_{n\to\infty} d^{-pn} \int_U (f^n)^*(\omega_{\mathrm{FS}}^p) \wedge \omega_{\mathrm{FS}}^{k-p} \neq 0;$$

3. *For every neighbourhood U of x, we have*

$$\limsup_{n\to\infty} d^{-(p-1)n} \int_U (f^n)^*(\omega_{\mathrm{FS}}^p) \wedge \omega_{\mathrm{FS}}^{k-p} = +\infty.$$

Proof. We have seen in Theorem 1.16 that $d^{-n}(f^n)^*(\omega_{\mathrm{FS}})$ converges to T when n goes to infinity. Moreover, $d^{-n}(f^n)^*(\omega_{\mathrm{FS}})$ admits a quasi-potential which converges uniformly to a quasi-potential of T. It follows that $\lim d^{-pn}(f^n)^*(\omega_{\mathrm{FS}}^p) = T^p$. We deduce that Properties 1) and 2) are equivalent. Since 2) implies 3), it remains to show that 3) implies 1). For this purpose, it is enough to show that for any open set V with $\overline{V} \cap \mathcal{J}_p = \varnothing$,

$$\int_V (f^n)^*(\omega_{\mathrm{FS}}^p) \wedge \omega_{\mathrm{FS}}^{k-p} = O(d^{(p-1)n}).$$

This is a consequence of a more general inequality in Theorem 1.112 below. We give here a direct proof.

Since $(\omega_{\mathrm{FS}}+dd^c\mathrm{g})^p=0$ on $\mathbb{P}^k\setminus\mathscr{J}_p$, we can write there $\omega_{\mathrm{FS}}^p=dd^c\mathrm{g}\wedge(S^+-S^-)$ where $S^\pm$ are positive closed $(p-1,p-1)$-currents on $\mathbb{P}^k$. Let χ be a cut-off function with compact support in $\mathbb{P}^k\setminus\mathscr{J}_p$ and equal to 1 on V. The above integral is bounded by

$$\int_{\mathbb{P}^k}\chi(f^n)^*\big(dd^c\mathrm{g}\wedge(S^+-S^-)\big)\wedge\omega_{\mathrm{FS}}^{k-p}=\int_{\mathbb{P}^k}dd^c\chi\wedge(\mathrm{g}\circ f^n)(f^n)^*(S^+-S^-)\wedge\omega_{\mathrm{FS}}^{k-p}.$$

Since g is bounded, the last integral is bounded by a constant times $\|(f^n)^*(S^+)\|+\|(f^n)^*(S^-)\|$. We conclude using the identity $\|(f^n)^*(S^\pm)\|=d^{(p-1)n}\|S^\pm\|$. □

The previous proposition suggests a notion of *local dynamical degree*. Define

$$\delta_p(x,r):=\limsup_{n\to\infty}\Big(\int_{B(x,r)}(f^n)^*(\omega_{\mathrm{FS}}^p)\wedge\omega_{\mathrm{FS}}^{k-p}\Big)^{1/n}$$

and

$$\delta_p(x):=\inf_{r>0}\delta_p(x,r)=\lim_{r\to 0}\delta_p(x,r).$$

It follows from the above proposition that $\delta_p(x)=d^p$ for $x\in\mathscr{J}_p$ and $\delta_p(x)=0$ for $x\not\in\mathscr{J}_p$. This notion can be extended to general meromorphic maps or correspondences and the sub-level sets $\{\delta_p(x)\geq c\}$ can be seen as a kind of Julia sets.

Exercise 1.27. Let f be an endomorphism of algebraic degree $d\geq 2$ of $\mathbb{P}^k$. Suppose a subsequence (f^{n_i}) is equicontinuous on an open set U. Show that U is contained in the Fatou set.

Exercise 1.28. Let f and g be two commuting holomorphic endomorphisms of $\mathbb{P}^k$, i.e. $f\circ g=g\circ f$. Show that f and g have the same Green currents. Deduce that they have the same Julia and Fatou sets.

Exercise 1.29. Determine the Green $(1,1)$-current and the Green measure for the map f in Example 1.11. Study the lamination on $\mathrm{supp}(T^p)\setminus\mathrm{supp}(T^{p+1})$. Express the current T^p on that set as an integral on appropriate manifolds.

Exercise 1.30. Let f be an endomorphism of algebraic degree $d\geq 2$ of $\mathbb{P}^k$ and T its Green $(1,1)$-current. Consider the family of maps $\tau:\Delta\to\mathbb{P}^k$ such that $\tau^*(T)=0$. The last equation means that if u is a local potential of T, i.e. $dd^cu=T$ on some open set, then $dd^cu\circ\tau=0$ on its domain of definition. Show that the sequence $(f^n_{|\tau(\Delta)})_{n\geq 1}$ is equicontinuous. Prove that there is a constant $c>0$ such that $\|D\tau(0)\|\leq c$ for every τ as above (this property holds for any positive closed $(1,1)$-current T with continuous potentials). Find the corresponding discs for f as in Exercise 1.29.

Exercise 1.31. Let f be an endomorphism of algebraic degree $d \geq 2$ of $\mathbb{P}^k$. Let X be an analytic set of pure dimension p in an open set $U \subset \mathbb{P}^k$. Show that for every compact $K \subset U$

$$\limsup_{n\to\infty} \frac{1}{n} \log \text{volume}(f^n(X \cap K)) \leq p \log d.$$

Hint. For an appropriate cut-off function χ, estimate $\int_X \chi (f^n)^*(\omega_{\text{FS}}^p)$.

1.3 Other Constructions of the Green Currents

In this paragraph, we give other methods, introduced and developped by the authors, in order to construct the Green currents and Green measures for meromorphic maps. We obtain estimates on the Perron-Frobenius operator and on the thickness of the Green measure, that will be applied in the stochastic study of the dynamical system. A key point here is the use of d.s.h. functions as observables.

We first present a recent direct construction of Green (p,p)-currents using super-potentials[1]. Super-potentials are a tool in order to compute with positive closed (p,p)-currents. They play the same role as potentials for bidegree $(1,1)$ currents. In dynamics, they are used in particular in the equidistribution problem for algebraic sets of arbitrary dimension and allow to get some estimates on the speed of convergence.

Theorem 1.32. *Let S be a positive closed (p,p)-current of mass 1 on $\mathbb{P}^k$. Assume that the super-potential of S is bounded. Then $d^{-pn}(f^n)^*(S)$ converge to the Green (p,p)-current T^p of f. Moreover, T^p has a Hölder continuous super-potential.*

Sketch of proof. We refer to Appendix A.2 and A.4 for an introduction to super-potentials and to the action of holomorphic maps on positive closed currents. Recall that f^* and f_* act on $H^{p,p}(\mathbb{P}^k,\mathbb{C})$ as the multiplications by d^p and d^{k-p} respectively. So, if S is a positive closed (p,p)-current of mass 1, then $\|f^*(S)\| = d^p$ and $\|f_*(S)\| = d^{k-p}$ since the mass can be computed cohomologically. Let Λ denote the operator $d^{-p+1}f_*$ acting on $\mathscr{C}_{k-p+1}(\mathbb{P}^k)$, the convex set of positive closed currents of bidegree $(k-p+1,k-p+1)$ and of mass 1. It is continuous and it takes values also in $\mathscr{C}_{k-p+1}(\mathbb{P}^k)$. Let $\mathscr{V}$, $\mathscr{U}$, $\mathscr{U}_n$ denote the super-potentials of $d^{-p}f^*(\omega_{\text{FS}}^p)$, S and $d^{-pn}(f^n)^*(S)$ respectively. Consider a quasi-potential U of mean 0 of S which is a DSH current satisfying $dd^cU = S - \omega_{\text{FS}}^p$. The following computations are valid for S smooth and can be extended to all currents S using a suitable regularization procedure.

By Theorem A.35 in the Appendix, the current $d^{-p}f^*(U)$ is DSH and satisfies

$$dd^c\big(d^{-p}f^*(U)\big) = d^{-p}f^*(S) - d^{-p}f^*(\omega_{\text{FS}}^p).$$

[1] These super-potentials correspond to super-potentials of mean 0 in [DS10].

If V is a smooth quasi-potential of mean 0 of $d^{-p}f^*(\omega_{\rm FS}^p)$, i.e. a smooth real $(p-1,p-1)$-form such that

$$dd^cV = d^{-p}f^*(\omega_{\rm FS}^p) - \omega_{\rm FS}^p \quad \text{and} \quad \langle \omega_{\rm FS}^{k-p+1}, V\rangle = 0,$$

then $V + d^{-p}f^*(U)$ is a quasi-potential of $d^{-p}f^*(S)$. Let m be the real number such that $V + d^{-p}f^*(U) + m\omega_{\rm FS}^{p-1}$ is a quasi-potential of mean 0 of $d^{-1}f^*(S)$. We have

$$\begin{aligned}\mathscr{U}_1(R) &= \langle V + d^{-p}f^*(U) + m\omega_{\rm FS}^{p-1}, R\rangle \\ &= \langle V, R\rangle + d^{-1}\langle U, \Lambda(R)\rangle + m \\ &= \mathscr{V}(R) + d^{-1}\mathscr{U}(\Lambda(R)) + m.\end{aligned}$$

By induction, we obtain

$$\begin{aligned}\mathscr{U}_n(R) = \mathscr{V}(R) + d^{-1}\mathscr{V}(\Lambda(R)) + \cdots + d^{-n+1}\mathscr{V}(\Lambda^{n-1}(R)) \\ + d^{-n}\mathscr{U}(\Lambda^n(R)) + m_n,\end{aligned}$$

where m_n is a constant depending on n and on S.

Since $d^{-p}f^*(\omega_{\rm FS}^p)$ is smooth, $\mathscr{V}$ is a Hölder continuous function. It is not difficult to show that Λ is Lipschitz with respect to the distance dist_α on $\mathscr{C}_{k-p+1}(\mathbb{P}^k)$. Therefore, by Lemma 1.19, the sum

$$\mathscr{V}(R) + d^{-1}\mathscr{V}(\Lambda(R)) + \cdots + d^{-n+1}\mathscr{V}(\Lambda^{n-1}(R))$$

converges uniformly to a Hölder continuous function $\mathscr{V}_\infty$ which does not depend on S. Recall that super-potentials vanish at $\omega_{\rm FS}^{k-p+1}$, in particular, $\mathscr{U}_n(\omega_{\rm FS}^{k-p+1}) = 0$. Since $\mathscr{U}$ is bounded, the above expression of $\mathscr{U}_n(R)$ for $R = \omega_{\rm FS}^{k-p+1}$ implies that m_n converge to $m_\infty := -\mathscr{V}_\infty(\omega_{\rm FS}^{k-p+1})$ which is independent of S. So, $\mathscr{U}_n$ converge uniformly to $\mathscr{V}_\infty + m_\infty$. We deduce that $d^{-pn}(f^n)^*(S)$ converge to a current T_p which does not depend on S. Moreover, the super-potential of T_p is the Hölder continuous function $\mathscr{V}_\infty + m_\infty$.

We deduce from the above discussion that $d^{-pn}(f^n)^*(\omega_{\rm FS}^p)$ converge in the Hartogs' sense to T_p. Theorem A.48 implies that $T_{p+q} = T_p \wedge T_q$ if $p+q \leq k$. Therefore, if T is the Green $(1,1)$-current, T_p is equal to T^p the Green (p,p)-current of f. □

Remark 1.33. Let S_n be positive closed (p,p)-currents of mass 1 on $\mathbb{P}^k$. Assume that their super-potentials $\mathscr{U}_{S_n}$ satisfy $\|\mathscr{U}_{S_n}\|_\infty = o(d^n)$. Then $d^{-pn}(f^n)^*(S_n)$ converge to T^p. If (f_s) is a family of maps depending holomorphically on s in a space of parameters Σ, then the Green super-functions are also locally Hölder continuous with respect to s and define a positive closed (p,p)-current on $\Sigma \times \mathbb{P}^k$. Its slice by $\{s\} \times \mathbb{P}^k$ is the Green (p,p)-current of f_s.

We now introduce two other constructions of the Green measure. The main point is the use of appropriate spaces of test functions adapted to complex analysis. Their norms take into account the complex structure of $\mathbb{P}^k$. The reason to introduce these spaces is that they are invariant under the push-forward by a holomorphic map. This is not the case for spaces of smooth forms because of the critical set. Moreover, we will see that there is a spectral gap for the action of endomorphisms of $\mathbb{P}^k$ which is a useful property in the stochastic study of the dynamical system. The first method, called *the dd^c-method*, was introduced in [DS1] and developped in [DS6]. It can be extended to Green currents of any bidegree. We show a convergence result for PB measures ν towards the Green measure. PB measures are diffuse in some sense; we will study equidistribution of Dirac masses in the next paragraph.

Recall that f is an endomorphism of $\mathbb{P}^k$ of algebraic degree $d \geq 2$. Define the *Perron-Frobenius operator* Λ on test functions φ by $\Lambda(\varphi) := d^{-k} f_*(\varphi)$. More precisely, we have

$$\Lambda(\varphi)(z) := d^{-k} \sum_{w \in f^{-1}(z)} \varphi(w),$$

where the points in $f^{-1}(z)$ are counted with multiplicity. The following proposition is crucial.

Proposition 1.34. *The operator $\Lambda : \mathrm{DSH}(\mathbb{P}^k) \to \mathrm{DSH}(\mathbb{P}^k)$ is well-defined, bounded and continuous with respect to the weak topology on $\mathrm{DSH}(\mathbb{P}^k)$. The operator $\widetilde{\Lambda} : \mathrm{DSH}(\mathbb{P}^k) \to \mathrm{DSH}(\mathbb{P}^k)$ defined by*

$$\widetilde{\Lambda}(\varphi) := \Lambda(\varphi) - \langle \omega_{\mathrm{FS}}^k, \Lambda(\varphi) \rangle$$

is contracting and satisfies the estimate

$$\|\widetilde{\Lambda}(\varphi)\|_{\mathrm{DSH}} \leq d^{-1} \|\varphi\|_{\mathrm{DSH}}.$$

Proof. We prove the first assertion. Let φ be a quasi-p.s.h. function such that $dd^c\varphi \geq -\omega_{\mathrm{FS}}$. We show that $\Lambda(\varphi)$ is d.s.h. Since φ is strongly upper semi-continuous, $\Lambda(\varphi)$ is strongly upper semi-continuous, see Appendix A.2. If $dd^c\varphi = S - \omega_{\mathrm{FS}}$ with S positive closed, we have $dd^c\Lambda(\varphi) = d^{-k} f_*(S) - d^{-k} f_*(\omega_{\mathrm{FS}})$. Therefore, if u is a quasi-potential of $d^{-k} f_*(\omega_{\mathrm{FS}})$, then $u + \Lambda(\varphi)$ is strongly semi-continuous and is a quasi-potential of $d^{-k} f_*(S)$. So, this function is quasi-p.s.h. We deduce that $\Lambda(\varphi)$ is d.s.h., and hence $\Lambda : \mathrm{DSH}(\mathbb{P}^k) \to \mathrm{DSH}(\mathbb{P}^k)$ is well-defined.

Observe that $\Lambda : L^1(\mathbb{P}^k) \to L^1(\mathbb{P}^k)$ is continuous. Indeed, if φ is in $L^1(\mathbb{P}^k)$, we have

$$\|\Lambda(\varphi)\|_{L^1} = \langle \omega_{\mathrm{FS}}^k, d^{-k} |f_*(\varphi)| \rangle \leq \langle \omega_{\mathrm{FS}}^k, d^{-k} f_*(|\varphi|) \rangle = d^{-k} \langle f^*(\omega_{\mathrm{FS}}^k), |\varphi| \rangle \lesssim \|\varphi\|_{L^1}.$$

Therefore, $\Lambda : \mathrm{DSH}(\mathbb{P}^k) \to \mathrm{DSH}(\mathbb{P}^k)$ is continuous with respect to the weak topology. This and the estimates below imply that Λ is a bounded operator.

We prove now the last estimate in the proposition. Write $dd^c\varphi = S^+ - S^-$ with $S^\pm$ positive closed. We have

$$dd^c\widetilde{\Lambda}(\varphi) = dd^c\Lambda(\varphi) = d^{-k} f_*(S^+ - S^-) = d^{-k} f_*(S^+) - d^{-k} f_*(S^-).$$

Since $\|f_*(S^\pm)\| = d^{k-1}\|S^\pm\|$ and $\langle \omega_{\mathrm{FS}}^k, \widetilde{\Lambda}(\varphi) \rangle = 0$, we obtain that $\|\widetilde{\Lambda}(\varphi)\|_{\mathrm{DSH}} \leq d^{-1}\|S^\pm\|$. The result follows. □

Recall that if ν is a positive measure on $\mathbb{P}^k$, the pull-back $f^*(\nu)$ is defined by the formula $\langle f^*(\nu),\varphi\rangle = \langle \nu, f_*(\varphi)\rangle$ for φ continuous on $\mathbb{P}^k$. Observe that since f is a ramified covering, $f_*(\varphi)$ is continuous when φ is continuous, see Exercise A.11 in Appendix. So, the above definition makes sense. For $\varphi = 1$, we obtain that $\|f^*(\nu)\| = d^k\|\nu\|$, since the covering is of degree d^k. If ν is the Dirac mass at a point a, $f^*(\nu)$ is the sum of Dirac masses at the points in $f^{-1}(a)$.

Recall that a measure ν is PB if quasi-p.s.h. are ν-integrable and ν is PC if it is PB and acts continuously on DSH($\mathbb{P}^k$) with respect to the weak topology on this space, see Appendix A.4. We deduce from Proposition 1.34 the following result where the norm $\|\cdot\|_\mu$ on DSH($\mathbb{P}^k$) is defined by

$$\|\varphi\|_\mu := |\langle \mu,\varphi\rangle| + \inf\|S^\pm\|,$$

with $S^\pm$ positive closed such that $dd^c\varphi = S^+ - S^-$. We will see that μ is PB, hence this norm is equivalent to $\|\cdot\|_{\rm DSH}$, see Proposition A.43.

Theorem 1.35. *Let f be as above. If ν is a PB probability measure, then $d^{-kn}(f^n)^*(\nu)$ converge to a PC probability measure μ which is independent of ν and totally invariant under f. Moreover, if φ is a d.s.h. function and $c_\varphi := \langle \mu,\varphi\rangle$, then*

$$\|\Lambda^n(\varphi) - c_\varphi\|_\mu \le d^{-n}\|\varphi\|_\mu \quad and \quad \|\Lambda^n(\varphi) - c_\varphi\|_{\rm DSH} \le Ad^{-n}\|\varphi\|_{\rm DSH},$$

where $A > 0$ is a constant independent of φ and n. In particular, there is a constant $c > 0$ depending on ν such that

$$|\langle d^{-kn}(f^n)^*(\nu) - \mu,\varphi\rangle| \le cd^{-n}\|\varphi\|_{\rm DSH}.$$

Proof. Since ν is PB, d.s.h. functions are ν integrable. It follows that there is a constant $c > 0$ such that $|\langle \nu,\varphi\rangle| \le c\|\varphi\|_{\rm DSH}$. Otherwise, there are d.s.h. functions φ_n with $\|\varphi_n\|_{\rm DSH} \le 1$ and $\langle \nu,\varphi_n\rangle \ge 2^n$, hence the d.s.h. function $\sum 2^{-n}\varphi_n$ is not ν-integrable.

It follows from Proposition 1.34 that $f^*(\nu)$ is PB. So, $d^{-kn}(f^n)^*(\nu)$ is PB for every n. Define for φ d.s.h.,

$$c_0 := \langle \omega_{\rm FS}^k,\varphi\rangle \quad \text{and} \quad \varphi_0 := \varphi - c_0$$

and inductively

$$c_{n+1} := \langle \omega_{\rm FS}^k,\Lambda(\varphi_n)\rangle \quad \text{and} \quad \varphi_{n+1} := \Lambda(\varphi_n) - c_{n+1} = \widetilde{\Lambda}(\varphi_n).$$

A straighforward computation gives

$$\Lambda^n(\varphi) = c_0 + \cdots + c_n + \varphi_n.$$

Therefore,

$$\langle d^{-kn}(f^n)^*(\nu),\varphi\rangle = \langle \nu,\Lambda^n(\varphi)\rangle = c_0 + \cdots + c_n + \langle \nu,\varphi_n\rangle.$$

Proposition 1.34 applied inductively on n implies that $\|\varphi_n\|_{\rm DSH} \leq d^{-n}\|\varphi\|_{\rm DSH}$. Since Λ is bounded, it follows that $|c_n| \leq Ad^{-n}\|\varphi\|_{\rm DSH}$, where $A > 0$ is a constant. The property that ν is PB and the above estimate on φ_n imply that $\langle \nu, \varphi_n\rangle$ converge to 0.

We deduce that $\langle d^{-kn}(f^n)^*(\nu), \varphi\rangle$ converge to $c_\varphi := \sum_{n\geq 0} c_n$ and $|c_\varphi| \lesssim \|\varphi\|_{\rm DSH}$. Therefore, $d^{-kn}(f^n)^*(\nu)$ converge to a PB measure μ defined by $\langle \mu, \varphi\rangle := c_\varphi$. The constant c_φ does not depend on ν, hence the measure μ is independent of ν. The above convergence implies that μ is totally invariant, i.e. $f^*(\mu) = d^k\mu$. Finally, since c_n depends continuously on the d.s.h. function φ, the constant c_φ, which is defined by a normally convergent series, depends also continuously on φ. It follows that μ is PC.

We prove now the estimates in the theorem. The total invariance of μ implies that $\langle \mu, \Lambda^n(\varphi)\rangle = \langle \mu, \varphi\rangle = c_\varphi$. If $dd^c\varphi = S^+ - S^-$ with $S^\pm$ positive closed, we have $dd^c\Lambda^n(\varphi) = d^{-kn}(f^n)_*(S^+) - d^{-kn}(f^n)_*(S^-)$, hence

$$\|\Lambda^n(\varphi) - c_\varphi\|_\mu \leq d^{-kn}\|(f^n)_*(S^\pm)\| = d^{-n}\|S^\pm\|.$$

It follows that

$$\|\Lambda^n(\varphi) - c_\varphi\|_\mu \leq d^{-n}\|\varphi\|_\mu.$$

For the second estimate, we have

$$\|\Lambda^n(\varphi) - c_\varphi\|_{\rm DSH} = \|\varphi_n\|_{\rm DSH} + \sum_{i\geq n} c_i.$$

The last sum is clearly bounded by a constant times $d^{-n}\|\varphi\|_{\rm DSH}$. This together with the inequality $\|\varphi_n\|_{\rm DSH} \lesssim d^{-n}\|\varphi\|_{\rm DSH}$ implies $\|\Lambda^n(\varphi) - c_\varphi\|_{\rm DSH} \lesssim d^{-n}\|\varphi\|_{\rm DSH}$. We can also use that $\|\ \|_\mu$ and $\|\ \|_{\rm DSH}$ are equivalent, see Proposition A.43.

The last inequality in the theorem is then deduced from the identity

$$\langle d^{-kn}(f^n)^*(\nu) - \mu, \varphi\rangle = \langle \nu, \Lambda^n(\varphi) - c_\varphi\rangle$$

and the fact that ν is PB. □

Remark 1.36. In the present case, the dd^c-method is quite simple. The function φ_n is the normalized solution of the equation $dd^c\psi = dd^c\Lambda^n(\varphi)$. It satisfies automatically good estimates. The other solutions differ from φ_n by constants. We will see that for polynomial-like maps, the solutions differ by pluriharmonic functions and the estimates are less straightforward. In the construction of Green (p,p)-currents with $p > 1$, φ is replaced with a test form of bidegree $(k-p,k-p)$ and φ_n is a solution of an appropriate dd^c-equation. The constants c_n will be replaced with dd^c-closed currents with a control of their cohomology class.

The second construction of the Green measure follows the same lines but we use the complex Sobolev space $W^*(\mathbb{P}^k)$ instead of DSH($\mathbb{P}^k$). We obtain that the Green measure μ is WPB, see Appendix A.4 for the terminology. We only mention here the result which replaces Proposition 1.34.

Proposition 1.37. *The operator $\Lambda : W^*(\mathbb{P}^k) \to W^*(\mathbb{P}^k)$ is well-defined, bounded and continuous with respect to the weak topology on $W^*(\mathbb{P}^k)$. The operator $\widetilde{\Lambda} : W^*(\mathbb{P}^k) \to W^*(\mathbb{P}^k)$ defined by*

$$\widetilde{\Lambda}(\varphi) := \Lambda(\varphi) - \langle \omega_{\rm FS}^k, \Lambda(\varphi)\rangle$$

is contracting and satisfies the estimate

$$\|\widetilde{\Lambda}(\varphi)\|_{W^*} \leq d^{-1/2}\|\varphi\|_{W^*}.$$

Sketch of proof. As in Proposition 1.34, since φ is in $L^1(\mathbb{P}^k)$, $\Lambda(\varphi)$ is also in $L^1(\mathbb{P}^k)$ and the main point here is to estimate $\partial\varphi$. Let S be a positive closed $(1,1)$-current on $\mathbb{P}^k$ such that $\sqrt{-1}\partial\varphi \wedge \overline{\partial}\varphi \leq S$. We show that $\sqrt{-1}\partial f_*(\varphi)\wedge\overline{\partial} f_*(\varphi) \leq d^k f_*(S)$, in particular, the Poincaré differential $d\Lambda(\varphi)$ of $\Lambda(\varphi)$ is in $L^2(\mathbb{P}^k)$.

If a is not a critical value of f and U a small neighbourhood of a, then $f^{-1}(U)$ is the union of d^k open sets U_i which are sent bi-holomorphically on U. Let $g_i : U \to U_i$ be the inverse branches of f. On U, we obtain using Schwarz's inequality that

$$\begin{aligned}\sqrt{-1}\partial f_*(\varphi)\wedge\overline{\partial} f_*(\varphi) &= \sqrt{-1}\Big(\sum \partial g_i^*(\varphi)\Big)\wedge\Big(\sum\overline{\partial} g_i^*(\varphi)\Big)\\ &\leq d^k\sum\sqrt{-1}\partial g_i^*(\varphi)\wedge\overline{\partial} g_i^*(\varphi)\\ &= d^k f_*\big(\sqrt{-1}\partial\varphi\wedge\overline{\partial}\varphi\big).\end{aligned}$$

Therefore, we have $\sqrt{-1}\partial f_*(\varphi)\wedge\overline{\partial} f_*(\varphi)\leq d^k f_*(S)$ out of the critical values of f which is a manifold of real codimension 2.

Recall that $f_*(\varphi)$ is in $L^1(\mathbb{P}^k)$. It is a classical result in Sobolev spaces theory that an L^1 function whose gradient out of a submanifold of codimension 2 is in L^2, is in fact in the Sobolev space $W^{1,2}(\mathbb{P}^k)$. We deduce that the inequality $\sqrt{-1}\partial f_*(\varphi)\wedge\overline{\partial} f_*(\varphi)\leq d^k f_*(S)$ holds on $\mathbb{P}^k$, because the left hand side term is an L^1 form and has no mass on critical values of f. Finally, we have

$$\sqrt{-1}\partial\Lambda(\varphi)\wedge\overline{\partial}\Lambda(\varphi)\leq d^{-k}f_*(S).$$

This, together with the identity $\|f_*(S)\| = d^{k-1}\|S\|$, implies that $\|\widetilde{\Lambda}(\varphi)\|_{W^*}\leq d^{-1/2}\|S\|$. The proposition follows. □

In the rest of this paragraph, we show that the Green measure μ is moderate, see Appendix A.4. Recall that a positive measure ν on $\mathbb{P}^k$ is *moderate* if there are constants $\alpha > 0$ and $c > 0$ such that

$$\|e^{-\alpha\varphi}\|_{L^1(\nu)} \leq c$$

for φ quasi-p.s.h. such that $dd^c\varphi \geq -\omega_{\rm FS}$ and $\langle\omega_{\rm FS}^k,\varphi\rangle = 0$. Moderate measures are PB and by linearity, if ν is moderate and $\mathscr{D}$ is a bounded set of d.s.h. functions then there are constants $\alpha > 0$ and $c > 0$ such that

$$\|e^{\alpha|\varphi|}\|_{L^1(\nu)} \leq c \quad \text{for} \quad \varphi\in\mathscr{D}.$$

Moderate measures were introduced in [DS1]. The fundamental estimate in Theorem A.22 in Appendix implies that smooth measures are moderate. So, when we use test d.s.h. functions, several estimates for the Lebesgue measure can be extended to moderate measures. For example, we will see that quasi-p.s.h. functions have comparable repartition functions with respect to the Lebesgue measure $\omega_{\rm FS}^k$ and to the equilibrium measure μ.

It is shown in [DNS] that measures which are wedge-products of positive closed $(1,1)$-currents with Hölder continuous potentials, are moderate. In particular, the Green measure μ is moderate. We will give here another proof of this result using the following criterion. Since DSH($\mathbb{P}^k$) is a subspace of $L^1(\mathbb{P}^k)$, the L^1-norm induces a distance on DSH($\mathbb{P}^k$) that we denote by dist_{L^1}.

Proposition 1.38. *Let ν be a PB positive measure on $\mathbb{P}^k$. Assume that ν restricted to any bounded subset of* DSH($\mathbb{P}^k$) *is Hölder continuous*[2] *with respect to* dist_{L^1}. *Then ν is moderate.*

Proof. Let φ be a quasi-p.s.h. function such that $dd^c\varphi \geq -\omega_{\rm FS}$ and $\langle \omega_{\rm FS}^k, \varphi\rangle = 0$. We want to prove that $\langle \nu, e^{-\alpha\varphi}\rangle \leq c$ for some positive constants α, c. For this purpose, we only have to show that $\nu\{\varphi \leq -M\} \lesssim e^{-\alpha M}$ for some constant $\alpha > 0$ and for $M \geq 1$. Define for $M > 0$, $\varphi_M := \max(\varphi, -M)$. These functions φ_M belong to a compact family $\mathscr{D}$ of d.s.h. functions. Observe that $\varphi_{M-1} - \varphi_M$ is positive with support in $\{\varphi \leq -M+1\}$. It is bounded by 1 and equal to 1 on $\{\varphi < -M\}$. Therefore, the Hölder continuity of ν on $\mathscr{D}$ implies that there is a constant $\lambda > 0$ such that

$$\begin{aligned} \nu\{\varphi < -M\} &\leq \langle \nu, \varphi_{M-1} - \varphi_M\rangle = \nu(\varphi_{M-1}) - \nu(\varphi_M) \\ &\lesssim \mathrm{dist}_{L^1}(\varphi_{M-1}, \varphi_M)^\lambda \leq \mathrm{volume}\{\varphi < -M+1\}^\lambda. \end{aligned}$$

Since the Lebesgue measure is moderate, the last expression is $\lesssim e^{-\alpha M}$ for some positive constant α. The proposition follows. □

We have the following result obtained in [DNS]. It will be used to establish several stochastic properties of d.s.h. observables for the equilibrium measure.

Theorem 1.39. *Let f be an endomorphism of algebraic degree $d \geq 2$ of $\mathbb{P}^k$. Then, the Green measure μ of f is Hölder continuous on bounded subsets of* DSH($\mathbb{P}^k$). *In particular, it is moderate.*

Proof. Let $\mathscr{D}$ be a bounded set of d.s.h. functions. We have to show that μ is Hölder continuous on $\mathscr{D}$ with respect to dist_{L^1}. By linearity, since μ is PC, it is enough to consider the case where $\mathscr{D}$ is the set of d.s.h. functions φ such that $\langle \mu, \varphi\rangle \geq 0$ and $\|\varphi\|_\mu \leq 1$. Let $\widetilde{\mathscr{D}}$ denote the set of d.s.h. functions $\varphi - \Lambda(\varphi)$ with $\varphi \in \mathscr{D}$. By Proposition 1.34, $\widetilde{\mathscr{D}}$ is a bounded family of d.s.h. functions. We claim that $\widetilde{\mathscr{D}}$ is invariant under $\widetilde{\Lambda} := d\Lambda$. Observe that if φ is in $\mathscr{D}$, then $\widetilde{\varphi} := \varphi - \langle \mu, \varphi\rangle$ is also in $\mathscr{D}$. Since $\langle \mu, \varphi\rangle = \langle \mu, \Lambda(\varphi)\rangle$, we have $\|\widetilde{\Lambda}(\widetilde{\varphi})\|_\mu \leq 1$ and

$$\widetilde{\Lambda}(\varphi - \Lambda(\varphi)) = \widetilde{\Lambda}(\widetilde{\varphi} - \Lambda(\widetilde{\varphi})) = \widetilde{\Lambda}(\widetilde{\varphi}) - \Lambda(\widetilde{\Lambda}(\widetilde{\varphi})).$$

[2] This property is close to the property that ν has a Hölder continuous super-potential.

By Theorem 1.35, $\widetilde{\Lambda}(\widetilde{\varphi})$ belongs to $\mathscr{D}$. This proves the claim. So, the crucial point is that Λ is contracting on an appropriate hyperplane.

For φ, ψ in $L^1(\mathbb{P}^k)$ we have

$$\|\widetilde{\Lambda}(\varphi)-\widetilde{\Lambda}(\psi)\|_{L^1} \leq \int \widetilde{\Lambda}(|\varphi-\psi|)\omega_{\rm FS}^k = d^{1-k}\int |\varphi-\psi| f^*(\omega_{\rm FS}^k) \lesssim \|\varphi-\psi\|_{L^1}.$$

So, the map $\widetilde{\Lambda}$ is Lipschitz with respect to dist_{L^1}. In particular, the map $\varphi \mapsto \varphi - \Lambda(\varphi)$ is Lipschitz with respect to this distance. Now, we have for $\varphi \in \mathscr{D}$

$$\begin{aligned}\langle \mu, \varphi\rangle &= \lim_{n\to\infty}\langle d^{-kn}(f^n)^*(\omega_{\rm FS}^k), \varphi\rangle = \lim_{n\to\infty}\langle \omega_{\rm FS}^k, \Lambda^n(\varphi)\rangle \\ &= \langle \omega_{\rm FS}^k, \varphi\rangle - \sum_{n\geq 0}\langle \omega_{\rm FS}^k, \Lambda^n(\varphi)-\Lambda^{n+1}(\varphi)\rangle \\ &= \langle \omega_{\rm FS}^k, \varphi\rangle - \sum_{n\geq 0} d^{-n}\langle \omega_{\rm FS}^k, \widetilde{\Lambda}^n(\varphi-\Lambda(\varphi))\rangle.\end{aligned}$$

By Lemma 1.19, the last series defines a function on $\widetilde{\mathscr{D}}$ which is Hölder continuous with respect to dist_{L^1}. Therefore, $\varphi \mapsto \langle \mu, \varphi\rangle$ is Hölder continuous on $\mathscr{D}$. □

Remark 1.40. Let f_s be a family of endomorphisms of algebraic degree $d \geq 2$, depending holomorphically on a parameter $s \in \Sigma$. Let μ_s denote its equilibrium measure. We get that $(s, \varphi) \mapsto \mu_s(\varphi)$ is Hölder continuous on bounded subsets of $\Sigma \times \mathrm{DSH}(\mathbb{P}^k)$.

The following results are useful in the stochastic study of the dynamical system.

Corollary 1.41. *Let f, μ and Λ be as above. There are constants $c > 0$ and $\alpha > 0$ such that if ψ is d.s.h. with $\|\psi\|_{\rm DSH} \leq 1$, then*

$$\big\langle \mu, e^{\alpha d^n|\Lambda^n(\psi)-\langle\mu,\psi\rangle|}\big\rangle \leq c \quad \textit{and} \quad \|\Lambda^n(\psi)-\langle\mu,\psi\rangle\|_{L^q(\mu)} \leq cqd^{-n}$$

for every $n \geq 0$ and every $1 \leq q < +\infty$.

Proof. By Theorem 1.35, $d^n(\Lambda^n(\psi) - \langle\mu,\psi\rangle)$ belong to a compact family in $\mathrm{DSH}(\mathbb{P}^k)$. The first inequality in the corollary follows from Theorem 1.39. For the second one, we can assume that q is integer and we easily deduce the result from the first inequality and the inequalities $x^q \leq q!e^x \leq q^q e^x$ for $x \geq 0$. □

Corollary 1.42. *Let $0 < \nu \leq 2$ be a constant. There are constants $c > 0$ and $\alpha > 0$ such that if ψ is a ν-Hölder continuous function with $\|\psi\|_{\mathscr{C}^\nu} \leq 1$, then*

$$\big\langle \mu, e^{\alpha d^{n\nu/2}|\Lambda^n(\psi)-\langle\mu,\psi\rangle|}\big\rangle \leq c \quad \textit{and} \quad \|\Lambda^n(\psi)-\langle\mu,\psi\rangle\|_{L^q(\mu)} \leq cq^{\nu/2}d^{-n\nu/2}$$

for every $n \geq 0$ and every $1 \leq q < +\infty$.

Proof. By Corollary 1.41, since $\|\cdot\|_{\rm DSH} \lesssim \|\cdot\|_{\mathscr{C}^2}$, we have

$$\|\Lambda^n(\psi)-\langle\mu,\psi\rangle\|_{L^q(\mu)} \leq cqd^{-n}\|\psi\|_{\mathscr{C}^2},$$

with $c > 0$ independent of q and of ψ. On the other hand, by definition of Λ, we have

$$\|\Lambda^n(\psi) - \langle\mu,\psi\rangle\|_{L^q(\mu)} \leq \|\Lambda^n(\psi) - \langle\mu,\psi\rangle\|_{L^\infty(\mu)} \leq 2\|\psi\|_{\mathscr{C}^0}.$$

The theory of interpolation between the Banach spaces $\mathscr{C}^0$ and $\mathscr{C}^2$ [T1], applied to the linear operator $\psi \mapsto \Lambda^n(\psi) - \langle\mu,\psi\rangle$, implies that

$$\|\Lambda^n(\psi) - \langle\mu,\psi\rangle\|_{L^q(\mu)} \leq A_\nu 2^{1-\nu/2}[cqd^{-n}]^{\nu/2}\|\psi\|_{\mathscr{C}^\nu},$$

for some constant $A_\nu > 0$ depending only on ν and on $\mathbb{P}^k$. This gives the second inequality in the corollary.

Recall that if L is a linear continuous functional on the space $\mathscr{C}^0$ of continuous functions, then we have for every $0 < \nu < 2$

$$\|L\|_{\mathscr{C}^\nu} \leq A_\nu \|L\|_{\mathscr{C}^0}^{1-\nu/2}\|L\|_{\mathscr{C}^2}^{\nu/2}$$

for some constant $A_\nu > 0$ independent of L (in our case, the functional is with values in $L^q(\mu)$).

For the first inequality, we have for a fixed constant $\alpha > 0$ small enough,

$$\left\langle\mu, e^{\alpha d^{n\nu/2}|\Lambda^n(\psi)-\langle\mu,\psi\rangle|}\right\rangle = \sum_{q\geq 0}\frac{1}{q!}\langle\mu, |\alpha d^{n\nu/2}(\Lambda^n(\psi)-\langle\mu,\psi\rangle)|^q\rangle \leq \sum_{q\geq 0}\frac{1}{q!}\alpha^q c^q q^q.$$

By Stirling's formula, the last sum converges. The result follows. □

Exercise 1.43. Let φ be a smooth function and φ_n as in Theorem 1.35. Show that we can write $\varphi_n = \varphi_n^+ - \varphi_n^-$ with $\varphi_n^\pm$ quasi-p.s.h. such that $\|\varphi_n^\pm\|_{\mathrm{DSH}} \lesssim d^{-n}$ and $dd^c\varphi_n^\pm \gtrsim -d^{-n}\omega_{\mathrm{FS}}$. Prove that φ_n converge pointwise to 0 out of a pluripolar set. Deduce that if ν is a probability measure with no mass on pluripolar sets, then $d^{-kn}(f^n)^*(\nu)$ converge to μ.

Exercise 1.44. Let $\mathrm{DSH}_0(\mathbb{P}^k)$ be the space of d.s.h. functions φ such that $\langle\mu,\varphi\rangle = 0$. Show that $\mathrm{DSH}_0(\mathbb{P}^k)$ is a closed subspace of $\mathrm{DSH}(\mathbb{P}^k)$, invariant under Λ, and that the spectral radius of Λ on this space is equal to $1/d$. Note that 1 is an eigenvalue of Λ on $\mathrm{DSH}(\mathbb{P}^k)$, so, Λ has a spectral gap on $\mathrm{DSH}(\mathbb{P}^k)$. Prove a similar result for $W^*(\mathbb{P}^k)$.

1.4 Equidistribution of Points

In this paragraph, we show that the preimages of a generic point by f^n are equidistributed with respect to the Green measure μ when n goes to infinity. The proof splits in two parts. First, we prove that there is a maximal proper algebraic set $\mathscr{E}$ which is totally invariant, then we show that for $a \notin \mathscr{E}$, the preimages of a are equidistributed. We will also prove that the convex set of probability measures ν, which are totally invariant, i.e. $f^*(\nu) = d^k\nu$, is finite dimensional. The equidistribution for a out of an algebraic set is reminiscent of the main questions in value distribution theory (we will see in the next paragraph that using super-potentials we can get an estimate on the speed of convergence towards μ, at least for generic maps). Finally, we prove a theorem due to Briend-Duval on the equidistribution of the repelling periodic points. The following result was obtained by the authors in [DS1], see also [DS9] and [BH, FL, LY] for the case of dimension 1.

Theorem 1.45. *Let f be an endomorphism of $\mathbb{P}^k$ of algebraic degree $d \geq 2$ and μ its Green measure. Then there is a proper algebraic set $\mathscr{E}$ of $\mathbb{P}^k$, possibly empty, such that $d^{-kn}(f^n)^*(\delta_a)$ converge to μ if and only if $a \notin \mathscr{E}$. Here, δ_a denotes the Dirac mass at a. Moreover, $\mathscr{E}$ is totally invariant: $f^{-1}(\mathscr{E}) = f(\mathscr{E}) = \mathscr{E}$ and is maximal in the sense that if E is a proper algebraic set of $\mathbb{P}^k$ such that $f^{-n}(E) \subset E$ for some $n \geq 1$, then E is contained in $\mathscr{E}$.*

Briend-Duval proved in [BD2] the above convergence for a outside the orbit of the critical set. They announced the property for a out of an algebraic set, but there is a problem with the counting of multiplicity in their lemma in [BD2, p.149].

We also have the following earlier result due to Fornæss-Sibony [FS1].

Proposition 1.46. *There is a pluripolar set $\mathscr{E}'$ such that if a is out of $\mathscr{E}'$, then $d^{-kn}(f^n)^*(\delta_a)$ converge to μ.*

Sketch of proof. We use here a version of the above dd^c-method which is given in [DS6] in a more general setting. Let φ be a smooth function and φ_n as in Theorem 1.35. Then, the functions φ_n are continuous. The estimates on φ_n imply that the series $\sum \varphi_n$ converges in $\mathrm{DSH}(\mathbb{P}^k)$, hence converges pointwise out of a pluripolar set. Therefore, $\varphi_n(a)$ converge to 0 for a out of some pluripolar set E_φ, see Exercise 1.43. If $c_n := \langle \omega_{\mathrm{FS}}^k, \Lambda(\varphi_n)\rangle$, we have as in Theorem 1.35

$$\langle d^{-kn}(f^n)^*(\delta_a), \varphi\rangle = c_0 + \cdots + c_n + \langle \delta_a, \varphi_n\rangle = c_0 + \cdots + c_n + \varphi_n(a).$$

Therefore, $\langle d^{-kn}(f^n)^*(\delta_a), \varphi\rangle$ converge to $c_\varphi = \langle \mu, \varphi\rangle$, for a out of E_φ.

Now, consider φ in a countable family $\mathscr{F}$ which is dense in the space of smooth functions. If a is not in the union $\mathscr{E}'$ of the pluripolar sets E_φ, the above convergence of $\langle d^{-kn}(f^n)^*(\delta_a), \varphi\rangle$ together with the density of $\mathscr{F}$ implies that $d^{-kn}(f^n)^*(\delta_a)$ converge to μ. Finally, $\mathscr{E}'$ is pluripolar since it is a countable union of such sets. □

For the rest of the proof, we follow a geometric method introduced by Lyubich [LY] in dimension one and developped in higher dimension by Briend-Duval and Dinh-Sibony. We first prove the existence of the exceptional set and give several characterizations in the following general situation. Let X be an analytic set of pure dimension p in $\mathbb{P}^k$ invariant under f, i.e. $f(X) = X$. Let $g : X \to X$ denote the restriction of f to X. The following result can be deduced from Section 3.4 in [DS1], see also [D5, DS9].

Theorem 1.47. *There is a proper analytic set $\mathscr{E}_X$ of X, possibly empty, totally invariant under g, which is maximal in the following sense. If E is an analytic set of X, of* *dimension* $< \dim X$, *such that* $g^{-s}(E) \subset E$ *for some* $s \geq 1$, *then* $E \subset \mathscr{E}_X$. *Moreover, there are at most finitely many analytic sets in* X *which are totally invariant under* g.

Since g permutes the irreducible components of X, we can find an integer $m \geq 1$ such that g^m fixes the components of X.

Lemma 1.48. *The topological degree of* g^m *is equal to* d^{mp}. *More precisely, there is a hypersurface* Y *of* X *containing* $\mathrm{sing}(X) \cup g^m(\mathrm{sing}(X))$ *such that for* $x \in X$ *out of* Y, *the fiber* $g^{-m}(x)$ *has exactly* d^{mp} *points.*

Proof. Since g^m fixes the components of X, we can assume that X is irreducible. It follows that g^m defines a covering over some Zariski dense open set of X. We want to prove that δ, the degree of this covering, is equal to d^{mp}. Consider the positive measure $(f^m)^*(\omega_{\mathrm{FS}}^p) \wedge [X]$. Since $(f^m)^*(\omega_{\mathrm{FS}}^p)$ is cohomologous to $d^{mp}\omega_{\mathrm{FS}}^p$, this measure is of mass $d^{mp}\deg(X)$. Observe that $(f^m)_*$ preserves the mass of positive measures and that we have $(f^m)_*[X] = \delta[X]$. Hence,

$$\begin{aligned} d^{mp}\deg(X) &= \|(f^m)^*(\omega_{\mathrm{FS}}^p) \wedge [X]\| = \|(f^m)_*((f^m)^*(\omega_{\mathrm{FS}}^p) \wedge [X])\| \\ &= \|\omega_{\mathrm{FS}}^p \wedge (f^m)_*[X]\| = \delta\|\omega_{\mathrm{FS}}^p \wedge [X]\| = \delta\deg(X). \end{aligned}$$

It follows that $\delta = d^{mp}$. So, we can take for Y, a hypersurface which contains the ramification values of g^m and the set $\mathrm{sing}(X) \cup g^m(\mathrm{sing}(X))$. □

Let Y be as above. Observe that if $g^m(x) \notin Y$ then g^m defines a bi-holomorphic map between a neighbourhood of x and a neighbourhood of $g^m(x)$ in X. Let $[Y]$ denote the $(k-p+1,k-p+1)$-current of integration on Y in $\mathbb{P}^k$. Since $(f^{mn})_*[Y]$ is a positive closed $(k-p+1,k-p+1)$-current of mass $d^{mn(p-1)}\deg(Y)$, we can define the following ramification current

$$R = \sum_{n \geq 0} R_n := \sum_{n \geq 0} d^{-mnp}(f^{mn})_*[Y].$$

Let $\nu(R,x)$ denote the Lelong number of R at x. By Theorem A.14, for $c > 0$, $E_c := \{\nu(R,x) \geq c\}$ is an analytic set of dimension $\leq p-1$ contained in X. Observe that E_1 contains Y. We will see that R measures the obstruction for constructing good backwards orbits.

For any point $x \in X$ let $\lambda_n'(x)$ denote the number of distinct orbits

$$x_{-n}, x_{-n+1}, \ldots, x_{-1}, x_0$$

such that $g^m(x_{-i-1}) = x_{-i}$, $x_0 = x$ and $x_{-i} \in X \setminus Y$ for $0 \leq i \leq n-1$. These are the "good" orbits. Define $\lambda_n := d^{-mpn}\lambda_n'$. The function λ_n is lower semi-continuous with respect to the Zariski topology on X. Moreover, by Lemma 1.48, we have $0 \leq \lambda_n \leq 1$ and $\lambda_n = 1$ out of the analytic set $\cup_{i=0}^{n-1} g^{mi}(Y)$. The sequence (λ_n) decreases to a function λ, which represents the asymptotic proportion of backwards orbits in $X \setminus Y$.

Lemma 1.49. *There is a constant* $\gamma > 0$ *such that* $\lambda \geq \gamma$ *on* $X \setminus E_1$.

Proof. We deduce from Theorem A.14, the existence of a constant $0 < \gamma < 1$ satisfying $\{\nu(R,x) > 1-\gamma\} = E_1$. Indeed, the sequence of analytic sets $\{\nu(R,x) \geq 1 - 1/i\}$ is decreasing, hence stationary. Consider a point $x \in X \setminus E_1$. We have $x \notin Y$ and if $\nu_n := \nu(R_n, x)$, then $\sum \nu_n \leq 1 - \gamma$. Since E_1 contains Y, $\nu_0 = 0$ and $F_1 := g^{-m}(x)$ contains exactly d^{mp} points. The definition of ν_1, which is "the multiplicity" of $d^{-mp}(f^m)_*[Y]$ at x, implies that $g^{-m}(x)$ contains at most $\nu_1 d^{mp}$ points in Y. Then

$$\#g^{-m}(F_1 \setminus Y) = d^{mp}\#(F_1 \setminus Y) \geq (1-\nu_1)d^{2mp}.$$

Define $F_2 := g^{-m}(F_1 \setminus Y)$. The definition of ν_2 implies that F_2 contains at most $\nu_2 d^{2mp}$ points in Y. Hence, $F_3 := g^{-m}(F_2 \setminus Y)$ contains at least $(1-\nu_1-\nu_2)d^{3mp}$ points. In the same way, we define $F_4, \ldots, F_n$ with $\#F_n \geq (1 - \sum \nu_i)d^{mpn}$. Hence, for every n we get the following estimate:

$$\lambda_n(x) \geq d^{-mpn}\#F_n \geq 1 - \sum \nu_i \geq \gamma.$$

This proves the lemma. □

End of the proof of Theorem 1.47. Let $\mathscr{E}_X^n$ denote the set of $x \in X$ such that $g^{-ml}(x) \subset E_1$ for $0 \leq l \leq n$ and define $\mathscr{E}_X := \cap_{n \geq 0} \mathscr{E}_X^n$. Then, $(\mathscr{E}_X^n)$ is a decreasing sequence of analytic subsets of E_1. It should be stationary. So, there is $n_0 \geq 0$ such that $\mathscr{E}_X^n = \mathscr{E}_X$ for $n \geq n_0$.

By definition, $\mathscr{E}_X$ is the set of $x \in X$ such that $g^{-mn}(x) \subset E_1$ for every $n \geq 0$. Hence, $g^{-m}(\mathscr{E}_X) \subset \mathscr{E}_X$. It follows that the sequence of analytic sets $g^{-mn}(\mathscr{E}_X)$ is decreasing and there is $n \geq 0$ such that $g^{-m(n+1)}(\mathscr{E}_X) = g^{-mn}(\mathscr{E}_X)$. Since g^{mn} is surjective, we deduce that $g^{-m}(\mathscr{E}_X) = \mathscr{E}_X$ and hence $\mathscr{E}_X = g^m(\mathscr{E}_X)$.

Assume as in the theorem that E is analytic with $g^{-s}(E) \subset E$. Define $E' := g^{-s+1}(E) \cup \ldots \cup E$. We have $g^{-1}(E') \subset E'$ which implies $g^{-n-1}(E') \subset g^{-n}(E')$ for every $n \geq 0$. Hence, $g^{-n-1}(E') = g^{-n}(E')$ for n large enough. This and the surjectivity of g imply that $g^{-1}(E') = g(E') = E'$. By Lemma 1.48, the topological degree of $(g^{m'})_{|E'}$ is at most $d^{m'(p-1)}$ for some $m' \geq 1$. This, the identity $g^{-1}(E') = g(E') = E'$ together with Lemma 1.49 imply that $E' \subset E_1$. Hence, $E' \subset \mathscr{E}_X$ and $E \subset \mathscr{E}_X$.

Define $\mathscr{E}_X' := g^{-m+1}(\mathscr{E}_X) \cup \ldots \cup \mathscr{E}_X$. We have $g^{-1}(\mathscr{E}_X') = g(\mathscr{E}_X') = \mathscr{E}_X'$. Applying the previous assertion to $E := \mathscr{E}_X'$ yields $\mathscr{E}_X' \subset \mathscr{E}_X$. Therefore, $\mathscr{E}_X' = \mathscr{E}_X$ and $g^{-1}(\mathscr{E}_X) = g(\mathscr{E}_X) = \mathscr{E}_X$. So, $\mathscr{E}_X$ is the maximal proper analytic set in X which is totally invariant under g.

We prove now that there are only finitely many totally invariant algebraic sets. We only have to consider totally invariant sets E of pure dimension q. The proof is by induction on the dimension p of X. The case $p = 0$ is clear. Assume that the assertion is true for X of dimension $\leq p-1$ and consider the case of dimension p. If $q = p$ then E is a union of components of X. There are only a finite number of such analytic sets. If $q < p$, we have seen that E is contained in $\mathscr{E}_X$. Applying the induction hypothesis to the restriction of f to $\mathscr{E}_X$ gives the result. □

We now give another characterization of $\mathscr{E}_X$. Observe that if X is not locally irreducible at a point x then $g^{-m}(x)$ may contain more than d^{mp} points. Let $\pi : \widetilde{X} \to X$

be the normalization of X, see Appendix A.1. By Theorem A.4 applied to $g\circ\pi$, g can be lifted to a map $\widetilde{g}:\widetilde{X}\to\widetilde{X}$ such that $g\circ\pi=\pi\circ\widetilde{g}$. Since g is finite, $\widetilde{g}$ is also finite. We deduce that $\widetilde{g}^m$ defines ramified coverings of degree d^{mp} on each component of $\widetilde{X}$. In particular, any fiber of $\widetilde{g}^m$ contains at most d^{mp} points. Observe that if $g^{-1}(E)\subset E$ then $\widetilde{g}^{-1}(\widetilde{E})\subset\widetilde{E}$ where $\widetilde{E}:=\pi^{-1}(E)$. Theorem 1.47 can be extended to $\widetilde{g}$. For simplicity, we consider the case where X is itself a normal analytic space. If X is not normal, one should work with its normalization.

Let Z be a hypersurface of X containing E_1. Let $N_n(a)$ denote the number of orbits of g^m

$$a_{-n},\ldots,a_{-1},a_0$$

with $g^m(a_{-i-1})=a_{-i}$ and $a_0=a$ such that $a_{-i}\in Z$ for every i. Here, the orbits are counted with multiplicity. So, $N_n(a)$ is the number of negative orbits of order n of a which stay in Z. Observe that the sequence of functions $\tau_n:=d^{-pmn}N_n$ decreases to some function τ. Since τ_n are upper semi-continuous with respect to the Zariski topology and $0\leq\tau_n\leq 1$ (we use here the assumption that X is normal), the function τ satisfies the same properties. Note that $\tau(a)$ is the probability that an infinite negative orbit of a stays in Z.

Proposition 1.50. *Assume that X is normal. Then, τ is the characteristic function of $\mathscr{E}_X$, that is, $\tau=1$ on $\mathscr{E}_X$ and $\tau=0$ on $X\setminus\mathscr{E}_X$.*

Proof. Since $\mathscr{E}_X\subset Z$ and $\mathscr{E}_X$ is totally invariant by g, we have $\mathscr{E}_X\subset\{\tau=1\}$. Let $\theta\geq 0$ denote the maximal value of τ on $X\setminus\mathscr{E}_X$. This value exists since τ is upper semi-continuous with respect to the Zariski topology (indeed, it is enough to consider the algebraic subsets $\{\tau\geq\theta'\}$ of X which decrease when θ' increases; the family is stationary). We have to check that $\theta=0$. Assume in order to obtain a contradiction that $\theta>0$. Since $\tau\leq 1$, we always have $\theta\leq 1$. Consider the non-empty analytic set $E:=\tau^{-1}(\theta)\setminus\mathscr{E}_X$ in $Z\setminus\mathscr{E}_X$. Let a be a point in E. Since $\mathscr{E}_X$ is totally invariant, we have $g^{-m}(a)\cap\mathscr{E}_X=\varnothing$. Hence, $\tau(b)\leq\theta$ for every $b\in g^{-m}(a)$. We deduce from the definition of τ and θ that

$$\theta=\tau(a)\leq d^{-pm}\sum_{b\in g^{-m}(a)}\tau(b)\leq\theta.$$

It follows that $g^{-m}(a)\subset E$. Therefore, the analytic subset $\overline{E}$ of Z satisfies $g^{-m}(\overline{E})\subset\overline{E}$. This contradicts the maximality of $\mathscr{E}_X$. □

We continue the proof of Theorem 1.45. We will use the above results for $X=\mathbb{P}^k$, Y the set of critical values of f. Let R be the ramification current defined as above by

$$R=\sum_{n\geq 0}R_n:=\sum_{n\geq 0}d^{-kn}(f^n)_*[Y].$$

The following proposition was obtained in [DS1], a weaker version was independently obtained by Briend-Duval [BD3]. Here, an inverse branch on B for f^n is a bi-holomorphic map $g_i:B\to U_i$ such that $g_i\circ f^n$ is identity on U_i.

Proposition 1.51. *Let ν be a strictly positive constant. Let a be a point in $\mathbb{P}^k$ such that the Lelong number $\nu(R,a)$ of R at a is strictly smaller than ν. Then, there is a ball B centered at a such that f^n admits at least $(1-\sqrt{\nu})d^{kn}$ inverse branches $g_i : B \to U_i$ where U_i are open sets in $\mathbb{P}^k$ of diameter $\leq d^{-n/2}$. In particular, if μ' is a limit value of the measures $d^{-kn}(f^n)^*(\delta_a)$ then $\|\mu' - \mu\| \leq 2\sqrt{\nu(R,a)}$.*

Given a local coordinate system at a, let $\mathscr{F}$ denote the family of complex lines passing through a. For such a line Δ denote by Δ_r the disc of center a and of radius r. The family $\mathscr{F}$ is parametrized by $\mathbb{P}^{k-1}$ where the probability measure (the volume form) associated to the Fubini-Study metric is denoted by $\mathscr{L}$. Let B_r denote the ball of center a and of radius r.

Lemma 1.52. *Let S be a positive closed $(1,1)$-current on a neighbourhood of a. Then for any $\delta > 0$ there is an $r > 0$ and a family $\mathscr{F}' \subset \mathscr{F}$, such that $\mathscr{L}(\mathscr{F}') \geq 1-\delta$ and for every Δ in $\mathscr{F}'$, the measure $S \wedge [\Delta_r]$ is well-defined and of mass $\leq \nu(S,a)+\delta$, where $\nu(S,a)$ is the Lelong number of S at a.*

Proof. Let $\pi : \widehat{\mathbb{P}}^k \to \mathbb{P}^k$ be the blow-up of $\mathbb{P}^k$ at a and E the exceptional hypersurface. Then, we can write $\pi^*(S) = \nu(S,a)[E] + S'$ with S' a current having no mass on E, see Exercise A.39. It is clear that for almost every Δ_r, the restriction of the potentials of S to Δ_r is not identically $-\infty$, so, the measure $S \wedge [\Delta_r]$ is well-defined. Let $\widehat{\Delta}_r$ denote the strict transform of Δ_r by π, i.e. the closure of $\pi^{-1}(\Delta_r \setminus \{a\})$. Then, the $\widehat{\Delta}_r$ define a smooth holomorphic fibration over E. The measure $S \wedge [\Delta_r]$ is equal to the push-forward of $\pi^*(S) \wedge [\widehat{\Delta}_r]$ by π. Observe that $\pi^*(S) \wedge [\widehat{\Delta}_r]$ is equal to $S' \wedge [\widehat{\Delta}_r]$ plus $\nu(S,a)$ times the Dirac mass at $\widehat{\Delta}_r \cap E$. Therefore, we only have to consider the Δ_r such that $S' \wedge [\widehat{\Delta}_r]$ are of mass $\leq \delta$.

Since S' have no mass on E, its mass on $\pi^{-1}(B_r)$ tends to 0 when r tends to 0. It follows from Fubini's theorem that when r is small enough the mass of the slices $S' \wedge [\widehat{\Delta}_r]$ is $\leq \delta$ except for a small family of Δ. This proves the lemma. □

Lemma 1.53. *Let U be a neighbourhood of $\overline{B}_r$. Let S be a positive closed $(1,1)$-current on U. Then, for every $\delta > 0$, there is a family $\mathscr{F}' \subset \mathscr{F}$ with $\mathscr{L}(\mathscr{F}') > 1-\delta$, such that for Δ in $\mathscr{F}'$, the measure $S \wedge [\Delta_r]$ is well-defined and of mass $\leq A\|S\|$, where $A > 0$ is a constant depending on δ but independent of S.*

Proof. We can assume that $\|S\| = 1$. Let π be as in Lemma 1.52. Then, by continuity of π^*, the mass of $\pi^*(S)$ on $\pi^{-1}(B_r)$ is bounded by a constant. It is enough to apply Fubini's theorem in order to estimate the mass of $\pi^*(S) \wedge [\widehat{\Delta}_r]$. □

Recall the following theorem due to Sibony-Wong [SW].

Theorem 1.54. *Let $m > 0$ be a positive constant. Let $\mathscr{F}' \subset \mathscr{F}$ be such that $\mathscr{L}(\mathscr{F}') \geq m$ and let Σ denote the intersection of the family $\mathscr{F}'$ with B_r. Then any holomorphic function h on a neighbourhood of Σ can be extended to a holomorphic function on $B_{\lambda r}$ where $\lambda > 0$ is a constant depending on m but independent of $\mathscr{F}'$ and r. Moreover, we have*

$$\sup_{B_{\lambda r}} |h| \leq \sup_{\Sigma} |h|.$$

We will use the following version of a lemma due to Briend-Duval [BD2]. Their proof uses the theory of moduli of annuli.

Lemma 1.55. *Let $g : \Delta_r \to \mathbb{P}^k$ be a holomorphic map from a disc of center 0 and of radius r in $\mathbb{C}$. Assume that* $\mathrm{area}(g(\Delta_r))$ *counted with multiplicity, is smaller than* $1/2$. *Then for any $\varepsilon > 0$ there is a constant $\lambda > 0$ independent of g, r such that the diameter of $g(\Delta_{\lambda r})$ is smaller than $\varepsilon\sqrt{\mathrm{area}(g(\Delta_r))}$.*

Proof. Observe that the lemma is an easy consequence of the Cauchy formula if g has values in a compact set of $\mathbb{C}^k \subset \mathbb{P}^k$. In order to reduce the problem to this case, it is enough to prove that given an $\varepsilon_0 > 0$, there is a constant $\lambda_0 > 0$ such that $\mathrm{diam}(g(\Delta_{\lambda_0 r})) \leq \varepsilon_0$. For ε_0 small enough, we can apply the above case to g restricted to $\Delta_{\lambda_0 r}$.

By hypothesis, the graphs Γ_g of g in $\Delta_r \times \mathbb{P}^k$ have bounded area. So, according to Bishop's theorem [BS], these graphs form a relatively compact family of analytic sets, that is, the limits of these graphs in the Hausdorff sense, are analytic sets. Since $\mathrm{area}(g(\Delta_r))$ is bounded by $1/2$, the limits have no compact components. So, they are also graphs and the family of the maps g is compact. We deduce that $\mathrm{diam}(g(\Delta_{\lambda_0 r})) \leq \varepsilon_0$ for λ_0 small enough. □

Sketch of the proof of Proposition 1.51. The last assertion in the proposition is deduced from the first one and Proposition 1.46 applied to a generic point in B. We obtain that $\|\mu' - \mu\| \leq 2\sqrt{\nu}$ for every ν strictly larger than $\nu(R,a)$ which implies the result.

For the first assertion, the idea is to construct inverse branches for many discs passing through a and then to apply Theorem 1.54 in order to construct inverse branches on balls. We can assume that ν is smaller than 1. Choose constants $\delta > 0$, $\varepsilon > 0$ small enough and then a constant $\kappa > 0$ large enough; all independent of n. Fix now the integer n. Recall that $\|(f^n)_*(\omega_{\mathrm{FS}})\| = d^{(k-1)n}$. By Lemmas 1.52 and 1.53, there is a family $\mathscr{F}' \subset \mathscr{F}$ and a constant $r > 0$ such that $\mathscr{L}(\mathscr{F}') > 1 - \delta$ and for any Δ in $\mathscr{F}'$, the mass of $R \wedge [\Delta_{\kappa^2 r}]$ is smaller than ν and the mass of $(f^n)_*(\omega_{\mathrm{FS}}) \wedge [\Delta_{\kappa r}]$ is smaller than $Ad^{(k-1)n}$ with $A > 0$.

Claim. For each Δ in $\mathscr{F}'$, f^n admits at least $(1 - 2\nu)d^{kn}$ inverse branches $g_i : \Delta_{\kappa^2 r} \to V_i$ with $\mathrm{area}(V_i) \leq A\nu^{-1}d^{-n}$. The inverse branches g_i can be extended to a neighbourhood of $\Delta_{\kappa^2 r}$.

Assuming the claim, we complete the proof of the proposition. Let $a_1, \ldots, a_l$ be the points in $f^{-n}(a)$, with $l \leq d^{kn}$, and $\mathscr{F}'_s \subset \mathscr{F}'$ the family of Δ's such that one of the previous inverse branches $g_i : \Delta_{\kappa^2 r} \to V_i$ passes through a_s, that is, V_i contains a_s. The above claim implies that $\sum \mathscr{L}(\mathscr{F}'_s) \geq (1-\delta)(1-2\nu)d^{kn}$. There are at most d^{kn} terms in this sum. We only consider the family $\mathscr{S}$ of the indices s such that $\mathscr{L}(\mathscr{F}'_s) \geq 1 - 3\sqrt{\nu}$. Since $\mathscr{L}(\mathscr{F}'_s) \leq 1$ for every s, we have

$$\#\mathscr{S} + (d^{kn} - \#\mathscr{S})(1 - 3\sqrt{\nu}) \geq \sum \mathscr{L}(\mathscr{F}'_s) \geq (1-\delta)(1-2\nu)d^{kn}.$$

Therefore, since δ is small, we have $\#\mathscr{S} \geq (1 - \sqrt{\nu})d^{kn}$. For any index $s \in \mathscr{S}$ and for Δ in $\mathscr{F}'_s$, by Lemma 1.55, the corresponding inverse branch on $\Delta_{\kappa r}$, which passes through a_s, has diameter $\leq \varepsilon d^{-n/2}$. By Theorem 1.54, f^n admits an inverse branch defined on the ball B_r and passing through a_s, with diameter $\leq d^{-n/2}$. This implies the result.

Proof of the claim. Let v_l denote the mass of $R_l \wedge [\Delta_{\kappa^2 r}]$. Then, $\sum v_l$ is the mass of $R \wedge [\Delta_{\kappa^2 r}]$. Recall that this mass is smaller than v. By definition, $v_l d^{kl}$ is the number of points in $f^l(Y) \cap \Delta_{\kappa^2 r}$, counted with multiplicity. We only have to consider the case $v < 1$. So, we have $v_0 = 0$ and $\Delta_{\kappa^2 r}$ does not intersect Y, the critical values of f. It follows that $\Delta_{\kappa^2 r}$ admits d^k inverse branches for f. By definition of v_1, there are at most $v_1 d^k$ such inverse branches which intersect Y, i.e. the images intersect Y. So, $(1 - v_1)d^k$ of them do not meet Y and the image of such a branch admits d^k inverse branches for f. We conclude that $\Delta_{\kappa^2 r}$ admits at least $(1 - v_1)d^{2k}$ inverse branches for f^2. By induction, we construct for f^n at least $(1 - v_1 - \cdots - v_{n-1})d^{kn}$ inverse branches on $\Delta_{\kappa^2 r}$.

Now, observe that the mass of $(f^n)_*(\omega_{\rm FS}) \wedge [\Delta_{\kappa r}]$ is exactly the area of $f^{-n}(\Delta_{\kappa r})$. We know that it is smaller than $Ad^{(k-1)n}$. It is not difficult to see that $\Delta_{\kappa^2 r}$ has at most $v d^{kn}$ inverse branches with area $\geq A v^{-1} d^{-n}$. This completes the proof. □

End of the proof of Theorem 1.45. Let a be a point out of the exceptional set $\mathscr{E}$ defined in Theorem 1.47 for $X = \mathbb{P}^k$. Fix $\varepsilon > 0$ and a constant $\alpha > 0$ small enough. If μ' is a limit value of $d^{-kn}(f^n)^*(\delta_a)$, it is enough to show that $\|\mu' - \mu\| \leq 2\alpha + 2\varepsilon$. Consider $Z := \{v(R,z) > \varepsilon\}$ and τ as in Proposition 1.50 for $X = \mathbb{P}^k$. We have $\tau(a) = 0$. So, for r large enough we have $\tau_r(a) \leq \alpha$. Consider all the negative orbits $\mathscr{O}_j$ of order $r_j \leq r$

$$\mathscr{O}_j = \left\{a_{-r_j}^{(j)}, \ldots, a_{-1}^{(j)}, a_0^{(j)}\right\}$$

with $f(a_{-i-1}^{(j)}) = a_{-i}^{(j)}$ and $a_0^{(j)} = a$ such that $a_{-r_j}^{(j)} \notin Z$ and $a_{-i}^{(j)} \in Z$ for $i \neq r_j$. Each orbit is repeated according to its multiplicity. Let S_r denote the family of points $b \in f^{-r}(a)$ such that $f^i(b) \in Z$ for $0 \leq i \leq r$. Then $f^{-r}(a) \setminus S_r$ consists of the preimages of the points $a_{-r_j}^{(j)}$. So, by definition of τ_r, we have

$$d^{-kr}\#S_r = \tau_r(a) \leq \alpha$$

and

$$d^{-kr}\sum_j d^{k(r-r_j)} = d^{-kr}\#(f^{-r}(a) \setminus S_r) = 1 - \tau_r(a) \geq 1 - \alpha.$$

We have for $n \geq r$

$$d^{-kn}(f^n)^*(\delta_a) = \quad d^{-kn}\sum_{b\in S_r}(f^{(n-r)})^*(\delta_b) + d^{-kn}\sum_j (f^{(n-r_j)})^*\left(\delta_{a_{-r_j}^{(j)}}\right).$$

Since $d^{-kn}(f^n)^*$ preserves the mass of any measure, the first term in the last sum is of mass $d^{-kr}\#S_r = \tau_r(a) \leq \alpha$ and the second term is of mass $\geq 1 - \alpha$. We apply Proposition 1.51 to the Dirac masses at $a_{-r_j}^{(j)}$. We deduce that if μ' is a limit value of $d^{-kn}(f^n)^*(\delta_a)$ then

$$\|\mu' - \mu\| \leq 2\alpha + 2\varepsilon.$$

This completes the proof of the theorem. □

We have the following more general result. When X is not normal, one has an analogous result for the lift of g to the normalization of X.

Theorem 1.56. *Let X be an irreducible analytic set of dimension p, invariant under f. Let g denote the restriction of f to X and $\mathscr{E}_X$ the exceptional set of g. Assume that X is a normal analytic space. Then $d^{-pn}(g^n)^*(\delta_a)$ converge to $\mu_X := (\deg X)^{-1}T^p \wedge [X]$ if and only if a is out of $\mathscr{E}_X$. Moreover, the convex set of probability measures on X which are totally invariant under g, is of finite dimension.*

Proof. The proof of the first assertion follows the same lines as in Theorem 1.45. We use the fact that g is the restriction of a holomorphic map in $\mathbb{P}^k$ in order to define the ramification current R. The assumption that X is normal allows to define $d^{-pn}(g^n)^*(\delta_a)$. We prove the second assertion. Observe that an analytic set, totally invariant by g^n, is not necessarily totally invariant by g, but it is a union of components of such sets, see Theorem 1.47. Therefore, we can replace g with an iterate g^n in order to assume that g fixes all the components of all the totally invariant analytic sets. Let μ' be an extremal element in the convex set of totally invariant probability measures and X' the smallest totally invariant analytic set such that $\mu'(X') = 1$. The first assertion applied to X' implies that $\mu' = \mu_{X'}$. Hence, the set of such μ' is finite. We use a normalization of X' if necessary. □

The following result due to Briend-Duval [BD1], shows that repelling periodic points are equidistributed on the support of the Green measure.

Theorem 1.57. *Let P_n denote the set of repelling periodic points of period n on the support of μ. Then the sequence of measures*

$$\mu_n := d^{-kn} \sum_{a \in P_n} \delta_a$$

converges to μ.

Proof. By Proposition 1.3, the number of periodic points of period n of f, counted with multiplicity, is equal to $(d^n - 1)^{-1}(d^{k(n+1)} - 1)$. Therefore, any limit value μ' of μ_n is of mass ≤ 1. Fix a small constant $\varepsilon > 0$. It is enough to check that for μ-almost every point $a \in \mathbb{P}^k$, there is a ball B centered at a, arbitrarily small, such that $\#P_n \cap B \geq (1-\varepsilon)d^{kn}\mu(B)$ for large n. We will use in particular a trick due to X. Buff, which simplifies the original proof.

Since μ is PC, it has no mass on analytic sets. So, it has no mass on the orbit $\mathscr{O}_Y$ of Y, the set of critical values of f. Fix a point a on the support of μ and out of $\mathscr{O}_Y$. We have $\nu(R,a) = 0$. By Proposition 1.51, there is a ball B of center a, with sufficiently small radius, which admits $(1-\varepsilon^2)d^{kn}$ inverse branches of diameter $\leq d^{-n/2}$ for f^n when n is large enough. Choose a finite family of such balls B_i of center b_i with $1 \leq i \leq m$ such that $\mu(B_1 \cup \ldots \cup B_m) > 1 - \varepsilon^2\mu(B)$ and each B_i admits $(1-\varepsilon^2\mu(B))d^{kn}$ inverse branches of diameter $\leq d^{-n/2}$ for f^n when n is large enough. Choose balls $B_i' \Subset B_i$ such that $\mu(B_1' \cup \ldots \cup B_m') > 1 - \varepsilon^2\mu(B)$.

Fix a constant N large enough. Since $d^{-kn}(f^n)^*(\delta_a)$ converge to μ, there are at least $(1-2\varepsilon^2)d^{kN}$ inverse branches for f^N whose image intersects $\cup B_j'$ and then with image contained in one of the B_j. In the same way, we show that for n large enough, each B_j admits $(1-2\varepsilon^2)\mu(B)d^{k(n-N)}$ inverse branches for f^{n-N} with images in B. Therefore, B admits at least $(1-2\varepsilon^2)^2\mu(B)d^{kn}$ inverse branches $g_i : B \to U_i$ for f^n with image $U_i \Subset B$. Observe that every holomorphic map $g : B \to U \Subset B$ contracts the Kobayashi metric and then admits an attractive fixed point z. Moreover, g^l converges uniformly to z and $\cap_l g^l(\overline{B}) = \{z\}$. Therefore, each g_i admits a fixed attractive point a_i. This point is fixed and repelling for f^n. They are different since the U_i are disjoint. Finally, since μ is totally invariant, its support is also totally invariant under f. Hence, a_i, which is equal to $\cap_l g_i^l(\mathrm{supp}(\mu)\cap\overline{B})$, is necessarily in $\mathrm{supp}(\mu)$. We deduce that

$$\#P_n \cap B \geq (1-2\varepsilon^2)^2\mu(B)d^{kn} \geq (1-\varepsilon)d^{kn}\mu(B).$$

This completes the proof. □

Note that the periodic points a_i, constructed above, satisfy $\|(Df^n)^{-1}(a_i)\| \lesssim d^{-n/2}$. Note also that in the previous theorem, one can replace P_n with the set of all periodic points counting with multiplicity or not.

Exercise 1.58. Let f be an endomorphism of algebraic degree $d \geq 2$ of $\mathbb{P}^k$ as above. Let K be a compact set such that $f^{-1}(K) \subset K$. Show that either K contains $\mathscr{J}_k$, the Julia set of order k, or K is contained in the exceptional set $\mathscr{E}$. Prove that $a \notin \mathscr{E}$ if and only if $\cup f^{-n}(a)$ is Zariski dense.

Exercise 1.59. Let f be as above and U an open set which intersects the Julia set $\mathscr{J}_k$. Show that $\cup_{n\geq 0} f^n(U)$ is a Zariski dense open set of $\mathbb{P}^k$. Deduce that f is topologically transitive on $\mathscr{J}_k$, that is, for any given non-empty open sets V,W on $\mathscr{J}_k$, there is an integer $n \geq 0$ such that $f^n(V)\cap W \neq \varnothing$. If $\mathscr{E} = \varnothing$, show that $f^n(U) = \mathbb{P}^k$ for n large enough. If $\mathscr{E} \cap \mathscr{J}_k = \varnothing$, show that $f^n(V) = \mathscr{J}_k$ for n large enough.

Exercise 1.60. Assume that p is a repelling fixed point in $\mathscr{J}_k$. If g is another endomorphism close enough to f in $\mathscr{H}_d(\mathbb{P}^k)$ such that $g(p) = p$, show that p belongs also to the Julia set of order k of g. Hint: use that $g \mapsto \mu_g$ is continuous.

Exercise 1.61. Using Example 1.10, construct a map f in $\mathscr{H}_d(\mathbb{P}^k)$, $d \geq 2$, such that for n large enough, every fiber of f^n contains more than $d^{(k-1/2)n}$ points. Deduce that there is Zariski dense open set in $\mathscr{H}_d(\mathbb{P}^k)$ such that if f is in that Zariski open set, its exceptional set is empty.

Exercise 1.62. Let ε be a fixed constant such that $0 < \varepsilon < 1$. Let P_n' the set of repelling periodic points a of prime period n on the support of μ such that all the eigenvalues of Df^n at a are of modulus $\geq (d-\varepsilon)^{n/2}$. Show that $d^{-kn}\sum_{a\in P_n'}\delta_a$ converges to μ.

Exercise 1.63. Let g be as in Theorem 1.56. Show that repelling periodic points on $\mathrm{supp}(\mu_X)$ are equidistributed with respect to μ_X. In particular, they are Zariski dense.

1.5 Equidistribution of Varieties

In this paragraph, we consider the inverse images by f^n of varieties in $\mathbb{P}^k$. The geometrical method in the last paragraph is quite difficult to apply here. Indeed, the inverse image of a generic variety of codimension $p < k$ is irreducible of degree $O(d^{pn})$. The pluripotential method that we introduce here is probably the right method for the equidistribution problem. Moreover, it should give some precise estimates on the convergence, see Remark 1.71.

The following result, due to the authors, gives a satisfactory solution in the case of hypersurfaces. It was proved for Zariski generic maps by Fornæss-Sibony in [FS3, S3] and for maps in dimension 2 by Favre-Jonsson in [FJ]. More precise results are given in [DS9] and in [FJ, FJ1] when $k = 2$. The proof requires some self-intersection estimates for currents, due to Demailly-Méo.

Theorem 1.64. *Let f be an endomorphism of algebraic degree $d \geq 2$ of $\mathbb{P}^k$. Let $\mathscr{E}_m$ denote the union of the totally invariant proper analytic sets in $\mathbb{P}^k$ which are minimal, i.e. do not contain smaller ones. Let S be a positive closed $(1,1)$-current of mass 1 on $\mathbb{P}^k$ whose local potentials are not identically $-\infty$ on any component of $\mathscr{E}_m$. Then, $d^{-n}(f^n)^*(S)$ converge weakly to the Green $(1,1)$-current T of f.*

The following corollary gives a solution to the equidistribution problem for hypersurfaces: the exceptional hypersurfaces belong to a proper analytic set in the parameter space of hypersurfaces of a given degree.

Corollary 1.65. *Let f, T and $\mathscr{E}_m$ be as above. If H is a hypersurface of degree s in $\mathbb{P}^k$, which does not contain any component of $\mathscr{E}_m$, then $s^{-1}d^{-n}(f^n)^*[H]$ converge to T in the sense of currents.*

Note that $(f^n)^*[H]$ is the current of integration on $f^{-n}(H)$ where the components of $f^{-n}(H)$ are counted with multiplicity.

Sketch of the proof of Theorem 1.64. We can write $S = T + dd^c u$ where u is a p.s.h. function modulo T, that is, the difference of quasi-potentials of S and of T. Subtracting from u a constant allows to assume that $\langle \mu, u \rangle = 0$. We call *$u$ the dynamical quasi-potential* of S. Since T has continuous quasi-potentials, u satisfies analogous properties that quasi-p.s.h. functions do. We are mostly concerned with the singularities of u.

The total invariance of T and μ implies that the dynamical quasi-potential of $d^{-n}(f^n)^*(S)$ is equal to $u_n := d^{-n} u \circ f^n$. We have to show that this sequence of functions converges to 0 in $L^1(\mathbb{P}^k)$. Since u is bounded from above, we have $\limsup u_n \leq 0$. Assume that u_n do not converge to 0. By Hartogs' lemma, see Proposition A.20, there is a ball B and a constant $\lambda > 0$ such that $u_n \leq -\lambda$ on B for infinitely many indices n. It follows that $u \leq -\lambda d^n$ on $f^n(B)$ for such an index n. On the other hand, the exponential estimate in Theorem A.22 implies that $\|e^{\alpha|u|}\|_{L^1} \leq A$ for some positive constants α and A independent of u. If the multiplicity of f at every point is $\leq d-1$, then a version of Lojasiewicz's theorem implies that $f^n(B)$ contains a ball of radius $\simeq e^{-c(d-1)^n}$, $c>0$. Therefore, we have

$$e^{-2kc(d-1)^n}e^{\lambda d^n\alpha} \lesssim \int_{f^n(B)} e^{\lambda d^n\alpha}\omega_{\rm FS}^k \leq \int_{\mathbb{P}^k} e^{\alpha|u|}\omega_{\rm FS}^k.$$

This contradicts the above exponential estimate.

In general, by Lemma 1.66 below, $f^n(B)$ contains always a ball of radius $\simeq e^{-cd^n}$. So, a slightly stronger version of the above exponential estimate will be enough to get a contradiction. We may improve this exponential estimate: if the Lelong numbers of S are small, we can increase the constant α and get a contradiction; if the Lelong numbers of S_n are small, we replace S with S_n.

The assumption $u_n < -\lambda d^n$ on $f^n(B)$ allows to show that all the limit currents of the sequence $d^{-n}(f^n)^*(S)$ have Lelong numbers larger than some constant $\nu>0$. If S' is such a current, there are other currents S'_n such that $S' = d^{-n}(f^n)^*(S'_n)$. Indeed, if S' is the limit of $d^{-n_i}(f^{n_i})^*(S)$ one can take S'_n a limit value of $d^{-n_i+n}(f^{n_i-n})^*(S)$.

Let a be a point such that $\nu(S'_n,a)\geq\nu$. The assumption on the potentials of S allows to prove by induction on the dimension of the totally invariant analytic sets that u_n converge to 0 on the maximal totally invariant set $\mathscr{E}$. So, a is out of $\mathscr{E}$. Lemma 1.49 allows to construct many distinct points in $f^{-n}(a)$. The identity $S' = d^{-n}(f^n)^*(S'_n)$ implies an estimate from below of the Lelong numbers of S' on $f^{-n}(a)$. This holds for every n. Finally, this permits to construct analytic sets of large degrees on which we have estimates on the Lelong numbers of S'. Therefore, S' has a too large self-intersection. This contradicts an inequality of Demailly-Méo [DE,ME3] and completes the proof. Note that the proof of Demailly-Méo inequality uses Hörmander's L^2 estimates for the $\overline{\partial}$-equation. □

The following lemma is proved in [DS9]. It also holds for meromorphic maps. Some earlier versions were given in [FS3] and in terms of Lebesgue measure in [FJ, G2].

Lemma 1.66. *There is a constant $c>0$ such that if B is a ball of radius r, $0<r<1$, in $\mathbb{P}^k$, then $f^n(B)$ contains a ball B_n of radius* $\exp(-cr^{-2k}d^n)$ *for any $n\geq 0$.*

The ball B_n is centered at $f^n(a_n)$ for some point $a_n\in B$ which is not necessarily the center of B. The key point in the proof of the lemma is to find a point a_n with an estimate from below on the Jacobian of f^n at a_n. If u is a quasi-potential of the current of integration on the critical set, the logarithm of this Jacobian is essentially the value of $u+u\circ f+\cdots+u\circ f^{n-1}$ at a_n. So, in order to prove the existence of a point a_n with a good estimate, it is enough to bound the L^1 norm of the last function. One easily obtains the result using the operator $f^*:\mathrm{DSH}(\mathbb{P}^k)\to\mathrm{DSH}(\mathbb{P}^k)$ and its iterates, as it is done for f_* in Proposition 1.34.

Remark 1.67. Let $\mathscr{C}$ denote the convex compact set of totally invariant positive closed $(1,1)$-currents of mass 1 on $\mathbb{P}^k$, i.e. currents S such that $f^*(S)=dS$. Define an operator $\vee$ on $\mathscr{C}$. If S_1, S_2 are elements of $\mathscr{C}$ and u_1,u_2 their dynamical quasi-potentials, then $u_i\leq 0$. Since $\langle\mu,u_i\rangle=0$ and u_i are upper semi-continuous, we deduce that $u_i = 0$ on $\mathrm{supp}(\mu)$. Define $S_1 \vee S_2 := T + dd^c \max(u_1, u_2)$. It is easy to check that $S_1 \vee S_2$ is an element of $\mathscr{C}$. An element S is said to be *minimal* if $S = S_1 \vee S_2$ implies $S_1 = S_2 = S$. It is clear that T is not minimal if $\mathscr{C}$ contains other currents. In fact, for S in $\mathscr{C}$, we have $T \vee S = T$. A current of integration on a totally invariant hypersurface is a minimal element. It is likely that $\mathscr{C}$ is generated by a finite number of currents, the operation $\vee$, convex hulls and limits.

Example 1.68. If f is the map given in Example 1.11, the exceptional set $\mathscr{E}_m$ is the union of the $k+1$ attractive fixed points

$$[0 : \cdots : 0 : 1 : 0 : \cdots : 0].$$

The convergence of $s^{-1} d^{-n} (f^n)^*[H]$ towards T holds for hypersurfaces H of degree s which do not contain these points. If $\pi : \mathbb{C}^{k+1} \setminus \{0\} \to \mathbb{P}^k$ is the canonical projection, the Green $(1,1)$-current T of f is given by $\pi^*(T) = dd^c(\max_i \log|z_i|)$, or equivalently $T = \omega_{\mathrm{FS}} + dd^c v$ where

$$v[z_0 : \cdots : z_k] := \max_{0 \leq i \leq k} \log|z_i| - \frac{1}{2}\log(|z_0|^2 + \cdots + |z_k|^2).$$

The currents T_i of integration on $\{z_i = 0\}$ belong to $\mathscr{C}$ and $T_i = T + dd^c u_i$ with $u_i := \log|z_i| - \max_j \log|z_j|$. These currents are minimal. If $\alpha_0, \ldots, \alpha_k$ are positive real numbers such that $\alpha := 1 - \sum \alpha_i$ is positive, then $S := \alpha T + \sum \alpha_i T_i$ is an element of $\mathscr{C}$. We have $S = T + dd^c u$ with $u := \sum \alpha_i u_i$. The current S is minimal if and only if $\alpha = 0$. One can obtain other elements of $\mathscr{C}$ using the operator $\vee$. We show that $\mathscr{C}$ is infinite dimensional. Define for $A := (\alpha_0, \ldots, \alpha_k)$ with $0 \leq \alpha_i \leq 1$ and $\sum \alpha_i = 1$ the p.s.h. function v_A by

$$v_A := \sum \alpha_i \log|z_i|.$$

If $\mathscr{A}$ is a family of such $(k+1)$-tuples A, define

$$v_{\mathscr{A}} := \sup_{A \in \mathscr{A}} v_A.$$

Then, we can define a positive closed $(1,1)$-currents $S_{\mathscr{A}}$ on $\mathbb{P}^k$ by $\pi^*(S_{\mathscr{A}}) = dd^c v_{\mathscr{A}}$. It is clear that $S_{\mathscr{A}}$ belongs to $\mathscr{C}$ and hence $\mathscr{C}$ is of infinite dimension.

The equidistribution problem in higher codimension is much more delicate and is still open for general maps. We first recall the following lemma.

Lemma 1.69. *For every $\delta > 1$, there is a Zariski dense open set $\mathscr{H}_d^*(\mathbb{P}^k)$ in $\mathscr{H}_d(\mathbb{P}^k)$ and a constant $A > 0$ such that for f in $\mathscr{H}_d^*(\mathbb{P}^k)$, the maximal multiplicity δ_n of f^n at a point in $\mathbb{P}^k$ is at most equal to $A\delta^n$. In particular, the exceptional set of such a map f is empty when $\delta < d$.*

Proof. Let X be a component of a totally invariant analytic set E of pure dimension $p \leq k-1$. Then, f permutes the components of E. We deduce that X is totally invariant under f^n for some $n \geq 1$. Lemma 1.48 implies that the maximal multiplicity of f^n at a point in X is at least equal to $d^{(k-p)n}$. Therefore, the second assertion in the lemma is a consequence of the first one.

Fix an N large enough such that $\delta^N > 2^k k!$. Let $\mathscr{H}_d^*(\mathbb{P}^k)$ be the set of f such that $\delta_N \leq 2^k k!$. This set is Zariski open in $\mathscr{H}_d(\mathbb{P}^k)$. Since the sequence (δ_n) is sub-multiplicative, i.e. $\delta_{n+m} \leq \delta_n \delta_m$ for $n,m \geq 0$, if f is in $\mathscr{H}_d^*(\mathbb{P}^k)$, we have $\delta_N < \delta^N$, hence $\delta_n \leq A\delta^n$ for A large enough and for all n. It remains to show that $\mathscr{H}_d^*(\mathbb{P}^k)$ is not empty. Choose a rational map $h : \mathbb{P}^1 \to \mathbb{P}^1$ of degree d whose critical points are simple and have disjoint infinite orbits. Observe that the multiplicity of h^N at every point is at most equal to 2. We construct the map f using the method described in Example 1.10. We have $f^N \circ \Pi = \Pi \circ \widehat{f}^N$. Consider a point x in $\mathbb{P}^k$ and a point $\widehat{x}$ in $\Pi^{-1}(x)$. The multiplicity of $\widehat{f}^N$ at $\widehat{x}$ is at most equal to 2^k. It follows that the multiplicity of f^N at x is at most equal to $2^k k!$ since Π has degree $k!$. Therefore, f satisfies the desired inequality. □

We have the following result due to the authors [DS10].

Theorem 1.70. *There is a Zariski dense open set $\mathscr{H}_d^*(\mathbb{P}^k)$ in $\mathscr{H}_d(\mathbb{P}^k)$ such that if f is in $\mathscr{H}_d^*(\mathbb{P}^k)$, then $d^{-pn}(f^n)^*(S) \to T^p$ uniformly on positive closed (p,p)-currents S of mass 1 on $\mathbb{P}^k$. In particular, the Green (p,p)-current T^p is the unique positive closed (p,p)-current of mass 1 which is totally invariant. If V is an analytic set of pure codimension p and of degree s in $\mathbb{P}^k$, then $s^{-1}d^{-pn}(f^n)^*[V]$ converge to T^p in the sense of currents.*

Sketch of proof. The proof uses the super-potentials of currents. In order to simplify the notation, introduce the dynamical super-potential $\mathscr{V}$ of S. Define $\mathscr{V} := \mathscr{U}_S - \mathscr{U}_{T^p} + c$ where $\mathscr{U}_S, \mathscr{U}_{T^p}$ are super-potentials of S, T^p and the constant c is chosen so that $\mathscr{V}(T^{k-p+1}) = 0$. Using a computation as in Theorem 1.32, we obtain that the dynamical super-potential of $d^{-pn}(f^n)^*(S)$ is equal to $d^{-n}\mathscr{V} \circ \Lambda^n$ where $\Lambda : \mathscr{C}_{k-p+1}(\mathbb{P}^k) \to \mathscr{C}_{k-p+1}(\mathbb{P}^k)$ is the operator $d^{-p+1}f_*$. Observe that the dynamical super-potential of T^p is identically 0. In order to prove the convergence $d^{-pn}(f^n)^*(S) \to T^p$, we only have to check that $d^{-n}\mathscr{V}(\Lambda^n(R)) \to 0$ for R smooth in $\mathscr{C}_{k-p+1}(\mathbb{P}^k)$. Since T^p has a continuous super-potential, $\mathscr{V}$ is bounded from above. Therefore, $\limsup d^{-n}\mathscr{V}(\Lambda^n(R)) \leq 0$.

Recall that $\mathscr{U}_S(R) = \mathscr{U}_R(S)$. So, in order to prove that $\liminf d^{-n}\mathscr{V}(\Lambda^n(R)) \geq 0$, it is enough to estimate $\inf_S \mathscr{U}_S(\Lambda^n(R))$, or equivalently, we have to estimate the capacity of $\Lambda^n(R)$ from below. Assume in order to explain the idea that the support of R is contained in a compact set K such that $f(K) \subset K$ and K does not intersect the critical set of f (this is possible only when $p = 1$). We easily obtain that $\|\Lambda^n(R)\|_\infty \lesssim A^n$ for some constant $A > 0$. The estimate in Theorem A.47 implies the result. In the general case, if $\mathscr{H}_d^*(\mathbb{P}^k)$ is chosen as in Lemma 1.69 for δ small enough and if f is in $\mathscr{H}_d^*(\mathbb{P}^k)$, we can prove the estimate $\mathrm{cap}(\Lambda^n(R)) \lesssim d'^n$ for any fixed constant d' such that $1 < d' < d$. This implies the desired convergence of super-potentials. The main technical difficulty is that when R hits the critical set, then $\Lambda(R)$ is not bounded. The estimates requires a smoothing and precise evaluation of the error. □

Remark 1.71. The above estimate on $\mathrm{cap}(\Lambda^n(R))$ can be seen as a version of Lojasiewicz's inequality for currents. It is quite delicate to obtain. We also have an explicit estimate on the speed of convergence. Indeed, we have for an appropriate $d' < d$:

$$\mathrm{dist}_2\big(d^{-pn}(f^n)^*(S), T^p\big) := \sup_{\|\Phi\|_{\mathscr{C}^2}\leq 1} |\langle d^{-pn}(f^n)^*(S) - T^p, \Phi\rangle| \lesssim d'^n d^{-n}.$$

The theory of interpolation between Banach spaces [T1] implies a similar estimate for Φ Hölder continuous.

Exercise 1.72. If $f := [z_0^d : \cdots : z_k^d]$, show that $\{z_1^p = z_2^q\}$, for arbitrary p, q, is invariant under f. Show that a curve invariant under an endomorphism is an image of $\mathbb{P}^1$ or a torus, possibly singular.

Exercise 1.73. Let f be as in Example 1.68. Let S be a (p,p)-current with strictly positive Lelong number at $[1:0:\cdots:0]$. Show that any limit of $d^{-pn}(f^n)^*(S)$ has a strictly positive Lelong number at $[1:0:\cdots:0]$ and deduce that $d^{-pn}(f^n)^*(S)$ do not converge to T^p. Note that for f generic, the multiplicity of the set of critical values of f^N at every point is smaller than δ^N.

Exercise 1.74. Let f be as in Theorem 1.70 for $p = k$ and Λ the associated Perron-Frobenius operator. If φ is a $\mathscr{C}^2$ function on $\mathbb{P}^k$, show that

$$\|\Lambda^n(\varphi) - \langle \mu, \varphi\rangle\|_\infty \leq c d'^n d^{-n}$$

for some constant $c > 0$. Deduce that $\Lambda^n(\varphi)$ converge uniformly to $\langle\mu,\varphi\rangle$. Give an estimate of $\|\Lambda^n(\varphi) - \langle\mu,\varphi\rangle\|_\infty$ for φ Hölder continuous.

Exercise 1.75. Let f be an endomorphism of algebraic degree $d \geq 2$ of $\mathbb{P}^k$. Assume that V is a totally invariant hypersurface, i.e. $f^{-1}(V) = V$. Let V_i denote the irreducible components of V and h_i minimal homogeneous polynomials such that $V_i = \{h_i = 0\}$. Define $h = \prod h_i$. Show that $h \circ f = ch^d$ where c is a constant. If F is a lift of f to $\mathbb{C}^{k+1}$, prove that $\mathrm{Jac}(F)$ contains $(\prod h_i)^{d-1}$ as a factor. Show that V is contained in the critical set of f and deduce[3] that $\deg V \leq k+1$. Assume now that V is reducible. Find a totally invariant positive closed $(1,1)$-current of mass 1 which is not the Green current nor the current associated to an analytic set.

Exercise 1.76. Let u be a p.s.h. function in $\mathbb{C}^k$, such that for $\lambda \in \mathbb{C}^*$, $u(\lambda z) = \log|\lambda| + u(z)$. If $\{u < 0\}$ is bounded in $\mathbb{C}^k$, show that $dd^c u^+$ is a positive closed current on $\mathbb{P}^k$ which is extremal in the cone of positive closed $(1,1)$-currents[4]. Deduce that the Green $(1,1)$-current of a polynomial map of $\mathbb{C}^k$ which extends holomorphically to $\mathbb{P}^k$, is extremal.

Exercise 1.77. Let v be a subharmonic function on $\mathbb{C}$. Suppose $v(e^{i\theta}z) = v(z)$ for every $z \in \mathbb{C}$ and for every $\theta \in \mathbb{R}$ such that $e^{i\theta d^n} = 1$ for some integer n. Prove

[3] It is known that in dimension $k = 2$, V is a union of at most 3 lines, [CL, FS7, SSU].

[4] Unpublished result by Berndtsson-Sibony.

that $v(z) = v(|z|)$ for $z \in \mathbb{C}$. Hint: use the Laplacian of v. Let f be as in Example 1.68, R a current in $\mathbb{C}^{k+1}$, and v a p.s.h. function on $\mathbb{C}^{k+1}$ such that $R = dd^c v$ and $v(F(z)) = dv(z)$, where $F(z) := (z_0^d, \ldots, z_k^d)$ is a lift of f to $\mathbb{C}^{k+1}$. Show that v is invariant under the action of the unit torus $\mathbb{T}^{k+1}$ in $\mathbb{C}^{k+1}$. Determine such functions v. Recall that $\mathbb{T}$ is the unit circle in $\mathbb{C}$ and $\mathbb{T}^{k+1}$ acts on $\mathbb{C}^{k+1}$ by multiplication.

Exercise 1.78. Define the Desboves map f_0 in $\mathscr{M}_4(\mathbb{P}^2)$ as

$$f_0[z_0 : z_1 : z_2] := [z_0(z_1^3 - z_2^3) : z_1(z_2^3 - z_0^3) : z_2(z_0^3 - z_1^3)].$$

Prove that f_0 has 12 indeterminacy points. If σ is a permutation of coordinates, compare $f_0 \circ \sigma$ and $\sigma \circ f_0$. Define

$$\Phi_\lambda(z_0, z_1, z_2) := z_0^3 + z_1^3 + z_2^3 - 3\lambda z_0 z_1 z_2, \quad \lambda \in \mathbb{C}$$

and

$$L[z_0 : z_1 : z_2] := [az_0 : bz_1 : cz_2], \quad a, b, c \in \mathbb{C}.$$

Show that for Zariski generic L, $f_L := f_0 + \Phi_\lambda L$ is in $\mathscr{H}_4(\mathbb{P}^2)$. Show that on the curve $\{\Phi_\lambda = 0\}$ in $\mathbb{P}^2$, f_L coincides with f_0, and that f_0 maps the cubic $\{\Phi_\lambda = 0\}$ onto itself.[5]

Exercise 1.79. Let f be an endomorphism of algebraic degree $d \geq 2$ of $\mathbb{P}^k$. Show that there is a finite invariant set E, possibly empty, such that if H is a hypersurface such that $H \cap E = \varnothing$, then $d^{-n}\deg(H)^{-1}(f^n)^*[H]$ converge to the Green $(1,1)$-current T of f.

1.6 Stochastic Properties of the Green Measure

In this paragraph, we are concerned with the stochastic properties of the equilibrium measure μ associated to an endomorphism f. If φ is an observable, $(\varphi \circ f^n)_{n \geq 0}$ can be seen as a sequence of dependent random variables. Since the measure is invariant, these variables are identically distributed, i.e. the Borel sets $\{\varphi \circ f^n < t\}$ have the same μ measure for any fixed constant t. The idea is to show that the dependence is weak and then to extend classical results in probability theory to our setting. One of the key point is the spectral study of the Perron-Frobenius operator $\Lambda := d^{-k} f_*$. It allows to prove the exponential decay of correlations for d.s.h. and Hölder continuous observables, the central limit theorem, the large deviation theorem, etc. An important point is to use the space of d.s.h. functions as a space of observables. For the reader's convenience, we recall few general facts from ergodic theory and probability theory. We refer to [KH, W] for the general theory.

[5] This example was considered in [BO]. It gives maps in $\mathscr{H}_4(\mathbb{P}^2)$ which preserves a cubic. The cubic is singular if $\lambda = 1$, non singular if $\lambda \neq 1$. In higher dimension, Beauville proved that a smooth hypersurface of $\mathbb{P}^k$, $k \geq 3$, of degree > 1 does not have an endomorphism with $d_t > 1$, unless the degree is 2, $k = 3$ and the hypersurface is isomorphic to $\mathbb{P}^1 \times \mathbb{P}^1$ [BV].

Consider a dynamical system associated to a map $g : X \to X$ which is measurable with respect to a σ-algebra $\mathscr{F}$ on X. The direct image of a probability measure ν by g is the probability measure $g_*(\nu)$ defined by

$$g_*(\nu)(A) := \nu(g^{-1}(A))$$

for every measurable set A. Equivalently, for any positive measurable function φ, we have

$$\langle g_*(\nu), \varphi \rangle := \langle \nu, \varphi \circ g \rangle.$$

The measure ν is *invariant* if $g_*(\nu) = \nu$. When X is a compact metric space and g is continuous, the set $\mathscr{M}(g)$ of invariant probability measures is convex, compact and non-empty: for any sequence of probability measures ν_N, the cluster points of

$$\frac{1}{N} \sum_{n=0}^{N-1} (g^n)_*(\nu_N)$$

are invariant probability measures.

A measurable set A is *totally invariant* if $\nu(A \setminus g^{-1}(A)) = \nu(g^{-1}(A) \setminus A) = 0$. An invariant probability measure ν is *ergodic* if any totally invariant set is of zero or full ν-measure. It is easy to show that ν is ergodic if and only if when $\varphi \circ g = \varphi$, for $\varphi \in L^1(\nu)$, then φ is constant. Here, we can replace $L^1(\nu)$ with $L^p(\nu)$ with $1 \leq p \leq +\infty$. The ergodicity of ν is also equivalent to the fact that it is extremal in $\mathscr{M}(g)$. We recall Birkhoff's ergodic theorem, which is the analogue of the law of large numbers for independent random variables [W].

Theorem 1.80 (Birkhoff). *Let $g : X \to X$ be a measurable map as above. Assume that ν is an ergodic invariant probability measure. Let φ be a function in $L^1(\nu)$. Then*

$$\frac{1}{N} \sum_{n=0}^{N-1} \varphi(g^n(x)) \to \langle \nu, \varphi \rangle$$

almost everywhere with respect to ν.

When X is a compact metric space, we can apply Birkhoff's theorem to continuous functions φ and deduce that for ν almost every x

$$\frac{1}{N} \sum_{n=0}^{N-1} \delta_{g^n(x)} \to \nu,$$

where δ_x denotes the Dirac mass at x. The sum

$$\mathrm{S}_N(\varphi) := \sum_{n=0}^{N-1} \varphi \circ g^n$$

is called *Birkhoff's sum*. So, Birkhoff's theorem describes the behavior of $\frac{1}{N}\mathrm{S}_N(\varphi)$ for an observable φ. We will be concerned with the precise behavior of $\mathrm{S}_N(\varphi)$ for various classes of functions φ.

A notion stronger than ergodicity is the notion of mixing. An invariant probability measure ν is *mixing* if for every measurable sets A, B

$$\lim_{n\to\infty} \nu(g^{-n}(A)\cap B) = \nu(A)\nu(B).$$

Clearly, mixing implies ergodicity. It is not difficult to see that ν is mixing if and only if for any test functions φ, ψ in $L^\infty(\nu)$ or in $L^2(\nu)$, we have

$$\lim_{n\to\infty} \langle \nu, (\varphi\circ g^n)\psi\rangle = \langle \nu, \varphi\rangle\langle \nu, \psi\rangle.$$

The quantity

$$I_n(\varphi,\psi) := |\langle \nu, (\varphi\circ g^n)\psi\rangle - \langle \nu, \varphi\rangle\langle \nu, \psi\rangle|$$

is called *the correlation at time n* of φ and ψ. So, mixing is equivalent to the convergence of $I_n(\varphi,\psi)$ to 0. We say that ν is *K-mixing* if for every $\psi\in L^2(\nu)$

$$\sup_{\|\varphi\|_{L^2(\nu)}\leq 1} I_n(\varphi,\psi) \to 0.$$

Note that K-mixing is equivalent to the fact that the σ-algebra $\mathscr{F}_\infty := \cap g^{-n}(\mathscr{F})$ contains only sets of zero and full measure. This is the strongest form of mixing for observables in $L^2(\nu)$. It is however of interest to get a quantitative information on the mixing speed for more regular observables like smooth or Hölder continuous functions.

Consider now an endomorphism f of algebraic degree $d\geq 2$ of $\mathbb{P}^k$ as above and its equilibrium measure μ. We know that μ is totally invariant: $f^*(\mu) = d^k\mu$. If φ is a continuous function, then

$$\langle \mu, \varphi\circ f\rangle = \langle d^{-k}f^*(\mu), \varphi\circ f\rangle = \langle \mu, d^{-k}f_*(\varphi\circ f)\rangle = \langle \mu, \varphi\rangle.$$

We have used the obvious fact that $f_*(\varphi\circ f) = d^k\varphi$. So, μ is invariant. We have the following proposition.

Proposition 1.81. *The Perron-Frobenius operator $\Lambda := d^{-k}f_*$ has a continuous extension of norm* 1 *to $L^2(\mu)$. Moreover, the adjoint of Λ satisfies ${}^t\Lambda(\varphi) = \varphi\circ f$ and $\Lambda\circ{}^t\Lambda = \mathrm{id}$. Let $L^2_0(\mu)$ denote the hyperplane of $L^2(\mu)$ defined by $\langle \mu, \varphi\rangle = 0$. Then, the spectral radius of Λ on $L^2_0(\mu)$ is also equal to* 1.

Proof. Schwarz's inequality implies that

$$|f_*(\varphi)|^2 \leq d^k f_*(|\varphi|^2).$$

Using the total invariance of μ, we get

$$\langle \mu, |\Lambda(\varphi)|^2\rangle \leq \langle \mu, \Lambda(|\varphi|^2)\rangle = \langle \mu, |\varphi|^2\rangle.$$

Therefore, Λ has a continuous extension to $L^2(\mu)$, with norm ≤ 1. Since $\Lambda(1) = 1$, the norm of this operator is equal to 1. The properties on the adjoint of Λ are easily deduced from the total invariance of μ.

Let φ be a function in $L^2_0(\mu)$ of norm 1. Then, $\varphi \circ f^n$ is also in $L^2(\mu)$ and of norm 1. Moreover, $\Lambda^n(\varphi \circ f^n)$, which is equal to φ, is of norm 1. So, the spectral radius of Λ on $L^2_0(\mu)$ is also equal to 1. □

Mixing for the measure μ was proved in [FS1]. We give in this paragraph two proofs of K-mixing. The first one is from [DS1] and does not use that μ is moderate.

Theorem 1.82. *Let f be an endomorphism of algebraic degree $d \geq 2$ of $\mathbb{P}^k$. Then, its Green measure μ is K-mixing.*

Proof. Let $c_\psi := \langle \mu, \psi \rangle$. Since μ is totally invariant, the correlations between two observables φ and ψ satisfy

$$\begin{aligned} I_n(\varphi,\psi) &= |\langle \mu, (\varphi \circ f^n)\psi \rangle - \langle \mu, \varphi \rangle \langle \mu, \psi \rangle| \\ &= |\langle \mu, \varphi \Lambda^n(\psi) \rangle - c_\psi \langle \mu, \varphi \rangle| \\ &= |\langle \mu, \varphi(\Lambda^n(\psi) - c_\psi) \rangle|. \end{aligned}$$

Hence,

$$\sup_{\|\varphi\|_{L^2(\mu)} \leq 1} I_n(\varphi,\psi) \leq \|\Lambda^n(\psi) - c_\psi\|_{L^2(\mu)}.$$

Since $\|\Lambda\|_{L^2(\mu)} \leq 1$, in order to show that $\|\Lambda^n(\psi) - c_\psi\|_{L^2(\mu)} \to 0$ for every $\psi \in L^2(\mu)$, it is enough to show that $\|\Lambda^n(\psi) - c_\psi\|_{L^2(\mu)} \to 0$ for a dense family of functions $\psi \in L^2(\mu)$. So, we can assume that ψ is a d.s.h. function such that $|\psi| \leq 1$. We have $|c_\psi| \leq 1$ and $\|\Lambda^n(\psi) - c_\psi\|_{L^\infty(\mu)} \leq 2$. Since μ is PB, we deduce from Theorem 1.35 and Cauchy-Schwarz's inequality that

$$\|\Lambda^n(\psi) - c_\psi\|_{L^2(\mu)} \lesssim \|\Lambda^n(\psi) - c_\psi\|^{1/2}_{L^1(\mu)} \lesssim \|\Lambda^n(\psi) - c_\psi\|^{1/2}_{\mathrm{DSH}} \lesssim d^{-n/2}.$$

This completes the proof. For the last argument, we can also use continuous test functions ψ and apply Proposition 1.46. We then obtain that $\Lambda^n(\psi) - c_\psi$ converges to 0 pointwise out of a pluripolar set. Lebesgue's convergence theorem and the fact that μ has no mass on pluripolar sets imply the result. □

In what follows, we show that the equilibrium measure μ satisfies remarkable stochastic properties which are quite hard to obtain in the setting of real dynamical systems. We will see the effectiveness of the pluripotential methods which replace the delicate estimates, used in some real dynamical systems. The following result was recently obtained by Nguyen and the authors [DNS]. It shows that the equilibrium measure is exponentially mixing and generalizes earlier results of [DS1, DS6, FS1]. Note that d.s.h. observables may be everywhere discontinuous.

Theorem 1.83. *Let f be a holomorphic endomorphism of algebraic degree $d \geq 2$ on $\mathbb{P}^k$. Let μ be the Green measure of f. Then for every $1 < p \leq +\infty$ there is a constant $c > 0$ such that*

$$|\langle \mu, (\varphi \circ f^n)\psi \rangle - \langle \mu, \varphi \rangle \langle \mu, \psi \rangle| \leq c d^{-n} \|\varphi\|_{L^p(\mu)} \|\psi\|_{\mathrm{DSH}}$$

for $n \geq 0$, φ in $L^p(\mu)$ and ψ d.s.h. Moreover, for $0 \leq \nu \leq 2$ there is a constant $c > 0$ such that

$$|\langle \mu, (\varphi \circ f^n)\psi \rangle - \langle \mu, \varphi \rangle \langle \mu, \psi \rangle| \leq c d^{-n\nu/2} \|\varphi\|_{L^p(\mu)} \|\psi\|_{\mathscr{C}^\nu}$$

for $n \geq 0$, φ in $L^p(\mu)$ and ψ of class $\mathscr{C}^\nu$.

Proof. We prove the first assertion. Observe that the correlations

$$I_n(\varphi, \psi) := |\langle \mu, (\varphi \circ f^n)\psi \rangle - \langle \mu, \varphi \rangle \langle \mu, \psi \rangle|$$

vanish if ψ is constant. Therefore, we can assume that $\langle \mu, \psi \rangle = 0$. In which case, we have

$$I_n(\varphi, \psi) = |\langle \mu, \varphi \Lambda^n(\psi) \rangle|,$$

where Λ denotes the Perron-Frobenius operator associated to f.

We can also assume that $\|\psi\|_{\mathrm{DSH}} \leq 1$. Corollary 1.41 implies that for $1 \leq q < \infty$,

$$\|\Lambda^n(\psi)\|_{L^q(\mu)} \leq cqd^{-n}$$

where $c > 0$ is a constant independent of n, q and ψ. Now, if q is chosen so that $p^{-1} + q^{-1} = 1$, we obtain using Hölder's inequality that

$$I_n(\varphi, \psi) \leq \|\varphi\|_{L^p(\mu)} \|\Lambda^n(\psi)\|_{L^q(\mu)} \leq cq \|\varphi\|_{L^p(\mu)} d^{-n}.$$

This completes the proof of the first assertion. The second assertion is proved in the same way using Corollary 1.42. □

Observe that the above estimates imply that for ψ smooth

$$\lim_{n \to \infty} \sup_{\|\varphi\|_{L^2(\mu)} \leq 1} I_n(\varphi, \psi) = 0.$$

Since smooth functions are dense in $L^2(\mu)$, the convergence holds for every ψ in $L^2(\mu)$ and gives another proof of the K-mixing. The following result [DNS] gives estimates for the exponential mixing of any order. It can be extended to Hölder continuous observables using the second assertion in Theorem 1.83.

Theorem 1.84. *Let f, d, μ be as in Theorem 1.83 and $r \geq 1$ an integer. Then there is a constant $c > 0$ such that*

$$\Big|\langle \mu, \psi_0(\psi_1 \circ f^{n_1}) \dots (\psi_r \circ f^{n_r}) \rangle - \prod_{i=0}^{r} \langle \mu, \psi_i \rangle \Big| \leq c d^{-n} \prod_{i=0}^{r} \|\psi_i\|_{\mathrm{DSH}}$$

for $0 = n_0 \leq n_1 \leq \cdots \leq n_r$, $n := \min_{0 \leq i < r}(n_{i+1} - n_i)$ and ψ_i d.s.h.

Proof. The proof is by induction on r. The case $r = 1$ is a consequence of Theorem 1.83. Suppose the result is true for $r - 1$, we have to check it for r. Without loss of generality, assume that $\|\psi_i\|_{\mathrm{DSH}} \leq 1$. This implies that $m := \langle \mu, \psi_0 \rangle$ is bounded. The invariance of μ and the induction hypothesis imply that

$$\Big| \langle \mu, m(\psi_1 \circ f^{n_1}) \ldots (\psi_r \circ f^{n_r}) \rangle - \prod_{i=0}^{r} \langle \mu, \psi_i \rangle \Big|$$
$$= \Big| \langle \mu, m\psi_1(\psi_2 \circ f^{n_2-n_1}) \ldots (\psi_r \circ f^{n_r-n_1}) \rangle - m\prod_{i=1}^{r} \langle \mu, \psi_i \rangle \Big| \leq cd^{-n}$$

for some constant $c > 0$. In order to get the desired estimate, it is enough to show that

$$\Big| \langle \mu, (\psi_0 - m)(\psi_1 \circ f^{n_1}) \ldots (\psi_r \circ f^{n_r}) \rangle \Big| \leq cd^{-n}.$$

Observe that the operator $(f^n)^*$ acts on $L^p(\mu)$ for $p \geq 1$ and its norm is bounded by 1. Using the invariance of μ and Hölder's inequality, we get for $p := r+1$

$$\Big| \langle \mu, (\psi_0 - m)(\psi_1 \circ f^{n_1}) \ldots (\psi_r \circ f^{n_r}) \rangle \Big|$$
$$= \Big| \langle \mu, \Lambda^{n_1}(\psi_0 - m)\psi_1 \ldots (\psi_r \circ f^{n_r-n_1}) \rangle \Big|$$
$$\leq \|\Lambda^{n_1}(\psi_0 - m)\|_{L^p(\mu)} \|\psi_1\|_{L^p(\mu)} \cdots \|\psi_r \circ f^{n_r-n_1}\|_{L^p(\mu)}$$
$$\leq cd^{-n_1} \|\psi_1\|_{L^p(\mu)} \cdots \|\psi_r\|_{L^p(\mu)},$$

for some constant $c > 0$. Since $\|\psi_i\|_{L^p(\mu)} \lesssim \|\psi_i\|_{\mathrm{DSH}}$, the previous estimates imply the result. Note that as in Theorem 1.83, it is enough to assume that ψ_i is d.s.h. for $i \leq r-1$ and ψ_r is in $L^p(\mu)$ for some $p > 1$. □

The mixing of μ implies that for any measurable observable φ, the times series $\varphi \circ f^n$, behaves like independent random variables with the same distribution. For example, the dependence of $\varphi \circ f^n$ and φ is weak when n is large: if a, b are real numbers, then the measure of $\{\varphi \circ f^n \leq a$ and $\varphi \leq b\}$ is almost equal to $\mu\{\varphi \circ f^n \leq a\}\mu\{\varphi \leq b\}$. Indeed, it is equal to

$$\langle \mu, (\mathbf{1}_{]-\infty,a]} \circ \varphi \circ f^n)(\mathbf{1}_{]-\infty,b]} \circ \varphi) \rangle,$$

and when n is large, mixing implies that the last integral is approximatively equal to

$$\langle \mu, \mathbf{1}_{]-\infty,a]} \circ \varphi \rangle \langle \mu, \mathbf{1}_{]-\infty,b]} \circ \varphi \rangle = \mu\{\varphi \leq a\}\mu\{\varphi \leq b\} = \mu\{\varphi \circ f^n \leq a\}\mu\{\varphi \leq b\}.$$

The estimates on the decay of correlations obtained in the above results, give at which speed the observables become "almost independent". We are going to show that under weak assumptions on the regularity of observables φ, the times series $\varphi \circ f^n$, satisfies the Central Limit Theorem (CLT for short). We recall the classical CLT for independent random variables. In what follows, $\mathrm{E}(\cdot)$ denotes expectation, i.e. the mean, of a random variable.

Theorem 1.85. *Let $(X, \mathscr{F}, \nu)$ be a probability space. Let $Z_1, Z_2, \ldots$ be independent identically distributed (i.i.d. for short) random variables with values in $\mathbb{R}$, and of* *mean zero, i.e.* $\mathrm{E}(Z_n) = 0$. *Assume also that* $0 < \mathrm{E}(Z_n^2) < \infty$. *Then for any open interval* $I \subset \mathbb{R}$ *and for* $\sigma := \mathrm{E}(Z_n^2)^{1/2}$, *we have*

$$\lim_{N\to\infty} \nu\left\{\frac{1}{\sqrt{N}}\sum_{n=0}^{N-1} Z_n \in I\right\} = \frac{1}{\sqrt{2\pi}\sigma}\int_I e^{-\frac{t^2}{2\sigma^2}}\,dt.$$

The important hypothesis here is that the variables have the same distribution (this means that for every interval $I \subset \mathbb{R}$, $\langle \nu, \mathbf{1}_I \circ Z_n\rangle$ is independent of n, where $\mathbf{1}_I$ is the characteristic function of I) and that they are independent. The result can be phrased as follows. If we define the random variables $\widehat{Z}_N$ by

$$\widehat{Z}_N := \frac{1}{\sqrt{N}}\sum_{n=0}^{N-1} Z_n,$$

then the sequence of probability measures $(\widehat{Z}_N)_*(\nu)$ on $\mathbb{R}$ converges to the probability measure of density $\frac{1}{\sqrt{2\pi}\sigma}e^{-\frac{t^2}{2\sigma^2}}$. This is also called *the convergence in law.*

We want to replace the random variables Z_n with the functions $\varphi \circ f^n$ on the probability space $(\mathbb{P}^k, \mathscr{B}, \mu)$ where $\mathscr{B}$ is the Borel σ-algebra. The fact that μ is invariant means exactly that $\varphi \circ f^n$ are identically distributed. We state first a central limit theorem due to Gordin [G], see also [V]. For simplicity, we consider a measurable map $g : (X, \mathscr{F}) \to (X, \mathscr{F})$ as above. Define $\mathscr{F}_n := g^{-n}(\mathscr{F})$, $n \geq 0$, the σ-algebra of $g^{-n}(A)$, with $A \in \mathscr{F}$. This sequence is non-increasing. Denote by $\mathrm{E}(\varphi|\mathscr{F}')$ the conditional expectation of φ with respect to a σ-algebra $\mathscr{F}' \subset \mathscr{F}$. We say that φ is *a coboundary* if $\varphi = \psi \circ g - \psi$ for some function $\psi \in L^2(\nu)$.

Theorem 1.86 (Gordin). *Let* ν *be an ergodic invariant probability measure on* X. *Let* φ *be an observable in* $L^1(\nu)$ *such that* $\langle \nu, \varphi\rangle = 0$. *Suppose*

$$\sum_{n\geq 0} \|\mathrm{E}(\varphi|\mathscr{F}_n)\|^2_{L^2(\nu)} < \infty.$$

Then $\langle \nu, \varphi^2\rangle + 2\sum_{n\geq 1}\langle \nu, \varphi(\varphi\circ g^n)\rangle$ *is a finite positive number which vanishes if and only if* φ *is a coboundary. Moreover, if*

$$\sigma := \Big[\langle \nu, \varphi^2\rangle + 2\sum_{n\geq 1}\langle \nu, \varphi(\varphi\circ g^n)\rangle\Big]^{1/2}$$

is strictly positive, then φ *satisfies the central limit theorem with variance* σ*: for any interval* $I \subset \mathbb{R}$

$$\lim_{N\to\infty} \nu\left\{\frac{1}{\sqrt{N}}\sum_{n=0}^{N-1} \varphi\circ g^n \in I\right\} = \frac{1}{\sqrt{2\pi}\sigma}\int_I e^{-\frac{t^2}{2\sigma^2}}\,dt.$$

It is not difficult to see that a function u is $\mathscr{F}_n$-measurable if and only if $u = u' \circ g^n$ with u' $\mathscr{F}$-measurable. Let $L^2(\nu, \mathscr{F}_n)$ denote the space of $\mathscr{F}_n$-measurable functions which are in $L^2(\nu)$. Then, $\mathrm{E}(\varphi|\mathscr{F}_n)$ is the orthogonal projection of $\varphi \in L^2(\nu)$ into $L^2(\nu, \mathscr{F}_n)$.

A straighforward computation using the invariance of ν gives that the variance σ in the above theorem is equal to

$$\sigma = \lim_{n\to\infty} n^{-1/2}\|\varphi + \cdots + \varphi\circ g^{n-1}\|_{L^2(\nu)}.$$

When φ is orthogonal to all $\varphi\circ g^n$, we find that $\sigma = \|\varphi\|_{L^2(\mu)}$. So, Gordin's theorem assumes a weak dependence and concludes that the observables satisfy the central limit theorem.

Consider now the dynamical system associated to an endomorphism f of $\mathbb{P}^k$ as above. Let $\mathscr{B}$ denote the Borel σ-algebra on $\mathbb{P}^k$ and define $\mathscr{B}_n := f^{-n}(\mathscr{B})$. Since the measure μ satisfies $f^*(\mu) = d^k\mu$, the norms $\|\mathrm{E}(\cdot|\mathscr{B}_n)\|_{L^2(\mu)}$ can be expressed in terms of the operator Λ. We have the following lemma.

Lemma 1.87. *Let φ be an observable in $L^2(\mu)$. Then*

$$\mathrm{E}(\varphi|\mathscr{F}_n) = \Lambda^n(\varphi)\circ f^n \quad \textit{and} \quad \|\mathrm{E}(\varphi|\mathscr{B}_n)\|_{L^p(\mu)} = \|\Lambda^n(\varphi)\|_{L^p(\mu)},$$

for $1\leq p\leq 2$.

Proof. We have

$$\begin{aligned}\langle\mu,\varphi(\psi\circ f^n)\rangle &= \langle d^{-kn}(f^n)^*(\mu),\varphi(\psi\circ f^n)\rangle = \langle\mu, d^{-kn}(f^n)_*[\varphi(\psi\circ f^n)]\rangle\\ &= \langle\mu,\Lambda^n(\varphi)\psi\rangle = \langle\mu,[\Lambda^n(\varphi)\circ f^n][\psi\circ f^n]\rangle.\end{aligned}$$

This proves the first assertion. The invariance of μ implies that $\|\psi\circ f^n\|_{L^p(\mu)} = \|\psi\|_{L^p(\mu)}$. Therefore, the second assertion is a consequence of the first one. □

Gordin's Theorem 1.86, Corollaries 1.41 and 1.42, applied to $q = 2$, give the following result.

Corollary 1.88. *Let f be an endomorphism of algebraic degree $d\geq 2$ of $\mathbb{P}^k$ and μ its equilibrium measure. Let φ be a d.s.h. function or a Hölder continuous function on $\mathbb{P}^k$, such that $\langle\mu,\varphi\rangle = 0$. Assume that φ is not a coboundary. Then φ satisfies the central limit theorem with the variance $\sigma > 0$ given by*

$$\sigma^2 := \langle\mu,\varphi^2\rangle + 2\sum_{n\geq 1}\langle\mu,\varphi(\varphi\circ f^n)\rangle.$$

We give an interesting decomposition of the space $L^2_0(\mu)$ which shows that Λ, acts like a "generalized shift". Recall that $L^2_0(\mu)$ is the space of functions $\psi\in L^2(\mu)$ such that $\langle\mu,\psi\rangle = 0$. Corollary 1.88 can also be deduced from the following result.

Proposition 1.89. *Let f be an endomorphism of algebraic degree $d\geq 2$ of $\mathbb{P}^k$ and μ the corresponding equilibrium measure. Define*

$$V_n := \{\psi\in L^2_0(\mu),\ \Lambda^n(\psi) = 0\}.$$

Then, we have $V_{n+1} = V_n\oplus V_1\circ f^n$ as an orthogonal sum and $L^2_0(\mu) = \oplus_{n=0}^\infty V_1\circ f^n$ as a Hilbert sum. Let $\psi = \sum\psi_n\circ f^n$, with $\psi_n\in V_1$, be a function in $L^2_0(\mu)$.

Then, ψ satisfies the Gordin's condition, see Theorem 1.86, if and only if the sum $\sum_{n\geq 1} n\|\psi_n\|^2_{L^2(\mu)}$ is finite. Moreover, if ψ is d.s.h. (resp. of class $\mathscr{C}^\nu$, $0<\nu\leq 2$) with $\langle\mu,\psi\rangle=0$, then $\|\psi_n\|_{L^2(\mu)}\lesssim d^{-n}$ (resp. $\|\psi_n\|_{L^2(\mu)}\lesssim d^{-n\nu/2}$).

Proof. It is easy to check that $V_n^\perp=\{\theta\circ f^n,\ \theta\in L^2(\mu)\}$. Let W_{n+1} denote the orthogonal complement of V_n in V_{n+1}. Suppose $\theta\circ f^n$ is in W_{n+1}. Then, $\Lambda(\theta)=0$. This gives the first decomposition in the proposition.

For the second decomposition, observe that $\oplus_{n=0}^\infty V_1\circ f^n$ is a direct orthogonal sum. We only have to show that $\cup V_n$ is dense in $L^2_0(\mu)$. Let θ be an element in $\cap V_n^\perp$. We have to show that $\theta=0$. For every n, $\theta=\theta_n\circ f^n$ for appropriate θ_n. Hence, θ is measurable with respect to the σ-algebra $\mathscr{B}_\infty:=\cap_{n\geq 0}\mathscr{B}_n$. We show that $\mathscr{B}_\infty$ is the trivial algebra. Let A be an element of $\mathscr{B}_\infty$. Define $A_n=f^n(A)$. Since A is in $\mathscr{B}_\infty$, $\mathbf{1}_A=\mathbf{1}_{A_n}\circ f^n$ and $\Lambda^n(\mathbf{1}_A)=\mathbf{1}_{A_n}$. K-mixing implies that $\Lambda^n(\mathbf{1}_A)$ converges in $L^2(\mu)$ to a constant, see Theorem 1.82. So, $\mathbf{1}_{A_n}$ converges to a constant which is necessarily 0 or 1. We deduce that $\mu(A_n)$ converges to 0 or 1. On the other hand, we have

$$\mu(A_n)=\langle\mu,\mathbf{1}_{A_n}\rangle=\langle\mu,\mathbf{1}_{A_n}\circ f^n\rangle=\langle\mu,\mathbf{1}_A\rangle=\mu(A).$$

Therefore, A is of measure 0 or 1. This implies the decomposition of $L^2_0(\mu)$.

Suppose now that $\psi:=\sum\psi_n\circ f^n$ with $\Lambda(\psi_n)=0$, is an element of $L^2_0(\mu)$. We have $\mathrm{E}(\psi|\mathscr{B}_n)=\sum_{i\geq n}\psi_i\circ f^i$. So,

$$\sum_{n\geq 0}\|\mathrm{E}(\psi|\mathscr{B}_n)\|^2_{L^2(\mu)}=\sum_{n\geq 0}(n+1)\|\psi_n\|^2_{L^2(\mu)},$$

and ψ satisfies Gordin's condition if and only if the last sum is finite.

Let ψ be a d.s.h. function with $\langle\mu,\psi\rangle=0$. It follows from Theorem 1.83 that

$$\|\mathrm{E}(\psi|\mathscr{B}_n)\|_{L^2(\mu)}=\sup_{\|\varphi\|_{L^2(\mu)}\leq 1}|\langle\mu,(\varphi\circ f^n)\psi\rangle|\lesssim d^{-n}.$$

Since $\psi_n\circ f^n=\mathrm{E}(\psi|\mathscr{B}_n)-\mathrm{E}(\psi|\mathscr{B}_{n+1})$, the above estimate implies that

$$\|\psi_n\|_{L^2(\mu)}=\|\psi_n\circ f^n\|_{L^2(\mu)}\lesssim d^{-n}.$$

The case of $\mathscr{C}^\nu$ observables is proved in the same way.

Observe that if $(\psi_n\circ f^n)_{n\geq 0}$ is the sequence of projections of ψ on the factors of the direct sum $\oplus_{n=0}^\infty V_1\circ f^n$, then the coordinates of $\Lambda(\psi)$ are $(\psi_n\circ f^{n-1})_{n\geq 1}$. □

We continue the study with other types of convergence. Let us recall the almost sure version of the central limit theorem in probability theory. Let Z_n be random variables, identically distributed in $L^2(X,\mathscr{F},\nu)$, such that $\mathrm{E}(Z_n)=0$ and $\mathrm{E}(Z_n^2)=\sigma^2$, $\sigma>0$. We say that *the almost sure central limit theorem* holds if at ν-almost every point in X, the sequence of measures

$$\frac{1}{\log N}\sum_{n=1}^N\frac{1}{n}\delta_{n^{-1/2}\sum_{i=0}^{n-1}Z_i}$$

converges in law to the normal distribution of mean 0 and variance σ. In particular, ν-almost surely

$$\frac{1}{\log N}\sum_{n=1}^{N}\frac{1}{n}\mathbf{1}_{\{n^{-1/2}\sum_{i=0}^{n-1}Z_i\leq t_0\}}\to\frac{1}{\sqrt{2\pi\sigma}}\int_{-\infty}^{t_0}e^{-\frac{t^2}{2\sigma^2}}dt,$$

for any $t_0\in\mathbb{R}$. In the central limit theorem, we only get the ν-measure of the set $\{N^{-1/2}\sum_{n=0}^{N-1}Z_n<t_0\}$ when N goes to infinity. Here, we get an information at ν-almost every point for the logarithmic averages.

The almost sure central limit theorem can be deduced from the so-called almost sure invariance principle (ASIP for short). In the case of i.i.d. random variables as above, this principle compares the variables $\widehat{Z}_N$ with Brownian motions and gives some information about the fluctuations of $\widehat{Z}_N$ around 0.

Theorem 1.90. *Let $(X,\mathscr{F},\nu)$ be a probability space. Let (Z_n) be a sequence of i.i.d. random variables with mean 0 and variance $\sigma>0$. Assume that there is an $\alpha>0$ such that Z_n is in $L^{2+\alpha}(\nu)$. Then, there is another probability space $(X',\mathscr{F}',\nu')$ with a sequence of random variables $\mathtt{S}'_N$ on X' which has the same joint distribution as $\mathtt{S}_N:=\sum_{n=0}^{N-1}Z_n$, and a Brownian motion $\mathtt{B}$ of variance σ on X' such that*

$$|\mathtt{S}'_N-\mathtt{B}(N)|\leq cN^{1/2-\delta},$$

for some positive constants c,δ. It follows that

$$|N^{-1/2}\mathtt{S}'_N-\mathtt{B}(1)|\leq cN^{-\delta}.$$

For weakly dependent variables, this type of result is a consequence of a theorem due to Philipp-Stout [PS]. It gives conditions which imply that the ASIP holds. Lacey-Philipp proved in [LP] that the ASIP implies the almost sure central limit theorem. For holomorphic endomorphisms of $\mathbb{P}^k$, we have the following result due to Dupont which holds in particular for Hölder continuous observables [DP2].

Theorem 1.91. *Let f be an endomorphism of algebraic degree $d\geq 2$ as above and μ its equilibrium measure. Let φ be an observable with values in $\mathbb{R}\cup\{-\infty\}$ such that e^φ is Hölder continuous, $H:=\{\varphi=-\infty\}$ is an analytic set and $|\varphi|\lesssim|\log\mathrm{dist}(\cdot,H)|^\rho$ near H for some $\rho>0$. If $\langle\mu,\varphi\rangle=0$ and φ is not a coboundary, then the almost sure invariance principle holds for φ. In particular, the almost sure central limit theorem holds for such observables.*

The ASIP in the above setting says that there is a probability space $(X',\mathscr{F}',\nu')$ with a sequence of random variables $\mathtt{S}'_N$ on X which has the same joint distribution as $\mathtt{S}_N:=\sum_{n=0}^{N-1}\varphi\circ f^n$, and a Brownian motion $\mathtt{B}$ of variance σ on X' such that

$$|\mathtt{S}'_N-\mathtt{B}(N)|\leq cN^{1/2-\delta},$$

for some positive constants c,δ.

The ASIP implies other stochastic results, see [PS], in particular, *the law of the iterated logarithm.* With our notations, it implies that for φ as above

$$\limsup_{N\to\infty} \frac{\mathrm{S}_N(\varphi)}{\sigma\sqrt{N\log\log(N\sigma^2)}} = 1 \quad \mu\text{-almost everywhere.}$$

Dupont's approach is based on the Philipp-Stout's result applied to a Bernoulli system and a quantitative Bernoulli property of the equilibrium measure of f, i.e. a construction of a coding tree. We refer to Dupont and Przytycki-Urbanski-Zdunik [PU] for these results.

Recall the Bernoulli property of μ which was proved by Briend in [BJ1]. The dimension one case is due to Mañé [MA] and Heicklen-Hoffman [HH]. Denote by (Σ,ν,σ) the *one-sided d^k-shift*, where $\Sigma := \{1,\ldots,d^k\}^{\mathbb{N}}$, ν is the probability measure on Σ induced by the equilibrium probability measure on $\{1,\ldots,d^k\}$ and $\sigma:\Sigma\to\Sigma$ is the d^k to 1 map defined by $\sigma(\alpha_0,\alpha_1,\ldots) = (\alpha_1,\alpha_2,\ldots)$.

Theorem 1.92. *Let f and μ be as above. Then $(\mathbb{P}^k,\mu,f)$ is measurably conjugated to (Σ,ν,σ). More precisely, there is a measurable map $\pi:\Sigma\to\mathbb{P}^k$, defined out of a set of zero ν-measure, which is invertible out of a set of zero μ-measure, such that $\pi_*(\nu)=\mu$ and $f=\pi\circ\sigma\circ\pi^{-1}$ μ-almost everywhere.*

The proof uses a criterion, the so called *tree very weak Bernoulli* property (tree-vwB for short) due to Hoffman-Rudolph [HR]. One can use Proposition 1.51 in order to check this criterion.

The last stochastic property we consider here is the large deviations theorem. As above, we first recall the classical result in probability theory.

Theorem 1.93. *Let $Z_1, Z_2, \ldots$ be independent random variables on $(X,\mathscr{F},\nu)$, identically distributed with values in $\mathbb{R}$, and of mean zero, i.e.* $\mathrm{E}(Z_1) = 0$. *Assume also that for $t\in\mathbb{R}$,* $\exp(tZ_n)$ *is integrable. Then, the limit*

$$I(\varepsilon) := -\lim_{N\to\infty} \log\nu\left\{\left|\frac{Z_1+\cdots+Z_N}{N}\right| > \varepsilon\right\}.$$

exists and $I(\varepsilon)>0$ for $\varepsilon>0$.

The theorem estimates the size of the set where the average is away from zero, the expected value. We have

$$\nu\left\{\left|\frac{Z_1+\cdots+Z_N}{N}\right| > \varepsilon\right\} \sim e^{-NI(\varepsilon)}.$$

Our goal is to give an analogue for the equilibrium measure of endomorphisms of $\mathbb{P}^k$. We first prove an abstract result corresponding to the above Gordin's result for the central limit theorem.

Consider a dynamical system $g:(X,\mathscr{F},\nu)\to(X,\mathscr{F},\nu)$ as above where ν is an invariant probability measure. So, g^* defines a linear operator of norm 1 from $L^2(\nu)$ into itself. We say that g has *bounded Jacobian* if there is a constant $\kappa>0$ such that $\nu(g(A))\leq\kappa\nu(A)$ for every $A\in\mathscr{F}$. The following result was obtained in [DNS].

Theorem 1.94. *Let $g : (X,\mathscr{F},\nu) \to (X,\mathscr{F},\nu)$ be a map with bounded Jacobian as above. Define $\mathscr{F}_n := g^{-n}(\mathscr{F})$. Let ψ be a bounded real-valued measurable function. Assume there are constants $\delta > 1$ and $c > 0$ such that*

$$\langle \nu, e^{\delta^n |\mathrm{E}(\psi|\mathscr{F}_n) - \langle \nu,\psi\rangle|}\rangle \leq c \quad \textit{for every} \quad n \geq 0.$$

Then ψ satisfies a weak large deviations theorem. More precisely, for every $\varepsilon > 0$, there exists a constant $h_\varepsilon > 0$ such that

$$\nu\left\{x \in X : \left|\frac{1}{N}\sum_{n=0}^{N-1} \psi \circ g^n(x) - \langle \nu,\psi\rangle\right| > \varepsilon\right\} \leq e^{-N(\log N)^{-2} h_\varepsilon}$$

for all N large enough[6].

We first prove some preliminary lemmas. The following one is a version of the classical Bennett's inequality see [DZ, Lemma 2.4.1].

Lemma 1.95. *Let ψ be an observable such that $\|\psi\|_{L^\infty(\nu)} \leq b$ for some constant $b \geq 0$, and $\mathrm{E}(\psi) = 0$. Then*

$$\mathrm{E}(e^{\lambda\psi}) \leq \frac{e^{-\lambda b} + e^{\lambda b}}{2}$$

for every $\lambda \geq 0$.

Proof. We can assume $\lambda = 1$. Consider first the case where there is a measurable set A such that $\nu(A) = 1/2$. Let ψ_0 be the function which is equal to $-b$ on A and to b on $X \setminus A$. We have $\psi_0^2 = b^2 \geq \psi^2$. Since $\nu(A) = 1/2$, we have $\mathrm{E}(\psi_0) = 0$. Let $g(t) = a_0 t^2 + a_1 t + a_2$, be the unique quadratic function such that $h(t) := g(t) - e^t$ satisfies $h(b) = 0$ and $h(-b) = h'(-b) = 0$. We have $g(\psi_0) = e^{\psi_0}$.

Since $h''(t) = 2a_0 - e^t$ admits at most one zero, h' admits at most two zeros. The fact that $h(-b) = h(b) = 0$ implies that h' vanishes in $]-b,b[$. Hence h' admits exactly one zero at $-b$ and another one in $]-b,b[$. We deduce that h'' admits a zero. This implies that $a_0 > 0$. Moreover, h vanishes only at $-b$, b and $h'(b) \neq 0$. It follows that $h(t) \geq 0$ on $[-b,b]$ because h is negative near $+\infty$. Thus, $e^t \leq g(t)$ on $[-b,b]$ and then $e^\psi \leq g(\psi)$.

Since $a_0 > 0$, if an observable ϕ satisfies $\mathrm{E}(\phi) = 0$, then $\mathrm{E}(g(\phi))$ is an increasing function of $\mathrm{E}(\phi^2)$. Now, using the properties of ψ and ψ_0, we obtain

$$\mathrm{E}(e^\psi) \leq \mathrm{E}(g(\psi)) \leq \mathrm{E}(g(\psi_0)) = \mathrm{E}(e^{\psi_0}) = \frac{e^{-b} + e^b}{2}.$$

This completes the proof under the assumption that $\nu(A) = 1/2$ for some measurable set A.

The general case is deduced from the previous particular case. Indeed, it is enough to apply the first case to the disjoint union of $(X,\mathscr{F},\nu)$ with a copy

[6] In the LDT for independent random variables, there is no factor $(\log N)^{-2}$ in the estimate.

$(X',\mathscr{F}',\nu')$ of this space, i.e. to the space $(X\cup X',\mathscr{F}\cup\mathscr{F}',\frac{\nu}{2}+\frac{\nu'}{2})$, and to the function equal to ψ on X and on X'. □

Lemma 1.96. *Let ψ be an observable such that $\|\psi\|_{L^\infty(\nu)}\leq b$ for some constant $b\geq 0$, and $\mathrm{E}(\psi|\mathscr{F}_1)=0$. Then*

$$\mathrm{E}(e^{\lambda\psi}|\mathscr{F}_1)\leq\frac{e^{-\lambda b}+e^{\lambda b}}{2}$$

for every $\lambda\geq 0$.

Proof. We consider the desintegration of ν with respect to g. For ν-almost every $x\in X$, there is a positive measure ν_x on $g^{-1}(x)$ such that if φ is a function in $L^1(\nu)$ then

$$\langle\nu,\varphi\rangle=\int_X\langle\nu_x,\varphi\rangle d\nu(x).$$

Since ν is g-invariant, we have

$$\langle\nu,\varphi\rangle=\langle\nu,\varphi\circ g\rangle=\int_X\langle\nu_x,\varphi\circ g\rangle d\nu(x)=\int_X\|\nu_x\|\varphi(x)d\nu(x).$$

Therefore, ν_x is a probability measure for ν-almost every x. Using also the invariance of ν, we obtain for φ and ϕ in $L^2(\nu)$ that

$$\begin{aligned}\langle\nu,\varphi(\phi\circ g)\rangle&=\int_X\langle\nu_x,\varphi(\phi\circ g)\rangle d\nu(x)=\int_X\langle\nu_x,\varphi\rangle\phi(x)d\nu(x)\\&=\int_X\langle\nu_{g(x)},\varphi\rangle\phi(g(x))d\nu(x).\end{aligned}$$

We deduce that

$$\mathrm{E}(\varphi|\mathscr{F}_1)(x)=\langle\nu_{g(x)},\varphi\rangle.$$

So, the hypothesis in the lemma is that $\langle\nu_x,\psi\rangle=0$ for ν-almost every x. It suffices to check that

$$\langle\nu_x,e^{\lambda\psi}\rangle\leq\frac{e^{-\lambda b}+e^{\lambda b}}{2}.$$

But this is a consequence of Lemma 1.95 applied to ν_x instead of ν. □

We continue the proof of Theorem 1.94. Without loss of generality we can assume that $\langle\nu,\psi\rangle=0$ and $|\psi|\leq 1$. The general idea is to write $\psi=\psi'+(\psi''-\psi''\circ g)$ for functions ψ' and ψ'' in $L^2(\nu)$ such that

$$\mathrm{E}(\psi'\circ g^n|\mathscr{F}_{n+1})=0,\qquad n\geq 0.$$

In the language of probability theory, these identities mean that $(\psi'\circ g^n)_{n\geq 0}$ is a *reversed martingale difference* as in Gordin's approach, see also [V]. The strategy is to prove the weak LDT for ψ' and for the coboundary $\psi''-\psi''\circ g$. Theorem 1.94 is then a consequence of Lemmas 1.99 and 1.101 below.

Let Λ_g denote the adjoint of the operator $\varphi \mapsto \varphi \circ g$ on $L^2(\nu)$. These operators are of norm 1. The computation in Lemma 1.96 shows that $\mathrm{E}(\varphi|\mathscr{F}_1) = \Lambda_g(\varphi)\circ g$. We obtain in the same way that $\mathrm{E}(\varphi|\mathscr{F}_n) = \Lambda_g^n(\varphi)\circ g^n$. Define

$$\psi'' := -\sum_{n=1}^{\infty} \Lambda_g^n(\psi), \qquad \psi' := \psi - (\psi'' - \psi''\circ g).$$

Using the hypotheses in Theorem 1.94, we see that ψ' and ψ'' are in $L^2(\nu)$ with norms bounded by some constant. However, we loose the uniform boundedness: these functions are not necessarily in $L^\infty(\nu)$.

Lemma 1.97. *We have* $\Lambda_g^n(\psi') = 0$ *for* $n \geq 1$ *and* $\mathrm{E}(\psi'\circ g^n|\mathscr{F}_m) = 0$ *for* $m > n \geq 0$.

Proof. Clearly $\Lambda_g(\psi''\circ g) = \psi''$. We deduce from the definition of ψ'' that

$$\Lambda_g(\psi') = \Lambda_g(\psi) - \Lambda_g(\psi'') + \Lambda_g(\psi''\circ g) = \Lambda_g(\psi) - \Lambda_g(\psi'') + \psi'' = 0.$$

Hence, $\Lambda_g^n(\psi') = 0$ for $n \geq 1$. For every function ϕ in $L^2(\nu)$, since ν is invariant, we have for $m > n$

$$\langle \nu, (\psi'\circ g^n)(\phi\circ g^m)\rangle = \langle \nu, \psi'(\phi\circ g^{m-n})\rangle = \langle \nu, \Lambda_g^{m-n}(\psi')\phi\rangle = 0.$$

It follows that $\mathrm{E}(\psi'\circ g^n|\mathscr{F}_m) = 0$. □

Lemma 1.98. *There are constants* $\delta_0 > 1$ *and* $c > 0$ *such that*

$$\nu\{|\psi'| > b\} \leq ce^{-\delta_0^b} \quad \text{and} \quad \nu\{|\psi''| > b\} \leq ce^{-\delta_0^b}$$

for any $b \geq 0$*. In particular,* $t\psi'$ *and* $t\psi''$ *are* ν*-integrable for every* $t \geq 0$.

Proof. Since $\psi' := \psi - (\psi'' - \psi''\circ g)$ and ψ is bounded, it is enough to prove the estimate on ψ''. Indeed, the invariance of ν implies that $\psi''\circ g$ satisfies a similar inequality.

Fix a positive constant δ_1 such that $1 < \delta_1^2 < \delta$, where δ is the constant in Theorem 1.94. Define $\varphi := \sum_{n\geq 1} \delta_1^{2n}|\Lambda_g^n(\psi)|$. We first show that there is a constant $\alpha > 0$ such that $\nu\{\varphi \geq b\} \lesssim e^{-\alpha b}$ for every $b \geq 0$. Recall that $\mathrm{E}(\psi|\mathscr{F}_n) = \Lambda_g^n(\psi)\circ g^n$. Using the hypothesis of Theorem 1.94, the inequality $\sum \frac{1}{2n^2} \leq 1$ and the invariance of ν, we obtain for $b \geq 0$

$$\begin{aligned}\nu\{\varphi \geq b\} &\leq \sum_{n\geq 1} \nu\Big\{|\Lambda_g^n(\psi)| \geq \frac{\delta_1^{-2n}b}{2n^2}\Big\} \leq \sum_{n\geq 1} \nu\Big\{|\mathrm{E}(\psi|\mathscr{F}_n)| \geq \frac{\delta_1^{-2n}b}{2n^2}\Big\} \\ &= \sum_{n\geq 1} \nu\Big\{\delta^n|\mathrm{E}(\psi|\mathscr{F}_n)| \geq \frac{\delta^n\delta_1^{-2n}b}{2n^2}\Big\} \lesssim \sum_{n\geq 1} \exp\Big(\frac{-\delta^n\delta_1^{-2n}b}{2n^2}\Big).\end{aligned}$$

It follows that $\nu\{\varphi \geq b\} \lesssim e^{-\alpha b}$ for some constant $\alpha > 0$.

We prove now the estimate $\nu\{|\psi''| > b\} \leq ce^{-\delta_0^b}$. It is enough to consider the case where $b = 2l$ for some positive integer l. Recall that for simplicity we assumed $|\psi| \leq 1$. It follows that $|\mathrm{E}(\psi|\mathscr{F}_n)| \leq 1$ and hence $|\Lambda_g^n(\psi)| \leq 1$. We have

$$|\psi''| \leq \sum_{n\geq 1} |\Lambda_g^n(\psi)| \leq \delta_1^{-2l} \sum_{n\geq 1} \delta_1^{2n} |\Lambda_g^n(\psi)| + \sum_{1\leq n\leq l} |\Lambda_g^n(\psi)| \leq \delta_1^{-2l}\varphi + l.$$

Consequently,

$$\nu\{|\psi''| > 2l\} \leq \nu\{\varphi > \delta_1^{2l}\} \lesssim e^{-\alpha\delta_1^{2l}}.$$

It is enough to choose $\delta_0 < \delta_1$ and c large enough. □

Lemma 1.99. *The coboundary $\psi'' - \psi'' \circ g$ satisfies the LDT.*

Proof. Given a function $\phi \in L^1(\mu)$, recall that Birkhoff's sum $\mathrm{S}_N(\phi)$ is defined by

$$\mathrm{S}_0(\phi) := 0 \qquad \text{and} \qquad \mathrm{S}_N(\phi) := \sum_{n=0}^{N-1} \phi \circ g^n \quad \text{for } N \geq 1.$$

Observe that $\mathrm{S}_N(\psi'' - \psi'' \circ g) = \psi'' - \psi'' \circ g^N$. Consequently, for a given $\varepsilon > 0$, using the invariance of ν, we have

$$\begin{aligned}\nu\{|\mathrm{S}_N(\psi'' - \psi'' \circ g)| > N\varepsilon\} &\leq \nu\left\{|\psi'' \circ g^N| > \frac{N\varepsilon}{2}\right\} + \mu\left\{|\psi''| > \frac{N\varepsilon}{2}\right\} \\ &= 2\nu\left\{|\psi''| > \frac{N\varepsilon}{2}\right\}.\end{aligned}$$

Lemma 1.98 implies that the last expression is smaller than e^{-Nh_ε} for some $h_\varepsilon > 0$ and for N large enough. This completes the proof. □

It remains to show that ψ' satisfies the weak LDT. We use the following lemma.

Lemma 1.100. *For every $b \geq 1$, there are Borel sets W_N such that $\nu(W_N) \leq cNe^{-\delta_0^b}$ and*

$$\int_{X\setminus W_N} e^{\lambda \mathrm{S}_N(\psi')} d\nu \leq 2\left[\frac{e^{-\lambda b} + e^{\lambda b}}{2}\right]^N,$$

where $c > 0$ is a constant independent of b.

Proof. For $N = 1$, define $W := \{|\psi'| > b\}$, $W' := g(W)$ and $W_1 := g^{-1}(W')$. Recall that the Jacobian of ν is bounded by some constant κ. This and Lemma 1.98 imply that

$$\nu(W_1) = \nu(W') = \nu(g(W)) \leq \kappa\nu(W) \leq ce^{-\delta_0^b}$$

for some constant $c > 0$. We also have

$$\int_{X\setminus W_1} e^{\lambda \mathrm{S}_1(\psi')} d\nu = \int_{X\setminus W_1} e^{\lambda\psi'} d\nu \leq e^{\lambda b} \leq 2\left[\frac{e^{-\lambda b} + e^{\lambda b}}{2}\right].$$

So, the lemma holds for $N = 1$.

Suppose the lemma for $N \geq 1$, we prove it for $N+1$. Define

$$W_{N+1} := g^{-1}(W_N) \cup W_1 = g^{-1}(W_N \cup W').$$

We have

$$\nu(W_{N+1}) \leq \nu(g^{-1}(W_N)) + \nu(W_1) = \nu(W_N) + \nu(W_1) \leq c(N+1)e^{-\delta_0^b}.$$

We will apply Lemma 1.96 to the function ψ^* such that $\psi^* = \psi'$ on $X \setminus W_1$ and $\psi^* = 0$ on W_1. By Lemma 1.97, we have $\mathrm{E}(\psi^*|\mathscr{F}_1) = 0$ since W_1 is an element of $\mathscr{F}_1$. The choice of W_1 gives that $|\psi^*| \leq b$. By Lemma 1.96, we have

$$\mathrm{E}(e^{\lambda\psi^*}|\mathscr{F}_1) \leq \frac{e^{-\lambda b} + e^{\lambda b}}{2} \quad \text{on} \quad X \quad \text{for} \quad \lambda \geq 0.$$

It follows that

$$\mathrm{E}(e^{\lambda\psi'}|\mathscr{F}_1) \leq \frac{e^{-\lambda b} + e^{\lambda b}}{2} \quad \text{on} \quad X \setminus W_1 \quad \text{for} \quad \lambda \geq 0.$$

Now, using the fact that W_{N+1} and $e^{\lambda \mathrm{S}_N(\psi' \circ g)}$ are $\mathscr{F}_1$-measurable, we can write

$$\begin{aligned}\int_{X\setminus W_{N+1}} e^{\lambda \mathrm{S}_{N+1}(\psi')} d\nu &= \int_{X\setminus W_{N+1}} e^{\lambda\psi'} e^{\lambda \mathrm{S}_N(\psi'\circ g)} d\nu \\ &= \int_{X\setminus W_{N+1}} \mathrm{E}(e^{\lambda\psi'}|\mathscr{F}_1) e^{\lambda \mathrm{S}_N(\psi'\circ g)} d\nu.\end{aligned}$$

Since $W_{N+1} = g^{-1}(W_N) \cup W_1$, the last integral is bounded by

$$\begin{aligned}&\sup_{X\setminus W_1} \mathrm{E}(e^{\lambda\psi'}|\mathscr{F}_1) \int_{X\setminus g^{-1}(W_N)} e^{\lambda \mathrm{S}_N(\psi'\circ g)} d\nu \\ &\leq \left[\frac{e^{-\lambda b} + e^{\lambda b}}{2}\right] \int_{X\setminus W_N} e^{\lambda \mathrm{S}_N(\psi')} d\nu \\ &\leq 2\left[\frac{e^{-\lambda b} + e^{\lambda b}}{2}\right]^{N+1},\end{aligned}$$

where the last inequality follows from the induction hypothesis. So, the lemma holds for $N+1$. □

The following lemma, together with Lemma 1.99, implies Theorem 1.94.

Lemma 1.101. *The function ψ' satisfies the weak LDT.*

Proof. Fix an $\varepsilon > 0$. By Lemma 1.100, we have, for every $\lambda \geq 0$

$$\begin{aligned}\nu\{\mathrm{S}_N(\psi') \geq N\varepsilon\} &\leq \nu(W_N) + e^{-\lambda N\varepsilon} \int_{X\setminus W_N} e^{\lambda \mathrm{S}_N(\psi')} d\nu \\ &\leq cNe^{-\delta_0^b} + 2e^{-\lambda N\varepsilon}\left[\frac{e^{-\lambda b} + e^{\lambda b}}{2}\right]^N.\end{aligned}$$

Let $b := \log N (\log \delta_0)^{-1}$. We have

$$cNe^{-\delta_0^b} = cNe^{-N} \leq e^{-N/2}$$

for N large. We also have

$$\frac{e^{-\lambda b} + e^{\lambda b}}{2} = \sum_{n \geq 0} \frac{\lambda^{2n} b^{2n}}{(2n)!} \leq e^{\lambda^2 b^2}.$$

Therefore, if $\lambda := u\varepsilon b^{-2}$ with a fixed $u > 0$ small enough

$$2e^{-\lambda N\varepsilon} \left[\frac{e^{-\lambda b} + e^{\lambda b}}{2} \right]^N \leq 2e^{-\varepsilon^2 b^{-2}(1-u)Nu} = 2e^{-2N(\log N)^{-2} h_\varepsilon}$$

for some constant $h_\varepsilon > 0$. We deduce from the previous estimates that

$$\nu\{S_N(\psi') \geq N\varepsilon\} \leq e^{-N(\log N)^{-2} h_\varepsilon}$$

for N large. A similar estimate holds for $-\psi'$. So, ψ' satisfies the weak LDT. □

We deduce from Theorem 1.94, Corollaries 1.41 and 1.42 the following result [DNS].

Theorem 1.102. *Let f be a holomorphic endomorphism of $\mathbb{P}^k$ of algebraic degree $d \geq 2$. Then the equilibrium measure μ of f satisfies the weak large deviations theorem for bounded d.s.h. observables and also for Hölder continuous observables. More precisely, if a function ψ is bounded d.s.h. or Hölder continuous, then for every $\varepsilon > 0$ there is a constant $h_\varepsilon > 0$ such that*

$$\mu \left\{ z \in \mathbb{P}^k : \left| \frac{1}{N} \sum_{n=0}^{N-1} \psi \circ f^n(z) - \langle \mu, \psi \rangle \right| > \varepsilon \right\} \leq e^{-N(\log N)^{-2} h_\varepsilon}$$

for all N large enough.

The exponential estimate on $\Lambda^n(\psi)$ is crucial in the proofs of the previous results. It is nearly an estimate in sup-norm. Note that if $\|\Lambda^n(\psi)\|_{L^\infty(\mu)}$ converge exponentially fast to 0 then ψ satisfies the LDT. This is the case for Hölder continuous observables in dimension 1, following a result by Drasin-Okuyama [DO], and when f is a generic map in higher dimension, see Remark 1.71. The LDT was recently obtained in dimension 1 by Xia-Fu in [X] for Lipschitz observables.

Exercise 1.103. Let $g : X \to X$ be a continuous map on a compact metric space X. Deduce from Birkhoff's theorem that any ergodic invariant measure of g is a limit of

$$\frac{1}{N} \sum_{n=0}^{N-1} \delta_{g^n(x)}$$

for an appropriate x.

Exercise 1.104. Show that if a Borel set A satisfies $\mu(A) > 0$, then $\mu(f^n(A))$ converges to 1.

Exercise 1.105. Show that $\sup_\psi I_n(\varphi,\psi)$ with ψ smooth $\|\psi\|_\infty \leq 1$ is equal to $\|\varphi\|_{L^1(\mu)}$. Deduce that there is no decay of correlations which is uniform on $\|\psi\|_\infty$.

Exercise 1.106. Let $V_1 := \{\psi \in L^2_0(\mu),\ \Lambda(\psi) = 0\}$. Show that V_1 is infinite dimensional and that bounded functions in V_1 are dense in V_1 with respect to the $L^2(\mu)$-topology. Using Theorem 1.82, show that the only eigenvalues of Λ are 0 and 1.

Exercise 1.107. Let φ be a d.s.h. function as in Corollary 1.88. Show that

$$\|\varphi+\cdots+\varphi\circ f^{n-1}\|^2_{L^2(\mu)} - n\sigma^2 + \gamma = O(d^{-n}),$$

where $\gamma := 2\sum_{n\geq 1} n\langle\mu,\varphi(\varphi\circ f^n)\rangle$ is a finite constant. Prove an analogous property for φ Hölder continuous.

1.7 Entropy, Hyperbolicity and Dimension

There are various ways to describe the complexity of a dynamical system. A basic measurement is the entropy which is closely related to the volume growth of the images of subvarieties. We will compute the topological entropy and the metric entropy of holomorphic endomorphisms of $\mathbb{P}^k$. We will also estimate the Lyapounov exponents with respect to the measure of maximal entropy and the Hausdorff dimension of this measure.

We recall few notions. Let (X,dist) be a compact metric space where dist is a distance on X. Let $g : X \to X$ be a continuous map. We introduce *the Bowen metric* associated to g. For a positive integer n, define the distance dist_n on X by

$$\mathrm{dist}_n(x,y) := \sup_{0\leq i\leq n-1} \mathrm{dist}(g^i(x),g^i(y)).$$

We have $\mathrm{dist}_n(x,y) > \varepsilon$ if the orbits $x,g(x),g^2(x),\ldots$ of x and $y,g(y),g^2(y),\ldots$ of y are distant by more than ε at a time i less than n. In which case, we say that x,y are *(n,ε)-separated.*

The topological entropy measures the rate of growth in function of time n, of the number of orbits that can be distinguished at ε-resolution. In other words, it measures the divergence of the orbits. More precisely, for $K\subset X$, not necessarily invariant, let $N(K,n,\varepsilon)$ denote the maximal number of points in K which are pairwise (n,ε)-separated. This number increases as ε decreases. *The topological entropy* of g on K is

$$h_t(g,K) := \sup_{\varepsilon>0}\limsup_{n\to\infty}\frac{1}{n}\log N(K,n,\varepsilon).$$

The topological entropy of g is the entropy on X and is denoted by $h_t(g)$. The reader can check that if g is an isometry, then $h_t(g) = 0$. In complex dynamics, we often have that for ε small enough, $\frac{1}{n}\log N(X,n,\varepsilon)$ converge to $h_t(g)$.

Let f be an endomorphism of algebraic degree $d \geq 2$ of $\mathbb{P}^k$as above. As we have seen, the iterate f^n of f has algebraic degree d^n. If Z is an algebraic set in $\mathbb{P}^k$ of codimension p then the degree of $f^{-n}(Z)$, counted with multiplicity, is equal to $d^{pn}\deg(Z)$ and the degree of $f^n(Z)$, counting with multiplicity, is equal to $d^{(k-p)n}\deg(Z)$. This is a consequence of Bézout's theorem. Recall that the degree of an algebraic set of codimension p in $\mathbb{P}^k$ is the number of points of intersection with a generic projective subspace of dimension p.

The pull-back by f induces a linear map $f^* : H^{p,p}(\mathbb{P}^k,\mathbb{C}) \to H^{p,p}(\mathbb{P}^k,\mathbb{C})$ which is just the multiplication by d^p. The constant d^p is *the dynamical degree of order p of f*. Dynamical degrees were considered by Gromov in [GR] where he introduced a method to bound the topological entropy from above. We will see that they measure the volume growth of the graphs. The degree of maximal order d^k is also called *the topological degree*. It is equal to the number of points in a fiber counting with multiplicity. The push-forward by f^n induces a linear map $f_* : H^{p,p}(\mathbb{P}^k,\mathbb{C}) \to H^{p,p}(\mathbb{P}^k,\mathbb{C})$ which is the multiplication by d^{k-p}. These operations act continuously on positive closed currents and hence, the actions are compatible with cohomology, see Appendix A.1.

We have the following result due to Gromov [GR] for the upper bound and to Misiurewicz-Przytycky [MP] for the lower bound of the entropy.

Theorem 1.108. *Let f be a holomorphic endomorphism of algebraic degree d on $\mathbb{P}^k$. Then the topological entropy $h_t(f)$ of f is equal to $k\log d$, i.e. to the logarithm of the maximal dynamical degree.*

The inequality $h_t(f) \geq k\log d$ is a consequence of the following result which is valid for arbitrary $\mathscr{C}^1$ maps [MP].

Theorem 1.109 (Misuriewicz-Przytycki). *Let X be a compact smooth orientable manifold and $g : X \to X$ a $\mathscr{C}^1$ map. Then*

$$h_t(g) \geq \log|\deg(g)|.$$

Recall that *the degree* of g is defined as follows. Let Ω be a continuous form of maximal degree on X such that $\int_X \Omega \neq 0$. Then

$$\deg(g) := \frac{\int_X g^*(\Omega)}{\int_X \Omega}.$$

The number is independent of the choice of Ω. When X is a complex manifold, it is necessarily orientable and $\deg(g)$ is just the generic number of preimages of a point, i.e. the topological degree of g. In our case, the topological degree of f is equal to d^k. So, $h_t(f) \geq k\log d$.

Instead of using Misuriewicz-Przytycki theorem, it is also possible to apply the following important result due to Yomdin [YO].

Theorem 1.110 (Yomdin). *Let X be a compact smooth manifold and $g : X \to X$ a smooth map. Let Y be a manifold in X smooth up to the boundary, then*

$$\limsup_{n\to\infty} \frac{1}{n} \log \text{volume}(g^n(Y)) \leq h_t(g),$$

where the volume of $g^n(Y)$ is counted with multiplicity.

In our situation, when $Y = \mathbb{P}^k$, we have volume$(g^n(Y)) \simeq d^{kn}$. Therefore, $h_t(f) \geq k \log d$. We can also deduce this inequality from Theorem 1.118 and the variational principle below.

End of the proof of Theorem 1.108. It remains to prove that $h_t(f) \leq k \log d$. Let Γ_n denote the graph of $(f, f^2, \ldots, f^{n-1})$ in $(\mathbb{P}^k)^n$, i.e. the set of points

$$(z, f(z), f^2(z), \ldots, f^{n-1}(z))$$

with z in $\mathbb{P}^k$. This is a manifold of dimension k. Let Π_i, $i = 0, \ldots, n-1$, denote the projections from $(\mathbb{P}^k)^n$ onto the factors $\mathbb{P}^k$. We use on $(\mathbb{P}^k)^n$ the metric and the distance associated to the Kähler form $\omega_n := \sum \Pi_i^*(\omega_{\text{FS}})$ induced by the Fubini-Study metrics ω_{FS} on the factors $\mathbb{P}^k$, see Appendix A.1. The following indicator lov(f) was introduced by Gromov, it measures the growth rate of the volume of Γ_n,

$$\text{lov}(f) := \lim_{n\to\infty} \frac{1}{n} \log \text{volume}(\Gamma_n).$$

The rest of the proof splits into two parts. We first show that the previous limit exists and is equal to $k \log d$ and then we prove the inequality $h_t(f) \leq \text{lov}(f)$.

Using that $\Pi_0 : \Gamma_n \to \mathbb{P}^k$ is a bi-holomorphic map and that $f^i = \Pi_i \circ (\Pi_{0|\Gamma_n})^{-1}$, we obtain

$$\begin{aligned} k!\text{volume}(\Gamma_n) &= \int_{\Gamma_n} \omega_n^k = \sum_{0\leq i_s\leq n-1} \int_{\Gamma_n} \Pi_{i_1}^*(\omega_{\text{FS}}) \wedge \ldots \wedge \Pi_{i_k}^*(\omega_{\text{FS}}) \\ &= \sum_{0\leq i_s\leq n-1} \int_{\mathbb{P}^k} (f^{i_1})^*(\omega_{\text{FS}}) \wedge \ldots \wedge (f^{i_k})^*(\omega_{\text{FS}}). \end{aligned}$$

The last sum contains n^k integrals that we can compute cohomologically. The above discussion on the action of f^n on cohomology implies that the last integral is equal to $d^{i_1+\cdots+i_k} \leq d^{kn}$. So, the sum is bounded from above by $n^k d^{kn}$. When $i_1 = \cdots = i_k = n-1$, we see that $k!\text{volume}(\Gamma_n) \geq d^{(n-1)k}$. Therefore, the limit in the definition of lov(f) exists and is equal to $k \log d$.

For the second step, we need the following classical estimate due to Lelong [LE], see also Appendix A.2.

Lemma 1.111 (Lelong). *Let A be an analytic set of pure dimension k in a ball B_r of radius r in $\mathbb{C}^N$. Assume that A contains the center of B_r. Then the $2k$-dimensional volume of A is at least equal to the volume of a ball of radius r in $\mathbb{C}^k$. In particular, we have*

$$\text{volume}(A) \geq c_k r^{2k},$$

where $c_k > 0$ is a constant independent of N and of r.

We prove now the inequality $h_t(f) \leq \mathrm{lov}(f)$. Consider an (n,ε)-separated set $\mathscr{F}$ in $\mathbb{P}^k$. For each point $a \in \mathscr{F}$, let $a^{(n)}$ denote the corresponding point $(a, f(a), \ldots, f^{n-1}(a))$ in Γ_n and $B_{a,n}$ the ball of center $a^{(n)}$ and of radius $\varepsilon/2$ in $(\mathbb{P}^k)^n$. Since $\mathscr{F}$ is (n,ε)-separated, these balls are disjoint. On the other hand, by Lelong's inequality, volume$(\Gamma_n \cap B_{a,n}) \geq c'_k \varepsilon^{2k}$, $c'_k > 0$. Note that Lelong's inequality is stated in the Euclidean metric. We can apply it using a fixed atlas of $\mathbb{P}^k$ and the corresponding product atlas of $(\mathbb{P}^k)^n$, the distortion is bounded. So, $\#\mathscr{F} \leq c_k'^{-1} \varepsilon^{-2k}$volume$(\Gamma_n)$ and hence,

$$\frac{1}{n}\log \#\mathscr{F} \leq \frac{1}{n}\log(\mathrm{volume}(\Gamma_n)) + O\Big(\frac{1}{n}\Big).$$

It follows that $h_t(f) \leq \mathrm{lov}(f) = k\log d$. □

We study the entropy of f on some subsets of $\mathbb{P}^k$. The following result is due to de Thélin and Dinh [DT3, D3].

Theorem 1.112. *Let f be a holomorphic endomorphism of $\mathbb{P}^k$ of algebraic degree $d \geq 2$ and $\mathscr{J}_p$ its Julia set of order p, $1 \leq p \leq k$. If K is a subset of $\mathbb{P}^k$ such that $\overline{K} \cap \mathscr{J}_p = \varnothing$, then $h_t(f, K) \leq (p-1)\log d$.*

Proof. The proof is based on Gromov's idea as in Theorem 1.108 and on the speed of convergence towards the Green current. Recall that $\mathscr{J}_p$ is the support of the Green (p,p)-current T^p of f. Fix an open neighbourhood W of $\overline{K}$ such that $W \Subset \mathbb{P}^k \setminus \mathrm{supp}(T^p)$. Using the notations in Theorem 1.108, we only have to prove that

$$\mathrm{lov}(f, W) := \limsup_{n\to\infty} \frac{1}{n} \log \mathrm{volume}(\Pi_0^{-1}(W) \cap \Gamma_n) \leq (p-1)\log d.$$

It is enough to show that volume$(\Pi_0^{-1}(W) \cap \Gamma_n) \lesssim n^k d^{(p-1)n}$. As in Theorem 1.108, it is sufficient to check that for $0 \leq n_i \leq n$

$$\int_W (f^{n_1})^*(\omega_{\mathrm{FS}}) \wedge \ldots \wedge (f^{n_k})^*(\omega_{\mathrm{FS}}) \lesssim d^{(p-1)n}.$$

To this end, we prove by induction on (r,s), $0 \leq r \leq p$ and $0 \leq s \leq k-p+r$, that

$$\|T^{p-r} \wedge (f^{n_1})^*(\omega_{\mathrm{FS}}) \wedge \ldots \wedge (f^{n_s})^*(\omega_{\mathrm{FS}})\|_{W_{r,s}} \leq c_{r,s} d^{n(r-1)},$$

where $W_{r,s}$ is a neighbourhood of $\overline{W}$ and $c_{r,s} \geq 0$ is a constant independent of n and of n_i. We obtain the result by taking $r = p$ and $s = k$.

It is clear that the previous inequality holds when $r = 0$ and also when $s = 0$. In both cases, we can take $W_{r,s} = \mathbb{P}^k \setminus \mathrm{supp}(T^p)$ and $c_{r,s} = 1$. Assume the inequality for $(r-1, s-1)$ and $(r, s-1)$. Let $W_{r,s}$ be a neighbourhood of $\overline{W}$ strictly contained in $W_{r-1,s-1}$ and $W_{r,s-1}$. Let $\chi \geq 0$ be a smooth cut-off function with support in $W_{r-1,s-1} \cap W_{r,s-1}$ which is equal to 1 on $W_{r,s}$. We only have to prove that

$$\int T^{p-r} \wedge (f^{n_1})^*(\omega_{\mathrm{FS}}) \wedge \ldots \wedge (f^{n_s})^*(\omega_{\mathrm{FS}}) \wedge \chi\, \omega_{\mathrm{FS}}^{k-p+r-s} \leq c_{r,s} d^{n(r-1)}.$$

If g is the Green function of f, we have

$$(f^{n_1})^*(\omega_{\rm FS}) = d^{n_1}T - dd^c(g\circ f^{n_1}).$$

The above integral is equal to the sum of the following integrals

$$d^{n_1}\int T^{p-r+1}\wedge (f^{n_2})^*(\omega_{\rm FS})\wedge\ldots\wedge (f^{n_s})^*(\omega_{\rm FS})\wedge\chi\,\omega_{\rm FS}^{k-p+r-s}$$

and

$$-\int T^{p-r}\wedge dd^c(g\circ f^{n_1})\wedge (f^{n_2})^*(\omega_{\rm FS})\wedge\ldots\wedge (f^{n_s})^*(\omega_{\rm FS})\wedge\chi\,\omega_{\rm FS}^{k-p+r-s}.$$

Using the case of $(r-1,s-1)$ we can bound the first integral by $cd^{n(r-1)}$. Stokes' theorem implies that the second integral is equal to

$$-\int T^{p-r}\wedge (f^{n_2})^*(\omega_{\rm FS})\wedge\ldots\wedge (f^{n_s})^*(\omega_{\rm FS})\wedge (g\circ f^{n_1}) dd^c\chi\wedge\omega_{\rm FS}^{k-p+r-s}$$

which is bounded by

$$\|g\|_\infty\|\chi\|_{\mathscr{C}^2}\|T^{p-r}\wedge (f^{n_2})^*(\omega_{\rm FS})\wedge\ldots\wedge (f^{n_s})^*(\omega_{\rm FS})\|_{W_{r,s-1}}$$

since χ has support in $W_{r,s-1}$. We obtain the result using the $(r,s-1)$ case. □

The above result suggests a local indicator of volume growth. Define for $a\in\mathbb{P}^k$

$$\mathrm{lov}(f,a) := \inf_{r>0}\limsup_{n\to\infty}\frac{1}{n}\log\mathrm{volume}(\Pi_0^{-1}(B_r)\cap\Gamma_n),$$

where B_r is the ball of center a and of radius r. We can show that if $a\in\mathscr{J}_p\setminus\mathscr{J}_{p+1}$ and if $\overline{B}_r$ does not intersect $\mathscr{J}_{p+1}$, the above limsup is in fact a limit and is equal to $p\log d$. One can also consider the graph of f^n instead of Γ_n. The notion can be extended to meromorphic maps and its sub-level sets are analogues of Julia sets.

We discuss now the metric entropy, i.e. the entropy of an invariant measure, a notion due to Kolgomorov-Sinai. Let $g: X\to X$ be map on a space X which is measurable with respect to a σ-algebra $\mathscr{F}$. Let ν be an invariant probability measure for g. Let $\xi=\{\xi_1,\ldots,\xi_m\}$ be a measurable partition of X. The entropy of ν with respect to ξ is a measurement of the information we gain when we know that a point x belongs to a member of the partition generated by $g^{-i}(\xi)$ with $0\leq i\leq n-1$.

The information we gain when we know that a point x belongs to ξ_i is a positive function $I(x)$ which depends only on $\nu(\xi_i)$, i.e. $I(x)=\varphi(\nu(\xi_i))$. The information given by independent events should be additive. In other words, we have

$$\varphi(\nu(\xi_i)\nu(\xi_j)) = \varphi(\nu(\xi_i)) + \varphi(\nu(\xi_j))$$

for $i\neq j$. Hence, $\varphi(t)=-c\log t$ with $c>0$. With the normalization $c=1$, the information function for the partition ξ is defined by

$$I_\xi(x) := \sum -\log\nu(\xi_i)\mathbf{1}_{\xi_i}(x).$$

The entropy of ξ is the average of I_ξ:

$$H(\xi) := \int I_\xi(x)d\nu(x) = -\sum \nu(\xi_i)\log\nu(\xi_i).$$

It is useful to observe that the function $t \mapsto -t\log t$ is concave on $]0,1]$ and has the maximal value e^{-1} at e^{-1}.

Consider now the information obtained if we measure the position of the orbit $x, g(x), \ldots, g^{n-1}(x)$ relatively to ξ. By definition, this is the measure of the entropy of the partition generated by $\xi, g^{-1}(\xi), \ldots, g^{-n+1}(\xi)$, which we denote by $\bigvee_{i=0}^{n-1} g^{-i}(\xi)$. The elements of this partition are $\xi_{i_1} \cap g^{-1}(\xi_{i_2}) \cap \ldots \cap g^{-n+1}(\xi_{i_{n-1}})$. It can be shown [W] that

$$h_\nu(g,\xi) := \lim_{n\to\infty} \frac{1}{n} H\left(\bigvee_{i=0}^{n-1} g^{-i}(\xi)\right)$$

exists. *The entropy of the measure* ν is defined as

$$h_\nu(g) := \sup_\xi h_\nu(g,\xi).$$

Two measurable dynamical systems g on $(X, \mathscr{F}, \nu)$ and g' on $(X', \mathscr{F}', \nu')$ are said to be *measurably conjugate* if there is a measurable invertible map $\pi : X \to X'$ such that $\pi \circ g = g' \circ \pi$ and $\pi_*(\nu) = \nu'$. In that case, we have $h_\nu(g) = h_{\nu'}(g')$. So, entropy is a conjugacy invariant. Note also that $h_\nu(g^n) = nh_\nu(g)$ and if g is invertible, $h_\nu(g^n) = |n|h_\nu(g)$ for $n \in \mathbb{Z}$. Moreover, if g is a continuous map of a compact metric space, then $\nu \mapsto h_\nu(g)$ is an affine function on the convex set of g-invariant probability measures [KH, p.164].

We say that a measurable partition ξ is *a generator* if up to sets of measure zero, $\mathscr{F}$ is the smallest σ-algebra containing ξ which is invariant under $g^n, n \in \mathbb{Z}$. A finite partition ξ is called *a strong generator* for a measure preserving dynamical system $(X, \mathscr{F}, \nu, g)$ as above, if up to sets of zero ν-measure, $\mathscr{F}$ generated by $\bigcup_{n=0}^\infty g^{-n}(\xi)$. The following result of Kolmogorov-Sinai is useful in order to compute the entropy [W].

Theorem 1.113 (Kolmogorov-Sinai). *Let ξ be a strong generator for the dynamical system $(X, \mathscr{F}, \nu, g)$ as above. Then*

$$h_\nu(g) = h_\nu(g,\xi).$$

We recall another useful theorem due to Brin-Katok [BK] which is valid for continuous maps $g : X \to X$ on a compact metric space. Let $B_n^g(x,\delta)$ denote the ball of center x and of radius δ with respect to the Bowen distance dist_n. We call $B_n^g(x,\delta)$ *the Bowen (n,δ)-ball.* Define local entropies of an invariant probability measure ν by

$$h_\nu(g,x) := \sup_{\delta>0} \limsup_{n\to\infty} -\frac{1}{n}\log\nu(B_n^g(x,\delta))$$

and

$$h_\nu^-(g,x) := \sup_{\delta>0} \liminf_{n\to\infty} -\frac{1}{n} \log \nu(B_n^g(x,\delta)).$$

Theorem 1.114 (Brin-Katok). *Let $g: X \to X$ be a continuous map on a compact metric space. Let ν be an invariant probability measure of finite entropy. Then, $h_\nu(g,x) = h_\nu^-(g,x)$ and $h_\nu(g,g(x)) = h_\nu(g,x)$ for ν-almost every x. Moreover, $\langle \nu, h_\nu(g,\cdot)\rangle$ is equal to the entropy $h_\nu(g)$ of ν. In particular, if ν is ergodic, we have $h_\nu(g,x) = h_\nu(g)$ ν-almost everywhere.*

One can roughly say that $\nu(B_n^g(x,\delta))$ goes to zero at the exponential rate $e^{-h_\nu(g)}$ for δ small. We can deduce from the above theorem that if $Y \subset X$ is a Borel set with $\nu(Y) > 0$, then $h_t(g,Y) \geq h_\nu(g)$. The comparison with the topological entropy is given by the variational principle [KH, W].

Theorem 1.115 (variational principle). *Let $g: X \to X$ be a continuous map on a compact metric space. Then*

$$\sup h_\nu(g) = h_t(g),$$

where the supremum is taken over the invariant probability measures ν.

Newhouse proved in [NE] that if g is a smooth map on a smooth compact manifold, there is always a measure ν of maximal entropy, i.e. $h_\nu(g) = h_t(g)$. One of the natural question in dynamics is to find the measures which maximize entropy. Their supports are in some sense the most chaotic parts of the system. The notion of Jacobian of a measure is useful in order to estimate the metric entropy.

Let $g: X \to X$ be a measurable map as above which preserves a probability measure ν. Assume there is a countable partition (ξ_i) of X, such that the map g is injective on each ξ_i. *The Jacobian* $J_\nu(g)$ of g with respect to ν is defined as the Radon-Nikodym derivative of $g^*(\nu)$ with respect to ν on each ξ_i. Observe that $g^*(\nu)$ is well-defined on ξ_i since g restricted to ξ_i is injective. We have the following theorem due to Parry [P].

Theorem 1.116 (Parry). *Let g, ν be as above and $J_\nu(g)$ the Jacobian of g with respect to ν. Then*

$$h_\nu(g) \geq \int \log J_\nu(g) d\nu.$$

We now discuss the metric entropy of holomorphic maps on $\mathbb{P}^k$. The following result is a consequence of the variational principle and Theorems 1.108 and 1.112.

Corollary 1.117. *Let f be an endomorphism of algebraic degree $d \geq 2$ of $\mathbb{P}^k$. Let ν be an invariant probability measure. Then $h_\nu(f) \leq k \log d$. If the support of ν does not intersect the Julia set $\mathcal{J}_p$ of order p, then $h_\nu(f) \leq (p-1)\log d$. In particular, if ν is ergodic and $h_\nu(f) > (p-1)\log d$, then ν is supported on $\mathcal{J}_p$.*

In the following result, the value of the metric entropy was obtained in [BD2, S3] and the uniqueness was obtained by Briend-Duval in [BD2]. The case of dimension 1 is due to Freire-Lopès-Mañé [FL] and Lyubich [LY].

Theorem 1.118. *Let f be an endomorphism of algebraic degree $d \geq 2$ of $\mathbb{P}^k$. Then the equilibrium measure μ of f is the unique invariant measure of maximal entropy $k \log d$.*

Proof. We have seen in Corollary 1.117 that $h_\mu(f) \leq k \log d$. Moreover, μ has no mass on analytic sets, in particular on the critical set of f. Therefore, if f is injective on a Borel set K, then $f_*(\mathbf{1}_K) = \mathbf{1}_{f(K)}$ and the total invariance of μ implies that $\mu(f(K)) = d^k \mu(K)$. So, μ is a measure of constant Jacobian d^k. It follows from Theorem 1.116 that its entropy is at least equal to $k \log d$. So, $h_\mu(f) = k \log d$.

Assume now that there is another invariant probability measure ν of entropy $k \log d$. We are looking for a contradiction. Since entropy is an affine function on ν, we can assume that ν is ergodic. This measure has no mass on proper analytic sets of $\mathbb{P}^k$ since otherwise its entropy is at most equal to $(k-1)\log d$, see Exercise 1.122 below. By Theorem 1.45, ν is not totally invariant, so it is not of constant Jacobian. Since μ has no mass on critical values of f, there is a simply connected open set U, not necessarily connected, such that $f^{-1}(U)$ is a union $U_1 \cup \ldots \cup U_{d^k}$ of disjoint open sets such that $f : U_i \to U$ is bi-holomorphic. One can choose U and U_i such that the U_i do not have the same ν-measure, otherwise $\mu = \nu$. So, we can assume that $\nu(U_1) > d^{-k}$. This is possible since two ergodic measures are mutually singular. Here, it is necessarily to chose U so that $\mu(\mathbb{P}^k \setminus U)$ is small.

Choose an open set $W \Subset U_1$ such that $\nu(W) > \sigma$ for some constant $\sigma > d^{-k}$. Let m be a fixed integer and let Y be the set of points x such that for every $n \geq m$, there are at least $n\sigma$ points $f^i(x)$ with $0 \leq i \leq n-1$ which belong to W. If m is large enough, Birkhoff's theorem implies that Y has positive ν-measure. By Brin-Katok's theorem 1.114, we have $h_t(f,Y) \geq h_\nu(f) = k \log d$.

Consider the open sets $\mathscr{U}_\alpha := U_{\alpha_0} \times \cdots \times U_{\alpha_{n-1}}$ in $(\mathbb{P}^k)^n$ such that there are at least $n\sigma$ indices α_i equal to 1. A straighforward computation shows that the number of such open sets is $\leq d^{k\rho n}$ for some constant $\rho < 1$. Let $\mathscr{V}_n$ denote the union of these $\mathscr{U}_\alpha$. Using the same arguments as in Theorem 1.108, we get that

$$k \log d \leq h_t(f,Y) \leq \lim_{n\to\infty} \frac{1}{n} \log \operatorname{volume}(\Gamma_n \cap \mathscr{V}_n)$$

and

$$k!\operatorname{volume}(\Gamma_n \cap \mathscr{V}_n) = \sum_{0\leq i_s \leq n-1} \sum_\alpha \int_{\Gamma_n \cap \mathscr{U}_\alpha} \Pi_{i_1}^*(\omega_{\mathrm{FS}}) \wedge \ldots \wedge \Pi_{i_k}^*(\omega_{\mathrm{FS}}).$$

Fix a constant λ such that $\rho < \lambda < 1$. Let I denote the set of multi-indices $i = (i_1, \ldots, i_k)$ in $\{0, \ldots, n-1\}^k$ such that $i_s \geq n\lambda$ for every s. We distinguish two cases where $i \notin I$ or $i \in I$. In the first case, we have

$$\begin{aligned}\sum_\alpha \int_{\Gamma_n \cap \mathscr{U}_\alpha} \Pi_{i_1}^*(\omega_{\mathrm{FS}}) \wedge \ldots \wedge \Pi_{i_k}^*(\omega_{\mathrm{FS}}) &\leq \int_{\Gamma_n} \Pi_{i_1}^*(\omega_{\mathrm{FS}}) \wedge \ldots \wedge \Pi_{i_k}^*(\omega_{\mathrm{FS}}) \\ &= \int_{\mathbb{P}^k} (f^{i_1})^*(\omega_{\mathrm{FS}}) \wedge \ldots \wedge (f^{i_k})^*(\omega_{\mathrm{FS}}) \\ &= d^{i_1+\cdots+i_k} \leq d^{(k-1+\lambda)n},\end{aligned}$$

since $i_1 + \cdots + i_k \leq (k-1+\lambda)n$.

Consider the second case with multi-indices $i \in I$. Let q denote the integer part of λn and W_α the projection of $\Gamma_n \cap \mathscr{U}_\alpha$ on $\mathbb{P}^k$ by Π_0. Observe that the choice of the open sets U_i implies that f^q is injective on W_α. Therefore,

$$\begin{aligned}
&\sum_\alpha \int_{\Gamma_n \cap \mathscr{U}_\alpha} \Pi_{i_1}^*(\omega_{\text{FS}}) \wedge \ldots \wedge \Pi_{i_k}^*(\omega_{\text{FS}}) \\
&\leq \sum_\alpha \int_{W_\alpha} (f^{i_1})^*(\omega_{\text{FS}}) \wedge \ldots \wedge (f^{i_k})^*(\omega_{\text{FS}}) \\
&= \sum_\alpha \int_{W_\alpha} (f^q)^* \big[(f^{i_1-q})^*(\omega_{\text{FS}}) \wedge \ldots \wedge (f^{i_k-q})^*(\omega_{\text{FS}})\big] \\
&\leq \sum_\alpha \int_{\mathbb{P}^k} (f^{i_1-q})^*(\omega_{\text{FS}}) \wedge \ldots \wedge (f^{i_k-q})^*(\omega_{\text{FS}}).
\end{aligned}$$

Recall that the number of open sets $\mathscr{U}_\alpha$ is bounded by $d^{k\rho n}$. So, the last sum is bounded by

$$d^{k\rho n} d^{(i_1-q)+\cdots+(i_k-q)} \leq d^{k\rho n} d^{k(n-q)} \lesssim d^{k(1+\rho-\lambda)n}.$$

Finally, since the number of multi-indices i is less than n^k, we deduce from the above estimates that

$$k!\text{volume}(\Gamma_n \cap \mathscr{V}_n) \lesssim n^k d^{(k-1+\lambda)n} + n^k d^{k(1+\rho-\lambda)n}.$$

This contradicts the above bound from below of volume$(\Gamma_n \cap \mathscr{V}_n)$. □

The remaining part of this paragraph deals with Lyapounov exponents associated to the measure μ and their relations with the Hausdorff dimension of μ. Results in this direction give some information about the rough geometrical behaviour of the dynamical system on the most chaotic locus. An abstract theory was developed by Oseledec and Pesin, see e.g. [KH]. However, it is often difficult to show that a given dynamical system has non-vanishing Lyapounov exponents. In complex dynamics as we will see, the use of holomorphicity makes the goal reachable. We first introduce few notions.

Let A be a linear endomorphism of $\mathbb{R}^k$. We can write $\mathbb{R}^k$ as the direct sum $\oplus E_i$ of invariant subspaces on which all the complex eigenvalues of A have the same modulus. This decomposition of $\mathbb{R}^k$ describes clearly the geometrical behaviour of the dynamical system associated to A. An important part in the dynamical study with respect to an invariant measure is to describe geometrical aspects following the directional dilation or contraction indicators.

Consider a smooth dynamical system $g : X \to X$ and an invariant ergodic probability measure ν. The map g induces a linear map from the tangent space at x to the tangent space at $g(x)$. This linear map is given by a square matrix when we fix local coordinates near x and $g(x)$. Then, we obtain a function on X with values in $\text{GL}(\mathbb{R}, k)$ where k denotes the real dimension of X. We will study the sequence of such functions associated to the sequence of iterates (g^n) of g.

Consider a more abstract setting. Let $g : X \to X$ be a measurable map and ν an invariant probability measure. Let $A : X \to \mathrm{GL}(\mathbb{R}, k)$ be a measurable function. Define for $n \geq 0$

$$A_n(x) := A(g^{n-1}(x)) \dots A(x).$$

These functions satisfy the identity

$$A_{n+m}(x) = A_n(g^m(x))A_m(x)$$

for $n, m \geq 0$. We say that the sequence (A_n) is *the multiplicative cocycle* over X *generated by* A.

The following Oseledec's multiplicative ergodic theorem is related to the Kingman's sub-multiplicative ergodic theorem [KH, W]. It can be seen as a generalization of the above property of a single square matrix A.

Theorem 1.119 (Oseledec). *Let $g : X \to X$, ν and the cocycle (A_n) be as above. Assume that ν is ergodic and that $\log^+ \|A^{\pm 1}(x)\|$ are in $L^1(\nu)$. Then there is an integer m, real numbers $\chi_1 < \cdots < \chi_m$, and for ν-almost every x, a unique decomposition of $\mathbb{R}^k$ into a direct sum of linear subspaces*

$$\mathbb{R}^k = \bigoplus_{i=1}^{m} E_i(x)$$

such that

1. *The dimension of $E_i(x)$ does not depend on x.*
2. *The decomposition is invariant, that is, $A(x)$ sends $E_i(x)$ to $E_i(g(x))$.*
3. *We have locally uniformly on vectors v in $E_i(x) \setminus \{0\}$*

$$\lim_{n \to \infty} \frac{1}{n} \log \|A_n(x) \cdot v\| = \chi_i.$$

4. *For $S \subset \{1, \dots, m\}$, define $E_S(x) := \oplus_{i \in S} E_i(x)$. If S, S' are disjoint, then the angle between $E_S(x)$ and $E_{S'}(x)$ is a tempered function, that is,*

$$\lim_{n \to \infty} \frac{1}{n} \log \sin \left| \angle\big(E_S(g^n(x)), E_{S'}(g^n(x))\big) \right| = 0.$$

The result is still valid for non-ergodic systems but the constants m and χ_i should be replaced with invariant functions. If g is invertible, the previous decomposition is the same for g^{-1} where the exponents χ_i are replaced with $-\chi_i$. The result is also valid in the complex setting where we replace $\mathbb{R}$ with $\mathbb{C}$ and $\mathrm{GL}(\mathbb{R}, k)$ by $\mathrm{GL}(\mathbb{C}, k)$. In this case, the subspaces $E_i(x)$ are complex.

We now come back to a smooth dynamical system $g : X \to X$ on a compact manifold. We assume that the Jacobian $J(g)$ of g associated to a smooth volume form satisfies $\langle \nu, \log J(g) \rangle > -\infty$. Under this hypothesis, we can apply Oseledec's theorem to the cocycle induced by g on the tangent bundle of X; this allows to decompose, ν-almost everywhere, the tangent bundle into invariant sub-bundles.

The corresponding constants χ_i are called *Lyapounov exponents* of g with respect to ν. The dimension of E_i is *the multiplicity* of χ_i. These notions do not depend on the choice of local coordinates on X. The Lyapounov exponents of g^n are equal to $n\chi_i$. We say that the measure ν is *hyperbolic* if no Lyapounov exponent is zero. It is not difficult to deduce from the Oseledec's theorem that the sum of Lyapounov exponents of ν is equal to $\langle \nu, \log J(g)\rangle$. The reader will find in [KH] a theorem due to Pesin, called *the ε-reduction theorem*, which generalizes Theorem 1.119. It gives some coordinate changes on $\mathbb{R}^k$ which allow to write $A(x)$ in the form of a diagonal block matrix with explicit estimates on the distortion.

The following result due to Briend-Duval [BD1], shows that endomorphisms in $\mathbb{P}^k$ are expansive with respect to the equilibrium measures. We give here a new proof using Proposition 1.51. Note that there are k Lyapounov exponents counted with multiplicity. If we consider these endomorphism as real maps, we should count twice the Lyapounov exponents.

Theorem 1.120. *Let f be a holomorphic endomorphism of algebraic degree $d \geq 2$ of $\mathbb{P}^k$. Then the equilibrium measure μ of f is hyperbolic. More precisely, its Lyapounov exponents are larger or equal to $\frac{1}{2}\log d$.*

Proof. Since the measure μ is PB, quasi-p.s.h. functions are μ-integrable. It is not difficult to check that if $J(f)$ is the Jacobian of f with respect to the Fubini-Study metric, then $\log J(f)$ is a quasi-p.s.h. function. Therefore, we can apply Oseledec's theorem 1.119. We deduce from this result that the smallest Lyapounov exponent of μ is equal to

$$\chi := \lim_{n\to\infty} -\frac{1}{n}\log\|Df^n(x)^{-1}\|$$

for μ-almost every x. By Proposition 1.51, there is a ball B of positive μ measure which admits at least $\frac{1}{2}d^{kn}$ inverse branches $g_i : B \to U_i$ for f^n with U_i of diameter $\leq d^{-n/2}$. If we slightly reduce the ball B, we can assume that $\|Dg_i\| \leq Ad^{-n/2}$ for some constant $A > 0$. This is a simple consequence of Cauchy's formula. It follows that $\|(Df^n)^{-1}\| \leq Ad^{-n/2}$ on U_i. The union V_n of the U_i is of measure at least equal to $\frac{1}{2}\mu(B)$. Therefore, by Fatou's lemma,

$$\frac{1}{2}\mu(B) \leq \limsup_{n\to\infty}\langle \mu, \mathbf{1}_{V_n}\rangle \leq \langle \mu, \limsup \mathbf{1}_{V_n}\rangle = \langle \mu, \mathbf{1}_{\limsup V_n}\rangle.$$

Hence, there is a set $K := \limsup V_n$ of positive measure such that if x is in K, we have $\|Df^n(x)^{-1}\| \leq Ad^{-n/2}$ for infinitely many of n. The result follows. □

Note that in a recent work [DT2], de Thélin proved that if ν is an invariant measure of entropy strictly larger than $(k-1)\log d$ such that the logarithm of the Jacobian is integrable, then the associated Lyapounov exponents are strictly positive. He also obtained for these exponents some explicit estimates from below. His method is more powerful and is valid in a very general setting.

The Hausdorff dimension $\dim_H(\nu)$ of a probability measure ν on $\mathbb{P}^k$ is the infimum of the numbers $\alpha \geq 0$ such that there is a Borel set K of Hausdorff dimension α of full measure, i.e. $\nu(K) = 1$. Hausdorff dimension says how the measure fills out its support. The following result was obtained by Binder-DeMarco [BI] and Dinh-Dupont [DD]. The fact that μ has positive dimension has been proved in [S3]; indeed, a lower bound is given in terms of the Hölder continuity exponent of the Green function g.

Theorem 1.121. *Let f be an endomorphism of algebraic degree $d \geq 2$ of $\mathbb{P}^k$ and μ its equilibrium measure. Let $\chi_1, \ldots, \chi_k$ denote the Lyapounov exponents of μ ordered by $\chi_1 \geq \cdots \geq \chi_k$ and Σ their sum. Then*

$$\frac{k \log d}{\chi_1} \leq \dim_H(\mu) \leq 2k - \frac{2\Sigma - k \log d}{\chi_1}.$$

The proof is quite technical. It is based on a delicate study of the inverse branches of balls along a generic negative orbit. We will not give the proof here. A better estimate in dimension 2, probably the sharp one, was recently obtained by Dupont. Indeed, Binder-DeMarco conjecture that the Hausdorff dimension of μ satisfies

$$\dim_H(\mu) = \frac{\log d}{\chi_1} + \cdots + \frac{\log d}{\chi_k}.$$

Dupont gives in [DP3] results in this direction.

Exercise 1.122. Let X be an analytic subvariety of pure dimension p in $\mathbb{P}^k$. Let f be an endomorphism of algebraic degree $d \geq 2$ of $\mathbb{P}^k$. Show that $h_t(f,X) \leq p \log d$.

Exercise 1.123. Let $f : X \to X$ be a smooth map and K an invariant compact subset of X. Assume that K is hyperbolic, i.e. there is a continuously varying decomposition $TX_{|K} = E \oplus F$ of the tangent bundle of X restricted to K, into the sum of two invariant vector bundles such that $\|Df\| < 1$ on E and $\|(Df)^{-1}\| < 1$ on F for some smooth metric near K. Show that f admits a hyperbolic ergodic invariant measure supported on K.

Exercise 1.124. Let $f : X \to X$ be a holomorphic map on a compact complex manifold and let ν be an ergodic invariant measure. Show that in Theorem 1.119 applied to the action of f on the complex tangent bundle, the spaces $E_i(x)$ are complex.

Exercise 1.125. Let $\alpha > 0$ be a constant. Show that there is an endomorphism f of $\mathbb{P}^k$ such that the Hausdorff dimension of the equilibrium measure of f is smaller than α. Show that there is an endomorphism f such that its Green function g is not α-Hölder continuous.

Notes. We do not give here results on local dynamics near a fixed point. If this point is non-critical attractive or repelling, a theorem of Poincaré says that the map is locally conjugated to a polynomial map [ST]. Maps which are tangent to the identity or are semi-attractive at a fixed point, were studied by Abate and Hakim [A, H1, H2, H3], see also Abate, Bracci and Tovena [ABT]. Dynamics near a super-attractive fixed point in dimension $k = 2$ was studied by Favre-Jonsson using a theory of valuations in [FJ1].

The study of the dynamical system outside the support of the equilibrium measure is not yet developped. Some results on attracting sets, attracting currents, etc. were obtained by de Thélin, Dinh, Fornæss, Jonsson, Sibony, Weickert [DT1, DT2, D3, FS6, FW, JW], see also Bonifant, Dabija, Milnor and Taflin [BO, T], Mihailescu and Urbański [MI, MIH].

In dimension 1, Fatou and Julia considered their theory as an investigation to solve some functional equations. In particular, they found all the commuting pairs of polynomials [F, JU], see also Ritt [R] and Eremenko [E] for the case of rational maps. Commuting endomorphisms of $\mathbb{P}^k$ were studied by the authors in [DS0]. A large family of solutions are Lattès maps. We refer to Berteloot, Dinh, Dupont, Loeb, Molino [BR, BDM, BL, D4, DP1] for a study of this class of maps, see also Milnor [MIL] for the dimension 1 case.

We do not consider here bifurcation problems for families of maps and refer to Bassanelli-Berteloot [BB] and Pham [PH] for this subject. Some results will be presented in the next chapter.

In [Z], Zhang considers some links between complex dynamics and arithmetic questions. He is interested in polarized holomorphic maps on Kähler varieties, i.e. maps which multiply a Kähler class by an integer. If the Kähler class is integral, the variety can be embedded into a projective space $\mathbb{P}^k$ and the maps extend to endomorphisms of $\mathbb{P}^k$. So, several results stated above can be directely applied to that situation. In general, most of the results for endomorphisms in $\mathbb{P}^k$ can be easily extended to general polarized maps. In the unpublished preprint [DS14], the authors considered the situation of smooth compact Kähler manifolds. We recall here the main result.

Let (X,ω) be an arbitrary compact Kähler manifold of dimension k. Let f be a holomorphic endomorphism of X. We assume that f is open. The spectral radius of f^* acting on $H^{p,p}(X,\mathbb{C})$ is called *the dynamical degree of order p* of f. It can be computed by the formula

$$d_p := \lim_{n\to\infty}\Big(\int_X (f^n)^*(\omega^p)\wedge\omega^{k-p}\Big)^{1/n}.$$

The last degree d_k is the topological degree of f, i.e. equal to the number of points in a generic fiber of f. We also denote it by d_t.

Assume that $d_t > d_p$ for $1\leq p\leq k-1$. Then, there is a maximal proper analytic subset $\mathscr{E}$ of X which is totally invariant by f, i.e. $f^{-1}(\mathscr{E}) = f(\mathscr{E}) = \mathscr{E}$. If δ_a is a Dirac mass at $a\not\in\mathscr{E}$, then $d_t^{-n}(f^n)^*(\delta_a)$ converge to a probability measure μ, which does not depend on a. This is *the equilibrium measure* of f. It satisfies $f^*(\mu) = d_t\mu$ and $f_*(\mu) = \mu$. If J is the Jacobian of f with respect to ω^k then $\langle\mu,\log J\rangle \geq \log d_t$. The measure μ is K-mixing and hyperbolic with Lyapounov exponents larger or equal to $\frac{1}{2}\log(d_t/d_{k-1})$. Moreover, there are sets $\mathscr{P}_n$ of repelling periodic points of order n, on $\mathrm{supp}(\mu)$ such that the probability measures equidistributed on $\mathscr{P}_n$ converge to μ, as n goes to infinity. If the periodic points of period n are isolated for every n, an estimate on the norm of $(f^n)^*$ on $H^{p,q}(X,\mathbb{C})$ obtained in [D22], implies that the number of these periodic points is $d_t^n + o(d_t^n)$. Therefore, periodic points are equidistributed with respect to μ. We can prove without difficulty that μ is the unique invariant measure of maximal entropy $\log d_t$ and is moderate. Then, we can extend the stochastic properties obtained for $\mathbb{P}^k$ to this more general setting.

When f is polarized by the cohomology class $[\omega]$ of a Kähler form ω, there is a constant $\lambda\geq 1$ such that $f^*[\omega] = \lambda[\omega]$. It is not difficult to check that $d_p = \lambda^p$. The above results can be applied for such a map when $\lambda > 1$. In which case, periodic points of a given period are isolated. Note also that Theorem 1.108 can be extended to this case.

2 Polynomial-like Maps in Higher Dimension

In this section we consider a large family of holomorphic maps in a semi-local setting: the polynomial-like maps. They can appear as a basic block in the study of some meromorphic maps on compact manifolds. The main reference for this section is our article [DS1] where the dd^c-method in dynamics was introduced. Endomorphisms of $\mathbb{P}^k$ can be considered as a special case of polynomial-like maps. However, in general, there is no Green $(1,1)$-current for such maps. The notion of dynamical degrees for polynomial-like maps replaces the algebraic degree. Under natural assumptions on dynamical degrees, we prove that the measure of maximal entropy is non-uniformly hyperbolic and we study its sharp ergodic properties.

2.1 Examples, Degrees and Entropy

Let V be a convex open set in $\mathbb{C}^k$ and $U \Subset V$ an open subset. A proper holomorphic map $f : U \to V$ is called *a polynomial-like map*. Recall that a map $f : U \to V$ is proper if $f^{-1}(K) \Subset U$ for every compact subset K of V. The map f sends the boundary of U to the boundary of V; more precisely, the points near ∂U are sent to points near ∂V. So, polynomial-like maps are somehow expansive in all directions, but the expansion is in the geometrical sense. In general, they may have a non-empty critical set. A polynomial-like mapping $f : U \to V$ defines a ramified covering over V. The degree d_t of this covering is also called *the topological degree*. It is equal to the number of points in a generic fiber, or in any fiber if we count the multiplicity.

Polynomial-like maps are characterized by the property that their graph Γ in $U \times V$ is in fact a submanifold of $V \times V$, that is, Γ is closed in $V \times V$. So, any small perturbation of f is polynomial-like of the same topological degree d_t, provided that we reduce slightly the open set V. We will construct large families of polynomial-like maps. In dimension one, it was proved by Douady-Hubbard [DH] that such a map is conjugated to a polynomial via a Hölder continuous homeomorphism. Many dynamical properties can be deduced from the corresponding properties of polynomials. In higher dimension, the analogous statement is not valid. Some new dynamical phenomena appear for polynomial-like mappings, that do not exist for polynomial maps. We use here an approach completely different from the one dimensional case, where the basic tool is the Riemann measurable mapping theorem.

Let $f : \mathbb{C}^k \to \mathbb{C}^k$ be a holomorphic map such that the hyperplane at infinity is attracting in the sense that $\|f(z)\| \geq A\|z\|$ for some constant $A > 1$ and for $\|z\|$ large enough. If V is a large ball centered at 0, then $U := f^{-1}(V)$ is strictly contained in V. Therefore, $f : U \to V$ is a polynomial-like map. Small transcendental perturbations of f, as we mentioned above, give a large family of polynomial-like maps. Observe also that the dynamical study of holomorphic endomorphisms on $\mathbb{P}^k$ can be reduced to polynomial-like maps by lifting to a large ball in $\mathbb{C}^{k+1}$. We give now other explicit examples.

Example 2.1. Let $f = (f_1, \dots, f_k)$ be a polynomial map in $\mathbb{C}^k$, with $\deg f_i = d_i \geq 2$. Using a conjugacy with a permutation of coordinates, we can assume that $d_1 \geq \dots \geq d_k$. Let f_i^+ denote the homogeneous polynomial of highest degree in f_i. If $\{f_1^+ = \dots = f_k^+ = 0\}$ is reduced to $\{0\}$, then f is polynomial-like in any large ball of center 0. Indeed, define $d := d_1 \dots d_k$ and $\pi(z_1, \dots, z_k) := (z_1^{d/d_1}, \dots, z_k^{d/d_k})$. Then, $\pi \circ f$ is a polynomial map of algebraic degree d which extends holomorphically at infinity to an endomorphism of $\mathbb{P}^k$. Therefore, $\|\pi \circ f(z)\| \gtrsim \|z\|^d$ for $\|z\|$ large enough. The estimate $\|f(z)\| \gtrsim \|z\|^{d_k}$ near infinity follows easily. If we consider the extension of f to $\mathbb{P}^k$, we obtain in general a meromorphic map which is not holomorphic. Small pertubations f_ε of f may have indeterminacy points in $\mathbb{C}^k$ and a priori, indeterminacy points of the sequence $(f_\varepsilon^n)_{n\geq 1}$ may be dense in $\mathbb{P}^k$.

Example 2.2. The map $(z_1,z_2) \mapsto (z_1^2 + az_2, z_1)$, $a \neq 0$, is not polynomial-like. It is invertible and the point $[0:0:1]$ at infinity, in homogeneous coordinates $[z_0:z_1:z_2]$, is an attractive fixed point for f^{-1}. Hence, the set $\mathscr{K}$ of points $z \in \mathbb{C}^2$ with bounded orbit, clusters at $[0:0:1]$.

The map $f(z_1,z_2) := (z_2^d, 2z_1)$, $d \geq 2$, is polynomial-like in any large ball of center 0. Considered as a map on $\mathbb{P}^2$, it is only meromorphic with an indeterminacy point $[0:1:0]$. On a fixed large ball of center 0, the perturbed maps $f_\varepsilon := (z_2^d + \varepsilon e^{z_1}, 2z_1 + \varepsilon e^{z_2})$ are polynomial-like, in an appropriate open set U.

Consider a general polynomial-like map $f: U \to V$ of topological degree $d_t \geq 2$. We introduce several growth indicators of the action of f on forms or currents. Define $f^n := f \circ \cdots \circ f$, n times, *the iterate of order* n of f. This map is only defined on $U_{-n} := f^{-n}(V)$. The sequence (U_{-n}) is decreasing: we have $U_{-n-1} = f^{-1}(U_{-n}) \Subset U_{-n}$. Their intersection $\mathscr{K} := \cap_{n\geq 0} U_{-n}$ is a non-empty compact set that we call *the filled Julia set* of f. The filled Julia set is totally invariant: we have $f^{-1}(\mathscr{K}) = \mathscr{K}$ which implies that $f(\mathscr{K}) = \mathscr{K}$. Only for x in $\mathscr{K}$, the infinite orbit $x, f(x), f^2(x), \ldots$ is well-defined. The preimages $f^{-n}(x)$ by f^n are defined for every $n \geq 0$ and every x in V.

Let $\omega := dd^c\|z\|^2$ denote the standard Kähler form on $\mathbb{C}^k$. Recall that the mass of a positive (p,p)-current S on a Borel set K is given by $\|S\|_K := \int_K S \wedge \omega^{k-p}$. Define the *dynamical degree of order* p of f, for $0 \leq p \leq k$, by

$$d_p(f) := \limsup_{n\to\infty} \|(f^n)_*(\omega^{k-p})\|_W^{1/n} = \limsup_{n\to\infty} \|(f^n)^*(\omega^p)\|_{f^{-n}(W)}^{1/n},$$

where $W \Subset V$ is a neighbourhood of $\mathscr{K}$. For simplicity, when there is no confusion, this degree is also denoted by d_p. We have the following lemma.

Lemma 2.3. *The degrees d_p do not depend on the choice of W. Moreover, we have $d_0 \leq 1$, $d_k = d_t$ and the dynamical degree of order p of f^m is equal to d_p^m.*

Proof. Let $W' \subset W$ be another neighbourhood of $\mathscr{K}$. For the first assertion, we only have to show that

$$\limsup_{n\to\infty} \|(f^n)^*(\omega^p)\|_{f^{-n}(W)}^{1/n} \leq \limsup_{n\to\infty} \|(f^n)^*(\omega^p)\|_{f^{-n}(W')}^{1/n}.$$

By definition of $\mathscr{K}$, there is an integer N such that $f^{-N}(V) \Subset W'$. Since $(f^N)^*(\omega^p)$ is smooth on $f^{-N}(V)$, we can find a constant $A > 0$ such that $(f^N)^*(\omega^p) \leq A\omega^p$ on $f^{-N}(W)$. We have

$$\begin{aligned}
\limsup_{n\to\infty} \|(f^n)^*(\omega^p)\|_{f^{-n}(W)}^{1/n} &= \limsup_{n\to\infty} \|(f^{n-N})^*((f^N)^*(\omega^p))\|_{f^{-n+N}(f^{-N}(W))}^{1/n} \\
&\leq \limsup_{n\to\infty} \|A(f^{n-N})^*(\omega^p)\|_{f^{-n+N}(W')}^{1/n} \\
&= \limsup_{n\to\infty} \|(f^n)^*(\omega^p)\|_{f^{-n}(W')}^{1/n}.
\end{aligned}$$

This proves the first assertion.

It is clear from the definition that $d_0 \leq 1$. Since f has topological degree d_t the pull-back of a positive measure multiplies the mass by d_t. Therefore, $d_k = d_t$. For the last assertion of the lemma, we only have to check that

$$\limsup_{n\to\infty} \|(f^n)^*(\omega^p)\|^{1/n}_{f^{-n}(W)} \leq \limsup_{s\to\infty} \|(f^{ms})^*(\omega^p)\|^{1/ms}_{f^{-ms}(W)}.$$

To this end, we proceed as above. Write $n = ms + r$ with $0 \leq r \leq m-1$. We obtain the result using that $(f^r)^*(\omega^p) \leq A\omega^p$ on a fixed neighbourhood of $\mathscr{K}$ for $0 \leq r \leq m-1$. □

The main result of this paragraph is the following formula for the entropy.

Theorem 2.4. *Let $f : U \to V$ be a polynomial-like map of topological degree $d_t \geq 2$. Let $\mathscr{K}$ be the filled Julia set of f. Then, the topological entropy of f on $\mathscr{K}$ is equal to $h_t(f, \mathscr{K}) = \log d_t$. Moreover, all the dynamical degrees d_p of f are smaller or equal to d_t.*

We need the following lemma where we use standard metrics on Euclidean spaces.

Lemma 2.5. *Let V be an open set of $\mathbb{C}^k$, U a relatively compact subset of V and L a compact subset of $\mathbb{C}$. Let π denote the canonical projection from $\mathbb{C}^m \times V$ onto V. Suppose Γ is an analytic subset of pure dimension k of $\mathbb{C}^m \times V$ contained in $L^m \times V$. Assume also that $\pi : \Gamma \to V$ defines a ramified covering of degree d_Γ. Then, there exist constants $c > 0$, $s > 0$, independent of Γ and m, such that*

$$\text{volume}(\Gamma \cap \mathbb{C}^m \times U) \leq cm^s d_\Gamma.$$

Proof. Since the problem is local on V, we can assume that V is the unit ball of $\mathbb{C}^k$ and U is the closed ball of center 0 and of radius $1/2$. We can also assume that L is the closed unit disc in $\mathbb{C}$. Denote by $x = (x_1, \ldots, x_m)$ and $y = (y_1, \ldots, y_k)$ the coordinates on $\mathbb{C}^m$ and on $\mathbb{C}^k$. Let ε be a $k \times m$ matrix whose entries have modulus bounded by $1/8mk$. Define $\pi_\varepsilon(x,y) := y + \varepsilon x$, $\Gamma_\varepsilon := \Gamma \cap \{\|\pi_\varepsilon\| < 3/4\}$ and $\Gamma^* := \Gamma \cap (L^m \times U)$.

We first show that $\Gamma^* \subset \Gamma_\varepsilon$. Consider a point $(x,y) \in \Gamma^*$. We have $|x_i| < 1$ and $\|y\| \leq 1/2$. Hence,

$$\|\pi_\varepsilon(x,y)\| \leq \|y\| + \|\varepsilon x\| < 3/4.$$

This implies that $(x,y) \in \Gamma_\varepsilon$.

Now, we prove that for every $a \in \mathbb{C}^k$ with $\|a\| < 3/4$, we have $\#\pi_\varepsilon^{-1}(a) \cap \Gamma = d_\Gamma$, where we count the multiplicities of points. To this end, we show that $\#\pi_{t\varepsilon}^{-1}(a) \cap \Gamma$ does not depend on $t \in [0,1]$. So, it is sufficient to check that the union of the sets $\pi_{t\varepsilon}^{-1}(a) \cap \Gamma$ is contained in the compact subset $\Gamma \cap \{\|\pi\| \leq 7/8\}$ of Γ. Let $(x,y) \in \Gamma$ and $t \in [0,1]$ such that $\pi_{t\varepsilon}(x,y) = a$. We have

$$3/4 > \|a\| = \|\pi_{t\varepsilon}(x,y)\| \geq \|y\| - t\|\varepsilon x\|.$$

It follows that $\|y\| < 7/8$ and hence $(x,y) \in \Gamma \cap \{\|\pi\| \leq 7/8\}$.

Let B denote the ball of center 0 and of radius $3/4$ in $\mathbb{C}^k$. We have for some constant $c' > 0$

$$\int_{\Gamma^*} \pi_\varepsilon^*(\omega^k) \le \int_{\Gamma_\varepsilon} \pi_\varepsilon^*(\omega^k) = d_\Gamma \int_B \omega^k = c' d_\Gamma.$$

Let $\Theta := dd^c\|x\|^2 + dd^c\|y\|^2$ be the standard Kähler $(1,1)$-form on $\mathbb{C}^m \times \mathbb{C}^k$. We have

$$\text{volume}(\Gamma \cap \mathbb{C}^m \times U) = \int_{\Gamma\cap\mathbb{C}^m\times U} \Theta^k.$$

It suffices to bound Θ by a linear combination of $2m+1$ forms of type $\pi_\varepsilon^*(\omega)$ with coefficients of order $\simeq m^2$ and then to use the previous estimates. Recall that $\omega = dd^c\|y\|^2$. So, we only have to bound $\sqrt{-1}dx_i \wedge d\overline{x}_i$ by a combination of $(1,1)$-forms of type $\pi_\varepsilon^*(\omega)$. Consider $\delta := 1/8mk$ and $\pi_\varepsilon(x,y) := (y_1 + \delta x_i, y_2, \ldots, y_k)$. We have

$$\begin{aligned}\sqrt{-1}dx_i \wedge d\overline{x}_i &= \frac{4\sqrt{-1}}{3\delta^2}\Big[3dy_1 \wedge d\overline{y}_1 + d(y_1+\delta x_i)\wedge d(\overline{y}_1 + \delta\overline{x}_i) \\ &\quad - d(2y_1 + \delta x_i/2)\wedge d(2\overline{y}_1 + \delta \overline{x}_i/2)\Big] \\ &\le \frac{4\sqrt{-1}}{3\delta^2}\Big[3dy_1 \wedge d\overline{y}_1 + d(y_1+\delta x_i)\wedge d(\overline{y}_1 + \delta\overline{x}_i)\Big]\end{aligned}$$

The last form can be bounded by a combination of $\pi_0^*(\omega)$ and $\pi_\varepsilon^*(\omega)$. This completes the proof. □

Proof of Theorem 2.4. We prove that $h_t(f,\mathscr{K}) \le \log d_t$. We will prove in Paragraphs 2.2 and 2.4 that f admits a totally invariant measure of maximal entropy $\log d_t$ with support in the boundary of $\mathscr{K}$. The variational principle then implies that $h_t(f,\mathscr{K}) = h_t(f,\partial\mathscr{K}) = \log d_t$. We can also conclude using Misiurewicz-Przytycki's theorem 1.109 or Yomdin's theorem 1.110 which can be extended to this case.

Let Γ_n denote the graph of $(f,\ldots,f^{n-1})$ in $V^n \subset (\mathbb{C}^k)^{n-1} \times V$. Let $\pi : (\mathbb{C}^k)^{n-1} \times V \to V$ be the canonical projection. Since $f : U \to V$ is polynomial-like, it is easy to see that $\Gamma_n \subset U^{n-1} \times V$ and that $\pi : \Gamma_n \to V$ defines a ramified covering of degree d_t^n. As in Theorem 1.108, we have

$$h_t(f,\mathscr{K}) \le \text{lov}(f) := \limsup_{n\to\infty} \frac{1}{n} \log \text{volume}(\Gamma_n \cap \pi^{-1}(U)).$$

But, it follows from Lemma 2.5 that

$$\text{volume}(\Gamma_n \cap \pi^{-1}(U)) \le c(kn)^s d_t^n.$$

Hence, $\text{lov}(f) \le \log d_t$. This implies the inequality $h_t(f,\mathscr{K}) \le \log d_t$. Note that the limit in the definition of $\text{lov}(f)$ exists and we have $\text{lov}(f) = \log d_t$. Indeed, since Γ_n is a covering of degree d_t^n over V, we always have

$$\text{volume}(\Gamma_n \cap \pi^{-1}(U)) \ge d_t^n \text{volume}(U).$$

We show that $d_p \leq d_t$. Let Π_i, $0 \leq i \leq n-1$, denote the projection of V^n onto its factors. We have

$$\text{volume}(\Gamma_n \cap \pi^{-1}(U)) = \sum_{0 \leq i_s \leq n-1} \int_{\Gamma_n \cap \pi^{-1}(U)} \Pi_{i_1}^*(\omega) \wedge \ldots \wedge \Pi_{i_k}^*(\omega).$$

The last sum contains the term

$$\int_{\Gamma_n \cap \pi^{-1}(U)} \Pi_0^*(\omega^{k-p}) \wedge \Pi_{n-1}^*(\omega^p) = \int_{f^{-n+1}(U)} \omega^{k-p} \wedge (f^{n-1})^*(\omega^p).$$

We deduce from the estimate on volume$(\Gamma_n \cap \pi^{-1}(U))$ and from the definition of d_p that $d_p \leq \text{lov}(f) = d_t$. □

We introduce now others useful dynamical degrees. We call *dynamical $*$-degree of order p* of f the following limit

$$d_p^* := \limsup_{n \to \infty} \sup_S \|(f^n)_*(S)\|_W^{1/n},$$

where $W \Subset V$ is a neighbourhood of $\mathcal{K}$ and the supremum is taken over positive closed $(k-p, k-p)$-current of mass ≤ 1 on a fixed neighbourhood $W' \Subset V$ of $\mathcal{K}$. Clearly, $d_p^* \geq d_p$, since we can take $S = c\omega^{k-p}$ with $c > 0$ small enough.

Lemma 2.6. *The above definition does not depend on the choice of W, W'. Moreover, we have $d_0^* = 1$, $d_k^* = d_t$ and the dynamical $*$-degree of order p of f^n is equal to d_p^{*n}.*

Proof. If N is an integer large enough, the operator $(f^N)_*$ sends continuously positive closed currents on W' to the ones on V. Therefore, the independence of the definition on W' is clear. If S is a probability measure on $\mathcal{K}$, then $(f^n)_*$ is also a probability measure on $\mathcal{K}$. Therefore, $d_0^* = 1$. Observe that

$$\|(f^n)_*(S)\|_W^{1/n} = \Big[\int_{f^{-n}(W)} S \wedge (f^n)^*(\omega^p)\Big]^{1/n}.$$

So, for the other properties, it is enough to follow the arguments given in Lemma 2.3. □

Many results below are proved under the hypothesis $d_{k-1}^* < d_t$. The following proposition shows that this condition is stable under small perturbations on f. This gives large families of maps satisfying the hypothesis. Indeed, the condition is satisfied for polynomial maps in $\mathbb{C}^k$ which extend at infinity as endomorphisms of $\mathbb{P}^k$. For such maps, if d is the algebraic degree, one can check that $d_p^* \leq d^p$.

Proposition 2.7. *Let $f: U \to V$ be a polynomial-like map of topological degree d_t. Let V' be a convex open set such that $U \Subset V' \Subset V$. If $g: U \to \mathbb{C}^k$ is a holomorphic map, close enough to f and $U' := g^{-1}(V')$, then $g: U' \to V'$ is a polynomial-like map of topological degree d_t. If moreover, f satisfies the condition $d_p^* < d_t$ for some $1 \leq p \leq k-1$, then g satisfies the same property.*

Proof. The first assertion is clear, using the characterization of polynomial-maps by their graphs. We prove the second one. Fix a constant δ with $d_p^* < \delta < d_t$ and an open set W such that $U \Subset W \Subset V$. Fix an integer N large enough such that $\|(f^N)_*(S)\|_W \leq \delta^N$ for any positive closed $(k-p,k-p)$-current S of mass 1 on U. If g is close enough to f, we have $g^{-N}(U) \Subset f^{-N}(W)$ and

$$\|(g^N)^*(\omega^p) - (f^N)^*(\omega^p)\|_{L^\infty(g^{-N}(U))} \leq \varepsilon$$

with $\varepsilon > 0$ a small constant. We have

$$\begin{aligned}\|(g^N)_*(S)\|_U &= \int_{g^{-N}(U)} S \wedge (g^N)^*(\omega^p) \\ &\leq \int_{f^{-N}(W)} S \wedge (f^N)^*(\omega^p) + \int_{g^{-N}(U)} S \wedge \left[(g^N)^*(\omega^p) - (f^N)^*(\omega^p)\right] \\ &\leq \|(f^N)_*(S)\|_W + \varepsilon \leq \delta^N + \varepsilon < d_t^N.\end{aligned}$$

Therefore, the dynamical $*$-degree $d_p^*(g^N)$ of g^N is strictly smaller than d_t^N. Lemma 2.6 implies that $d_p^*(g) < d_t$. □

Remark 2.8. The proof gives that $g \mapsto d_p^*(g)$ is upper semi-continuous on g.

Consider a simple example. Let $f : \mathbb{C}^2 \to \mathbb{C}^2$ be the polynomial map $f(z_1,z_2) = (2z_1, z_2^2)$. The restriction of f to $V := \{|z_1| < 2, |z_2| < 2\}$ is polynomial-like and using the current $S = [z_1 = 0]$, it is not difficult to check that $d_1 = d_1^* = d_t = 2$. The example shows that in general one may have $d_{k-1}^* = d_t$.

Exercise 2.9. Let $f : \mathbb{C}^2 \to \mathbb{C}^2$ be the polynomial map defined by $f(z_1,z_2) := (3z_2, z_1^2 + z_2)$. Show that the hyperplane at infinity is attracting. Compute the topological degree of f. Compute the topological degree of the map in Example 2.1.

Exercise 2.10. Let f be a polynomial map on $\mathbb{C}^k$ of algebraic degree $d \geq 2$, which extends to a holomorphic endomorphism of $\mathbb{P}^k$. Let V be a ball large enough centered at 0 and $U := f^{-1}(V)$. Prove that the polynomial-like map $f : U \to V$ satisfies $d_p^* = d^p$ and $d_t = d^k$. Hint: use the Green function and Green currents.

2.2 Construction of the Green Measure

In this paragraph, we introduce the first version of the dd^c-method. It allows to construct for a polynomial-like map f a canonical measure which is totally invariant. As we have seen in the case of endomorphisms of $\mathbb{P}^k$, the method gives good estimates and allows to obtain precise stochastic properties. Here, we will see that it applies under a very weak hypothesis. The construction of the measure does not require any hypothesis on the dynamical degrees and give useful convergence results.

Consider a polynomial-like map $f : U \to V$ of topological degree $d_t > 1$ as above. Define the Perron-Frobenius operator Λ acting on test functions φ by

$$\Lambda(\varphi)(z) := d_t^{-1} f_*(\varphi)(z) := d_t^{-1} \sum_{w \in f^{-1}(z)} \varphi(w),$$

where the points in $f^{-1}(z)$ are counted with multiplicity. Since f is a ramified covering, $\Lambda(\varphi)$ is continuous when φ is continuous. If ν is a probability measure on V, define the measure $f^*(\nu)$ by

$$\langle f^*(\nu), \varphi \rangle := \langle \nu, f_*(\varphi) \rangle.$$

This is a positive measure of mass d_t supported on $f^{-1}(\text{supp}(\nu))$. Observe that the operator $\nu \mapsto d_t^{-1} f^*(\nu)$ is continuous on positive measures, see Exercise A.11.

Theorem 2.11. *Let $f: U \to V$ be a polynomial-like map as above. Let ν be a probability measure supported on V which is defined by an L^1 form. Then $d_t^{-n}(f^n)^*(\nu)$ converge to a probability measure μ which does not depend on ν. For φ p.s.h. on a neighbourhood of the filled Julia set $\mathscr{K}$, we have $\langle d_t^{-n}(f^n)^*(\nu), \varphi \rangle \to \langle \mu, \varphi \rangle$. The measure μ is supported on the boundary of $\mathscr{K}$ and is totally invariant: $d_t^{-1} f^*(\mu) = f_*(\mu) = \mu$. Moreover, if Λ is the Perron-Frobenius operator associated to f and φ is a p.s.h. function on a neighbourhood of $\mathscr{K}$, then $\Lambda^n(\varphi)$ converge to $\langle \mu, \varphi \rangle$.*

Note that in general $\langle \mu, \varphi \rangle$ may be $-\infty$. If $\langle \mu, \varphi \rangle = -\infty$, the above convergence means that $\Lambda^n(\varphi)$ tend locally uniformly to $-\infty$; otherwise, the convergence is in L^p_{loc} for $1 \leq p < +\infty$, see Appendix A.2. The above result still holds for measures ν which have no mass on pluripolar sets. The proof in that case is more delicate. We have the following lemma.

Lemma 2.12. *If φ is p.s.h. on a neighbourhood of $\mathscr{K}$, then $\Lambda^n(\varphi)$ converge to a constant c_φ in $\mathbb{R} \cup \{-\infty\}$.*

Proof. Observe that $\Lambda^n(\varphi)$ is defined on V for n large enough. It is not difficult to check that these functions are p.s.h. Indeed, when φ is a continuous p.s.h. function, $\Lambda^n(\varphi)$ is a continuous function, see Exercise A.11, and $dd^c \Lambda^n(\varphi) = d_t^{-n}(f^n)_*(dd^c\varphi) \geq 0$. So, $\Lambda^n(\varphi)$ is p.s.h. The general case is obtained using an approximation of φ by a decreasing sequence of smooth p.s.h. functions.

Consider ψ the upper semi-continuous regularization of $\limsup \Lambda^n(\varphi)$. We deduce from Proposition A.20 that ψ is a p.s.h. function. We first prove that ψ is constant. Assume not. By maximum principle, there is a constant δ such that $\sup_{\overline{U}} \psi < \delta < \sup_V \psi$. By Hartogs' lemma A.20, for n large enough, we have $\Lambda^n(\varphi) < \delta$ on U. Since the fibers of f are contained in U, we deduce from the definition of Λ that

$$\sup_V \Lambda^{n+1}(\varphi) = \sup_V \Lambda(\Lambda^n(\varphi)) \leq \sup_U \Lambda^n(\varphi) < \delta.$$

This implies that $\psi \leq \delta$ which contradicts the choice of δ. So ψ is constant.

Denote by c_φ this constant. If $c_\varphi = -\infty$, it is clear that $\Lambda^n(\varphi)$ converge to $-\infty$ uniformly on compact sets. Assume that c_φ is finite and $\Lambda^{n_i}(\varphi)$ does not converge to c_φ for some sequence (n_i). By Hartogs' lemma, we have $\Lambda^{n_i}(\varphi) \leq c_\varphi - \varepsilon$ for some constant $\varepsilon > 0$ and for i large enough. We deduce as above that $\Lambda^n(\varphi) \leq c_\varphi - \varepsilon$ for $n \geq n_i$. This contradicts the definition of c_φ. □

Proof of Theorem 2.11. We can replace ν with $d_t^{-1}f^*(\nu)$ in order to assume that ν is supported on U. The measure ν can be written as a finite or countable sum of bounded positive forms, we can assume that ν is a bounded form.

Consider a smooth p.s.h. function φ on a neighbourhood of $\mathscr{K}$. It is clear that $\Lambda^n(\varphi)$ are uniformly bounded for n large enough. Therefore, the constant c_φ is finite. We deduce from Lemma 2.12 that $\Lambda^n(\varphi)$ converge in $L^1_{loc}(V)$ to c_φ. It follows that

$$\langle d_t^{-n}(f^n)^*(\nu),\varphi\rangle = \langle \nu,\Lambda^n(\varphi)\rangle \to c_\varphi.$$

Let ϕ be a general smooth function on V. We can always write ϕ as a difference of p.s.h. functions on U. Therefore, $\langle d_t^{-n}(f^n)^*(\nu),\phi\rangle$ converge. It follows that the sequence of probability measures $d_t^{-n}(f^n)^*(\nu)$ converges to some probability measure μ. Since c_φ does not depend on ν, the measure μ does not depend on ν. Consider a measure ν supported on $U\setminus\mathscr{K}$. So, the limit μ of $d_t^{-n}(f^n)^*(\nu)$ is supported on $\partial\mathscr{K}$. The total invariance is a direct consequence of the above convergence.

For the rest of the theorem, assume that φ is a general p.s.h. function on a neighbourhood of $\mathscr{K}$. Since $\limsup\Lambda^n(\varphi)\leq c_\varphi$, Fatou's lemma implies that

$$\langle\mu,\varphi\rangle = \langle d_t^{-n}(f^n)^*(\mu),\varphi\rangle = \langle\mu,\Lambda^n(\varphi)\rangle \leq \langle\mu,\limsup_{n\to\infty}\Lambda^n(\varphi)\rangle = c_\varphi.$$

On the other hand, for ν smooth on U, we have since φ is upper semi-continuous

$$c_\varphi = \lim_{n\to\infty}\langle\nu,\Lambda^n(\varphi)\rangle = \lim_{n\to\infty}\langle d_t^{-n}(f^n)^*(\nu),\varphi\rangle \leq \langle\lim_{n\to\infty} d_t^{-n}(f^n)^*(\nu),\varphi\rangle = \langle\mu,\varphi\rangle.$$

Therefore, $c_\varphi=\langle\mu,\varphi\rangle$. Hence, $\Lambda^n(\varphi)$ converge to $\langle\mu,\varphi\rangle$ for an arbitrary p.s.h. function φ. This also implies that $\langle d_t^{-n}(f^n)^*(\nu),\varphi\rangle\to\langle\mu,\varphi\rangle$. □

The measure μ is called *the equilibrium measure* of f. We deduce from the above arguments the following result.

Proposition 2.13. *Let ν be a totally invariant probability measure. Then ν is supported on $\mathscr{K}$. Moreover, $\langle\nu,\varphi\rangle\leq\langle\mu,\varphi\rangle$ for every function φ which is p.s.h. in a neighbourhood of $\mathscr{K}$ and $\langle\nu,\varphi\rangle=\langle\mu,\varphi\rangle$ if φ is pluriharmonic in a neighbourhood of $\mathscr{K}$.*

Proof. Since $\nu=d_t^{-n}(f^n)^*(\nu)$, it is supported on $f^{-n}(V)$ for every $n\geq 0$. So, ν is supported on $\mathscr{K}$. We know that $\limsup\Lambda^n(\varphi)\leq c_\varphi$, then Fatou's lemma implies that

$$\langle\nu,\varphi\rangle = \lim_{n\to\infty}\langle d_t^{-n}(f^n)^*(\nu),\varphi\rangle = \lim_{n\to\infty}\langle\nu,\Lambda^n(\varphi)\rangle \leq c_\varphi.$$

When φ is pluriharmonic, the inequality holds for $-\varphi$; we then deduce that $\langle\nu,\varphi\rangle\geq c_\varphi$. The proposition follows. □

Corollary 2.14. *Let X_1,X_2 be two analytic subsets of V such that $f^{-1}(X_1)\subset X_1$ and $f^{-1}(X_2)\subset X_2$. Then $X_1\cap X_2\neq\varnothing$. In particular, f admits at most one point a such that $f^{-1}(a)=\{a\}$.*

Proof. Observe that there are totally invariant probability measures ν_1, ν_2 supported on X_1, X_2. Indeed, if ν is a probability measure supported on X_i, then any limit value of

$$\frac{1}{N}\sum_{n=0}^{N-1} d_t^{-n}(f^n)^*(\nu)$$

is supported on X_i and is totally invariant. We are using the continuity of the operator $\nu \mapsto d_t^{-1} f^*(\nu)$, for the weak topology on measures.

On the other hand, if X_1 and X_2 are disjoint, we can find a holomorphic function h on U such that $h = c_1$ on X_1 and $h = c_2$ on X_2, where c_1, c_2 are distinct constants. We consider the function defined on $X_1 \cup X_2$ as claimed and extend it as a holomorphic function in U. This is possible since V is convex [HO2]. Adding to h a constant allows to assume that h does not vanish on $\mathscr{K}$. Therefore, $\varphi := \log|h|$ is pluriharmonic on a neighbourhood of $\mathscr{K}$. We have $\langle \nu_1, \varphi\rangle \neq \langle \nu_2, \varphi\rangle$. This contradicts Proposition 2.13. □

When the test function is pluriharmonic, we have the following exponential convergence.

Proposition 2.15. *Let W be a neighbourhood of $\mathscr{K}$ and $\mathscr{F}$ a bounded family of pluriharmonic functions on W. There are constants $N \geq 0$, $c > 0$ and $0 < \lambda < 1$ such that if φ is a function in $\mathscr{F}$, then*

$$|\langle \Lambda^n(\varphi) - \langle \mu, \varphi\rangle| \leq c\lambda^n \quad \text{on} \quad V$$

for $n \geq N$.

Proof. Observe that if N is large enough, the functions $\Lambda^N(\varphi)$ are pluriharmonic and they absolute values are bounded on V by the same constant. We can replace φ with $\Lambda^N(\varphi)$ in order to assume that $W = V$, $N = 0$ and that $|\varphi|$ is bounded by some constant A. Then, $|\Lambda^n(\varphi)| \leq A$ for every n. Subtracting from φ the constant $\langle \mu, \varphi\rangle$ allows to assume that $\langle \mu, \varphi\rangle = 0$.

Let $\mathscr{F}_\alpha$ denote the family of pluriharmonic functions φ on V such that $|\varphi| \leq \alpha$ and $\langle \mu, \varphi\rangle = 0$. It is enough to show that Λ sends $\mathscr{F}_\alpha$ into $\mathscr{F}_{\lambda\alpha}$ for some constant $0 < \lambda < 1$. We can assume $\alpha = 1$. Since μ is totally invariant, Λ preserves the subspace $\{\varphi,\ \langle \mu, \varphi\rangle = 0\}$. The family $\mathscr{F}_1$ is compact and does not contain the function identically equal to 1. By maximum principle applied to $\pm\varphi$, there is a constant $0 < \lambda < 1$ such that $\sup_U |\varphi| \leq \lambda$ for φ in $\mathscr{F}_1$. We deduce that $\sup_V |\Lambda(\varphi)| \leq \lambda$. The result follows. □

The following result shows that the equilibrium measure μ satisfies the Oseledec's theorem hypothesis. It can be extended to a class of orientation preserving smooth maps on Riemannian manifolds [DS1].

Theorem 2.16. *Let $f : U \to V$ be a polynomial-like map as above. Let μ be the equilibrium measure and J the Jacobian of f with respect to the standard volume form on $\mathbb{C}^k$. Then*

$$\langle \mu, \log J\rangle \geq \log d_t.$$

In particular, μ has no mass on the critical set of f.

Proof. Let ν be the restriction of the Lebesgue measure to U multiplied by a constant so that $\|\nu\| = 1$. Define

$$\nu_n := d_t^{-n}(f^n)^*(\nu) \quad \text{and} \quad \mu_N := \frac{1}{N}\sum_{n=1}^{N} \nu_n.$$

By Theorem 2.11, μ_N converge to μ. Choose a constant $M > 0$ such that $J \leq M$ on U. For any constant $m > 0$, define

$$g_m(x) := \min\Big(\log\frac{M}{J(x)}, m + \log M\Big) = \min\Big(\log\frac{M}{J(x)}, m'\Big)$$

with $m' := m + \log M$. This is a family of continuous functions which are positive, bounded on U and which converge to $\log M/J$ when m goes to infinity. Define

$$s_N(x) := \frac{1}{N}\sum_{q=0}^{N-1} g_m(f^q(x)).$$

Using the definition of f^* on measures, we obtain

$$\begin{aligned}
\langle \nu_N, s_N\rangle &= \frac{1}{N}\sum_{q=0}^{N-1} d_t^{-N}\langle (f^N)^*(\nu), g_m\circ f^q\rangle \\
&= \frac{1}{N}\sum_{q=0}^{N-1} d_t^{-N+q}\langle (f^{N-q})^*(\nu), g_m\rangle \\
&= \frac{1}{N}\sum_{q=0}^{N-1} \langle \nu_{N-q}, g_m\rangle = \langle \mu_N, g_m\rangle.
\end{aligned}$$

In order to bound $\langle \mu, \log J\rangle$ from below, we will bound $\langle \mu_N, g_m\rangle$ from above.

For $\alpha > 0$, let U_N^α denote the set of points $x \in U$ such that $s_N(x) > \alpha$. Since $s_N(x) \leq m'$, we have

$$\begin{aligned}
\langle \mu_N, g_m\rangle = \langle \nu_N, s_N\rangle &\leq m'\nu_N(U_N^\alpha) + \alpha(1 - \nu_N(U_N^\alpha)) \\
&= \alpha + (m' - \alpha)\nu_N(U_N^\alpha).
\end{aligned}$$

If $\nu_N(U_N^\alpha)$ converge to 0 when $N \to \infty$, then

$$\langle \mu, g_m\rangle = \lim_{N\to\infty}\langle \mu_N, g_m\rangle \leq \alpha \quad \text{and hence} \quad \langle \mu, \log M/J\rangle \leq \alpha.$$

We determine a value of α such that $\nu_N(U_N^\alpha)$ tend to 0.

By definition of ν_N, we have

$$\nu_N(U_N^\alpha) = \int_{U_N^\alpha} d_t^{-N}(f^N)^*(\nu) = \int_{U_N^\alpha} d_t^{-N}\Big(\prod_{q=0}^{N-1} J\circ f^q\Big)d\nu.$$

Define for a given $\delta > 0$ and for any integer j,

$$W_j := \{\exp(-j\delta) < J \le \exp(-(j-1)\delta)\}$$

and

$$\tau_j(x) := \frac{1}{N}\#\{q, \quad f^q(x) \in W_j \text{ and } 0 \le q \le N-1\}.$$

We have $\sum \tau_j = 1$ and

$$\nu_N(U_N^\alpha) \le \int_{U_N^\alpha} \left[\frac{1}{d_t}\exp\left(\sum -(j-1)\delta\tau_j\right)\right]^N d\nu.$$

Using the inequality $g_m \le \log M/J$, we have on U_N^α

$$\alpha < s_N < \sum \tau_j(\log M + j\delta) = \sum j\delta\tau_j + \log M.$$

Therefore,

$$-\sum (j-1)\delta\tau_j < -\alpha + (\log M + \delta).$$

We deduce from the above estimate on $\nu_N(U_N^\alpha)$ that

$$\nu_N(U_N^\alpha) \le \int_{U_N^\alpha} \left[\frac{\exp(-\alpha)M\exp(\delta)}{d_t}\right]^N d\nu.$$

So, for every $\alpha > \log(M/d_t) + \delta$, we have $\nu_N(U_N^\alpha) \to 0$.

Choosing δ arbitrarily small, we deduce from the above discussion that

$$\lim_{N\to\infty} \langle \mu_N, g_m\rangle \le \log(M/d_t).$$

Since g_m is continuous and μ_N converge to μ, we have $\langle \mu, g_m\rangle \le \log(M/d_t)$. Letting m go to infinity gives $\langle \mu, \log J\rangle \ge \log d_t$. □

Exercise 2.17. Let φ be a strictly p.s.h. function on a neighbourhood of $\mathscr{K}$, i.e. a p.s.h. function satisfying $dd^c\varphi \ge c\, dd^c\|z\|^2$, with $c > 0$, in a neighbourhood of $\mathscr{K}$. Let ν be a probability measure such that $\langle d_t^{-n}(f^n)^*(\nu), \varphi\rangle$ converge to $\langle \mu, \varphi\rangle$ and that $\langle \mu, \varphi\rangle$ is finite. Show that $d_t^{-n}(f^n)^*(\nu)$ converge to μ.

Exercise 2.18. Using the test function $\varphi = \|z\|^2$, show that

$$\int_{f^{-n}(U)} (f^n)^*(\omega^{k-1}) \wedge \omega = o(d_t^n),$$

when n goes to infinity.

Exercise 2.19. Let Y denote the set of critical values of f. Show that the volume of $f^n(Y)$ in U satisfies volume$(f^n(Y) \cap U) = o(d_t^n)$ when n goes to infinity.

Exercise 2.20. Let f be a polynomial endomorphism of $\mathbb{C}^2$ of algebraic degree $d \geq 2$. Assume that f extends at infinity to an endomorphism of $\mathbb{P}^2$. Show that f admits at most three totally invariant points[7].

2.3 Equidistribution Problems

In this paragraph, we consider polynomial-like maps f satisfying the hypothesis that the dynamical degree d_{k-1}^* is strictly smaller than d_t. We say that f has a *large topological degree*. We have seen that this property is stable under small pertubations of f. Let Y denote the hypersurface of critical values of f. As in the case of endomorphisms of $\mathbb{P}^k$, define the ramification current R by

$$R := \sum_{n\geq 0} d_t^{-n}(f^n)_*[Y].$$

The following result is a version of Proposition 1.51.

Proposition 2.21. *Let $f : U \to V$ be a polynomial-like map as above with large topological degree. Let ν be a strictly positive constant and let a be a point in V such that the Lelong number $\nu(R,a)$ is strictly smaller than ν. Let δ be a constant such that $d_{k-1} < \delta < d_t$. Then, there is a ball B centered at a such that f^n admits at least $(1-\sqrt{\nu})d_t^n$ inverse branches $g_i : B \to W_i$ where W_i are open sets in V of diameter $\leq \delta^{n/2}d_t^{-n/2}$. In particular, if μ' is a limit value of the measures $d_t^{-n}(f^n)^*(\delta_a)$ then $\|\mu' - \mu\| \leq 2\sqrt{\nu(R,a)}$.*

Proof. Since $d_{k-1}^* < d_t$, the current R is well-defined and has locally finite mass. If ω is the standard Kähler form on $\mathbb{C}^k$, we also have $\|(f^n)_*(\omega)\|_{V'} \lesssim \delta^n$ for every open set $V' \Subset V$. So, for the first part of the proposition, it is enough to follow the arguments in Proposition 1.51. The proof there is written in such way that the estimates are local and can be extended without difficulty to the present situation. In particular, we did not use Bézout's theorem.

For the second assertion, we do not have yet the analogue of Proposition 1.46, but it is enough to compare $d_t^{-n}(f^n)^*(\delta_a)$ with the pull-backs of a smooth measure supported on B and to apply Theorem 2.11. □

We deduce the following result as in the case of endomorphisms of $\mathbb{P}^k$.

Theorem 2.22. *Let $f : U \to V$ be a polynomial-like map as above with large topological degree. Let P_n denote the set of repelling periodic points of period n on the support of μ. Then the sequence of measures*

[7] This result was proved in [DS1]. Amerik and Campana proved in [AC] for a general non-invertible endomorphism of $\mathbb{P}^2$ that the number of totally invariant points is at most equal to 9. The sharp bound (probably 3) is unknown. One deduces from Corollary 2.14 and Amerik-Campana result that any non-invertible polynomial map on $\mathbb{C}^3$ which extends at infinity to an endomorphism of $\mathbb{P}^3$, admits at most 10 totally invariant points.

$$\mu_n := d_t^{-n} \sum_{a \in P_n} \delta_a$$

converges to μ.

Proof. It is enough to repeat the proof of Theorem 1.57 and to show that f^n admits exactly d_t^n fixed points counted with multiplicity. We can assume $n = 1$. Let Γ denote the graph of f in $U \times V \subset V \times V$. The number of fixed points of f is the number of points in the intersection of Γ with the diagonal Δ of $\mathbb{C}^k \times \mathbb{C}^k$. Observe that this intersection is contained in the compact set $\mathscr{K} \times \mathscr{K}$. For simplicity, assume that V contains the point 0 in $\mathbb{C}^k$. Let (z,w) denote the standard coordinates on $\mathbb{C}^k \times \mathbb{C}^k$ where the diagonal is given by the equation $z = w$. Consider the deformations $\Delta_t := \{w = tz\}$ with $0 \leq t \leq 1$ of Δ. Since V is convex, it is not difficult to see that the intersection of Γ with this family stays in a compact subset of $V \times V$. Therefore, the number of points in $\Delta_t \cap \Gamma$, counted with multiplicity, does not depend on t. For $t = 0$, this is just the number of points in the fiber $f^{-1}(0)$. The result follows. □

The equidistribution of negative orbits of points is more delicate than in the case of endomorphisms of $\mathbb{P}^k$. It turns out that the exceptional set $\mathscr{E}$ does not satisfy in general $f^{-1}(\mathscr{E}) = \mathscr{E} \cap U$. We have the following result.

Theorem 2.23. *Let $f : U \to V$ be a polynomial-like map as above with large topological degree. Then there is a proper analytic subset $\mathscr{E}$ of V, possibly empty, such that $d_t^{-n}(f^n)^*(\delta_a)$ converge to the equilibrium measure μ if and only if a does not belong to the orbit of $\mathscr{E}$. Moreover, $\mathscr{E}$ satisfies $f^{-1}(\mathscr{E}) \subset \mathscr{E} \subset f(\mathscr{E})$ and is maximal in the sense that if E is a proper analytic subset of V contained in the orbit of critical values such that $f^{-n}(E) \subset E$ for some $n \geq 1$ then $E \subset \mathscr{E}$.*

The proof follows the main lines of the case of endomorphisms of $\mathbb{P}^k$ using the following proposition applied to Z the set of critical values of f. The set $\mathscr{E}$ will be defined as $\mathscr{E} := \mathscr{E}_Z$. Observe that unlike in the case of endomorphisms of $\mathbb{P}^k$, we need to assume that E is in the orbit of the critical values.

Let Z be an arbitrary analytic subset of V not necessarily of pure dimension. Let $N_n(a)$ denote the number of orbits

$$a_{-n}, \ldots, a_{-1}, a_0$$

with $f(a_{-i-1}) = a_{-i}$ and $a_0 = a$ such that $a_{-i} \in Z$ for every i. Here, the orbits are counted with multiplicity, i.e. we count the multiplicity of f^n at a_{-n}. So, $N_n(a)$ is the number of negative orbits of order n of a which stay in Z. Observe that the sequence of functions $\tau_n := d_t^{-n} N_n$ decreases to some function τ. Since τ_n are upper semi-continuous with respect to the Zariski topology and $0 \leq \tau_n \leq 1$, the function τ satisfies the same properties. Note that $\tau(a)$ is the proportion of infinite negative orbits of a staying in Z. Define $\mathscr{E}_Z := \{\tau = 1\}$. The Zariski upper semi-continuity of τ implies that $\mathscr{E}_Z$ is analytic. It is clear that $f^{-1}(\mathscr{E}_Z) \subset \mathscr{E}_Z$ which implies that $\mathscr{E}_Z \subset f(\mathscr{E}_Z)$.

Proposition 2.24. *If a point $a \in V$ does not belong to the orbit $\cup_{n \geq 0} f^n(\mathscr{E}_Z)$ of $\mathscr{E}_Z$, then $\tau(a) = 0$.*

Proof. Assume there is $\theta_0 > 0$ such that $\{\tau \geq \theta_0\}$ is not contained in the orbit of $\mathscr{E}_Z$. We claim that there is a maximal value θ_0 satisfying the above property. Indeed, by definition, $\tau(a)$ is smaller than or equal to the average of τ on the fiber of a. So, we only have to consider the components of $\{\tau \geq \theta_0\}$ which intersect U and there are only finitely many of such components, hence the maximal value exists.

Let E be the union of irreducible components of $\{\tau \geq \theta_0\}$ which are not contained in the orbit of $\mathscr{E}_Z$. Since $\theta_0 > 0$, we know that $E \subset Z$. We want to prove that E is empty. If a is a generic point in E, it does not belong to the orbit of $\mathscr{E}_Z$ and we have $\tau(a) = \theta_0$. If b is a point in $f^{-1}(a)$, then b is not in the orbit of $\mathscr{E}_Z$. Therefore, $\tau(b) \leq \theta_0$. Since $\tau(a)$ is smaller than or equal to the average of τ on $f^{-1}(a)$, we deduce that $\tau(b) = \theta_0$, and hence $f^{-1}(a) \subset E$. By induction, we obtain that $f^{-n}(a) \subset E \subset Z$ for every $n \geq 1$. Hence, $a \in \mathscr{E}_Z$. This is a contradiction. □

The following example shows that in general the orbit of $\mathscr{E}$ is not an analytic set. We deduce that in general polynomial-like maps are not homeomorphically conjugated to restrictions on open sets of endomorphisms of $\mathbb{P}^k$ (or polynomial maps of $\mathbb{C}^k$ such that the infinity is attractive) with the same topological degree.

Example 2.25. Denote by $D(a,R)$ the disc of radius R and of center a in $\mathbb{C}$. Observe that the polynomial $P(z) := 6z^2 + 1$ defines a ramified covering of degree 2 from $D := P^{-1}(D(0,4))$ to $D(0,4)$. The domain D is simply connected and is contained in $D(0,1)$. Let ψ be a bi-holomorphic map between $D(1,2)$ and $D(0,1)$ such that $\psi(0) = 0$. Define $h(z,w) := (P(z), 4\psi^m(w))$ with m large enough. This application is holomorphic and proper from $W := D \times D(1,2)$ to $V := D(0,4) \times D(0,4)$. Its critical set is given by $zw = 0$.

Define also

$$g(z,w) := 10^{-2}\big(\exp(z)\cos(\pi w/2), \exp(z)\sin(\pi w/2)\big).$$

One easily check that g defines a bi-holomorphic map between W and $U := g(W)$. Consider now the polynomial-like map $f : U \to V$ defined by $f = h \circ g^{-1}$. Its topological degree is equal to $2m$; its critical set C is equal to $g\{zw = 0\}$. The image of $C_1 := g\{z = 0\}$ by f is equal to $\{z = 1\}$ which is outside U. The image of $g\{w = 0\}$ by f is $\{w = 0\} \cap V$.

The intersection $\{w = 0\} \cap U$ contains two components $C_2 := g\{w = 0\}$ and $C_2' := g\{w = 2\}$. They are disjoint because g is bi-holomorphic. We also have $f(C_2) = \{w = 0\} \cap V$ and $f^{-1}\{w = 0\} = C_2$. Therefore, $\mathscr{E} = \{w = 0\} \cap V$, $f^{-1}(\mathscr{E}) \subset \mathscr{E}$ and $f^{-1}(\mathscr{E}) \neq \mathscr{E} \cap U$ since $f^{-1}(\mathscr{E})$ does not contain C_2'. The orbit of $\mathscr{E}$ is the union of C_2 and of the orbit of C_2'. Since m is large, the image of C_2' by f is a horizontal curve very close to $\{w = 0\}$. It follows that the orbit of C_2' is a countable union of horizontal curves close to $\{w = 0\}$ and it is not analytic.

It follows that f is not holomorphically conjugate to an endomorphism (or a polynomial map such that the hyperplane at infinity is attractive) with the same topological degree. If it were, the exceptional set would not have infinitely many components in a neighbourhood of $w = 0$.

Remark 2.26. Assume that f is not with large topological degree but that the series which defines the ramification current R converges. We can then construct inverse branches as in Proposition 2.21. To obtain the same exponential estimates on the diameter of W_i, it is enough to assume that $d_{k-1} < d_t$. In general, we only have that these diameters tend uniformly to 0 when n goes to infinity. Indeed, we can use the estimate in Exercise 2.18. The equidistribution of periodic points and of negative orbits still holds in this case.

Exercise 2.27. Let f be a polynomial-like map with large topological degree. Show that there is a small perturbation of f, arbitrarily close to f, whose exceptional set is empty.

2.4 Properties of the Green Measure

Several properties of the equilibrium measure of polynomial-like maps can be proved using the arguments that we introduced in the case of endomorphisms of $\mathbb{P}^k$. We have the following result for general polynomial-like maps.

Theorem 2.28. *Let $f : U \to V$ be a polynomial-like map of topological degree $d_t > 1$. Then its equilibrium measure μ is an invariant measure of maximal entropy* $\log d_t$. *Moreover, μ is K-mixing.*

Proof. By Theorem 2.16, μ has no mass on the critical set of f. Therefore, it is an invariant measure of constant Jacobian d_t in the sense that $\mu(f(A)) = d_t\mu(A)$ when f is injective on a Borel set A. We deduce from Parry's theorem 1.116 that $h_\mu(f) \geq \log d_t$. The variational principle and Theorem 2.4 imply that $h_\mu(f) = \log d_t$.

We prove the K-mixing property. As in the case of endomorphisms of $\mathbb{P}^k$, it is enough to show for φ in $L^2(\mu)$ that $\Lambda^n(\varphi) \to \langle \mu, \varphi \rangle$ in $L^2(\mu)$. Since $\Lambda : L^2(\mu) \to L^2(\mu)$ is of norm 1, it is enough to check the convergence for a dense family of φ. So, we only have to consider φ smooth. We can also assume that φ is p.s.h. because smooth functions can be written as a difference of p.s.h. functions. Assume also for simplicity that $\langle \mu, \varphi \rangle = 0$.

So, the p.s.h. functions $\Lambda^n(\varphi)$ converge to 0 in $L^p_{loc}(V)$. By Hartogs' lemma A.20, $\sup_U \Lambda^n(\varphi)$ converge to 0. This and the identity $\langle \mu, \Lambda^n(\varphi) \rangle = \langle \mu, \varphi \rangle = 0$ imply that $\mu\{\Lambda^n(\varphi) < -\delta\} \to 0$ for every fixed $\delta > 0$. On the other hand, by definition of Λ, $|\Lambda^n(\varphi)|$ is bounded by $\|\varphi\|_\infty$ which is a constant independent of n. Therefore, $\Lambda^n(\varphi) \to 0$ in $L^2(\mu)$ and K-mixing follows. □

The following result, due to Saleur [S], generalizes Theorem 1.118.

Theorem 2.29. *Let $f : U \to V$ be a polynomial-like map with large topological degree. Then its equilibrium measure is the unique invariant probability measure of maximal entropy.*

The proof follows the one of Theorem 1.118. The estimates of some integrals over $\mathbb{P}^k$ are replaced with the volume estimate in Lemma 2.5. A new difficulty here is to prove that if ν is an ergodic measure of maximal entropy, it has no mass on the set of critical values of f which may be singular. For this purpose, Saleur uses a version of Lemma 2.5 where V is replaced with an analytic set.

Theorem 2.30. *Let f and μ be as above. Then the sum of the Lyapounov exponents of μ is at least equal to $\frac{1}{2}\log d_t$. In particular, f admits a strictly positive Lyapounov exponent. If f is with large topological degree, then μ is hyperbolic and its Lyapounov exponents are at least equal to $\frac{1}{2}\log(d_t/d_{k-1})$.*

Proof. By Oseledec's theorem 1.119, applied in the complex setting, the sum of the Lyapounov exponents of μ (associated to complex linear spaces) is equal to $\frac{1}{2}\langle\mu,\log J\rangle$. Theorem 2.16 implies that this sum is at least equal to $\frac{1}{2}\log d_t$. The second assertion is proved as in Theorem 1.120 using Proposition 2.21. □

From now on, we only consider maps with large topological degree. The following result was obtained by Dinh-Dupont in [DD]. It generalizes Theorem 1.121.

Theorem 2.31. *Let f be a polynomial-like map with large topological degree as above. Let $\chi_1,\ldots,\chi_k$ denote the Lyapounov exponents of the equilibrium measure μ ordered by $\chi_1\geq\cdots\geq\chi_k$ and Σ their sum. Then the Hausdorff dimension of μ satisfies*

$$\frac{\log d_t}{\chi_1}\leq \dim_H(\mu)\leq 2k-\frac{2\Sigma-\log d_t}{\chi_1}.$$

We now prove some stochastic properties of the equilibrium measure. We first introduce some notions. Let V be an open subset of $\mathbb{C}^k$ and ν a probability measure with compact support in V. We consider ν as a function on the convex cone PSH(V) of p.s.h. functions on V, with the L^1_{loc}-topology. We say that ν is *PB* if this function is finite, i.e. p.s.h. functions on V are ν-integrable. We say that ν is *PC* if it is PB and defines a continuous functional on PSH(V). Recall that the weak topology on PSH(V) coincides with the L^p_{loc} topology for $1\leq p<+\infty$. In dimension 1, a measure is PB if it has locally bounded potentials, a measure is PC if it has locally continuous potentials. A measure ν is *moderate* if for any bounded subset $\mathscr{P}$ of PSH(V), there are constants $\alpha>0$ and $A>0$ such that

$$\langle\nu,e^{\alpha|\varphi|}\rangle\leq A \quad\text{for}\quad \varphi\in\mathscr{P}.$$

Let K be a compact subset of V. Define a *pseudo-distance* $\text{dist}_{L^1(K)}$ between φ,ψ in PSH(V) by

$$\text{dist}_{L^1(K)}(\varphi,\psi):=\|\varphi-\psi\|_{L^1(K)}.$$

Observe that if ν is continuous with respect to $\text{dist}_{L^1(K)}$ then ν is PC. The following proposition gives a criterion for a measure to be moderate.

Proposition 2.32. *If ν is Hölder continuous with respect to $\text{dist}_{L^1(K)}$ for some compact subset K of V, then ν is moderate. If ν is moderate, then p.s.h. functions on V are in $L^p(\nu)$ for every $1\leq p<+\infty$ and ν has positive Hausdorff dimension.*

Proof. We prove the first assertion. Assume that ν is Hölder continuous with respect to $\mathrm{dist}_{L^1(K)}$ for some compact subset K of V. Consider a bounded subset $\mathscr{P}$ of $\mathrm{PSH}(V)$. The functions in $\mathscr{P}$ are uniformly bounded above on K. Therefore, subtracting from these functions a constant allows to assume that they are negative on K. We want to prove the estimate $\langle \nu, e^{-\alpha\varphi}\rangle \leq A$ for $\varphi \in \mathscr{P}$ and for some constants α, A. It is enough to show that $\nu\{\varphi < -M\} \lesssim e^{-\alpha M}$ for some (other) constant $\alpha > 0$ and for $M \geq 1$. For $M \geq 0$ and $\varphi \in \mathscr{P}$, define $\varphi_M := \max(\varphi, -M)$. We replace $\mathscr{P}$ with the family of functions φ_M. This allows to assume that the family is stable under the operation $\max(\cdot, -M)$. Observe that $\varphi_{M-1} - \varphi_M$ is positive, supported on $\{\varphi < -M+1\}$, smaller or equal to 1, and equal to 1 on $\{\varphi < -M\}$. In order to obtain the above estimate, we only have to show that $\langle \nu, \varphi_{M-1} - \varphi_M\rangle \lesssim e^{-\alpha M}$ for some $\alpha > 0$.

Fix a constant $\lambda > 0$ small enough and a constant $A > 0$ large enough. Since ν is Hölder continuous and $\varphi_{M-1} - \varphi_M$ vanishes on $\{\varphi > -M+1\}$, we have

$$\begin{aligned}\nu\{\varphi < -M\} &\leq \langle \nu, \varphi_{M-1}\rangle - \langle \nu, \varphi_M\rangle \leq A\, \|\varphi_{M-1} - \varphi_M\|^{\lambda}_{L^1(K)} \\ &\leq A\mathrm{volume}\{\varphi \leq -M+1\}^{\lambda}.\end{aligned}$$

On the other hand, since $\mathscr{P}$ is a bounded family in $\mathrm{PSH}(V)$, by Theorem A.22, we have $\|e^{-\lambda\varphi}\|_{L^1(K)} \leq A$ for $\varphi \in \mathscr{P}$. Hence, we have on K

$$\mathrm{volume}\{\varphi \leq -M+1\} \leq Ae^{-\lambda(M-1)}.$$

This implies the desired estimate for $\alpha = \lambda^2$ and completes the proof of the first assertion.

Assume now that ν is moderate. Let φ be a p.s.h. function on V. Then $e^{\alpha|\varphi|}$ is in $L^1(\nu)$ for some constant $\alpha > 0$. Since $e^{\alpha x} \gtrsim x^p$ for $1 \leq p < +\infty$, we deduce that φ is in $L^p(\nu)$. For the last assertion in the proposition, it is enough to show that $\nu(B_r) \leq Ar^{\alpha}$ for any ball B_r of radius $r > 0$ where A, α are some positive constants. We can assume that B_r is a small ball centered at a point $a \in K$. Define $\varphi(z) := \log\|z-a\|$. This function belongs to a compact family of p.s.h. functions. Therefore, $\|e^{-\alpha\varphi}\|_{L^1(\nu)} \leq A$ for some positive constants A, α independent of B_r. Since $e^{-\alpha\varphi} \geq r^{-\alpha}$ on B_r, we deduce that $\nu(B_r) \leq Ar^{\alpha}$. It is well-known that in order to compute the Hausdorff dimension of a set, it is enough to use only coverings by balls. It follows easily that if a Borel set has positive measure, then its Hausdorff dimension is at least equal to α. This completes the proof. □

The following results show that the equilibrium measure of a polynomial-like map with large topological degree satisfies the above regularity properties.

Theorem 2.33. *Let $f : U \to V$ be a polynomial-like map. Then the following properties are equivalent:*

1. *The map f has large topological degree, i.e. $d_t > d^*_{k-1}$;*
2. *The measure μ is PB, i.e. p.s.h. functions on V are integrable with respect to μ;*
3. *The measure μ is PC, i.e. μ can be extended to a linear continuous form on the cone of p.s.h. functions on V;*

Moreover, if f is such a map, then there is a constant $0 < \lambda < 1$ *such that*

$$\sup_V \Lambda(\varphi) - \langle\mu,\varphi\rangle \leq \lambda\big[\sup_V \varphi - \langle\mu,\varphi\rangle\big],$$

for φ p.s.h. on V.

Proof. It is clear that 3) $\Rightarrow$ 2). We show that 1) $\Rightarrow$ 3) and 2) $\Rightarrow$ 1).

1) $\Rightarrow$ 3). Let φ be a p.s.h. function on V. Let W be a convex open set such that $U \Subset W \Subset V$. For simplicity, assume that $\|\varphi\|_{L^1(W)} \leq 1$. So, φ belongs to a compact subset of PSH(W). Therefore, $S := dd^c\varphi$ has locally bounded mass in W. Define $S_n := (f^n)_*(S)$. Fix a constant $\delta > 1$ such that $d_{k-1}^* < \delta < d_t$. Condition 1) implies that $\|(f^n)_*(S)\|_W \lesssim \delta^n$. By Proposition A.16, there are p.s.h. functions φ_n on U such that $dd^c\varphi_n = S_n$ and $\|\varphi_n\|_U \lesssim \delta^n$ on U.

Define $\psi_0 := \varphi - \varphi_0$ and $\psi_n := f_*(\varphi_{n-1}) - \varphi_n$. Observe that these functions are pluriharmonic on U and depend continuously on φ. Moreover, f_* sends continuously p.s.h. functions on U to p.s.h. functions on V. Hence,

$$\|\psi_n\|_{L^1(U)} \leq \|f_*(\varphi_{n-1})\|_{L^1(U)} + \|\varphi_n\|_{L^1(U)} \lesssim \delta^n.$$

We have

$$\begin{aligned}\Lambda^n(\varphi) &= \Lambda^n(\psi_0+\varphi_0) = \Lambda^n(\psi_0) + d_t^{-1}\Lambda^{n-1}(f_*(\varphi_0))\\ &= \Lambda^n(\psi_0) + d_t^{-1}\Lambda^{n-1}(\psi_1+\varphi_1) = \cdots\\ &= \Lambda^n(\psi_0) + d_t^{-1}\Lambda^{n-1}(\psi_1) + \cdots + d_t^{-n+1}\Lambda(\psi_{n-1}) + d_t^{-n}\psi_n + d_t^{-n}\varphi_n.\end{aligned}$$

The last term in the above sum converges to 0. The above estimate on ψ_n and their pluriharmonicity imply, by Proposition 2.15, that the sum

$$\Lambda^n(\psi_0) + d_t^{-1}\Lambda^{n-1}(\psi_1) + \cdots + d_t^{-n+1}\Lambda(\psi_{n-1}) + d_t^{-n}\psi_n$$

converges uniformly to the finite constant

$$\langle\mu,\psi_0\rangle + d_t^{-1}\langle\mu,\psi_1\rangle + \cdots + d_t^{-n}\langle\mu,\psi_n\rangle + \cdots$$

which depends continuously on φ. We used here the fact that when ψ is pluriharmonic, $\langle\mu,\psi\rangle$ depends continuously on ψ. By Theorem 2.11, the above constant is equal to $\langle\mu,\varphi\rangle$. Consequently, μ is PC.

2) $\Rightarrow$ 1). Let $\mathscr{F}$ be an L^1 bounded family of p.s.h. functions on a neighbourhood of $\mathscr{K}$. We first show that $\langle\mu,\varphi\rangle$ is uniformly bounded on $\mathscr{F}$. Since $\langle\mu,\Lambda^N(\varphi)\rangle = \langle\mu,\varphi\rangle$, we can replace φ with $\Lambda^N(\varphi)$, with N large enough, in order to assume that $\mathscr{F}$ is a bounded family of p.s.h. functions φ on V which are uniformly bounded above. Subtracting from φ a fixed constant allows to assume that these functions are negative. If $\langle\mu,\varphi\rangle$ is not uniformly bounded on $\mathscr{F}$, there are φ_n such that $\langle\mu,\varphi_n\rangle \leq -n^2$. It follows that the series $\sum n^{-2}\varphi_n$ decreases to a p.s.h. function which is not integrable with respect to μ. This contradicts that μ is PB.

We deduce that for any neighbourhood W of $\mathscr{K}$, there is a constant $c > 0$ such that $|\langle\mu, \varphi\rangle| \leq c\|\varphi\|_{L^1(W)}$ for φ p.s.h. on W.

We now show that there is a constant $0 < \lambda < 1$ such that $\sup_V \Lambda(\varphi) \leq \lambda$ if φ is a p.s.h. function on V, bounded from above by 1, such that $\langle\mu, \varphi\rangle = 0$. This property implies the last assertion in the proposition. Assume that the property is not satisfied. Then, there are functions φ_n such that $\sup_V \varphi_n = 1$, $\langle\mu, \varphi_n\rangle = 0$ and $\sup_V \Lambda(\varphi_n) \geq 1 - 1/n^2$. By definition of Λ, we have

$$\sup_U \varphi_n \geq \sup_V \Lambda(\varphi_n) \geq 1 - 1/n^2.$$

The submean value inequality for p.s.h. functions implies that φ_n converge to 1 in $L^1_{loc}(V)$. On the other hand, we have

$$1 = |\langle\mu, \varphi_n - 1\rangle| \leq c\|\varphi_n - 1\|_{L^1(W)}.$$

This is a contradiction.

Finally, consider a positive closed (1,1)-current S of mass 1 on W. By Proposition A.16, there is a p.s.h. function φ on a neighbourhood of $\overline{U}$ with bounded L^1 norm such that $dd^c\varphi = S$. The submean inequality for p.s.h functions implies that φ is bounded from above by a constant independent of S. We can after subtracting from φ a constant, assume that $\langle\mu, \varphi\rangle = 0$. The p.s.h. functions $\lambda^{-n}\Lambda^n(\varphi)$ are bounded above and satisfy $\langle\mu, \lambda^{-n}\Lambda^n(\varphi)\rangle = \langle\mu, \varphi\rangle = 0$. Hence, they belong to a compact subset of $\mathrm{PSH}(U)$ which is independent of S. If W' is a neighbourhood of $\mathscr{K}$ such that $W' \Subset U$, the mass of $dd^c\big[\lambda^{-n}\Lambda^n(\varphi)\big]$ on W' is bounded uniformly on n and on S. Therefore,

$$\|dd^c(f^n)_*(S)\|_{W'} \leq c\lambda^n d_t^n$$

for some constant $c > 0$ independent of n and of S. It follows that $d^*_{k-1} \leq \lambda d_t$. This implies property 1). □

Theorem 2.34. *Let $f : U \to V$ be a polynomial-like map with large topological degree. Let $\mathscr{P}$ be a bounded family of p.s.h. functions on V. Let K be a compact subset of V such that $f^{-1}(K)$ is contained in the interior of K. Then, the equilibrium measure μ of f is Hölder continuous on $\mathscr{P}$ with respect to* $\mathrm{dist}_{L^1(K)}$. *In particular, this measure is moderate.*

Let $\mathrm{DSH}(V)$ denote the space of d.s.h. functions on V, i.e. functions which are differences of p.s.h. functions. They are in particular in $L^p_{loc}(V)$ for every $1 \leq p < +\infty$. Consider on $\mathrm{DSH}(V)$ the following topology: a sequence (φ_n) converges to φ in $\mathrm{DSH}(V)$ if φ_n converge weakly to φ and if we can write $\varphi_n = \varphi_n^+ - \varphi_n^-$ with $\varphi_n^\pm$ in a compact subset of $\mathrm{PSH}(V)$, independent of n. We deduce from the compactness of bounded sets of p.s.h. functions that $\varphi_n \to \varphi$ in all $L^p_{loc}(V)$ with $1 \leq p < +\infty$. Since μ is PC, it extends by linearity to a continuous functional on $\mathrm{DSH}(V)$.

Proof of Theorem 2.34. Let $\mathscr{P}$ be a compact family of p.s.h. functions on V. We show that μ is Hölder continuous on $\mathscr{P}$ with respect to $\mathrm{dist}_{L^1(K)}$. We claim that Λ

is Lipschitz with respect to $\mathrm{dist}_{L^1(K)}$. Indeed, if φ, ψ are in $L^1(K)$, we have for the standard volume form Ω on $\mathbb{C}^k$

$$\|\Lambda(\varphi)-\Lambda(\psi)\|_{L^1(K)} = \int_K |\Lambda(\varphi-\psi)|\Omega \leq d_t^{-1}\int_{f^{-1}(K)} |\varphi-\psi| f^*(\Omega).$$

since $f^{-1}(K)\subset K$ and $f^*(\Omega)$ is bounded on $f^{-1}(K)$, this implies that

$$\|\Lambda(\varphi)-\Lambda(\psi)\|_{L^1(K)} \leq \mathrm{const}\|\varphi-\psi\|_{L^1(K)}.$$

Since $\mathscr{P}$ is compact, the functions in $\mathscr{P}$ are uniformly bounded above on U. Therefore, replacing $\mathscr{P}$ by the family of $\Lambda(\varphi)$ with $\varphi\in\mathscr{P}$ allows to assume that functions in $\mathscr{P}$ are uniformly bounded above on V. On the other hand, since μ is PC, μ is bounded on $\mathscr{P}$. Without loss of generality, we can assume that $\mathscr{P}$ is the set of functions φ such that $\langle\mu,\varphi\rangle\geq 0$ and $\varphi\leq 1$. In particular, $\mathscr{P}$ is invariant under Λ. Let $\mathscr{D}$ be the family of d.s.h. functions $\varphi-\Lambda(\varphi)$ with $\varphi\in\mathscr{P}$. This is a compact subset of $\mathrm{DSH}(V)$ which is invariant under Λ, and we have $\langle\mu,\varphi'\rangle=0$ for φ' in $\mathscr{D}$.

Consider a function $\varphi\in\mathscr{P}$. Observe that $\widetilde{\varphi}:=\varphi-\langle\mu,\varphi\rangle$ is also in $\mathscr{P}$. Define $\widetilde{\Lambda}:=\lambda^{-1}\Lambda$ with λ the constant in Theorem 2.33. We deduce from that theorem that $\widetilde{\Lambda}(\widetilde{\varphi})$ is in $\mathscr{P}$. Moreover,

$$\widetilde{\Lambda}(\varphi-\Lambda(\varphi))=\widetilde{\Lambda}(\widetilde{\varphi}-\Lambda(\widetilde{\varphi}))=\widetilde{\Lambda}(\widetilde{\varphi})-\Lambda(\widetilde{\Lambda}(\widetilde{\varphi})).$$

Therefore, $\mathscr{D}$ is invariant under $\widetilde{\Lambda}$. This is the key point in the proof. Observe that we can extend $\mathrm{dist}_{L^1(K)}$ to $\mathrm{DSH}(V)$ and that $\widetilde{\Lambda}$ is Lipschitz with respect to this pseudo-distance.

Let ν be a smooth probability measure with support in K. We have seen that $d_t^{-n}(f^n)^*(\nu)$ converge to μ. If φ is d.s.h. on V, then

$$\langle d_t^{-n}(f^n)^*(\nu),\varphi\rangle=\langle\nu,\Lambda^n(\varphi)\rangle.$$

Define for φ in $\mathscr{P}$, $\varphi':=\varphi-\Lambda(\varphi)$. We have

$$\begin{aligned}\langle\mu,\varphi\rangle &= \lim_{n\to\infty}\langle\nu,\Lambda^n(\varphi)\rangle\\ &= \langle\nu,\varphi\rangle-\sum_{n\geq 0}\langle\nu,\Lambda^n(\varphi)\rangle-\langle\nu,\Lambda^{n+1}(\varphi)\rangle\\ &= \langle\nu,\varphi\rangle-\sum_{n\geq 0}\lambda^n\langle\nu,\widetilde{\Lambda}^n(\varphi')\rangle.\end{aligned}$$

Since ν is smooth with support in K, it is Lipschitz with respect to $\mathrm{dist}_{L^1(K)}$. We deduce from Lemma 1.19 which is also valid for a pseudo-distance, that the last series defines a Hölder continuous function on $\mathscr{D}$. We use here the invariance of $\mathscr{D}$ under $\widetilde{\Lambda}$. Finally, since the map $\varphi\mapsto\varphi'$ is Lipschitz on $\mathscr{P}$, we conclude that μ is Hölder continuous on $\mathscr{P}$ with respect to $\mathrm{dist}_{L^1(K)}$. □

As in the case of endomorphisms of $\mathbb{P}^k$, we deduce from the above results the following fundamental estimates on the Perron-Frobenius operator Λ.

Corollary 2.35. *Let f be a polynomial-like map with large topological degree as above. Let μ be the equilibrium measure and Λ the Perron-Frobenius operator associated to f. Let $\mathscr{D}$ be a bounded subset of d.s.h. functions on V. There are constants $c > 0$, $\delta > 1$ and $\alpha > 0$ such that if ψ is in $\mathscr{D}$, then*

$$\big\langle \mu, e^{\alpha\delta^n|\Lambda^n(\psi)-\langle\mu,\psi\rangle|}\big\rangle \leq c \quad \text{and} \quad \|\Lambda^n(\psi)-\langle\mu,\psi\rangle\|_{L^q(\mu)} \leq cq\delta^{-n}$$

for every $n \geq 0$ and every $1 \leq q < +\infty$.

Corollary 2.36. *Let f, μ, Λ be as above. Let $0 < \nu \leq 2$ be a constant. There are constants $c > 0$, $\delta > 1$ and $\alpha > 0$ such that if ψ is a ν-Hölder continuous function on V with $\|\psi\|_{\mathscr{C}^\nu} \leq 1$, then*

$$\big\langle \mu, e^{\alpha\delta^{n\nu/2}|\Lambda^n(\psi)-\langle\mu,\psi\rangle|}\big\rangle \leq c \quad \text{and} \quad \|\Lambda^n(\psi)-\langle\mu,\psi\rangle\|_{L^q(\mu)} \leq cq^{\nu/2}\delta^{-n\nu/2}$$

for every $n \geq 0$ and every $1 \leq q < +\infty$. Moreover, δ is independent of ν.

The following results are deduced as in the case of endomorphisms of $\mathbb{P}^k$.

Theorem 2.37. *Let $f : U \to V$ be a polynomial-like map with large topological degree and μ the equilibrium measure of f. Then f is exponentially mixing. More precisely, there is a constant $0 < \lambda < 1$, such that if $1 < p \leq +\infty$, we have*

$$|\langle \mu, (\varphi \circ f^n)\psi\rangle - \langle\mu,\varphi\rangle\langle\mu,\psi\rangle| \leq c_p\lambda^n\|\varphi\|_{L^p(\mu)}\|\psi\|_{L^1(V)}$$

for φ in $L^p(\mu)$, ψ p.s.h. on V and $n \geq 0$, where $c_p > 0$ is a constant independent of φ, ψ. If ν is such that $0 \leq \nu \leq 2$, then there is a constant $c_{p,\nu} > 0$ such that

$$|\langle \mu, (\varphi \circ f^n)\psi\rangle - \langle\mu,\varphi\rangle\langle\mu,\psi\rangle| \leq c_{p,\nu}\lambda^{n\nu/2}\|\varphi\|_{L^p(\mu)}\|\psi\|_{\mathscr{C}^\nu}$$

for φ in $L^p(\mu)$, ψ a $\mathscr{C}^\nu$ function on V and $n \geq 0$.

The following result gives the exponential mixing of any order. It can be extended to Hölder continuous observables using the theory of interpolation between Banach spaces.

Theorem 2.38. *Let f, μ be as in Theorem 2.37 and $r \geq 1$ an integer. Then there are constants $c > 0$ and $0 < \lambda < 1$ such that*

$$\Big|\langle \mu, \psi_0(\psi_1 \circ f^{n_1})\ldots(\psi_r \circ f^{n_r})\rangle - \prod_{i=0}^r \langle\mu,\psi_i\rangle\Big| \leq c\lambda^n \prod_{i=0}^r \|\psi_i\|_{L^1(V)}$$

for $0 = n_0 \leq n_1 \leq \cdots \leq n_r$, $n := \min_{0\leq i<r}(n_{i+1}-n_i)$ and ψ_i p.s.h. on V.

As in Section 1, we deduce the following result, as a consequence of Gordin's theorem and of the exponential decay of correlations.

Theorem 2.39. *Let f be a polynomial-like map with large topological degree as above. Let φ be a test function which is $\mathscr{C}^\nu$ with $\nu > 0$, or is d.s.h. on V. Then, either φ is a coboundary or it satisfies the central limit theorem with the variance $\sigma > 0$ given by*

$$\sigma^2 := \langle \mu, \varphi^2 \rangle + 2 \sum_{n \geq 1} \langle \mu, \varphi(\varphi \circ f^n) \rangle.$$

The following result is obtained as in Theorem 1.102, as a consequence of the exponential estimates in Corollaries 2.35 and 2.36.

Theorem 2.40. *Let f be a polynomial-like map with large topological degree as above. Then, the equilibrium measure μ of f satisfies the weak large deviations theorem for bounded d.s.h. observables and for Hölder continuous observables. More precisely, if a function ψ is bounded d.s.h. or Hölder continuous then for every $\varepsilon > 0$ there is a constant $h_\varepsilon > 0$ such that*

$$\mu \left\{ z \in \mathrm{supp}(\mu) : \left| \frac{1}{N} \sum_{n=0}^{N-1} \varphi \circ f^n(z) - \langle \mu, \varphi \rangle \right| > \varepsilon \right\} \leq e^{-N(\log N)^{-2} h_\varepsilon}$$

for all N large enough.

Theorem 1.91 can be extended to polynomial-like maps with large topological degrees.

Theorem 2.41. *Let $f : U \to V$ be a polynomial-like map with large topological degree and μ its equilibrium measure. Let φ be an observable on V with values in $\mathbb{R} \cup \{-\infty\}$ such that e^φ is Hölder continuous, $H := \{\varphi = -\infty\}$ is an analytic subset of V and $|\varphi| \lesssim |\log \mathrm{dist}(\cdot, H)|^\rho$ near H for some $\rho > 0$. If $\langle \mu, \varphi \rangle = 0$ and φ is not a coboundary, then the almost sure invariance principle holds for φ. In particular, the almost sure central limit theorem holds for such observables.*

A technical point here is to prove that if H_t is the t-neighbourhood of an analytic set H, then $\mu(H_t) \lesssim t^\alpha$ for $t > 0$ and for some constant $\alpha > 0$. This property is a consequence of the fact that μ is moderate.

We also have the following version of Theorem 1.92. In the proof, one uses Proposition 2.21 instead of Proposition 1.51.

Theorem 2.42. *Let $f : U \to V$ be a polynomial-like map with large topological degree $d_t > 1$ and μ its equilibrium measure. Then (U, μ, f) is measurably conjugate to a one-sided d_t-shift.*

Exercise 2.43. Assume that for every positive closed $(1,1)$-current S on V we have $\limsup \|(f^n)_*(S)\|^{1/n} < d_t$. Show that μ is PB and deduce that $d_{k-1}^* < d_t$. Hint: write $S = dd^c \varphi$.

Exercise 2.44. Let ν be a positive measure with compact support in $\mathbb{C}$. Prove that ν is moderate if and only if there are positive constants α and c such that for every disc D of radius r, $\nu(D) \leq c r^\alpha$. Give an example showing that this condition is not sufficient in $\mathbb{C}^2$.

2.5 Holomorphic Families of Maps

In this paragraph, we consider polynomial-like maps $f_s : U_s \to V_s$ depending holomorphically on a parameter $s \in \Sigma$. We will show that the Green measure μ_s of f_s depends "holomorphically" on s and then we study the dependence of the Lyapounov exponents on the parameters. Since the problems are local, we assume for simplicity that Σ is a ball in $\mathbb{C}^l$. Of course, we assume that U_s and V_s depend continuously on s. Observe that if we replace V_s with a convex open set $V'_s \subset V_s$ and U_s by $f_s^{-1}(V'_s)$ with $V_s \setminus V'_s$ small enough, the map f_s is still polynomial-like. So, for simplicity, assume that $V := V_s$ is independent of s. Let $U_\Sigma := \cup_s \{s\} \times U_s$. This is an open set in $V_\Sigma := \Sigma \times V$. Define the holomorphic map $F : U_\Sigma \to V_\Sigma$ by $F(s,z) := (s, f_s(z))$. This map is proper. By continuity, the topological degree d_t of f_s is independent of s. So, the topological degree of F is also d_t. Define $\mathscr{K}_\Sigma := \cap_{n\geq 0} F^{-n}(V_\Sigma)$. Then $\mathscr{K}_\Sigma$ is closed in U_Σ. If $\pi : \Sigma \times \mathbb{C}^k \to \Sigma$ is the canonical projection, then π is proper on $\mathscr{K}_\Sigma$ and $\mathscr{K}_s := \mathscr{K}_\Sigma \cap \pi^{-1}(s)$ is the filled Julia set of f_s.

It is not difficult to show that $\mathscr{K}_s$ depends upper semi-continuously on s with respect to the Hausdorff metric on compact sets of V. This means that if W_{s_0} is a neighbourhood of $\mathscr{K}_{s_0}$, then $\mathscr{K}_s$ is contained in W_{s_0} for s closed enough to s_0. We will see that for maps with large topological degree, $s \mapsto \mu_s$ is continuous in a strong sense. However, in general, the Julia set $\mathscr{J}_s$, i.e. the support of the equilibrium measure μ_s, does not depend continuously on s.

In our context, the goal is to construct and to study currents which measure the bifurcation, i.e. the discontinuity of $s \mapsto \mathscr{J}_s$. We have the following result due to Pham [PH].

Proposition 2.45. *Let $(f_s)_{s\in\Sigma}$ be as above. Then, there is a positive closed current $\mathscr{R}$ of bidegree (k,k), supported on $\mathscr{K}_\Sigma$ such that the slice $\langle \mathscr{R}, \pi, s\rangle$ is equal to the equilibrium measure μ_s of f_s for $s \in \Sigma$. Moreover, if φ is a p.s.h. function on a neighbourhood of $\mathscr{K}_\Sigma$, then the function $s \mapsto \langle \mu_s, \varphi(s,\cdot)\rangle$ is either equal to $-\infty$ or is p.s.h. on Σ.*

Proof. Let Ω be a smooth probability measure with compact support in V. Define the positive closed (k,k)-current Θ on $\Sigma \times V$ by $\Theta := \tau^*(\Omega)$ where $\tau : \Sigma \times V \to V$ is the canonical projection. Observe that the slice $\langle \Theta, \pi, s\rangle$ coincides with Ω on $\{s\} \times V$, since Ω is smooth. Define $\Theta_n := d_t^{-n}(F^n)^*(\Theta)$. The slice $\langle \Theta_n, \pi, s\rangle$ can be identified with $d_t^{-n}(f_s^n)^*(\Omega)$ on $\{s\} \times V$. This is a smooth probability measure which tends to μ_s when n goes to infinity.

Since the problem is local for s, we can assume that all the forms Θ_n are supported on $\Sigma \times K$ for some compact subset K of V. As we mentioned in Appendix A.3, since these forms have slice mass 1, they belong to a compact family of currents. Therefore, we can extract a sequence Θ_{n_i} which converges to some current $\mathscr{R}$ with slice mass 1. We want to prove that $\langle \mathscr{R}, \pi, s\rangle = \mu_s$.

Let φ be a smooth p.s.h. function on a neighbourhood of $\mathscr{K}_\Sigma$. So, for n large enough, φ is defined on the support of Θ_n (we reduce Σ if necessary). By slicing theory, $\pi_*(\Theta_{n_i} \wedge \varphi)$ is equal to the p.s.h. function $\psi_{n_i}(s) := \langle \Theta_{n_i}, \pi, s\rangle(\varphi)$ and $\pi_*(\mathscr{R} \wedge \varphi)$ is equal to the p.s.h. function $\psi(s) := \langle \mathscr{R}, \pi, s\rangle(\varphi)$ in the sense of currents. By definition of $\mathscr{R}$, since π_* is continuous on currents supported on $\Sigma \times K$, ψ_{n_i} converge to ψ in $L^1_{loc}(\Sigma)$. On the other hand, $\langle \Theta_{n_i}, \pi, s\rangle$ converge to μ_s. So, the function $\psi'(s) := \lim \psi_{n_i}(s) = \langle \mu_s, \varphi\rangle$ is equal to $\psi(s)$ almost everywhere. Since ψ_{n_i} and ψ are p.s.h., the Hartogs' lemma implies that $\psi' \leq \psi$. We show the inequality $\psi'(s) \geq \psi(s)$.

The function ψ is p.s.h., hence it is strongly upper semi-continuous. Therefore, there is a sequence (s_n) converging to s such that $\psi'(s_n) = \psi(s_n)$ and $\psi(s_n)$ converge to $\psi(s)$. Up to extracting a subsequence, we can assume that μ_{s_n} converge to some probability measure μ'_s. By continuity, μ'_s is totally invariant under f_s. We deduce from Proposition 2.13 that $\langle \mu'_s, \varphi(s,\cdot)\rangle \leq \langle \mu_s, \varphi(s,\cdot)\rangle$. The first integral is equal to $\psi(s)$, the second one is equal to $\psi'(s)$. Therefore, $\psi(s) \leq \psi'(s)$. The identity $\langle \mathscr{R}, \pi, s\rangle = \mu_s$ follows.

The second assertion in the proposition is also a consequence of the above arguments. This is clear when φ is smooth. The general case is deduced using an approximation of φ by a decreasing sequence of smooth p.s.h. functions. □

Let $\mathrm{Jac}(F)$ denote the Jacobian of F with respect to the standard volume form on $\Sigma \times \mathbb{C}^k$. Its restriction to $\pi^{-1}(s)$ is the Jacobian $\mathrm{Jac}(f_s)$ of f_s. Since $\mathrm{Jac}(F)$ is a p.s.h. function, we can apply the previous proposition and deduce that the function $L_k(s) := \frac{1}{2}\langle \mu_s, \log \mathrm{Jac}(f_s)\rangle$ is p.s.h. on Σ. Indeed, by Theorem 2.16, this function is bounded from below by $\frac{1}{2}\log d_t$, hence it is not equal to $-\infty$. By Oseledec's theorem 1.119, $L_k(s)$ is the sum of the Lyapounov exponents of f_s. We deduce the following result of [DS1].

Corollary 2.46. *Let $(f_s)_{s\in\Sigma}$ be as above. Then, the sum of the Lyapounov exponents associated to the equilibrium measure μ_s of f_s is a p.s.h. function on s. In particular, it is upper semi-continuous.*

Pham defined in [PH] *the bifurcation (p,p)-currents* by $\mathscr{B}^p := (dd^c L_k)^p$ for $1 \leq p \leq \dim\Sigma$. The wedge-product is well-defined since L_k is locally bounded: it is bounded from below by $\frac{1}{2}\log d_t$. Very likely, these currents play a crucial role in the study of bifurcation as we see in the following observation. Assume that the critical set of f_{s_0} does not intersect the filled Julia set $\mathscr{K}_{s_0}$ for some $s_0 \in \Sigma$. Since the filled Julia sets $\mathscr{K}_s$ vary upper semi-continuously in the Hausdorff metric, $\log \mathrm{Jac}(F)$ is pluriharmonic near $\{s_0\} \times \mathscr{K}_{s_0}$. It follows that L_k is pluriharmonic and $\mathscr{B}^p = 0$ in a neighbourhood of s_0[8]. On the other hand, using Kobayashi metric, it is easy to show that f is uniformly hyperbolic on $\mathscr{K}_s$ for s close to s_0. It follows that $\mathscr{K}_s = \mathscr{J}_s$ and $s \mapsto \mathscr{J}_s$ is continuous near s_0, see [FS6].

Note that L_k is equal in the sense of currents to $\pi_*(\log \mathrm{Jac}(F) \wedge \mathscr{R})$, where $\mathscr{R}$ is the current in Proposition 2.45. Therefore, $\mathscr{B}$ can be obtained using the formula

$$\mathscr{B} = dd^c \pi_*(\log \mathrm{Jac}(F) \wedge \mathscr{R}) = \pi_*([\mathscr{C}_F] \wedge \mathscr{R}),$$

since $dd^c \log|\mathrm{Jac}(F)| = [\mathscr{C}_F]$, the current of integration on the critical set $\mathscr{C}_F$ of F. We also have the following property of the function L_k.

[8] This observation was made by the second author for the family $z^2 + c$, with $c \in \mathbb{C}$. He showed that the bifurcation measure is the harmonic measure associated to the Mandelbrot set [SI].

Theorem 2.47. *Let $(f_s)_{s\in\Sigma}$ be a family of polynomial-like maps as above. Assume that f_{s_0} has a large topological degree for some $s_0 \in \Sigma$. Then L_k is Hölder continuous in a neighbourhood of s_0. In particular, the bifurcation currents $\mathscr{B}^p$ are moderate for $1 \leq p \leq \dim \Sigma$.*

Let Λ_s denote the Perron-Frobenius operator associated to f_s. For any Borel set B, denote by Ω_B the standard volume form on $\mathbb{C}^k$ restricted to B. We first prove some preliminary results.

Lemma 2.48. *Let W be a neighbourhood of the filled Julia set $\mathscr{K}_{s_0}$ of f_{s_0}. Then, there is a neighbourhood Σ_0 of s_0 such that $\langle \mu_s, \varphi \rangle$ depends continuously on (s,φ) in $\Sigma_0 \times \mathrm{PSH}(W)$.*

Proof. We first replace Σ with a neighbourhood Σ_0 of s_0 small enough. So, for every $s \in \Sigma$, the filled Julia set of f_s is contained in $U := f_{s_0}^{-1}(V)$ and in W. We also reduce the size of V in order to assume that f_s is polynomial-like on a neighbourhood of $\overline{U}$ with values in a neighbourhood V' of $\overline{V}$. Moreover, since $\langle \mu_s, \varphi \rangle = \langle \mu_s, \Lambda_s(\varphi) \rangle$ and $\Lambda_s(\varphi)$ depends continuously on (s,φ) in $\Sigma \times \mathrm{PSH}(W)$, we can replace φ with $\Lambda_s^N(\varphi)$ with N large enough and $s \in \Sigma$, in order to assume that $W = V$. Finally, since $\Lambda_s(\varphi)$ is defined on V', it is enough to prove the continuity for φ p.s.h. on V such that $\varphi \leq 1$ and $\langle \Omega_U, \varphi \rangle \geq 0$. Denote by $\mathscr{P}$ the family of such functions φ. Since μ_{s_0} is PC, we have $|\langle \mu_{s_0}, \varphi \rangle| \leq A$ for some constant $A \geq 1$ and for $\varphi \in \mathscr{P}$. Let $\mathscr{P}'$ denote the family of p.s.h. functions ψ such that $\psi \leq 2A$ and $\langle \mu_{s_0}, \psi \rangle = 0$. The function $\varphi' := \varphi - \langle \mu_{s_0}, \varphi \rangle$ belongs to this family. Observe that $\mathscr{P}'$ is bounded and therefore if $A' \geq 1$ is a fixed constant large enough, we have $|\langle \Omega_U, \psi \rangle| \leq A'$ for $\psi \in \mathscr{P}'$.

Fix an integer N large enough. By Theorem 2.33, $\Lambda_{s_0}^N(\varphi') \leq 1/8$ on V' and $|\langle \Omega_U, \Lambda_{s_0}^N(\varphi') \rangle| \leq 1/8$ for φ' as above. We deduce that $2\Lambda_{s_0}^N(\varphi') - \langle \Omega_U, 2\Lambda_{s_0}^N(\varphi') \rangle$ is a function in $\mathscr{P}$, smaller than $1/2$ on V'. This function differs from $2\Lambda_{s_0}^N(\varphi)$ by a constant. So, it is equal to $2\Lambda_{s_0}^N(\varphi) - \langle \Omega_U, 2\Lambda_{s_0}^N(\varphi) \rangle$. When Σ_0 is small enough, by continuity, the operator $L_s(\varphi) := 2\Lambda_s^N(\varphi) - \langle \Omega_U, 2\Lambda_s^N(\varphi) \rangle$ preserves $\mathscr{P}$ for $s \in \Sigma_0$. Therefore, since Λ_s preserves constant functions, we have

$$\begin{aligned}
\Lambda_s^{mN}(\varphi) &= \Lambda_s^{(m-1)N}\big[\langle \Omega_U, \Lambda_s^N(\varphi) \rangle + 2^{-1} L_s(\varphi)\big] \\
&= \langle \Omega_U, \Lambda_s^N(\varphi) \rangle + 2^{-1} \Lambda_s^{(m-1)N}(L_s(\varphi)).
\end{aligned}$$

By induction, we obtain

$$\begin{aligned}
\Lambda_s^{mN}(\varphi) &= \langle \Omega_U, \Lambda_s^N(\varphi) \rangle + \cdots + 2^{-m+1} \langle \Omega_U, \Lambda_s^N(L_s^{m-1}(\varphi)) \rangle + 2^{-m} L_s^m(\varphi) \\
&= \big\langle \Omega_U, \Lambda_s^N\big[\varphi + \cdots + 2^{-m+1} L_s^{m-1}(\varphi)\big] \big\rangle + 2^{-m} L_s^m(\varphi) \\
&= \big\langle d_t^{-N} (f_s^N)^*(\Omega_U), \varphi + \cdots + 2^{-m+1} L_s^{m-1}(\varphi) \big\rangle + 2^{-m} L_s^m(\varphi).
\end{aligned}$$

We deduce from the above property of L_s that the last term converges uniformly to 0 when m goes to infinity. The sum in the first term converges normally to the p.s.h. function $\sum_{m\geq 1} 2^{-m+1} L_s^{m-1}(\varphi)$, which depends continuously on (s,φ). Therefore, $\Lambda_s^{mN}(\varphi)$ converge to a constant which depends continuously on (s,φ). But we know that the limit is $\langle \mu_s, \varphi \rangle$. The lemma follows. □

Using the same approach as in Theorem 2.34, we prove the following result.

Theorem 2.49. *Let f_s, s_0 and W be as in Theorem 2.47 and Lemma 2.48. Let K be a compact subset of W such that $f_{s_0}^{-1}(K)$ is contained in the interior of K. There is a neighbourhood Σ_0 of s_0 such that if $\mathscr{P}$ is a bounded family of p.s.h. functions on W, then $(s,\varphi)\mapsto\langle\mu_s,\varphi\rangle$ is Hölder continuous on $\Sigma_0\times\mathscr{P}$ with respect to the pseudo-distance $\mathrm{dist}_{L^1(K)}$ on $\mathscr{P}$.*

Proof. We replace Σ with Σ_0 as in Lemma 2.48. It is not difficult to check that $(s,\varphi)\mapsto(s,\Lambda_s(\varphi))$ is locally Lipschitz with respect to $\mathrm{dist}_{L^1(K)}$. So, replacing (s,φ) by $(s,\Lambda_s^N(\varphi))$ with N large enough allows to assume that $W=V$. Let $\widehat{\mathscr{P}}$ be the set of (s,φ) in $\Sigma\times\mathrm{PSH}(V)$ such that $\varphi\leq 1$ and $\langle\mu_s,\varphi\rangle\geq 0$. By Lemma 2.48, such functions φ belong to a compact subset of $\mathrm{PSH}(V)$. It is enough to prove that $(s,\varphi)\mapsto\langle\mu_s,\varphi\rangle$ is Hölder continuous on $\widehat{\mathscr{P}}$.

Let $\widehat{\mathscr{D}}$ denote the set of $(s,\varphi-\Lambda_s(\varphi))$ with $(s,\varphi)\in\widehat{\mathscr{P}}$. Consider the operator $\widehat{\Lambda}(s,\psi):=(s,\lambda^{-1}\Lambda_s(\psi))$ on $\mathscr{D}$ as in Theorem 2.34 where $\lambda<1$ is a fixed constant close enough to 1. Theorem 2.33 and the continuity in Lemma 2.48 imply that $\widehat{\Lambda}$ preserves $\widehat{\mathscr{D}}$. Therefore, we only have to follow the arguments in Theorem 2.34. □

Proof of Theorem 2.47. We replace Σ with a small neighbourhood of s_0. Observe that $\log\mathrm{Jac}(f_s)$, $s\in\Sigma$, is a bounded family of p.s.h. functions on U. By Theorem 2.49, it is enough to show that $s\mapsto\log\mathrm{Jac}(f_s)$ is Hölder continuous with respect to $\mathrm{dist}_{L^1(K)}$.

We also deduce from Theorem A.22 that $\langle\Omega_K,e^{\lambda|\log\mathrm{Jac}(f_s)|}\rangle\leq A$ for some positive constants λ and A. Reducing V and Σ allows to assume that $\mathrm{Jac}(F)$, their derivatives and the vanishing order of $\mathrm{Jac}(F)$ are bounded on $\Sigma\times U$ by some constant m.

Fix a constant $\alpha>0$ small enough and a constant $A>0$ large enough. Define $\psi(s):=\langle\Omega_K,\log\mathrm{Jac}(f_s)\rangle$. Consider s and t in Σ such that $r:=\|s-t\|$ is smaller than a fixed small constant. We will compare $|\psi(s)-\psi(t)|$ with $r^{\lambda\alpha}$ in order to show that ψ is Hölder continuous with exponent $\lambda\alpha$. Define $S:=\{z\in U,\ \mathrm{Jac}(f_s)<2r^{2\alpha}\}$. We will bound separately

$$\langle\Omega_{K\setminus S},\log\mathrm{Jac}(f_s)-\log\mathrm{Jac}(f_t)\rangle$$

and

$$\langle\Omega_{K\cap S},\log\mathrm{Jac}(f_s)-\log\mathrm{Jac}(f_t)\rangle.$$

Note that $\psi(s)-\psi(t)$ is the sum of the above two integrals.

Consider now the integral on $K\setminus S$. The following estimates are only valid on $K\setminus S$. Since the derivatives of $\mathrm{Jac}(F)$ is bounded, we have $\mathrm{Jac}(f_t)\geq r^{2\alpha}$. It follows that the derivatives on t of $\log\mathrm{Jac}(f_t)$ is bounded by $Ar^{-2\alpha}$. We deduce that

$$|\log\mathrm{Jac}(f_s)-\log\mathrm{Jac}(f_t)|\leq Ar^{1-2\alpha}.$$

Therefore,

$$|\langle\Omega_{K\setminus S},\log\mathrm{Jac}(f_s)-\log\mathrm{Jac}(f_t)\rangle|\leq\|\Omega_{K\setminus S}\|Ar^{1-2\alpha}\leq r^{\lambda\alpha}.$$

We now estimate the integral on $K\cap S$. Its absolute value is bounded by

$$\langle \Omega_{K\cap S}, |\log \mathrm{Jac}(f_s)|\rangle + \langle \Omega_{K\cap S}, |\log \mathrm{Jac}(f_t)|\rangle.$$

We deduce from the estimate $\langle \Omega_K, e^{\lambda|\log \mathrm{Jac}(f_s)|}\rangle \leq A$ that $\mathrm{volume}(K\cap S)\leq Ar^{2\lambda\alpha}$. Therefore, by Cauchy-Schwarz's inequality, we have

$$\begin{aligned}\langle \Omega_{K\cap S}, |\log \mathrm{Jac}(f_s)|\rangle &\lesssim \mathrm{volume}(K\cap S)^{1/2}\big\langle \Omega_K, |\log \mathrm{Jac}(f_s)|^2\big\rangle^{1/2}\\ &\lesssim r^{\lambda\alpha}\langle \Omega_K, e^{\lambda|\log \mathrm{Jac}(f_s)|}\rangle^{1/2}\lesssim r^{\lambda\alpha}.\end{aligned}$$

The estimate holds for f_t instead of f_s. Hence, ψ is Hölder continuous. The fact that $\mathscr{B}^p$ are moderate follows from Theorem A.34. □

The following result of Pham generalizes Corollary 2.46 and allows to define other bifurcation currents by considering $dd^c L_p$ or their wedge-products [PH].

Theorem 2.50. *Let $(f_s)_{s\in\Sigma}$ be a holomorphic family of polynomial-like maps as above. Let $\chi_1(s)\geq\cdots\geq\chi_k(s)$ be the Lyapounov exponents of the equilibrium measure μ_s of f_s. Then, for $1\leq p\leq k$, the function*

$$L_p(s):=\chi_1(s)+\cdots+\chi_p(s)$$

is p.s.h. on Σ. In particular, L_p is upper semi-continuous.

Proof. Observe that $L_p(s)\geq \frac{p}{k}L_k(s)\geq \frac{p}{2k}\log d_t$. We identify the tangent space of V at any point with $\mathbb{C}^k$. So, the differential $Df_s(z)$ of f_s at a point $z\in U_s$ is a linear self-map on $\mathbb{C}^k$ which depends holomorphically on (s,z). It induces a linear self-map on the exterior product $\bigwedge^p\mathbb{C}^k$ that we denote by $D^pf_s(z)$. This map depends holomorphically on (s,z). In the standard coordinate system on $\bigwedge^p\mathbb{C}^k$, the function $(s,z)\mapsto \log\|D^pf_s(z)\|$ is p.s.h. on U_Σ. By Proposition 2.45, the function $\psi_1(s):=\langle\mu_s,\log\|D^pf_s\|\rangle$ is p.s.h. or equal to $-\infty$ on Σ. Define in the same way the functions $\psi_n(s):=\langle\mu_s,\log\|D^pf_s^n\|\rangle$ associated to the iterate f_s^n of f_s. We have

$$D^pf_s^{n+m}(z)=D^pf_s^m(f_s^n(z))\circ D^pf_s^n(z).$$

Hence,

$$\|D^pf_s^{n+m}(z)\|\leq\|D^pf_s^m(f_s^n(z))\|\,\|D^pf_s^n(z)\|.$$

We deduce using the invariance of μ_s that

$$\psi_{m+n}(s)\leq\psi_m(s)+\psi_n(s).$$

Therefore, the sequence $n^{-1}\psi_n$ decreases to $\inf_n n^{-1}\psi_n$. So, the limit is p.s.h. or equal to $-\infty$. On the other hand, Oseledec's theorem 1.119 implies that the limit is equal to $L_p(s)$ which is a positive function. It follows that $L_p(s)$ is p.s.h. □

Consider now the family f_s of endomorphisms of algebraic degree $d\geq 2$ of $\mathbb{P}^k$ with $s\in\mathscr{H}_d(\mathbb{P}^k)$. We can lift f_s to polynomial-like maps on $\mathbb{C}^{k+1}$ and apply the above results. The construction of the bifurcation currents $\mathscr{B}^p$ can be obtained directly using the Green measures of f_s. This was done by Bassanelli-Berteloot in [BB]. They studied some properties of the bifurcation currents and obtained nice formulas for that currents in terms of the Green functions. We also refer to DeMarco, Dujardin-Favre, McMullen, Milnor, Sibony and Silverman [DM, DM1, DF, MM, MI1, SI, SJ] for results in dimension one.

Exercise 2.51. If f is an endomorphism in $\mathscr{H}_d(\mathbb{P}^k)$, denote by $L_k(f)$ the sum of the Lyapounov exponents of the equilibrium measure. Show that $f \mapsto L_k(f)$ is locally Hölder continuous on $\mathscr{H}_d(\mathbb{P}^k)$. Deduce that the bifurcation currents are moderate. Hint: use that the lift of f to $\mathbb{C}^{k+1}$ has always a Lyapounov exponent equal to $\log d$.

Exercise 2.52. Find a family $(f_s)_{s\in\Sigma}$ such that $\mathscr{J}_s$ does not vary continuously.

Exercise 2.53. A family $(X_s)_{s\in\Sigma}$ of compact subsets in V is lower semi-continuous at s_0 if for every $\varepsilon > 0$, X_{s_0} is contained in the ε-neighbourhood of X_s when s is close enough to s_0. If $(\nu_s)_{s\in\Sigma}$ is a continuous family of probability measures on V, show that $s \mapsto \text{supp}(\nu_s)$ is lower semi-continuous. If $(f_s)_{s\in\Sigma}$ is a holomorphic family of polynomial-like maps, deduce that $s \mapsto \mathscr{J}_s$ is lower semi-continuous. Show that if $\mathscr{J}_{s_0} = \mathscr{K}_{s_0}$, then $s \mapsto \mathscr{J}_s$ is continuous at s_0 for the Hausdorff metric.

Exercise 2.54. Assume that f_{s_0} is of large topological degree. Let $\delta > 0$ be a constant small enough. Using the continuity of $s \mapsto \mu_s$, show that if p_{s_0} is a repelling fixed point in $\mathscr{J}_{s_0}$ for f_{s_0}, there are repelling fixed points p_s in $\mathscr{J}_s$ for f_s, with $|s - s_0| < \delta$, such that $s \mapsto p_s$ is holomorphic. Suppose $s \mapsto \mathscr{J}_s$ is continuous with respect to the Hausdorff metric. Construct a positive closed current $\mathscr{R}$ supported on $\bigcup_{|s-s_0|<\delta}\{s\} \times \mathscr{J}_s$ with slices μ_s. Deduce that if $\mathscr{J}_{s_0}$ does not contain critical points of f_{s_0} then $s \mapsto L_k(s)$ is pluriharmonic near s_0.

Notes. Several results in this section still hold for larger classes of polynomial-like maps. For example, the construction of the equilibrium measure is valid for a manifold V admitting a smooth strictly p.s.h. function. The dd^c-method was originally introduced for polynomial-like maps. However, we have seen that it is also effective for endomorphisms of $\mathbb{P}^k$. In a forthcoming survey, we will show that the method can be extended to other dynamical systems. Several statistical properties obtained in this section are new.

Appendix: Currents and Pluripotential Theory

In this appendix, we recall some basic notions and results on complex geometry and on currents in the complex setting. Most of the results are classical and their proofs are not given here. In constrast, we describe in detail some notions in order to help the reader who are not familiar with complex geometry or currents. The main references for the abstract theory of currents are [CH, DR, FE, SC, WA]. The reader will find in [DEM, GU, HO, LE, N] the basics on currents on complex manifolds. We also refer to [DEM, GH, HB, VO] for the theory of compact Kähler manifolds.

A.1 Projective Spaces and Analytic Sets

In this paragraph, we recall the definition of complex projective spaces. We then discuss briefly compact Kähler manifolds, projective manifolds and analytic sets.

The complex projective space $\mathbb{P}^k$ is a compact complex manifold of dimension k. It is obtained as the quotient of $\mathbb{C}^{k+1}\setminus\{0\}$ by the natural multiplicative action of $\mathbb{C}^*$. In other words, $\mathbb{P}^k$ is the parameter space of the complex lines passing through 0 in $\mathbb{C}^{k+1}$. The image of a subspace of dimension $p+1$ of $\mathbb{C}^{k+1}$ is a submanifold of dimension p in $\mathbb{P}^k$, bi-holomorphic to $\mathbb{P}^p$, and is called *a projective subspace of dimension p*. *Hyperplanes* of $\mathbb{P}^k$ are projective subspaces of dimension $k-1$. The group $\mathrm{GL}(\mathbb{C},k+1)$ of invertible linear endomorphisms of $\mathbb{C}^{k+1}$ induces the group $\mathrm{PGL}(\mathbb{C},k+1)$ of automorphisms of $\mathbb{P}^k$. It acts transitively on $\mathbb{P}^k$ and sends projective subspaces to projective subspaces.

Let $z=(z_0,\ldots,z_k)$ denote the standard coordinates of $\mathbb{C}^{k+1}$. Consider the equivalence relation: $z\sim z'$ *if there is* $\lambda\in\mathbb{C}^*$ *such that* $z=\lambda z'$. The projective space $\mathbb{P}^k$ is the quotient of $\mathbb{C}^{k+1}\setminus\{0\}$ by this relation. We can cover $\mathbb{P}^k$ by open sets U_i associated to the open sets $\{z_i\neq 0\}$ in $\mathbb{C}^{k+1}\setminus\{0\}$. Each U_i is bi-holomorphic to $\mathbb{C}^k$ and $(z_0/z_i,\ldots,z_{i-1}/z_i,z_{i+1}/z_i,\ldots,z_k/z_i)$ is a coordinate system on this chart. The complement of U_i is the hyperplane defined by $\{z_i=0\}$. So, $\mathbb{P}^k$ can be considered as a natural compactification of $\mathbb{C}^k$. We denote by $[z_0:\cdots:z_k]$ the point of $\mathbb{P}^k$ associated to $(z_0,\ldots,z_k)$. This expression is *the homogeneous coordinates* on $\mathbb{P}^k$. Projective spaces are compact Kähler manifolds. We will describe this notion later.

Let X be a complex manifold of dimension k. Let φ be a differential l-form on X. In local holomorphic coordinates $z=(z_1,\ldots,z_k)$, it can be written as

$$\varphi(z)=\sum_{|I|+|J|=l}\varphi_{IJ}dz_I\wedge d\overline{z}_J,$$

where φ_{IJ} are complex-valued functions, $dz_I:=dz_{i_1}\wedge\ldots\wedge dz_{i_p}$ if $I=(i_1,\ldots,i_p)$, and $d\overline{z}_J:=d\overline{z}_{j_1}\wedge\ldots\wedge d\overline{z}_{j_q}$ if $J=(j_1,\ldots,j_q)$. The *conjugate* of φ is

$$\overline{\varphi}(z):=\sum_{|I|+|J|=l}\overline{\varphi}_{IJ}d\overline{z}_I\wedge dz_J.$$

The form φ is real if and only if $\varphi=\overline{\varphi}$.

We say that φ is a form of *of bidegree* (p,q) if $\varphi_{IJ}=0$ when $(|I|,|J|)\neq(p,q)$. The bidegree does not depend on the choice of local coordinates. Let $T_X^{\mathbb{C}}$ denote the complexification of the tangent bundle of X. The complex structure on X induces a linear endomorphism $\mathscr{J}$ on the fibers of $T_X^{\mathbb{C}}$ such that $\mathscr{J}^2=-\mathrm{id}$. This endomorphism induces a decomposition of $T_X^{\mathbb{C}}$ into the direct sum of two proper sub-bundles of dimension k: *the holomorphic part* $T_X^{1,0}$ associated to the eigenvalue i of $\mathscr{J}$, and the *anti-holomorphic part* $T_X^{0,1}$ associated to the eigenvalue $-i$. Let $\Omega_X^{1,0}$ and $\Omega_X^{0,1}$ denote the dual bundles of $T_X^{1,0}$ and $T_X^{0,1}$. Then, (p,q)-forms sections of the vector bundle $\bigwedge^p\Omega^{1,0}\otimes\bigwedge^q\Omega^{0,1}$.

If φ is a (p,q)-form then the differential $d\varphi$ is the sum of a $(p+1,q)$-form and a $(p,q+1)$-form. We then denote by $\partial\varphi$ the part of bidegree $(p+1,q)$ and $\overline{\partial}\varphi$ the the part of bidegree $(p,q+1)$. The operators ∂ and $\overline{\partial}$ extend linearly to arbitrary forms φ. The operator d is real, i.e. it sends real forms to real forms but ∂ and $\overline{\partial}$ are not real. The identity $d\circ d=0$ implies that $\partial\circ\partial=0$, $\overline{\partial}\circ\overline{\partial}=0$ and $\partial\overline{\partial}+\overline{\partial}\partial=0$. Define $d^c:=\frac{\sqrt{-1}}{2\pi}(\overline{\partial}-\partial)$. This operator is real and satisfies $dd^c=\frac{\sqrt{-1}}{\pi}\partial\overline{\partial}$. Note that the above operators commute with the pull-back by holomorphic maps. More precisely, if $\tau: X_1\to X_2$ is a holomorphic map between complex manifolds and φ is a form on X_2 then $df^*(\varphi)=f^*(d\varphi)$, $dd^cf^*(\varphi)=f^*(dd^c\varphi)$, etc. Recall that the form φ is *closed* (resp. ∂-closed, $\overline{\partial}$-closed, dd^c-closed) if $d\varphi$ (resp. $\partial\varphi$, $\overline{\partial}\varphi$, $dd^c\varphi$) vanishes. The form φ is *exact* (resp. ∂-exact, $\overline{\partial}$-exact, dd^c-exact) if it is equal to the differential $d\psi$ (resp. $\partial\psi$, $\overline{\partial}\psi$, $dd^c\psi$) of a form ψ. Clearly, exact forms are closed.

A smooth $(1,1)$-form ω on X is *Hermitian* if it can be written in local coordinates as

$$\omega(z)=\sqrt{-1}\sum_{1\leq i,j\leq k}\alpha_{ij}(z)dz_i\wedge d\overline{z}_j,$$

where α_{ij} are smooth functions such that the matrix (α_{ij}) is Hermitian. We consider a form ω such that the matrix (α_{ij}) is positive definite at every point. It is strictly positive in the sense that we will introduce later. If a is a point in X, we can find local coordinates z such that $z=0$ at a and ω is equal near 0 to the Euclidean form $dd^c\|z\|^2$ modulo a term of order $\|z\|$. The form ω is always real and induces a norm on the tangent spaces of X. So, it defines a Riemannian metric on X. We say that ω is *a Kähler form* if it is a closed positive definite Hermitian form. In this case, one can find local coordinates z such that $z=0$ at a and ω is equal near 0 to $dd^c\|z\|^2$ modulo a term of order $\|z\|^2$. So, at the infinitesimal level, a Kähler metric is close to the Euclidean one. This is a crucial property in Hodge theory in the complex setting.

Consider now a compact complex manifold X of dimension k. Assume that X is *a Kähler manifold*, i.e. it admits a Kähler form ω. Recall that *the de Rham cohomology group* $H^l(X,\mathbb{C})$ is the quotient of the space of closed l-forms by the subspace of exact l-forms. This complex vector space is of finite dimension. The real groups $H^l(X,\mathbb{R})$ are defined in the same way using real forms. We have

$$H^l(X,\mathbb{C})=H^l(X,\mathbb{R})\otimes_{\mathbb{R}}\mathbb{C}.$$

If α is a closed l-form, its class in $H^l(X,\mathbb{C})$ is denoted by $[\alpha]$. The group $H^0(X,\mathbb{C})$ is just the set of constant functions. So, it is isomorphic to $\mathbb{C}$. The group $H^{2k}(X,\mathbb{C})$ is also isomorphic to $\mathbb{C}$. The isomorphism is given by the canonical map $[\alpha]\mapsto\int_X\alpha$. For l,m such that $l+m\leq 2k$, *the cup-product*

$$\smile: H^l(X,\mathbb{C})\times H^m(X,\mathbb{C})\to H^{l+m}(X,\mathbb{C})$$

is defined by $[\alpha]\smile[\beta]:=[\alpha\wedge\beta]$. The Poincaré duality theorem says that the cup-product is a non-degenerated bilinear form when $l+m=2k$. So, it defines an isomorphism between $H^l(X,\mathbb{C})$ and the dual of $H^{2k-l}(X,\mathbb{C})$.

Let $H^{p,q}(X,\mathbb{C})$, $0 \leq p,q \leq k$, denote the subspace of $H^{p+q}(X,\mathbb{C})$ generated by the classes of closed (p,q)-forms. We call $H^{p,q}(X,\mathbb{C})$ the *Hodge cohomology group*. Hodge theory shows that

$$H^l(X,\mathbb{C}) = \bigoplus_{p+q=l} H^{p,q}(X,\mathbb{C}) \qquad \text{and} \qquad H^{q,p}(X,\mathbb{C}) = \overline{H^{p,q}(X,\mathbb{C})}.$$

This, together with the Poincaré duality, induces a canonical isomorphism between $H^{p,q}(X,\mathbb{C})$ and the dual space of $H^{k-p,k-q}(X,\mathbb{C})$. Define for $p = q$

$$H^{p,p}(X,\mathbb{R}) := H^{p,p}(X,\mathbb{C}) \cap H^{2p}(X,\mathbb{R}).$$

We have

$$H^{p,p}(X,\mathbb{C}) = H^{p,p}(X,\mathbb{R}) \otimes_{\mathbb{R}} \mathbb{C}.$$

Recall that *the Dolbeault cohomology group* $H^{p,q}_{\overline{\partial}}(X)$ is the quotient of the space of $\overline{\partial}$-closed (p,q)-forms by the subspace of $\overline{\partial}$-exact (p,q)-forms. Observe that a (p,q)-form is d-closed if and only if it is ∂-closed and $\overline{\partial}$-closed. Therefore, there is a natural morphism between the Hodge and the Dolbeault cohomology groups. Hodge theory asserts that this is in fact an isomorphism: we have

$$H^{p,q}(X,\mathbb{C}) \simeq H^{p,q}_{\overline{\partial}}(X).$$

The result is a consequence of the following theorem, the so-called *dd^c-lemma*, see e.g. [DEM, VO].

Theorem A.1. *Let φ be a smooth d-closed (p,q)-form on X. Then φ is dd^c-exact if and only if it is d-exact (or ∂-exact or $\overline{\partial}$-exact).*

The projective space $\mathbb{P}^k$ admits a Kähler form ω_{FS}, called *the Fubini-Study form*. It is defined on the chart U_i by

$$\omega_{\text{FS}} := dd^c \log\left(\sum_{j=0}^{k} \left|\frac{z_j}{z_i}\right|^2\right).$$

In other words, if $\pi : \mathbb{C}^{k+1} \setminus \{0\} \to \mathbb{P}^k$ is the canonical projection, then ω_{FS} is defined by

$$\pi^*(\omega_{\text{FS}}) := dd^c \log\left(\sum_{i=0}^{k} |z_i|^2\right).$$

One can check that ω_{FS}^k is a probability measure on $\mathbb{P}^k$. The cohomology groups of $\mathbb{P}^k$ are very simple. We have $H^{p,q}(\mathbb{P}^k,\mathbb{C}) = 0$ for $p \neq q$ and $H^{p,p}(\mathbb{P}^k,\mathbb{C}) \simeq \mathbb{C}$. The groups $H^{p,p}(\mathbb{P}^k,\mathbb{R})$ and $H^{p,p}(\mathbb{P}^k,\mathbb{C})$ are generated by the class of ω_{FS}^p. Submanifolds of $\mathbb{P}^k$ are Kähler, as submanifolds of a Kähler manifold. Chow's theorem says that such a manifold is algebraic, i.e. it is the set of common zeros of a finite family of homogeneous polynomials in z. A compact manifold is *projective* if it is bi-holomorphic to a submanifold of a projective space. Their cohomology groups are in general very rich and difficult to describe.

A useful result of Blanchard [BN] says that the blow-up of a compact Kähler manifold along a submanifold is always a compact Kähler manifold. The construction of the blow-up is as follows. Consider first the case of open sets in $\mathbb{C}^k$ with $k \geq 2$. Observe that $\mathbb{C}^k$ is the union of the complex lines passing through 0 which are parametrized by the projective space $\mathbb{P}^{k-1}$. *The blow-up* $\widehat{\mathbb{C}^k}$ of $\mathbb{C}^k$ at 0 is obtained by separating these complex lines, that is, we keep $\mathbb{C}^k \setminus \{0\}$ and replace 0 with a copy of $\mathbb{P}^{k-1}$. More precisely, if $z = (z_1, \ldots, z_k)$ denote the coordinates of $\mathbb{C}^k$ and $[w] = [w_1 : \cdots : w_k]$ are homogeneous coordinates of $\mathbb{P}^{k-1}$, then $\widehat{\mathbb{C}^k}$ is the submanifold of $\mathbb{C}^k \times \mathbb{P}^{k-1}$ defined by the equations $z_i w_j = z_j w_i$ for $1 \leq i, j \leq k$. If U is an open set in $\mathbb{C}^k$ containing 0, the blow-up $\widehat{U}$ of U at 0 is defined by $\pi^{-1}(U)$ where $\pi : \widehat{\mathbb{C}^k} \to \mathbb{C}^k$ is the canonical projection.

If U is a neighbourhood of 0 in $\mathbb{C}^{k-p}$, $p \leq k-2$, and V is an open set in $\mathbb{C}^p$, then the blow-up of $U \times V$ along $\{0\} \times V$ is equal to $\widehat{U} \times V$. Consider now a submanifold Y of X of dimension $p \leq k-2$. We cover X by charts which either do not intersect Y or are of the type $U \times V$, where Y is identified with $\{0\} \times V$. *The blow-up* $\widehat{X}$ is obtained by sticking the charts outside Y with the blow-ups of charts which intersect Y. The natural projection $\pi : \widehat{X} \to X$ defines a bi-holomorphic map between $\widehat{X} \setminus \pi^{-1}(Y)$ and $X \setminus Y$. The set $\pi^{-1}(Y)$ is a smooth hypersurface, i.e. submanifold of codimension 1; it is called *the exceptional hypersurface*. Blow-up may be defined using the local ideals of holomorphic functions vanishing on Y. The blow-up of a projective manifold along a submanifold is a projective manifold.

We now recall some facts on analytic sets, see [GU, N]. Let X be an arbitrary complex manifold of dimension k[9]. Analytic sets of X can be seen as submanifolds of X, possibly with singularities. Analytic sets of dimension 0 are locally finite subsets, those of dimension 1 are (possibly singular) Riemann surfaces. For example, $\{z_1^2 = z_2^3\}$ is an analytic set of $\mathbb{C}^2$ of dimension 1 with a singularity at 0. Chow's theorem holds for analytic sets: any analytic set in $\mathbb{P}^k$ is the set of common zeros of a finite family of homogeneous polynomials.

Recall that an *analytic set* Y of X is locally the set of common zeros of holomorphic functions: for every point $a \in X$ there is a neighbourhood U of a and holomorphic functions f_i on U such that $Y \cap U$ is the intersection of $\{f_i = 0\}$. We can choose U so that $Y \cap U$ is defined by a finite family of holomorphic functions. Analytic sets are closed for the usual topology on X. Local rings of holomorphic functions on X induce local rings of holomorphic functions on Y. An analytic set Y is *irreducible* if it is not a union of two different non-empty analytic sets of X. A general analytic set Y can be decomposed in a unique way into a union of irreducible analytic subsets $Y = \cup Y_i$, where no component Y_i is contained in another one. The decomposition is locally finite, that is, given a compact set K in X, only finitely many Y_i intersect K.

Any increasing sequence of irreducible analytic subsets of X is stationary. A decreasing sequence (Y_n) of analytic subsets of X is always locally stationary, that is, for any compact subset K of X, the sequence $(Y_n \cap K)$ is stationary. Here, we do not suppose Y_n irreducible. The topology on X whose closed sets are exactly the analytic sets, is called *the Zariski topology*. When X is connected, non-empty open

[9] We often assume implicitely that X is connected for simplicity.

Zariski sets are dense in X for the usual topology. The restriction of the Zariski topology on X to Y is also called the *Zariski topology* of Y. When Y is irreducible, the non-empty Zariski open subsets are also dense in Y but this is not the case for reducible analytic sets.

There is a minimal analytic subset $\mathrm{sing}(Y)$ in X such that $Y \setminus \mathrm{sing}(Y)$ is a (smooth) complex submanifold of $X \setminus \mathrm{sing}(Y)$, i.e. a complex manifold which is closed and without boundary in $X \setminus \mathrm{sing}(Y)$. The analytic set $\mathrm{sing}(Y)$ is the *singular part* of Y. The *regular part* of Y is denoted by $\mathrm{reg}(Y)$; it is equal to $Y \setminus \mathrm{sing}(Y)$. The manifold $\mathrm{reg}(Y)$ is not necessarily irreducible; it may have several components. We call *dimension of* Y, $\dim(Y)$, the maximum of the dimensions of these components; the *codimension* $\mathrm{codim}(Y)$ of Y in X is the integer $k - \dim(Y)$. We say that Y is a proper analytic set of X if it has positive codimension. When all the components of Y have the same dimension, we say that Y is *of pure dimension* or *of pure codimension*. When $\mathrm{sing}(Y)$ is non-empty, its dimension is always strictly smaller than the dimension of Y. We can again decompose $\mathrm{sing}(Y)$ into regular and singular parts. The procedure can be repeated less than k times and gives a stratification of Y into disjoint complex manifolds. Note that Y is irreducible if and only if $\mathrm{reg}(Y)$ is a connected manifold. The following result is due to Wirtinger.

Theorem A.2 (Wirtinger). *Let Y be analytic set of pure dimension p of a Hermitian manifold (X, ω). Then the $2p$-dimensional volume of Y on a Borel set K is equal to*

$$\mathrm{volume}(Y \cap K) = \frac{1}{p!} \int_{\mathrm{reg}(Y) \cap K} \omega^p.$$

Here, the volume is with respect to the Riemannian metric induced by ω.

Let D_k denote the unit polydisc $\{|z_1| < 1, \ldots, |z_k| < 1\}$ in $\mathbb{C}^k$. The following result describes the local structure of analytic sets.

Theorem A.3. *Let Y be an analytic set of pure dimension p of X. Let a be a point of Y. Then there is a holomorphic chart U of X, bi-holomorphic to D_k, with local coordinates $z = (z_1, \ldots, z_k)$, such that $z = 0$ at a, U is given by $\{|z_1| < 1, \ldots, |z_k| < 1\}$ and the projection $\pi : U \to D_p$, defined by $\pi(z) := (z_1, \ldots, z_p)$, is proper on $Y \cap U$. In this case, there is a proper analytic subset S of D_p such that $\pi : Y \cap U \setminus \pi^{-1}(S) \to D_p \setminus S$ is a finite covering and the singularities of Y are contained in $\pi^{-1}(S)$.*

Recall that a holomorphic map $\tau : X_1 \to X_2$ between complex manifolds of the same dimension is *a covering of degree d* if each point of X_2 admits a neighbourhood V such that $\tau^{-1}(V)$ is a disjoint union of d open sets, each of which is sent bi-holomorphically to V by τ. Observe the previous theorem also implies that the fibers of $\pi : Y \cap U \to D_p$ are finite and contain at most d points if d is the degree of the covering. We can reduce U in order to have that a is the unique point in the fiber $\pi^{-1}(0) \cap Y$. The degree d of the covering depends on the choice of coordinates and the smallest integer d obtained in this way is called *the multiplicity* of Y at a and is denoted by $\mathrm{mult}(Y, a)$. We will see that $\mathrm{mult}(Y, a)$ is the Lelong number at a of the positive closed current associated to Y. In other words, if B_r denotes the ball of center a and of radius r, then the ratio between the volume of $Y \cap B_r$ and the volume of a ball of radius r in $\mathbb{C}^p$ decreases to $\mathrm{mult}(Y, a)$ when r decreases to 0.

Let $\tau : X_1 \to X_2$ be an open holomorphic map between complex manifolds of the same dimension. Applying the above result to the graph of τ, we can show that for any point $a \in X_1$ and for a neighbourhood U of a small enough, if z is a generic point in X_2 close enough to $\tau(a)$, the number of points in $\tau^{-1}(z) \cap U$ does not depend on z. We call this number *the multiplicity* or *the local topological degree* of τ at a. We say that τ is *a ramified covering of degree d* if τ is open, proper and each fiber of τ contains exactly d points counted with multiplicity. In this case, if Σ_2 is the set of critical values of τ and $\Sigma_1 := \tau^{-1}(\Sigma_2)$, then $\tau : X_1 \setminus \Sigma_1 \to X_2 \setminus \Sigma_2$ is a covering of degree d.

We recall the notion of analytic space which generalizes complex manifolds and their analytic subsets. An *analytic space of dimension* $\leq p$ is defined as a complex manifold but a chart is replaced with an analytic subset of dimension $\leq p$ in an open set of a complex Euclidean space. As in the case of analytic subsets, one can decompose analytic spaces into irreducible components and into regular and singular parts. The notions of dimension, of Zariski topology and of holomorphic maps can be extended to analytic spaces. The precise definition uses the local ring of holomorphic functions, see [GU, N]. An analytic space is *normal* if the local ring of holomorphic functions at every point is integrally closed. This is equivalent to the fact that for U open in Z holomorphic functions on $\mathrm{reg}(Z) \cap U$ which are bounded near $\mathrm{sing}(Z) \cap U$, are holomorphic on U. In particular, normal analytic spaces are locally irreducible. A holomorphic map $f : Z_1 \to Z_2$ between complex spaces is a continuous map which induces morphisms from local rings of holomorphic functions on Z_2 to the ones on Z_1. The notions of ramified covering, of multiplicity and of open maps can be extended to normal analytic spaces. We have the following useful result where $\widetilde{Z}$ is called *normalization* of Z.

Theorem A.4. *Let Z be an analytic space. Then there is a unique, up to a bi-holomorphic map, normal analytic space $\widetilde{Z}$ and a finite holomorphic map $\pi : \widetilde{Z} \to Z$ such that*

1. *$\pi^{-1}(\mathrm{reg}(Z))$ is a dense Zariski open set of $\widetilde{Z}$ and π defines a bi-holomorphic map between $\pi^{-1}(\mathrm{reg}(Z))$ and $\mathrm{reg}(Z)$;*
2. *If $\tau : Z' \to Z$ is a holomorphic map between analytic spaces, then there is a unique holomorphic map $h : Z' \to \widetilde{Z}$ satisfying $\pi \circ h = \tau$.*

In particular, holomorphic self-maps of Z can be lifted to holomorphic self-maps of $\widetilde{Z}$.

Example A.5. Let $\pi : \mathbb{C} \to \mathbb{C}^2$ be the holomorphic map given by $\pi(t) = (t^2, t^3)$. This map defines a normalization of the analytic curve $\{z_1^3 = z_2^2\}$ in $\mathbb{C}^2$ which is singular at 0. The normalization of the analytic set $\{z_1 = 0\} \cup \{z_1^3 = z_2^2\}$ is the union of two disjoint complex lines. The normalization of a complex curve (an analytic set of pure dimension 1) is always smooth.

The following desingularization theorem, due to Hironaka, is very useful.

Theorem A.6. *Let Z be an analytic space. Then there is a smooth manifold $\widehat{Z}$, possibly reducible, and a holomorphic map $\pi : \widehat{Z} \to Z$ such that $\pi^{-1}(\mathrm{reg}(Z))$ is a dense Zariski open set of $\widehat{Z}$ and π defines a bi-holomorphic map between $\pi^{-1}(\mathrm{reg}(Z))$ and $\mathrm{reg}(Z)$.*

When Z is an analytic subset of a manifold X, then one can obtain a map $\pi : \widehat{X} \to X$ using a sequence of blow-ups along the singularities of Z. The manifold $\widehat{Z}$ is the strict transform of Z by π. The difference with the normalization of Z is that we do not have the second property in Theorem A.4 but $\widehat{Z}$ is smooth.

Exercise A.7. Let X be a compact Kähler manifold of dimension k. Show that the Betti number b_l, i.e. the dimension of $H^l(X,\mathbb{R})$, is even if l is odd and does not vanish if l is even.

Exercise A.8. Let $\mathrm{Grass}(l,k)$ denote the Grassmannian, i.e. the set of linear subspaces of dimension l of $\mathbb{C}^k$. Show that $\mathrm{Grass}(l,k)$ admits a natural structure of a projective manifold.

Exercise A.9. Let X be a compact complex manifold of dimension ≥ 2 and $\pi : \widehat{X \times X} \to X \times X$ the blow-up of $X \times X$ along the diagonal Δ. Let Π_1, Π_2 denote the natural projections from $\widehat{X \times X}$ onto the two factors X of $X \times X$. Show that Π_1, Π_2 and their restrictions to $\pi^{-1}(\Delta)$ are submersions.

Exercise A.10. Let E be a finite or countable union of proper analytic subsets of a connected manifold X. Show that $X \setminus E$ is connected and dense in X for the usual topology.

Exercise A.11. Let $\tau : X_1 \to X_2$ be a ramified covering of degree n. Let φ be a function on X_1. Define

$$\tau_*(\varphi)(z) := \sum_{w \in \tau^{-1}(z)} \varphi(w),$$

where the points in $\tau^{-1}(z)$ are counted with multiplicity. If φ is upper semi-continuous or continuous, show that $\tau_*(\varphi)$ is upper semi-continuous or continuous respectively. Show that the result still holds for a general open map τ between manifolds of the same dimension if φ has compact support in X_1.

A.2 Positive Currents and p.s.h. Functions

In this paragraph, we introduce positive forms, positive currents and plurisubharmonic functions on complex manifolds. The concept of positivity and the notion of plurisubharmonic functions are due to Lelong and Oka. The theory has many applications in complex algebraic geometry and in dynamics.

Let X be a complex manifold of dimension k and ω a Hermitian $(1,1)$-form on X which is positive definite at every point. Recall that a current S on X, of degree l and of dimension $2k-l$, is a continuous linear form on the space $\mathscr{D}^{2k-l}(X)$ of smooth $(2k-l)$-forms with compact support in X. Its value on a $(2k-l)$-form $\varphi \in \mathscr{D}^{2k-l}(X)$ is denoted by $S(\varphi)$ or more frequently by $\langle S,\varphi\rangle$. On a chart, S corresponds to a continuous linear form acting on the coefficients of φ. So, it can be represented as an l-form with distribution coefficients. A sequence (S_n) of l-currents converges to an l-current S if for every $\varphi \in \mathscr{D}^{2k-l}(X)$, $\langle S_n,\varphi\rangle$ converge to $\langle S,\varphi\rangle$. The conjugate of S is the l-current $\overline{S}$ defined by

$$\langle \overline{S},\varphi\rangle := \overline{\langle S,\overline{\varphi}\rangle},$$

for $\varphi \in \mathscr{D}^{2k-l}(X)$. The current S is real if and only if $\overline{S}=S$.

The support of S is the smallest closed subset $\mathrm{supp}(S)$ of X such that $\langle S,\varphi\rangle=0$ when φ is supported on $X\setminus\mathrm{supp}(S)$. The current S extends continuously to the space of smooth forms φ such that $\mathrm{supp}(\varphi)\cap\mathrm{supp}(S)$ is compact in X. If X' is a complex manifold of dimension k' with $2k'\geq 2k-l$, and if $\tau : X\to X'$ is a holomorphic map which is proper on the support of S, we can define *the push-forward* $\tau_*(S)$ of S by τ. This is a current $\tau_*(S)$ of the same dimension than S, i.e. of degree $2k'-2k+l$, which is supported on $\tau(\mathrm{supp}(S))$, it satisfies

$$\langle \tau_*(S),\varphi\rangle := \langle S,\tau^*(\varphi)\rangle$$

for $\varphi\in\mathscr{D}^{2k-l}(X')$. If X' is a complex manifold of dimension $k'\geq k$ and if $\tau : X'\to X$ is a submersion, we can define *the pull-back* $\tau^*(S)$ of S by τ. This is an l-current supported on $\tau^{-1}(\mathrm{supp}(S))$, it satisfies

$$\langle \tau^*(S),\varphi\rangle := \langle S,\tau_*(\varphi)\rangle$$

for $\varphi\in\mathscr{D}^{2k'-l}(X')$. Indeed, since τ is a submersion, the current $\tau_*(\varphi)$ is in fact a smooth form with compact support in X; it is given by an integral of φ on the fibers of τ.

Any smooth differential l-form ψ on X defines a current: it defines the continuous linear form $\varphi\mapsto\int_X\psi\wedge\varphi$ on $\varphi\in\mathscr{D}^{2k-l}(X)$. So, currents extend the notion of differential forms. The operators $d,\partial,\overline{\partial}$ on differential forms extend to currents. For example, we have that dS is an $(l+1)$-current defined by

$$\langle dS,\varphi\rangle := (-1)^{l+1}\langle S,d\varphi\rangle$$

for $\varphi\in\mathscr{D}^{2k-l-1}(X)$. One easily check that when S is a smooth form, the above identity is a consequence of the Stokes' formula. We say that S is of *bidegree* (p,q) and *of bidimension* $(k-p,k-q)$ if it vanishes on forms of bidegree $(r,s)\neq(k-p,k-q)$. The conjugate of a (p,q)-current is of bidegree (q,p). So, if such a current is real, we have necessarily $p=q$. Note that the push-forward and the pull-back by holomorphic maps commute with the above operators. They preserve real currents; the push-forward preserves the bidimension and the pull-back preserves the bidegree.

There are three notions of positivity which coincide for the bidegrees $(0,0)$, $(1,1)$, $(k-1,k-1)$ and (k,k). Here, we only use two of them. They are dual to each other. A (p,p)-form φ is *(strongly) positive* if at each point, it is equal to a combination with positive coefficients of forms of type

$$(\sqrt{-1}\alpha_1 \wedge \overline{\alpha}_1) \wedge \ldots \wedge (\sqrt{-1}\alpha_p \wedge \overline{\alpha}_p),$$

where α_i are $(1,0)$-forms. Any (p,p)-form can be written as a finite combination of positive (p,p)-forms. For example, in local coordinates z, a $(1,1)$-form ω is written as

$$\omega = \sum_{i,j=1}^{k} \alpha_{ij}\sqrt{-1}dz_i \wedge d\overline{z}_j,$$

where α_{ij} are functions. This form is positive if and only if the matrix (α_{ij}) is positive semi-definite at every point. In local coordinates z, the $(1,1)$-form $dd^c\|z\|^2$ is positive. One can write $dz_1 \wedge d\overline{z}_2$ as a combination of $dz_1 \wedge d\overline{z}_1$, $dz_2 \wedge d\overline{z}_2$, $d(z_1 \pm z_2) \wedge d\overline{(z_1 \pm z_2)}$ and $d(z_1 \pm \sqrt{-1}z_2) \wedge d\overline{(z_1 \pm \sqrt{-1}z_2)}$. Hence, positive forms generate the space of (p,p)-forms.

A (p,p)-current S is *weakly positive* if for every smooth positive $(k-p,k-p)$-form φ, $S \wedge \varphi$ is a positive measure and is *positive* if $S \wedge \varphi$ is a positive measure for every smooth weakly positive $(k-p,k-p)$-form φ. Positivity implies weak positivity. These properties are preserved under pull-back by holomorphic submersions and push-forward by proper holomorphic maps. Positive and weakly positive forms or currents are real. One can consider positive and weakly positive (p,p)-forms as sections of some bundles of salient convex closed cones which are contained in the real part of the vector bundle $\bigwedge^p \Omega^{1,0} \otimes \bigwedge^p \Omega^{0,1}$.

The wedge-product of a positive current with a positive form is positive. The wedge-product of a weakly positive current with a positive form is weakly positive. Wedge-products of weakly positive forms or currents are not always weakly positive. For real (p,p)-currents or forms S, S', we will write $S \geq S'$ and $S' \leq S$ if $S - S'$ is positive. A current S is *negative* if $-S$ is positive. A (p,p)-current or form S is *strictly positive* if in local coordinates z, there is a constant $\varepsilon > 0$ such that $S \geq \varepsilon(dd^c\|z\|^2)^p$. Equivalently, S is strictly positive if we have locally $S \geq \varepsilon\omega^p$ with $\varepsilon > 0$.

Example A.12. Let Y be an analytic set of pure codimension p of X. Using the local description of Y near a singularity in Theorem A.3 and Wirtinger's theorem A.2, one can prove that the $2(k-p)$-dimensional volume of Y is locally finite in X. This allows to define the following (p,p)-current $[Y]$ by

$$\langle [Y], \varphi \rangle := \int_{\mathrm{reg}(Y)} \varphi$$

for φ in $\mathscr{D}^{k-p,k-p}(X)$, the space of smooth $(k-p,k-p)$-forms with compact support in X. Lelong proved that this current is positive and closed [DEM, LE].

If S is a (weakly) positive (p,p)-current, it is of order 0, i.e. it extends continuously to the space of continuous forms with compact support in X. In other words, on a chart of X, the current S corresponds to a differential form with measure coefficients. We define the *mass* of S on a Borel set K by

$$\|S\|_K := \int_K S \wedge \omega^{k-p}.$$

When K is relatively compact in X, we obtain an equivalent norm if we change the Hermitian metric on X. This is a consequence of the property we mentioned above, which says that S takes values in salient convex closed cones. Note that the previous mass-norm is just defined by an integral, which is easier to compute or to estimate than the usual mass for currents on real manifolds.

Positivity implies an important compactness property. As for positive measures, any family of positive (p,p)-currents with locally uniformly bounded mass, is relatively compact in the cone of positive (p,p)-currents. For the current $[Y]$ in Example A.12, by Wirtinger's theorem, the mass on K is equal to $(k-p)!$ times the volume of $Y \cap K$ with respect to the considered Hermitian metric. If S is a negative (p,p)-current, its mass is defined by

$$\|S\|_K := -\int_K S \wedge \omega^{k-p}.$$

The following result is the complex version of the classical support theorem in the real setting, [BA, HP, FE].

Proposition A.13. *Let S be a (p,p)-current supported on a smooth complex submanifold Y of X. Let $\tau : Y \to X$ denote the inclusion map. Assume that S is $\mathbb{C}$-normal, i.e. S and $dd^c S$ are of order 0. Then, S is a current on Y. More precisely, there is a $\mathbb{C}$-normal (p,p)-current S' on Y such that $S = \tau_*(S')$. If S is positive closed and Y is of dimension $k-p$, then S is equal to a combination with positive coefficients of currents of integration on components of Y.*

The last property holds also when Y is a singular analytic set. Proposition A.13 applies to positive closed (p,p)-currents which play an important role in complex geometry and dynamics. These currents generalize analytic sets of dimension $k-p$, as we have seen in Example A.12. They have no mass on Borel sets of $2(k-p)$-dimensional Hausdorff measure 0. The proposition is used in order to develop a calculus on potentials of closed currents.

We introduce now the notion of Lelong number for such currents which generalizes the notion of multiplicity for analytic sets. Let S be a positive closed (p,p)-current on X. Consider local coordinates z on a chart U of X and the local Kähler form $dd^c\|z\|^2$. Let $B_a(r)$ denote the ball of center a and of radius r contained in U. Then, $S \wedge (dd^c\|z\|^2)^{k-p}$ is a positive measure on U. Define for $a \in U$

$$\nu(S,a,r) := \frac{\|S \wedge (dd^c\|z\|^2)^{k-p}\|_{B_a(r)}}{\pi^{k-p} r^{2(k-p)}}.$$

Note that $\pi^{k-p}r^{2(k-p)}$ is $(k-p)!$ times the volume of a ball in $\mathbb{C}^{k-p}$ of radius r, i.e. the mass of the current associated to this ball. When r decreases to 0, $\nu(S,a,r)$ is decreasing and the *Lelong number* of S at a is the limit

$$\nu(S,a) := \lim_{r\to 0} \nu(S,a,r).$$

It does not depend on the coordinates. So, we can define the Lelong number for currents on any manifold. Note that $\nu(S,a)$ is also the mass of the measure $S \wedge (dd^c \log \|z-a\|)^{k-p}$ at a. We will discuss the wedge-product (intersection) of currents in the next paragraph.

If S is the current of integration on an analytic set Y, by Thie's theorem, $\nu(S,a)$ is equal to the multiplicity of Y at a which is an integer. This implies the following Lelong's inequality: *the Euclidean $2(k-p)$-dimensional volume of Y in a ball $B_a(r)$ centered at a point $a \in Y$, is at least equal to $\frac{1}{(k-p)!}\pi^{k-p}r^{2(k-p)}$, the volume in $B_a(r)$ of a $(k-p)$-dimensional linear space passing through a.*

Positive closed currents generalize analytic sets but they are much more flexible. A remarkable fact is that the use of positive closed currents allows to construct analytic sets. The following theorem of Siu [SIU] is a beautiful application of the complex L^2 method.

Theorem A.14. *Let S be a positive closed (p,p)-current on X. Then, for $c>0$, the level set $\{\nu(S,a) \geq c\}$ of the Lelong number is an analytic set of X, of dimension $\leq k-p$. Moreover, there is a unique decomposition $S = S_1 + S_2$ where S_1 is a locally finite combination, with positive coefficients, of currents of integration on analytic sets of codimension p and S_2 is a positive closed (p,p)-current such that $\{\nu(S_2,z)>0\}$ is a finite or countable union of analytic sets of dimension $\leq k-p-1$.*

Calculus on currents is often delicate. However, the theory is well developped for positive closed $(1,1)$-currents thanks to the use of plurisubharmonic functions. Note that positive closed $(1,1)$-currents correspond to hypersurfaces (analytic sets of pure codimension 1) in complex geometry and working with (p,p)-currents, as with higher codimension analytic sets, is more difficult.

An upper semi-continuous function $u : X \to \mathbb{R} \cup \{-\infty\}$, not identically $-\infty$ on any component of X, is *plurisubharmonic* (p.s.h. for short) if it is subharmonic or identically $-\infty$ on any holomorphic disc in X. Recall that a *holomorphic disc* in X is a holomorphic map $\tau : \Delta \to X$ where Δ is the unit disc in $\mathbb{C}$. One often identifies this holomorphic disc with its image $\tau(\Delta)$. If u is p.s.h., then $u \circ \tau$ is subharmonic or identically $-\infty$ on Δ. As for subharmonic functions, we have the submean inequality: *in local coordinates, the value at a of a p.s.h. function is smaller or equal to the average of the function on a sphere centered at a.* Indeed, this average increases with the radius of the sphere. The submean inequality implies the maximum principle: *if a p.s.h. function on a connected manifold X has a maximum, it is constant.* The semi-continuity implies that p.s.h. functions are locally bounded from above. A function v is *pluriharmonic* if v and $-v$ are p.s.h. Pluriharmonic functions are locally real parts of holomorphic functions, in particular, they are real analytic. The following proposition is of constant use.

Proposition A.15. *A function $u : X \to \mathbb{R} \cup \{-\infty\}$ is p.s.h. if and only if the following conditions are satisfied*

1. *u is strongly upper semi-continuous, that is, for any subset A of full Lebesgue measure in X and for any point a in X, we have $u(a) = \limsup u(z)$ when $z \to a$ and $z \in A$.*
2. *u is locally integrable with respect to the Lebesgue measure on X and $dd^c u$ is a positive closed $(1,1)$-current.*

Conversely, any positive closed $(1,1)$-current can be locally written as $dd^c u$ where u is a (local) p.s.h. function. This function is called *a local potential* of the current. Two local potentials differ by a pluriharmonic function. So, there is almost a correspondence between positive closed $(1,1)$-currents and p.s.h. functions. We say that u is *strictly p.s.h.* if $dd^c u$ is strictly positive. The p.s.h. functions are defined at every point; this is a crucial property in pluripotential theory. Other important properties of this class of functions are some strong compactness properties that we state below.

If S is a positive closed (p,p)-current, one can write locally $S = dd^c U$ with U a $(p-1,p-1)$-current. We can choose the potential U negative with good estimates on the mass but the difference of two potentials may be very singular. The use of potentials U is much more delicate than in the bidegree $(1,1)$ case. We state here a useful local estimate, see e.g. [DN].

Proposition A.16. *Let V be convex open domain in $\mathbb{C}^k$ and W an open set with $W \Subset V$. Let S be a positive closed (p,p)-current on V. Then there is a negative L^1 form U of bidegree $(p-1,p-1)$ on W such that $dd^c U = S$ and $\|U\|_{L^1(W)} \leq c\|S\|_V$ where $c > 0$ is a constant independent of S. Moreover, U depends continuously on S, where the continuity is with respect to the weak topology on S and the $L^1(W)$ topology on U.*

Note that when $p = 1$, U is equal almost everywhere to a p.s.h. function u such that $dd^c u = S$.

Example A.17. Let f be a holomorphic function on X not identically 0 on any component of X. Then, $\log|f|$ is a p.s.h. function and we have $dd^c \log|f| = \sum n_i [Z_i]$ where Z_i are irreducible components of the hypersurface $\{f = 0\}$ and n_i their multiplicities. The last equation is called *Poincaré-Lelong equation*. Locally, the ideal of holomorphic functions vanishing on Z_i is generated by a holomorphic function g_i and f is equal to the product of $\prod g_i^{n_i}$ with a non-vanishing holomorphic function. In some sense, $\log|f|$ is one of the most singular p.s.h. functions. If X is a ball, the convex set generated by such functions is dense in the cone of p.s.h. functions [HO, GU] for the L^1_{loc} topology. If $f_1, \ldots, f_n$ are holomorphic on X, not identically 0 on a component of X, then $\log(|f_1|^2 + \cdots + |f_n|^2)$ is also a p.s.h. function.

The following proposition is useful in constructing p.s.h. functions.

Proposition A.18. *Let χ be a function defined on $(\mathbb{R} \cup \{-\infty\})^n$ with values in $\mathbb{R} \cup \{-\infty\}$, not identically $-\infty$, which is convex in all variables and increasing in* *each variable. Let $u_1,\dots,u_n$ be p.s.h. functions on X. Then $\chi(u_1,\dots,u_n)$ is p.s.h. In particular, the function* $\max(u_1,\dots,u_n)$ *is p.s.h.*

We call *complete pluripolar set* the pole set $\{u=-\infty\}$ of a p.s.h. function and *pluripolar set* a subset of a complete pluripolar one. Pluripolar sets are of Hausdorff dimension $\leq 2k-2$, in particular, they have zero Lebesgue measure. Finite and countable unions of (locally) pluripolar sets are (locally) pluripolar. In particular, finite and countable unions of analytic subsets are locally pluripolar.

Proposition A.19. *Let E be a closed pluripolar set in X and u a p.s.h. function on $X\setminus E$, locally bounded above near E. Then the extension of u to X given by*

$$u(z) := \limsup_{\substack{w\to z\\ w\in X\setminus E}} u(w) \quad \textit{for} \quad z\in E,$$

is a p.s.h. function.

The following result describes compactness properties of p.s.h. functions, see [HO].

Proposition A.20. *Let (u_n) be a sequence of p.s.h. functions on X, locally bounded from above. Then either it converges locally uniformly to $-\infty$ on a component of X or there is a subsequence (u_{n_i}) which converges in $L^p_{loc}(X)$ to a p.s.h. function u for every p with $1\leq p<\infty$. In the second case, we have $\limsup u_{n_i}\leq u$ with equality outside a pluripolar set. Moreover, if K is a compact subset of X and if h is a continuous function on K such that $u<h$ on K, then $u_{n_i}<h$ on K for i large enough.*

The last assertion is the classical Hartogs' lemma. It suggests the following notion of convergence introduced in [DS10]. Let (u_n) be a sequence of p.s.h. functions converging to a p.s.h. function u in $L^1_{loc}(X)$. We say that the sequence (u_n) *converges in the Hartogs' sense* or *is H-convergent* if for any compact subset K of X there are constants c_n converging to 0 such that $u_n+c_n\geq u$ on K. In this case, Hartogs' lemma implies that u_n converge pointwise to u. If (u_n) decreases to a function u, not identically $-\infty$, then u is p.s.h. and (u_n) converges in the Hartogs' sense. The following result is useful in the calculus with p.s.h. functions.

Proposition A.21. *Let u be a p.s.h. function on an open subset D of $\mathbb{C}^k$. Let $D'\Subset D$ be an open set. Then, there is a sequence of smooth p.s.h. functions u_n on D' which decreases to u.*

The functions u_n can be obtained as the standard convolution of u with some radial function ρ_n on $\mathbb{C}^k$. The submean inequality for u allows to choose ρ_n so that u_n decrease to u.

The following result, see [HO2], may be considered as the strongest compactness property for p.s.h. functions. The proof can be reduced to the one dimensional case by slicing.

Theorem A.22. *Let $\mathscr{F}$ be a family of p.s.h. functions on X which is bounded in $L^1_{loc}(X)$. Let K be a compact subset of X. Then there are constants $\alpha > 0$ and $A > 0$ such that*

$$\|e^{-\alpha u}\|_{L^1(K)} \leq A$$

for every function u in $\mathscr{F}$.

P.s.h. functions are in general unbounded. However, the last result shows that such functions are nearly bounded. The above family $\mathscr{F}$ is uniformly bounded from above on K. So, we also have the estimate

$$\|e^{\alpha|u|}\|_{L^1(K)} \leq A$$

for u in $\mathscr{F}$ and for some (other) constants α, A. More precise estimates can be obtained in terms of the maximal Lelong number of $dd^c u$ in a neighbourhood of K.

Define the *Lelong number* $\nu(u,a)$ of u at a as the Lelong number of $dd^c u$ at a. The following result describes the relation with the singularity of p.s.h. functions near a pole. We fix here a local coordinate system for X.

Proposition A.23. *The Lelong number $\nu(u,a)$ is the supremum of the number ν such that the inequality $u(z) \leq \nu \log \|z-a\|$ holds in a neighbourhood of a.*

If S is a positive closed (p,p)-current, the Lelong number $\nu(S,a)$ can be computed as the mass at a of the measure $S \wedge (dd^c \log \|z-a\|)^{k-p}$. This property allows to prove the following result, due to Demailly [DEM], which is useful in dynamics.

Proposition A.24. *Let $\tau : (\mathbb{C}^k, 0) \to (\mathbb{C}^k, 0)$ be a germ of an open holomorphic map with $\tau(0) = 0$. Let d denote the multiplicity of τ at 0. Let S be a positive closed (p,p)-current on a neighbourhood of 0. Then, the Lelong number of $\tau_*(S)$ at 0 satisfies the inequalities*

$$\nu(S,0) \leq \nu(\tau_*(S),0) \leq d^{k-p}\nu(S,0).$$

In particular, we have $\nu(\tau_*(S),0) = 0$ if and only if $\nu(S,0) = 0$.

Assume now that X is a compact Kähler manifold and ω is a Kähler form on X. If S is a dd^c-closed (p,p)-current, we can, using the dd^c-lemma, define a linear form on $H^{k-p,k-p}(X,\mathbb{C})$ by $[\alpha] \mapsto \langle S, \alpha \rangle$. Therefore, the Poincaré duality implies that S is canonically associated to a class $[S]$ in $H^{p,p}(X,\mathbb{C})$. If S is real then $[S]$ is in $H^{p,p}(X,\mathbb{R})$. If S is positive, its mass $\langle S, \omega^{k-p} \rangle$ depends only on the class $[S]$. So, the mass of positive dd^c-closed currents can be computed cohomologically. In $\mathbb{P}^k$, the mass of ω^p_{FS} is 1 since ω^k_{FS} is a probability measure. If H is a subspace of codimension p of $\mathbb{P}^k$, then the current associated to H is of mass 1 and it belongs to the class $[\omega^p_{\mathrm{FS}}]$. If Y is an analytic set of pure codimension p of $\mathbb{P}^k$, *the degree* $\deg(Y)$ of Y is by definition the number of points in its intersection with a generic projective space of dimension p. One can check that the cohomology class of Y is $\deg(Y)[\omega^p_{\mathrm{FS}}]$. The volume of Y, obtained using Wirtinger's theorem A.2, is equal to $\frac{1}{p!}\deg(Y)$.

Exercise A.25. With the notation of Exercise A.11, show that $\tau_*(\varphi)$ is p.s.h. if φ is p.s.h.

Exercise A.26. Using that $\nu(S,a,r)$ is decreasing, show that if (S_n) is a sequence of positive closed (p,p)-currents on X converging to a current S and (a_n) is a sequence in X converging to a, then $\limsup \nu(S_n,a_n) \leq \nu(S,a)$.

Exercise A.27. Let S and S' be positive closed $(1,1)$-currents such that $S' \leq S$. Assume that the local potentials of S are bounded or continuous. Show that the local potentials of S' are also bounded or continuous.

Exercise A.28. Let $\mathscr{F}$ be an L^1_{loc} bounded family of p.s.h. functions on X. Let K be a compact subset of X. Show that $\mathscr{F}$ is locally bounded from above and that there is $c > 0$ such that $\|dd^c u\|_K \leq c$ for every $u \in \mathscr{F}$. Prove that there is a constant $\nu > 0$ such that $\nu(u,a) \leq \nu$ for $u \in \mathscr{F}$ and $a \in K$.

Exercise A.29. Let Y_i, $1 \leq i \leq m$, be analytic sets of pure codimension p_i in $\mathbb{P}^k$. Assume $p_1 + \cdots + p_m \leq k$. Show that the intersection of the Y_i's is a non-empty analytic set of dimension $\geq k - p_1 - \cdots - p_m$.

A.3 Intersection, Pull-back and Slicing

We have seen that positive closed currents generalize differential forms and analytic sets. However, it is not always possible to extend the calculus on forms or on analytic sets to currents. We will give here some results which show how positive closed currents are flexible and how they are rigid.

The theory of intersection is much more developed in bidegree $(1,1)$ thanks to the use of their potentials which are p.s.h. functions. The case of continuous potentials was considered by Chern-Levine-Nirenberg [CLN]. Bedford-Taylor [BD] developed a nice theory when the potentials are locally bounded. The case of unbounded potentials was considered by Demailly [DE] and Fornæss-Sibony [FS2, S2]. We have the following general definition.

Let S be a positive closed (p,p)-current on X with $p \leq k-1$. If ω is a fixed Hermitian form on X as above, then $S \wedge \omega^{k-p}$ is a positive measure which is called *the trace measure* of S. In local coordinates, the coefficients of S are measures, bounded by a constant times the trace measure. Now, if u is a p.s.h function on X, locally integrable with respect to the trace measure of S, then uS is a current on X and we can define

$$dd^c u \wedge S := dd^c(uS).$$

Since u can be locally approximated by decreasing sequences of smooth p.s.h. functions, it is easy to check that the previous intersection is a positive closed $(p+1,p+1)$-current with support contained in $\mathrm{supp}(S)$. When u is pluriharmonic, $dd^c u \wedge S$ vanishes identically. So, the intersection depends only on $dd^c u$ and on S. If R is a positive closed $(1,1)$-current on X, one defines $R \wedge S$ as above using local potentials of R. In general, $dd^c u \wedge S$ does not depend continuously on u and S. The following proposition is a consequence of Hartogs' lemma.

Proposition A.30. *Let $u^{(n)}$ be p.s.h. functions on X which converge in the Hartogs' sense to a p.s.h. function u. If u is locally integrable with respect to the trace measure of S, then $dd^c u^{(n)} \wedge S$ are well-defined and converge to $dd^c u \wedge S$. If u is continuous and S_n are positive closed $(1,1)$-currents converging to S, then $dd^c u^{(n)} \wedge S_n$ converge to $dd^c u \wedge S$.*

If $u_1, \ldots, u_q$, with $q \leq k-p$, are p.s.h. functions, we can define by induction the wedge-product

$$dd^c u_1 \wedge \ldots \wedge dd^c u_q \wedge S$$

when some integrability conditions are satisfied, for example when the u_i are locally bounded. In particular, if $u_j^{(n)}$, $1 \leq j \leq q$, are continuous p.s.h. functions converging locally uniformly to continuous p.s.h. functions u_j and if S_n are positive closed converging to S, then

$$dd^c u_1^{(n)} \wedge \ldots \wedge dd^c u_q^{(n)} \wedge S_n \to dd^c u_1 \wedge \ldots \wedge dd^c u_q \wedge S$$

The following version of the Chern-Levine-Nirenberg inequality is a very useful result [CLN, DEM].

Theorem A.31. *Let S be a positive closed (p,p)-current on X. Let $u_1, \ldots, u_q$, $q \leq k-p$, be locally bounded p.s.h. functions on X and K a compact subset of X. Then there is a constant $c > 0$ depending only on K and X such that if v is p.s.h. on X then*

$$\|v dd^c u_1 \wedge \ldots \wedge dd^c u_q \wedge S\|_K \leq c \|v\|_{L^1(\sigma_S)} \|u_1\|_{L^\infty(X)} \cdots \|u_q\|_{L^\infty(X)},$$

where σ_S denotes the trace measure of S.

This inequality implies that p.s.h. functions are locally integrable with respect to the current $dd^c u_1 \wedge \ldots \wedge dd^c u_q$. We deduce the following corollary.

Corollary A.32. *Let $u_1, \ldots, u_p$, $p \leq k$, be locally bounded p.s.h. functions on X. Then, the current $dd^c u_1 \wedge \ldots \wedge dd^c u_p$ has no mass on locally pluripolar sets, in particular on proper analytic sets of X.*

We give now two other regularity properties of the wedge-product of currents with Hölder continuous local potentials.

Proposition A.33. *Let S be a positive closed (p,p)-current on X and q a positive integer such that $q \leq k-p$. Let u_i be Hölder continuous p.s.h. functions of Hölder exponents α_i with $0 < \alpha_i \leq 1$ and $1 \leq i \leq q$. Then, the current $dd^c u_1 \wedge \ldots \wedge dd^c u_q \wedge S$ has no mass on Borel sets with Hausdorff dimension less than or equal to $2(k-p-q) + \alpha_1 + \cdots + \alpha_q$.*

The proof of this result is given in [S3]. It is based on a mass estimate on a ball in term of the radius which is a consequence of the Chern-Levine-Nirenberg inequality.

We say that a positive measure ν in X is *locally moderate* if for any compact subset K of X and any compact family $\mathscr{F}$ of p.s.h. functions in a neighbourhood of K, there are positive constants α and c such that

$$\int_K e^{-\alpha u} d\nu \leq c$$

for u in $\mathscr{F}$. This notion was introduced in [DS1]. We say that a positive current is *locally moderate* if its trace measure is locally moderate. The following result was obtained in [DNS].

Theorem A.34. *Let S be a positive closed (p,p)-current on X and u a p.s.h. function on X. Assume that S is locally moderate and u is Hölder continuous. Then the current $dd^c u \wedge S$ is locally moderate. In particular, wedge-products of positive closed $(1,1)$-currents with Hölder continuous local potentials are locally moderate.*

Theorem A.22 implies that a measure defined by a smooth form is locally moderate. Theorem A.34 implies, by induction, that $dd^c u_1 \wedge \ldots \wedge dd^c u_p$ is locally moderate when the p.s.h. functions u_j are Hölder continuous. So, using p.s.h. functions as test functions, the previous currents satisfy similar estimates as smooth forms do. One may also consider that Theorem A.34 strengthens A.22 and gives a strong compactness property for p.s.h. functions. The estimate has many consequences in complex dynamics.

The proof of Theorem A.34 is based on a mass estimate of $dd^c u \wedge S$ on the sub-level set $\{v < -M\}$ of a p.s.h function v. Some estimates are easily obtained for u continuous using the Chern-Levine-Nirenberg inequality or for u of class $\mathscr{C}^2$. The case of Hölder continuous function uses arguments close to the interpolation between the Banach spaces $\mathscr{C}^0$ and $\mathscr{C}^2$. However, the non-linearity of the estimate and the positivity of currents make the problem more subtle.

We discuss now the pull-back of currents by holomorphic maps which are not submersions. The problem can be considered as a particular case of the general intersection theory, but we will not discuss this point here. The following result was obtained in [DS8].

Theorem A.35. *Let $\tau : X' \to X$ be an open holomorphic map between complex manifolds of the same dimension. Then the pull-back operator τ^* on smooth positive closed (p,p)-forms can be extended in a canonical way to a continuous operator on positive closed (p,p)-currents S on X. If S has no mass on a Borel set $K \subset X$, then $\tau^*(S)$ has no mass on $\tau^{-1}(K)$. The result also holds for negative currents S such that $dd^c S$ is positive.*

By canonical way, we mean that the extension is functorial. More precisely, one can locally approximate S by a sequence of smooth positive closed forms. The pull-back of these forms converge to some positive closed (p,p)-current which does not depend on the chosen sequence of forms. This limit defines the pull-back current $\tau^*(S)$. The result still holds when X' is singular. In the case of bidegree $(1,1)$, we have the following result due to Méo [ME].

Proposition A.36. *Let $\tau : X' \to X$ be a holomorphic map between complex manifolds. Assume that τ is dominant, that is, the image of τ contains an open subset of X. Then the pull-back operator τ^* on smooth positive closed $(1,1)$-forms can be extended in a canonical way to a continuous operator on positive closed $(1,1)$-currents S on X.*

Indeed, locally we can write $S = dd^c u$ with u p.s.h. The current $\tau^*(S)$ is then defined by $\tau^*(S) := dd^c(u \circ \tau)$. One can check that the definition does not depend on the choice of u.

The remaining part of this paragraph deals with the slicing of currents. We only consider a situation used in this course. Let $\pi : X \to V$ be a dominant holomorphic map from X to a manifold V of dimension l and S a current on X. Slicing theory allows to define the slice $\langle S, \pi, \theta \rangle$ of some currents S on X by the fiber $\pi^{-1}(\theta)$. Slicing theory generalizes the restriction of forms to fibers. One can also consider it as a generalization of Sard's and Fubini's theorems for currents or as a special case of intersection theory: the slice $\langle S, \pi, \theta \rangle$ can be seen as the wedge-product of S with the current of integration on $\pi^{-1}(\theta)$. We can consider the slicing of $\mathbb{C}$-flat currents, in particular, of (p,p)-currents such that S and $dd^c S$ are of order 0. The operation preserves positivity and commutes with ∂, $\overline{\partial}$. If φ is a smooth form on X then $\langle S \wedge \varphi, \pi, \theta \rangle = \langle S, \pi, \theta \rangle \wedge \varphi$. Here, we only consider positive closed $(k-l, k-l)$-currents S. In this case, the slices $\langle S, \pi, \theta \rangle$ are positive measures on X with support in $\pi^{-1}(\theta)$.

Let y denote the coordinates in a chart of V and $\lambda_V := (dd^c \|y\|^2)^l$ the Euclidean volume form associated to y. Let $\psi(y)$ be a positive smooth function with compact support such that $\int \psi \lambda_V = 1$. Define $\psi_\varepsilon(y) := \varepsilon^{-2l} \psi(\varepsilon^{-1} y)$ and $\psi_{\theta,\varepsilon}(y) := \psi_\varepsilon(y - \theta)$. The measures $\psi_{\theta,\varepsilon} \lambda_V$ approximate the Dirac mass at θ. For every smooth test function Φ on X, we have

$$\langle S, \pi, \theta \rangle (\Phi) = \lim_{\varepsilon \to 0} \langle S \wedge \pi^*(\psi_{\theta,\varepsilon} \lambda_V), \Phi \rangle$$

when $\langle S, \pi, \theta \rangle$ exists. This property holds for all choice of ψ. Conversely, when the previous limit exists and is independent of ψ, it defines the measure $\langle S, \pi, \theta \rangle$ and we say that $\langle S, \pi, \theta \rangle$ *is well-defined*. The slice $\langle S, \pi, \theta \rangle$ is well-defined for θ out of a set of Lebesgue measure zero in V and the following formula holds for smooth forms Ω of maximal degree with compact support in V:

$$\int_{\theta \in V} \langle S, \pi, \theta \rangle (\Phi) \Omega(\theta) = \langle S \wedge \pi^*(\Omega), \Phi \rangle.$$

We recall the following result which was obtained in [DS7].

Theorem A.37. *Let V be a complex manifold of dimension l and let π denote the canonical projection from $\mathbb{C}^k \times V$ onto V. Let S be a positive closed current of bidimension (l,l) on $\mathbb{C}^k \times V$, supported on $K \times V$ for a given compact subset K*

of $\mathbb{C}^k$. Then the slice $\langle S,\pi,\theta\rangle$ is well-defined for every θ in V and is a positive measure whose mass is independent of θ. Moreover, if Φ is a p.s.h. function in a neighbourhood of $\mathrm{supp}(S)$*, then the function $\theta\mapsto\langle S,\pi,\theta\rangle(\Phi)$ is p.s.h.*

The mass of $\langle S,\pi,\theta\rangle$ is called the *slice mass* of S. The set of currents S as above with bounded slice mass is compact for the weak topology on currents. In particular, their masses are locally uniformly bounded on $\mathbb{C}^k\times V$. In general, the slice $\langle S,\pi,\theta\rangle$ does not depend continuously on S nor on θ. The last property in Theorem A.37 shows that the dependence on θ satisfies a semi-continuity property. More generally, we have that $(\theta,S)\mapsto\langle S,\pi,\theta\rangle(\Phi)$ is upper semi-continuous for Φ p.s.h. We deduce easily from the above definition that the slice mass of S depends continuously on S.

Exercise A.38. Let X,X' be complex manifolds. Let ν be a positive measure with compact support on X such that p.s.h. functions on X are ν-integrable. If u is a p.s.h. function on $X\times X'$, show that $x'\mapsto\int u(x,x')d\nu(x)$ is a p.s.h function on X'. Show that if ν,ν' are positive measures on X,X' which are locally moderate, then $\nu\otimes\nu'$ is a locally moderate measure on $X\times X'$.

Exercise A.39. Let S be a positive closed $(1,1)$-current on the unit ball of $\mathbb{C}^k$. Let $\pi:\widehat{\mathbb{C}^k}\to\mathbb{C}^k$ be the blow-up of $\mathbb{C}^k$ at 0 and E the exceptional set. Show $\pi^*(S)$ is equal to $\nu[E]+S'$, where ν is the Lelong number of S at 0 and S' is a current without mass on E.

A.4 Currents on Projective Spaces

In this paragraph, we will introduce quasi-potentials of currents, the spaces of d.s.h. functions, DSH currents and the complex Sobolev space which are used as observables in complex dynamics. We also introduce PB, PC currents and the notion of super-potentials which are crucial in the calculus with currents in higher bidegree.

Recall that the Fubini-Study form ω_{FS} on $\mathbb{P}^k$ satisfies $\int_{\mathbb{P}^k}\omega_{\mathrm{FS}}^k=1$. If S is a positive closed (p,p)-current, the mass of S is given by by $\|S\|:=\langle S,\omega_{\mathrm{FS}}^{k-p}\rangle$. Since $H^{p,p}(\mathbb{P}^k,\mathbb{R})$ is generated by ω_{FS}^p, such a current S is cohomologous to $c\omega_{\mathrm{FS}}^p$ where c is the mass of S. So, $S-c\omega_{\mathrm{FS}}^p$ is exact and the dd^c-lemma, which also holds for currents, implies that there exists a $(p-1,p-1)$-current U, such that $S=c\omega_{\mathrm{FS}}^p+dd^cU$. We call U *a quasi-potential* of S. We have in fact the following more precise result [DS10].

Theorem A.40. *Let S be a positive closed (p,p)-current of mass 1 in $\mathbb{P}^k$. Then, there is a negative form U such that $dd^cU=S-\omega_{\mathrm{FS}}^p$. For r,s with $1\leq r<k/(k-1)$ and $1\leq s<2k/(2k-1)$, we have*

$$\|U\|_{L^r}\leq c_r \quad \textit{and} \quad \|\nabla U\|_{L^s}\leq c_s,$$

where c_r,c_s are constants independent of S. Moreover, U depends linearly and continuously on S with respect to the weak topology on the currents S and the L^r topology on U.

The construction of U uses a kernel constructed in Bost-Gillet-Soulé [BGS]. We call U *the Green quasi-potential* of S. When $p = 1$, two quasi-potentials of S differ by a constant. So, the solution is unique if we require that $\langle \omega_{\text{FS}}^k, U\rangle = 0$. In this case, we have a bijective and bi-continuous correspondence $S \leftrightarrow u$ between positive closed $(1,1)$-currents S and their normalized quasi-potentials u.

By maximum principle, p.s.h. functions on a compact manifold are constant. However, the interest of p.s.h. functions is their type of local singularities. S.T. Yau introduced in [YA] the useful notion of quasi-p.s.h. functions. A *quasi-p.s.h. function* is locally the difference of a p.s.h. function and a smooth one. Several properties of quasi-p.s.h. functions can be deduced from properties of p.s.h. functions. If u is a quasi-p.s.h. function on $\mathbb{P}^k$ there is a constant $c > 0$ such that $dd^c u \geq -c\omega_{\text{FS}}$. So, $dd^c u$ is the difference of a positive closed $(1,1)$-current and a smooth positive closed $(1,1)$-form: $dd^c u = (dd^c u + c\omega_{\text{FS}}) - c\omega_{\text{FS}}$. Conversely, if S is a positive closed $(1,1)$-current cohomologous to a real $(1,1)$-form α, there is a quasi-p.s.h. function u, unique up to a constant, such that $dd^c u = S - \alpha$. The following proposition is easily obtained using a convolution on the group of automorphisms of $\mathbb{P}^k$, see Demailly [DEM] for analogous results on compact Kähler manifolds.

Proposition A.41. *Let u be a quasi-p.s.h. function on $\mathbb{P}^k$ such that $dd^c u \geq -\omega_{\text{FS}}$. Then, there is a sequence (u_n) of smooth quasi-p.s.h. functions decreasing to u such that $dd^c u_n \geq -\omega_{\text{FS}}$. In particular, if S is a positive closed $(1,1)$-current on $\mathbb{P}^k$, then there are smooth positive closed $(1,1)$-forms S_n converging to S.*

A subset E of $\mathbb{P}^k$ is *pluripolar* if it is contained in $\{u = -\infty\}$ where u is a quasi-p.s.h. function. It is *complete pluripolar* if there is a quasi-p.s.h. function u such that $E = \{u = -\infty\}$. It is easy to check that analytic sets are complete pluripolar and that a countable union of pluripolar sets is pluripolar. The following capacity is close to a notion of capacity introduced by H. Alexander in [AL]. The interesting point here is that our definition extends to general compact Kähler manifold [DS6]. We will see that the same idea allows to define the capacity of a current. Let $\mathscr{P}_1$ denote the set of quasi-p.s.h. functions u on $\mathbb{P}^k$ such that $\max_{\mathbb{P}^k} u = 0$. *The capacity* of a Borel set E in $\mathbb{P}^k$ is

$$\text{cap}(E) := \inf_{u \in \mathscr{P}_1} \exp\Big(\sup_E u\Big).$$

The Borel set E is pluripolar if and only if $\text{cap}(E) = 0$. It is not difficult to show that when the volume of E tends to the volume of $\mathbb{P}^k$ then $\text{cap}(E)$ tends to 1.

The space of d.s.h. functions (differences of quasi-p.s.h. functions) and the complex Sobolev space of functions on compact Kähler manifolds were introduced by the authors in [DS6, DS11]. They satisfy strong compactness properties and are invariant under the action of holomorphic maps. Using them as test functions, permits to obtain several results in complex dynamics.

A function on $\mathbb{P}^k$ is called *d.s.h.* if it is equal outside a pluripolar set to the difference of two quasi-p.s.h. functions. We identify two d.s.h. functions if they are equal outside a pluripolar set. Let $\text{DSH}(\mathbb{P}^k)$ denote the space of d.s.h. functions on $\mathbb{P}^k$. We deduce easily from properties of p.s.h. functions that $\text{DSH}(\mathbb{P}^k)$ is contained in $L^p(\mathbb{P}^k)$ for $1 \leq p < \infty$. If u is d.s.h. then $dd^c u$ can be written as the difference of two positive closed $(1,1)$-currents which are cohomologous. Conversely, if $S^\pm$ are positive closed $(1,1)$-currents of the same mass, then there is a d.s.h. function u, unique up to a constant, such that $dd^c u = S^+ - S^-$.

We introduce several equivalent norms on $\mathrm{DSH}(\mathbb{P}^k)$. Define

$$\|u\|_{\mathrm{DSH}} := |\langle \omega_{\mathrm{FS}}^k, u\rangle| + \min \|S^\pm\|,$$

where the minimum is taken over positive closed $(1,1)$-currents $S^\pm$ such that $dd^c u = S^+ - S^-$. The term $|\langle \omega_{\mathrm{FS}}^k, u\rangle|$ may be replaced with $\|u\|_{L^p}$, $1 \leq p < \infty$; we then obtain equivalent norms. The space of d.s.h. functions endowed with the above norm is a Banach space. However, we will use on this space *a weaker topology*: we say that a sequence (u_n) converges to u in $\mathrm{DSH}(\mathbb{P}^k)$ if u_n converge to u in the sense of currents and if (u_n) is bounded with respect to $\|\cdot\|_{\mathrm{DSH}}$. Under the last condition on the DSH-norm, the convergence in the sense of currents of u_n is equivalent to the convergence in L^p for $1 \leq p < \infty$. We have the following proposition [DS6].

Proposition A.42. *Let u be a d.s.h. function on $\mathbb{P}^k$ such that $\|u\|_{\mathrm{DSH}} \leq 1$. Then there are negative quasi-p.s.h. function $u^\pm$ such that $u = u^+ - u^-$, $\|u^\pm\|_{\mathrm{DSH}} \leq c$ and $dd^c u^\pm \geq -c\omega_{\mathrm{FS}}$, where $c > 0$ is a constant independent of u.*

A positive measure on $\mathbb{P}^k$ is said to be *PC* [10] if it can be extended to a continuous linear form on $\mathrm{DSH}(\mathbb{P}^k)$. Here, the continuity is with respect to the weak topology on d.s.h. functions. A positive measure is *PB* [11] if quasi-p.s.h. functions are integrable with respect to this measure. PB measures have no mass on pluripolar sets and d.s.h. functions are integrable with respect to such measures. PC measures are always PB. Let μ be a non-zero PB positive measure on X. Define

$$\|u\|_\mu := |\langle \mu, u\rangle| + \min \|S^\pm\|,$$

with $S^\pm$ as above. We have the following useful property [DS6].

Proposition A.43. *The semi-norm $\|\cdot\|_\mu$ is in fact a norm on $\mathrm{DSH}(\mathbb{P}^k)$ which is equivalent to $\|\cdot\|_{\mathrm{DSH}}$.*

One can extend the above notions to currents but the definitions are slightly different. Let $\mathrm{DSH}^p(\mathbb{P}^k)$ denote the space generated by negative (p,p)-currents Φ such that $dd^c\Phi$ is the difference of two positive closed $(p+1,p+1)$-currents. A DSH (p,p)-current, i.e. a current in $\mathrm{DSH}^p(\mathbb{P}^k)$, is not an L^1 form in general. Define the $\|\Phi\|_{\mathrm{DSH}}$-norm of a negative current Φ in $\mathrm{DSH}^p(\mathbb{P}^k)$ by

$$\|\Phi\|_{\mathrm{DSH}} := \|\Phi\| + \min \|\Omega^\pm\|,$$

[10] In dimension 1, the measure is PC if and only if its local Potentials are Continuous.

[11] In dimension 1, the measure is PB if and only if its local Potentials are Bounded.

where $\Omega^\pm$ are positive closed such that $dd^c\Phi = \Omega^+ - \Omega^-$. For a general Φ in $\mathrm{DSH}^p(\mathbb{P}^k)$ define

$$\|\Phi\|_{\mathrm{DSH}} := \min(\|\Phi^+\|_{\mathrm{DSH}} + \|\Phi^-\|_{\mathrm{DSH}}),$$

where $\Phi^\pm$ are negative currents in $\mathrm{DSH}^p(X)$ such that $\Phi = \Phi^+ - \Phi^-$. We also consider on this space *the weak topology*: a sequence (Φ_n) converges to Φ in $\mathrm{DSH}^p(\mathbb{P}^k)$ if it converges to Φ in the sense of currents and if $(\|\Phi_n\|_{\mathrm{DSH}})$ is bounded. Using a convolution on the group of automorphisms of $\mathbb{P}^k$, we can show that smooth forms are dense in $\mathrm{DSH}^p(\mathbb{P}^k)$.

A positive closed (p,p)-current S is called *PB* if there is a constant $c>0$ such that $|\langle S,\Phi\rangle| \leq c\|\Phi\|_{\mathrm{DSH}}$ for any real smooth $(k-p,k-p)$-form Φ. The current S is *PC* if it can be extended to a linear continuous form on $\mathrm{DSH}^{k-p}(\mathbb{P}^k)$. The continuity is with respect to the weak topology we consider on $\mathrm{DSH}^{k-p}(\mathbb{P}^k)$. PC currents are PB. We will see that these notions correspond to currents with bounded or continuous super-potentials. As a consequence of Theorem A.35, we have the following useful result.

Proposition A.44. *Let $f:\mathbb{P}^k\to\mathbb{P}^k$ be a holomorphic surjective map. Then, the operator f^* on smooth forms has a continuous extension $f^*:\mathrm{DSH}^p(\mathbb{P}^k)\to\mathrm{DSH}^p(\mathbb{P}^k)$. If S is a current on $\mathrm{DSH}^p(\mathbb{P}^k)$ with no mass on a Borel set A, then $f^*(S)$ has no mass on $f^{-1}(A)$.*

Another useful functional space is *the complex Sobolev space* $W^*(\mathbb{P}^k)$. Its definition uses the complex structure of $\mathbb{P}^k$. In dimension one, $W^*(\mathbb{P}^1)$ coincides with the Sobolev space $W^{1,2}(\mathbb{P}^1)$ of real-valued functions in L^2 with gradient in L^2. In higher dimension, $W^*(\mathbb{P}^k)$ is the space of functions u in $W^{1,2}(\mathbb{P}^k)$ such that $i\partial u\wedge\overline{\partial}u$ is bounded by a positive closed $(1,1)$-current Θ. We define

$$\|u\|_{W^*} := |\langle\omega_{\mathrm{FS}}^k,u\rangle| + \min\|\Theta\|^{1/2}$$

with Θ as above, see [DS11, V3]. By Sobolev-Poincaré inequality, the term $|\langle\omega_{\mathrm{FS}}^k,u\rangle|$ may be replaced with $\|u\|_{L^1}$ or $\|u\|_{L^2}$; we then obtain equivalent norms. *The weak topology* on $W^*(\mathbb{P}^k)$ is defined as in the case of d.s.h. functions: a sequence (u_n) converges in $W^*(\mathbb{P}^k)$ to a function u if it converges to u in the sense of currents and if $(\|u_n\|_{W^*})$ is bounded. A positive measure μ is *WPC* if it can be extended to a linear continuous form on $W^*(\mathbb{P}^k)$. If u is a strictly negative quasi-p.s.h. function, one can prove that $\log(-u)$ is in $W^*(\mathbb{P}^k)$. This allows to show that WPC measures have no mass on pluripolar sets.

In the rest of the paragraph, we will introduce the notion of super-potentials associated to positive closed (p,p)-currents. They are canonical functions defined on infinite dimensional spaces and are, in some sense, quasi-p.s.h. functions there. Super-potentials were introduced by the authors in order to replace ordinary quasi-p.s.h. functions which are used as quasi-potentials for currents of bidegree $(1,1)$. The theory is satisfactory in the case of projective spaces [DS10] and can be easily extended to homogeneous manifolds.

Let $\mathscr{C}_{k-p+1}(\mathbb{P}^k)$ denote the convex set of positive closed currents of bidegree $(k-p+1,k-p+1)$ and of mass 1, i.e. currents cohomologous to $\omega_{\rm FS}^{k-p+1}$. Let S be a positive closed (p,p)-current on $\mathbb{P}^k$. We assume for simplicity that S is of mass 1; the general case can be deduced by linearity. The super-potential[12] $\mathscr{U}_S$ of S is a function on $\mathscr{C}_{k-p+1}(\mathbb{P}^k)$ with values in $\mathbb{R}\cup\{-\infty\}$. Let R be a current in $\mathscr{C}_{k-p+1}(\mathbb{P}^k)$ and U_R a potential of $R-\omega_{\rm FS}^{k-p+1}$. Subtracting from U_R a constant times $\omega_{\rm FS}^{k-p}$ allows to have $\langle U_R,\omega_{\rm FS}^p\rangle=0$. We say that U_R is a quasi-potential of mean 0 of R. Formally, i.e. in the case where R and U_R are smooth, the value of $\mathscr{U}_S$ at R is defined by

$$\mathscr{U}_S(R):=\langle S,U_R\rangle.$$

One easily check using Stokes' formula that formally if U_S is a quasi-potential of mean 0 of S, then $\mathscr{U}_S(R)=\langle U_S,R\rangle$. Therefore, the previous definition does not depend on the choice of U_R or U_S. By definition, we have $\mathscr{U}_S(\omega_{\rm FS}^{k-p+1})=0$. Observe also that when S is smooth, the above definition makes sense for every R and $\mathscr{U}_S$ is a continuous affine function on $\mathscr{C}_{k-p+1}(\mathbb{P}^k)$. It is also clear that if $\mathscr{U}_S=\mathscr{U}_{S'}$, then $S=S'$. The following theorem allows to define $\mathscr{U}_S$ in the general case.

Theorem A.45. *The above function $\mathscr{U}_S$, which is defined on smooth forms R in $\mathscr{C}_{k-p+1}(\mathbb{P}^k)$, can be extended to an affine function on $\mathscr{C}_{k-p+1}(\mathbb{P}^k)$ with values in $\mathbb{R}\cup\{-\infty\}$ by*

$$\mathscr{U}_S(R):=\limsup \mathscr{U}_S(R'),$$

where R' is smooth in $\mathscr{C}_{k-p+1}(\mathbb{P}^k)$ and converges to R. We have $\mathscr{U}_S(R)=\mathscr{U}_R(S)$. Moreover, there are smooth positive closed (p,p)-forms S_n of mass 1 and constants c_n converging to 0 such that $\mathscr{U}_{S_n}+c_n$ decrease to $\mathscr{U}_S$. In particular, $\mathscr{U}_{S_n}$ converge pointwise to $\mathscr{U}_S$.

For bidegree $(1,1)$, there is a unique quasi-p.s.h. function u_S such that $dd^cu_S=S-\omega_{\rm FS}$ and $\langle\omega_{\rm FS}^k,u_S\rangle=0$. If δ_a denotes the Dirac mass at a, we have $\mathscr{U}_S(\delta_a)=u_S(a)$. Dirac masses are extremal elements in $\mathscr{C}_k(\mathbb{P}^k)$. The super-potential $\mathscr{U}_S$ in this case is just the affine extension of u_S, that is, we have for any probability measure ν:

$$\mathscr{U}_S(\nu)=\int\mathscr{U}_S(\delta_a)d\nu(a)=\int u_S(a)d\nu(a).$$

The function $\mathscr{U}_S$ extends the action $\langle S,\Phi\rangle$ on smooth forms Φ to $\langle S,U\rangle$ where U is a quasi-potential of a positive closed current. Super-potentials satisfy analogous properties as quasi-p.s.h. functions do. They are upper semi-continuous and bounded from above by a universal constant. Note that we consider here the weak topology on $\mathscr{C}_{k-p+1}(\mathbb{P}^k)$. We have the following version of the Hartogs' lemma.

Proposition A.46. *Let S_n be positive closed (p,p)-currents of mass 1 on $\mathbb{P}^k$ converging to S. Then for every continuous function $\mathscr{U}$ on $\mathscr{C}_{k-p+1}$ with $\mathscr{U}_S<\mathscr{U}$, we have $\mathscr{U}_{S_n}<\mathscr{U}$ for n large enough. In particular,* $\limsup\mathscr{U}_{S_n}\leq\mathscr{U}_S$.

[12] The super-potential we consider here corresponds to the super-potential of mean 0 in [DS10]. The other super-potentials differ from $\mathscr{U}_S$ by constants.

We say that S_n converge to S *in the Hartogs' sense* if S_n converge to S and if there are constants c_n converging to 0 such that $\mathscr{U}_{S_n} + c_n \geq \mathscr{U}_S$. If $\mathscr{U}_{S_n}$ converge uniformly to $\mathscr{U}_S$, we say that S_n *converge SP-uniformly* to S.

One can check that PB and PC currents correspond to currents of bounded or continuous super-potential. In the case of bidegree $(1,1)$, they correspond to currents with bounded or continuous quasi-potential. We say that S' is *more diffuse than* S if $\mathscr{U}_{S'} - \mathscr{U}_S$ is bounded from below. So, PB currents are more diffuse than any other currents.

In order to prove the above results and to work with super-potentials, we have to consider a geometric structure on $\mathscr{C}_{k-p+1}(\mathbb{P}^k)$. In a weak sense, $\mathscr{C}_{k-p+1}(\mathbb{P}^k)$ can be seen as a space of infinite dimension which contains many "analytic" sets of finite dimension that we call *structural varieties*. Let V be a complex manifold and $\mathscr{R}$ a positive closed current of bidegree $(k-p+1, k-p+1)$ on $V \times \mathbb{P}^k$. Let π_V denote the canonical projection map from $V \times \mathbb{P}^k$ onto V. One can prove that the slice $\langle \mathscr{R}, \pi_V, \theta \rangle$ is defined for θ outside a locally pluripolar set of V. Each slice can be identified with a positive closed (p,p)-current R_θ on $\mathbb{P}^k$. Its mass does not depend on θ. So, multiplying $\mathscr{R}$ with a constant, we can assume that all the R_θ are in $\mathscr{C}_{k-p+1}(\mathbb{P}^k)$. The map $\tau(\theta) := R_\theta$ or the family (R_θ) is called *a structural variety* of $\mathscr{C}_{k-p+1}(\mathbb{P}^k)$. The restriction of $\mathscr{U}_S$ to this structural variety, i.e. $\mathscr{U}_S \circ \tau$, is locally a d.s.h. function or identically $-\infty$. When the structural variety is nice enough, this restriction is quasi-p.s.h. or identically $-\infty$. In practice, we often use some special structural discs parametrized by θ in the unit disc of $\mathbb{C}$. They are obtained by convolution of a given current R with a smooth probability measure on the group $\mathrm{PGL}(\mathbb{C}, k+1)$ of automorphisms of $\mathbb{P}^k$.

Observe that since the correspondence $S \leftrightarrow \mathscr{U}_S$ is $1:1$, the compactness on positive closed currents should induce some compactness on super-potentials. We have the following result.

Theorem A.47. *Let $W \subset \mathbb{P}^k$ be an open set and $K \subset W$ a compact set. Let S be a current in $\mathscr{C}_p(\mathbb{P}^k)$ with support in K and R a current in $\mathscr{C}_{k-p+1}(\mathbb{P}^k)$. Assume that the restriction of R to W is a bounded form. Then, the super-potential $\mathscr{U}_S$ of S satisfies*

$$|\mathscr{U}_S(R)| \leq A\big(1 + \log^+ \|R\|_{\infty,W}\big)$$

where $A > 0$ is a constant independent of S, R and $\log^+ := \max(0, \log)$.

This result can be applied to $K = W = \mathbb{P}^k$ and can be considered as a version of the exponential estimate in Theorem A.22. Indeed, the weaker estimate $|\mathscr{U}_S(R)| \lesssim 1 + \|R\|_\infty$ is easy to obtain. It corresponds to the L^1 estimate on the quasi-p.s.h. function u_S in the case of bidegree $(1,1)$.

Using the analogy with the bidegree $(1,1)$ case, we define *the capacity* of a current R as

$$\mathrm{cap}(R) := \inf_S \exp\big(\mathscr{U}_S(R) - \max \mathscr{U}_S\big).$$

This capacity describes rather the regularity of R: an R with big capacity is somehow more regular. Theorem A.47 implies that $\mathrm{cap}(R) \gtrsim \|R\|_\infty^{-\lambda}$ for some universal constant $\lambda > 0$. This property is close to the capacity estimate for Borel sets in term of volume.

Super-potentials allow to develop a theory of intersection of currents in higher bidegree. Here, the fact that $\mathscr{U}_S$ has a value at every point (i.e. at every current $R \in \mathscr{C}_{k-p+1}(\mathbb{P}^k)$) is crucial. Let S, S' be positive closed currents of bidegree (p,p) and (p',p') with $p+p' \leq k$. We assume for simplicity that their masses are equal to 1. We say that S and S' are *wedgeable* if $\mathscr{U}_S$ is finite at $S' \wedge \omega_{\rm FS}^{k-p-p'+1}$. This property is symmetric on S and S'. If $\widetilde{S}, \widetilde{S}'$ are more diffuse than S, S' and if S, S' are wedgeable, then $\widetilde{S}, \widetilde{S}'$ are wedgeable.

Let Φ be a real smooth form of bidegree $(k-p-p', k-p-p')$. Write $dd^c\Phi = c(\Omega^+ - \Omega^-)$ with $c \geq 0$ and $\Omega^\pm$ positive closed of mass 1. If S and S' are wedgeable, define the current $S \wedge S'$ by

$$\langle S \wedge S', \Phi \rangle := \langle S', \omega_{\rm FS}^p \wedge \Phi \rangle + c\mathscr{U}_S(S' \wedge \Omega^+) - c\mathscr{U}_S(S' \wedge \Omega^-).$$

A simple computation shows that the definition coincides with the usual wedge-product when S or S' is smooth. One can also prove that the previous definition does not depend on the choice of c, $\Omega^\pm$ and is symmetric with respect to S, S'. If S is of bidegree $(1,1)$, then S, S' are wedgeable if and only if the quasi-potentials of S are integrable with respect to the trace measure of S'. In this case, the above definition coincides with the definition in Appendix A.3. We have the following general result.

Theorem A.48. *Let S_i be positive closed currents of bidegree (p_i, p_i) on $\mathbb{P}^k$ with $1 \leq i \leq m$ and $p_1 + \cdots + p_m \leq k$. Assume that for $1 \leq i \leq m-1$, S_i and $S_{i+1} \wedge \ldots \wedge S_m$ are wedgeable. Then, this condition is symmetric on $S_1, \ldots, S_m$. The wedge-product $S_1 \wedge \ldots \wedge S_m$ is a positive closed current of mass $\|S_1\| \ldots \|S_m\|$ supported on $\mathrm{supp}(S_1) \cap \ldots \cap \mathrm{supp}(S_m)$. It depends linearly on each variable and is symmetric on the variables. If $S_i^{(n)}$ converge to S_i in the Hartogs' sense, then the $S_i^{(n)}$ are wedgeable and $S_1^{(n)} \wedge \ldots \wedge S_m^{(n)}$ converge in the Hartogs' sense to $S_1 \wedge \ldots \wedge S_m$.*

We deduce from this result that wedge-products of PB currents are PB. One can also prove that wedge-products of PC currents are PC. If S_n is defined by analytic sets, they are wedgeable if the intersection of these analytic sets is of codimension $p_1 + \cdots + p_m$. In this case, the intersection in the sense of currents coincides with the intersection of cycles, i.e. is equal to the current of integration on the intersection of cycles where we count the multiplicities. We have the following criterion of wedgeability which contains the case of cycles.

Proposition A.49. *Let S, S' be positive closed currents on $\mathbb{P}^k$ of bidegrees (p,p) and (p',p'). Let W, W' be open sets such that S restricted to W and S' restricted to W' are bounded forms. Assume that $W \cup W'$ is $(p+p')$-concave in the sense that there is a positive closed smooth form of bidegree $(k-p-p'+1, k-p-p'+1)$ with compact support in $W \cup W'$. Then S and S' are wedgeable.*

The following result can be deduced from Theorem A.48.

Corollary A.50. *Let S_i be positive closed $(1,1)$-currents on $\mathbb{P}^k$ with $1 \leq i \leq p$. Assume that for $1 \leq i \leq p-1$, S_i admits a quasi-potential which is integrable with* *respect to the trace measure of $S_{i+1}\wedge\ldots\wedge S_p$. Then, this condition is symmetric on $S_1,\ldots,S_p$. The wedge-product $S_1\wedge\ldots\wedge S_p$ is a positive closed (p,p)-current of mass $\|S_1\|\ldots\|S_p\|$ supported on $\text{supp}(S_1)\cap\ldots\cap\text{supp}(S_p)$. It depends linearly on each variable and is symmetric on the variables. If $S_i^{(n)}$ converge to S_i in the Hartogs' sense, then the $S_i^{(n)}$ are wedgeable and $S_1^{(n)}\wedge\ldots\wedge S_p^{(n)}$ converge to $S_1\wedge\ldots\wedge S_p$.*

We discuss now currents with Hölder continuous super-potential and moderate currents. The space $\mathscr{C}_{k-p+1}(\mathbb{P}^k)$ admits natural distances dist_α, with $\alpha>0$, defined by

$$\text{dist}_\alpha(R,R') := \sup_{\|\Phi\|_{\mathscr{C}^\alpha}\leq 1} |\langle R-R',\Phi\rangle|,$$

where Φ is a smooth $(p-1,p-1)$-form on $\mathbb{P}^k$. The norm $\mathscr{C}^\alpha$ on Φ is the sum of the $\mathscr{C}^\alpha$-norms of its coefficients for a fixed atlas of $\mathbb{P}^k$. The topology associated to dist_α coincides with the weak topology. Using the theory of interpolation between Banach spaces [T1], we obtain for $\beta>\alpha>0$ that

$$\text{dist}_\beta \leq \text{dist}_\alpha \leq c_{\alpha,\beta}[\text{dist}_\beta]^{\alpha/\beta}$$

where $c_{\alpha,\beta}>0$ is a constant. So, a function on $\mathscr{C}_{k-p+1}(\mathbb{P}^k)$ is Hölder continuous with respect to dist_α if and only if it is Hölder continuous with respect to dist_β. The following proposition is useful in dynamics.

Proposition A.51. *The wedge-product of positive closed currents on $\mathbb{P}^k$ with Hölder continuous super-potentials has a Hölder continuous super-potential. Let S be a positive closed (p,p)-current with a Hölder continuous super-potential. Then, the Hausdorff dimension of S is strictly larger than $2(k-p)$. Moreover, S is moderate, i.e. for any bounded family $\mathscr{F}$ of d.s.h. functions on $\mathbb{P}^k$, there are constants $c>0$ and $\alpha>0$ such that*

$$\int e^{\alpha|u|}d\sigma_S \leq c$$

for every u in $\mathscr{F}$, where σ_S is the trace measure of S.

Exercise A.52. Show that there is a constant $c>0$ such that

$$\text{cap}(E) \geq \exp(-c/\text{volume}(E)).$$

Hint: use the compactness of $\mathscr{P}_1$ in L^1.

Exercise A.53. Let (u_n) be a sequence of d.s.h. functions such that $\sum\|u_n\|_{\text{DSH}}$ is finite. Show that $\sum u_n$ converge pointwise out of a pluripolar set to a d.s.h. function. Hint: write $u_n=u_n^+-u_n^-$ with $u_n^\pm\leq 0$, $\|u_n^\pm\|_{\text{DSH}}\lesssim\|u_n\|_{\text{DSH}}$ and $dd^c u_n^\pm \geq -\|u_n\|_{\text{DSH}}\omega_{\text{FS}}$.

Exercise A.54. If χ is a convex increasing function on $\mathbb{R}$ with bounded derivative and u is a d.s.h. function, show that $\chi\circ u$ is d.s.h. If χ is Lipschitz and u is in $W^*(\mathbb{P}^k)$, show that $\chi\circ u$ is in $W^*(\mathbb{P}^k)$. Prove that bounded d.s.h. functions are in $W^*(\mathbb{P}^k)$. Show that $\text{DSH}(\mathbb{P}^k)$ and $W^*(\mathbb{P}^k)$ are stable under the max and min operations.

Exercise A.55. Let μ be a non-zero positive measure which is WPC. Define

$$\|u\|_\mu^* := |\langle \mu, u\rangle| + \min \|\Theta\|^{1/2}$$

with Θ as above. Show that $\|\cdot\|_\mu^*$ defines a norm which is equivalent to $\|\cdot\|_{W^*}$.

Exercise A.56. Show that the capacity of R is positive if and only if R is PB.

Exercise A.57. Let S be a positive closed (p,p)-current of mass 1 with positive Lelong number at a point a. Let H be a hyperplane containing a such that S and $[H]$ are wedgeable. Show that the Lelong number of $S \wedge [H]$ at a is the same if we consider it as a current on $\mathbb{P}^k$ or on H. If R is a positive closed current of bidimension $(p-1,p-1)$ on H, show that $\mathscr{U}_S(R) \leq \mathscr{U}_{S\wedge[H]}(R) + c$ where $c > 0$ is a constant independent of S, R and H. Deduce that PB currents have no positive Lelong numbers.

Exercise A.58. Let K be a compact subset in $\mathbb{C}^k \subset \mathbb{P}^k$. Let $S_1, \ldots, S_p$ be positive closed $(1,1)$-currents on $\mathbb{P}^k$. Assume that their quasi-potentials are bounded on $\mathbb{P}^k \setminus K$. Show that $S_1, \ldots, S_p$ are wedgeable. Show that the wedge-product $S_1 \wedge \ldots \wedge S_p$ is continuous for Hartogs' convergence.

Exercise A.59. Let S and S' be positive closed (p,p)-currents on $\mathbb{P}^k$ such that $S' \leq S$. Assume that S is PB (resp. PC). Show that S' is PB (resp. PC).

References

[A] Abate, M.: The residual index and the dynamics of holomorphic maps tangent to the identity. Duke Math. J. **107**(1), 173–207 (2001)

[ABT] Abate, M., Bracci, F., Tovena, F.: Index theorems for holomorphic self-maps. Ann. of Math. (2), **159**(2), 819–864 (2004)

[AL] Alexander, H.: Projective capacity. Recent developments in several complex variables. Ann. Math. Stud. **100**, Princeton Univerity Press, Princeton, N.J., 3–27 (1981)

[AC] Amerik, E., Campana, F.: Exceptional points of an endomorphism of the projective plane. Math. Z. **249**(4), 741–754 (2005)

[BA] Bassanelli, G.: A cut-off theorem for plurisubharmonic currents. Forum Math. **6**(5), 567–595 (1994)

[BB] Bassanelli, G., Berteloot, F.: Bifurcation currents in holomorphic dynamics on $\mathbb{P}^k$. J. Reinc Angew. Math. **608**, 201–235 (2007)

[BE] Beardon, A.: Iteration of rational functions. Complex analytic dynamical systems. Graduate Texts in Mathematics, **132**, Springer-Verlag, New York (1991)

[BV] Beauville, A.: Endomorphisms of hypersurfaces and other manifolds. Internat. Math. Res. Notices (1), 53–58 (2001)

[BD] Bedford, E., Taylor, B.A.: A new capacity for plurisubharmonic functions. Acta Math. **149**(1–2), 1–40 (1982)

[BC] Benedicks, M., Carleson, L.: The dynamics of the Hénon map. Ann. Math. (2), **133**(1), 73–169 (1991)

[BR] Berteloot, F., Dupont, C.: Une caractérisation des endomorphismes de Lattès par leur mesure de Green. Comment. Math. Helv. **80**(2), 433–454 (2005)

[BDM] Berteloot, F., Dupont, C., Molino, L.: Normalization of bundle holomorphic contractions and applications to dynamics. Ann. Inst. Fourier (Grenoble) **58**(6), 2137–2168 (2008)

[BL] Berteloot, F., Loeb, J.-J.: Une caractérisation géométrique des exemples de Lattès de $\mathbb{P}^k$. Bull. Soc. Math. France **129**(2), 175–188 (2001)

[BI] Binder, I., DeMarco, L.: Dimension of pluriharmonic measure and polynomial endomorphisms of $\mathbb{C}^n$. Int. Math. Res. Not. (11), 613–625 (2003)

[BS] Bishop, E.: Conditions for the analyticity of certain sets. Michigan Math. J. **11**, 289–304 (1964)

[BN] Blanchard, A.: Sur les variétés analytiques complexes. Ann. Sci. Ecole Norm. Sup. (3), **73**, 157–202 (1956)

[BO] Bonifant, A., Dabija, M., Milnor, J.: Elliptic curves as attractors in $\mathbb{P}^2$. I. Dynamics. Exp. Math. **16**(4), 385–420 (2007)

[BGS] Bost, J.-B., Gillet, H., Soulé, C.: Heights of projective varieties and positive Green forms. J. Am. Math. Soc.. **7**(4), 903–1027 (1994)

[BJ] Briend, J.-Y.: Exposants de Liapounoff et points périodiques d'endomorphismes holomorphes de $\mathbb{CP}^k$. PhD thesis (1997)

[BJ1] ——: La propriété de Bernoulli pour les endomorphismes de $\mathrm{P}^k(\mathbb{C})$. Ergod. Theor. Dyn. Syst. **22**(2), 323–327 (2002)

[BD1] Briend, J.-Y., Duval, J.: Exposants de Liapounoff et distribution des points périodiques d'un endomorphisme de $\mathbf{CP}^k$. Acta Math. **182**(2), 143–157 (1999)

[BD2] ——: Deux caractérisations de la mesure d'équilibre d'un endomorphisme de $\mathrm{P}^k(\mathbf{C})$. Publ. Math. Inst. Hautes Études Sci. **93**, 145–159 (2001)

[BD3] ——: Personal communication.

[BK] Brin, M., Katok, A.: On local entropy. Geometric Dynamics Lecture Notes in Mathematics, **1007**. Springer, Berlin, 30–38 (1983)

[BH] Brolin, H.: Invariant sets under iteration of rational functions. Ark. Mat. **6**, 103–144 (1965)

[CG] Carleson, L., Gamelin, T.W.: Complex dynamics. Universitext: Tracts in Mathematics. Springer-Verlag, New York (1993)

[CL] Cerveau, D., Lins Neto, A.: Hypersurfaces exceptionnelles des endomorphismes de $\mathrm{CP}(n)$. Bol. Soc. Brasil. Mat. (N.S.), **31**(2), 155–161 (2000)

[CH] Chemin, J.-Y.: Théorie des distributions et analyse de Fourier, cours à l'école polytechnique de Paris (2003)

[CLN] Chern, S.S., Levine, H.I., Nirenberg, L.: Intrinsic norms on a complex manifold, in Global Analysis (Papers in Honor of K. Kodaira), Univ. Tokyo Press, 119–139 (1969)

[DE] Demailly, J.-P.: Monge-Ampère Operators, Lelong numbers and Intersection theory in Complex Analysis and Geometry. In (Ancona, V., Silva, A. eds.) Plemum Press, 115–193 (1993)

[DEM] ——: Complex analytic geometry, available at `www.fourier.ujf-grenoble.fr/~demailly`

[DM] DeMarco, L.: Dynamics of rational maps: Lyapunov exponents, bifurcations, and capacity. Math. Ann. **326**(1), 43–73 (2003)

[DM1] ——: The moduli space of quadratic rational maps. J. Am. Math. Soc. **20**(2), 321–355 (2007)

[DZ] Dembo, A., Zeitouni, O.: Large deviations techniques and applications, Second edition, Applications of Mathematics (New York), **38**, Springer-Verlag, New York (1998)

[DR] de Rham, G.: Differentiable Manifolds. Forms, Currents, Harmonic Forms, **266**. Springer-Verlag, Berlin (1984)

[DT1] de Thélin, H.: Sur la laminarité de certains courants. Ann. Sci. École Norm. Sup. (4), **37**(2), 304–311 (2004)

[DT3] ——: Sur la construction de mesures selles. Ann. Inst. Fourier, **56**(2), 337–372 (2006)

[DT2] ——: Sur les exposants de Lyapounov des applications méromorphes. Invent. Math. **172**(1), 89–116 (2008)

[D4] Dinh, T.-C.: Sur les applications de Lattès de $\mathbb{P}^k$. J. Math. Pures Appl. (9), **80**(6), 577–592 (2001)

[D22] ——: Suites d'applications méromorphes multivaluées et courants laminaires. J. Geom. Anal. **15**(2), 207–227 (2005)

[D3] ——: Attracting current and equilibrium measure for attractors on $\mathbb{P}^k$. J. Geom. Anal. **17**(2), 227–244 (2007)

[D5] ——: Analytic multiplicative cocycles over holomorphic dynamical systems, special issue of Complex Variables. Complex Var. Elliptic Equ. **54**(3–4), 243–251 (2009)

[DD] Dinh, T.-C., Dupont, C.: Dimension de la mesure d'équilibre d'applications méromorphes. J. Geom. Anal. **14**(4), 613–627 (2004)

[DN] Dinh, T.-C., Nguyen, V.-A., Sibony, N.: Dynamics of horizontal-like maps in higher dimension. Adv. Math. **219**, 1689–1721 (2008)

[DNS] ——: Exponential estimates for plurisubharmonic functions and stochastic dynamics, Preprint (2008), to appear in J. Diff. Geometry (2010). `arXiv:0801.1983`

[DS14] Dinh, T.-C., Sibony, N.: Dynamique des endomorphismes holomorphes. prépublications d'Orsay, No. 2002-15 (2002)

[DS0] ——: Sur les endomorphismes holomorphes permutables de $\mathbb{P}^k$. Math. Ann. **324**(1), 33–70 (2002)

[DS1] ——: Dynamique des applications d'allure polynomiale. J. Math. Pures Appl. **82**, 367–423 (2003)

[DS4] ——: Green currents for holomorphic automorphisms of compact Kähler manifolds. J. Am. Math. Soc. **18**, 291–312 (2005)

[DS6] ——: Distribution des valeurs de transformations méromorphes et applications. Comment. Math. Helv. **81**(1), 221–258 (2006)

[DS11] ——: Decay of correlations and the central limit theorem for meromorphic maps. Comm. Pure Appl. Math. **59**(5), 754–768 (2006)

[DS7] ——: Geometry of currents, intersection theory and dynamics of horizontal-like maps. Ann. Inst. Fourier (Grenoble), **56**(2), 423–457 (2006)

[DS8] ——: Pull-back of currents by holomorphic maps. Manuscripta Math. **123**, 357–371 (2007)

[DS9] ——: Equidistribution towards the Green current for holomorphic maps. Ann. Sci. École Norm. Sup. **41**, 307–336 (2008)

[DS10] ——: Super-potentials of positive closed currents, intersection theory and dynamics. Acta Math. **203**, 1–82 (2009)

[DH] Douady, A., Hubbard, J.: On the dynamics of polynomial-like mappings. Ann. Sci. École Norm. Sup. (4), **18**(2), 287–343 (1985)

[DO] Drasin, D., Okuyama, Y.: Equidistribution and Nevanlinna theory. Bull. Lond. Math. Soc. **39**(4), 603–613 (2007)

[DF] Dujardin, R., Favre, C.: Distribution of rational maps with a preperiodic critical point. Am. Math. J. **130**(4), 979–1032 (2008)

[DP1] Dupont, C.: Exemples de Lattès et domaines faiblement sphériques de $\mathbb{C}^n$. Manuscripta Math. **111**(3), 357–378 (2003)

[DP2] ——: Bernoulli coding map and singular almost-sure invariance principle for endomorphisms of $\mathbb{P}^k$. Probability Theory & Related Field, **146**, 337–359 (2010)

[DP3] ——: On the dimension of the maximal entropy measure of endomorphisms of $\mathbb{P}^k$, preprint (2008)

[E] Eremenko, A.E.: On some functional equations connected with iteration of rational function. Leningrad. Math. J. **1**(4), 905–919 (1990)

[F] Fatou, P.: Sur l'itération analytique et les substitutions permutables. J. Math. **2**, 343 (1923)

[FJ] Favre, C., Jonsson, M.: Brolin's theorem for curves in two complex dimensions. Ann. Inst. Fourier (Grenoble), **53**(5), 1461–1501 (2003)

[FJ1] ——: Eigenvaluations. Ann. Sci. École Norm. Sup. (4), **40**(2), 309–349 (2007)

[FE] Federer, H.: Geometric Measure Theory, Die Grundlehren der mathematischen Wissenschaften, Band **153**. Springer-Verlag, New York Inc., New York (1969)

[FS] Fornæss, J.-E., Sibony, N.: Critically finite maps on $\mathbb{P}^2$. Contemp. Math. **137**, A.M.S, 245–260 (1992)

[FS1] ——: Complex dynamics in higher dimensions. Notes partially written by Estela A. Gavosto. NATO Adv. Sci. Inst. Ser. C Math. Phys. Sci. **439**, Complex potential theory (Montreal, PQ, 1993), 131–186, Kluwer Acadamic Publishing, Dordrecht (1994)

[FS7] ——: Complex dynamics in higher dimension. I., (IMPA, 1992). *Astérisque*, **222**, 201–231 (1994)

[FS3] ——: Complex dynamics in higher dimension. II. Modern methods in complex analysis (Princeton, NJ, 1992), 135–182, Ann. of Math. Stud. **137**, Princeton University Press, Princeton, NJ (1995)

[FS8] ——: Classification of recurrent domains for some holomorphic maps. Math. Ann. **301**, 813–820 (1995)

[FS2] ——: Oka's inequality for currents and applications. Math. Ann. **301**, 399–419 (1995)

[FS6] ——: Dynamics of $\mathbb{P}^2$ (examples). Laminations and foliations in dynamics, geometry and topology (Stony Brook, NY, 1998), 47–85. Contemp. Math. **269**. American Mathematical Society, Providence, RI (2001)

[FW] Fornæss, J.-E., Weickert, B.: Attractors in $\mathbb{P}^2$. Several complex variables (Berkeley, CA, 1995–1996), 297–307. Math. Sci. Res. Inst. Publ. **37**. Cambridge University Press, Cambridge (1999)

[FL] Freire, A., Lopès, A., Mañé, R.: An invariant measure for rational maps. Bol. Soc. Brasil. Mat. **14**(1), 45–62 (1983)

[G] Gordin M.I.: The central limit theorem for stationary processes. Dokl. Akad. Nauk SSSR, **188**, 739–741 (1969)

[GH] Griffiths, Ph., Harris, J.: Principles of Algebraic Geometry. Wiley, London (1994)

[GR] Gromov, M.: On the entropy of holomorphic maps. Enseign. Math. (2), **49**(3–4), 217–235 (2003). Manuscript (1977)

[G2] Guedj, V.: Equidistribution towards the Green current. Bull. Soc. Math. France, **131**(3), 359–372 (2003)

[GK] Guelfand, I.M., Kapranov, M.M., Zelevinsky, A.V.: Discriminants, Resultants, and Multidimensional Determinants. Mathematics: Theory & Applications. Birkhäuser Boston, Inc., Boston, MA (1994)

[GU] Gunning, R.C.: Introduction to holomorphic functions in several variables. Wadsworh and Brooks (1990)

[H1] Hakim, M.: Attracting domains for semi-attractive transformations of $\mathbf{C}^p$. Publ. Mat. **38**(2), 479–499 (1994)

[H2] ——: Analytic transformations of $(\mathbf{C}^p, 0)$ tangent to the identity. Duke Math. J. **92**(2), 403–428 (1998)

[H3] ——: Stable pieces in transformations tangent to the identity. Preprint Orsay (1998)

[HP] Harvey, R., Polking, J.: Extending analytic objects. Comm. Pure Appl. Math. **28**, 701–727 (1975)

[HH] Heicklen, D., Hoffman, C.: Rational maps are d-adic Bernoulli. Ann. Math., **156**, 103–114 (2002)

[HR] Hoffman, C., Rudolph, D.: Uniform endomorphisms which are isomorphic to a Bernoulli shift. Ann. Math. (2), **156**(1), 79–101 (2002)

[HO] Hörmander, L.: The Analysis of Linear Partial Differential Operators I, II. Springer-Verlag, Berlin (1983)

[HO2] ——: An Introduction to Complex Analysis in Several Variables, 3rd edn. North-Holland Mathematical Library, **7**, North-Holland Publishing Co., Amsterdam (1990)

[HJ] Hubbard, J., Papadopol, P.: Superattractive fixed points in $\mathbf{C}^n$. Indiana Univ. Math. J. **43**(1), 321–365 (1994)

[HB] Huybrechts, D.: Complex Geometry. An Introduction, Universitext. Springer-Verlag, Berlin (2005)

[JW] Jonsson, M., Weickert, B.: A nonalgebraic attractor in $\mathbf{P}^2$. Proc. Am. Math. Soc. **128**(10), 2999–3002 (2000)

[JU] Julia, G.: Mémoire sur la permutabilité des fractions rationnelles. Ann. Sci. École Norm. Sup. **39**, 131–215 (1922)

[KH] Katok, A., Hasselblatt, B.: Introduction to the modern theory of dynamical systems. Encyclopedia of Mathematics and its Applications, **54**. Cambridge University Press, Cambridge (1995)

[K] Kobayashi, S.: Hyperbolic complex spaces. Grundlehren der Mathematischen Wissenschaften, **318**. Springer-Verlag, Berlin (1998)

[KO] Kosek, M.: Hölder continuity property of filled-in Julia sets in $\mathbf{C}^n$. Proc. Am. Math. Soc. **125**(7), 2029–2032 (1997)

[KR] Krengel, U.: Ergodic Theorems, de Gruyter, Berlin, New-York (1985)

[LP] Lacey, M.T., Philipp, W.: A note on the almost sure central limit theorem. Statist. Probab. Lett. **9**(3), 201–205 (1990)

[LE] Lelong, P.: Fonctions plurisousharmoniques et formes différentielles positives, Dunod Paris (1968)

[LY] Lyubich, M.Ju.: Entropy properties of rational endomorphisms of the Riemann sphere. Ergod. Theor. Dyn. Syst. **3**(3), 351–385 (1983)

[MA] Mañé, R.: On the Bernoulli property of rational maps. Ergod. Theor. Dyn. Syst. **5**, 71–88 (1985)

[MM] McMullen, C.T.: Complex dynamics and renormalization. Annals of Mathematics Studies, **135**. Princeton University Press, Princeton, NJ (1994)

[ME] Méo, M.: Image inverse d'un courant positif fermé par une application surjective. C.R.A.S. **322**, 1141–1144 (1996)

[ME3] ——: Inégalités d'auto-intersection pour les courants positifs fermés définis dans les variétés projectives. Ann. Scuola Norm. Sup. Pisa Cl. Sci. (4), **26**(1), 161–184 (1998)

[MI] Mihailescu, E.: Unstable manifolds and Hölder structures associated with noninvertible maps. Discrete Contin. Dyn. Syst. **14**(3), 419–446 (2006)

[MIH] Mihailescu, E., Urbański, M.: Inverse pressure estimates and the independence of stable dimension for non-invertible maps. Can. J. Math. **60**(3), 658–684 (2008)

[MI1] Milnor, J.: Geometry and dynamics of quadratic rational maps, with an appendix by the author and Lei Tan. Exp. Math. **2**(1), 37–83 (1993)

[MIL] ——: On Lattès maps, Dynamics on the Riemann sphere, 9-43. Eur. Math. Soc. Zürich (2006)

[MP] Misiurewicz, M., Przytycki, F.: Topological entropy and degree of smooth mappings. Bull. Acad. Polon. Sci. Sér. Sci. Math. Astronom. Phys. **25**(6), 573–574 (1977)

[N] Narasimhan, R.: Introduction to the theory of analytic spaces, Lecture Notes in Mathematics, No. **25**. Springer-Verlag, Berlin-New York (1966)

[NE] Newhouse, S.E.: Continuity properties of entropy. Ann. Math. (2), **129**(2), 215–235 (1989)

[P] Parry, W.: Entropy and generators in ergodic theory. W.A. Benjamin, Inc., New York-Amsterdam (1969)

[PH] Pham, N.-M.: Lyapunov exponents and bifurcation current for polynomial-like maps. Preprint, 2005. arXiv:math/0512557

[PS] Philipp, W., Stout, W.: Almost sure invariance principles for partial sums of weakly dependent random variables. Mem. Am. Math. Soc. **2**(2), no. 161 (1975)

[PU] Przytycki, F., Urbański, M., Zdunik, A.: Harmonic, Gibbs and Hausdorff measures on repellers for holomorphic maps I. Ann. Math. (2), **130**(1), 1–40 (1989)

[R] Ritt, J.F.: Permutable rational functions. Trans. Am. Math. Soc. **25**, 399–448 (1923)

[S] Saleur, B.: Unicité de la mesure d'entropie maximale pour certaines applications d'allure polynomiale. Preprint (2008)

[SC] Schwartz, L.: Théorie des Distributions. Hermann, Paris (1966)

[SSU] Shiffman, B., Shishikura, M., Ueda, T.: On totally invariant varieties of holomorphic mappings of $\mathbb{P}^n$. Preprint (2000)

[SI] Sibony, N.: Seminar talk at Orsay, October (1981)

[S2] ——: Quelques problèmes de prolongement de courants en analyse complexe. Duke Math. J. **52**(1), 157–197 (1985)

[S3] ——: Dynamique des applications rationnelles de $\mathbb{P}^k$. Panoramas et Synthèses, **8**, 97–185 (1999)

[SW] Sibony, N., Wong, P.: Some results on global analytic sets, *Séminaire Pierre Lelong-Henri Skoda (Analyse), années 1978/79*, 221–237. Lecture Notes in Mathematics **822**. Springer, Berlin (1980)

[SJ] Silverman, J.: The space of rational maps on $\mathbf{P}^1$. Duke Math. J. **94**(1), 41–77 (1998)

[SIU] Siu, Y.-T.: Analyticity of sets associated to Lelong numbers and the extension of closed positive currents. Invent. Math. **27**, 53–156 (1974)

[ST] Sternberg, S.: Local contractions and a theorem of Poincaré. Am. J. Math. **79**, 809–824 (1957)

[SU] Sullivan, D.: Quasiconformal homeomorphisms and dynamics, I, Solution of the Fatou-Julia problem on wandering domains. Ann. Math. (2), **122**(3), 401–418 (1985)

[U] Ueda, T.: Fatou sets in complex dynamics on projective spaces. J. Math. Soc. Jpn. **46**(3), 545–555 (1994)

[U2] ——: Critical orbits of holomorphic maps on projective spaces. J. Geom. Anal. **8**(2), 319–334 (1998)

[V] Viana, M.: Stochastic dynamics of deterministic systems, vol. **21**, IMPA (1997)

[V3] Vigny, G.: Dirichlet-like space and capacity in complex analysis in several variables. J. Funct. Anal. **252**(1), 247–277 (2007)

[VO] Voisin, C. Théorie de Hodge et géométrie algébrique complexe, Cours Spécialisés, **10**, Société Mathématique de France, Paris (2002)

[T] Taflin, J.: Invariant elliptic curves as attractors in the projective plane. J. Geom. Anal. **20**(1), 219–225 (2010)

[T1] Triebel, H.: Interpolation theory, function spaces, differential operators, North-Holland (1978)

[W] Walters, P.: An introduction to ergodic theory. Graduate Texts in Mathematics, **79**. Springer-Verlag, New York-Berlin (1982)

[WA] Warner, F.W.: Foundations of Differentiable Manifolds and Lie Groups. Springer-Verlag, Berlin (1971)

[X] Xia, H., Fu, X.: Remarks on large deviation for rational maps on the Riemann sphere. Stochast. Dyn. **7**(3), 357–363 (2007)

[YA] Yau, S.-T.: On the Ricci curvature of a compact Kähler manifold and the complex Monge-Ampère equation. I. Comm. Pure Appl. Math., **31**(3), 339–411 (1978)

[YO] Yomdin, Y.: Volume growth and entropy. Israel J. Math. **57**(3), 285–300 (1987)

[Y] Young, L.-S.: Statistical properties of dynamical systems with some hyperbolicity. Ann. Math. (2), **147**(3), 585–650 (1998)

[Y1] ——: Ergodic theory of chaotic dynamical systems. XIIth International Congress of Mathematical Physics (ICMP '97) (Brisbane), 131–143. Int. Press, Cambridge, MA (1999)

[Z] Zhang, S.-W.: Distributions in Algebraic Dynamics. Survey in Differential Geometry, vol 10, 381–430. International Press, Somerville, MA (2006)

Dynamics of Entire Functions

Dierk Schleicher

Abstract Complex dynamics of iterated entire holomorphic functions is an active and exciting area of research. This manuscript collects known background in this field and describes several of the most active research areas within the dynamics of entire functions.

Complex dynamics, in the sense of holomorphic iteration theory, has been a most active research area for the last three decades. A number of interesting developments have taken place during this time. After the foundational work by Fatou and Julia in the early 20th century, which developed much of the basic theory of iterated general rational (and also transcendental maps), the advent of computer graphics

D. Schleicher
School of Engineering and Science, Research I, Jacobs University,
Postfach 750 561, D-28725 Bremen, Germany
e-mail: dierk@jacobs-university.de

G. Gentili et al. (eds.), *Holomorphic Dynamical Systems*,
Lecture Notes in Mathematics 1998, DOI 10.1007/978-3-642-13171-4_5,

made possible detailed studies of particular maps; most often, quadratic polynomials as the simplest non-trivial holomorphic mappings were studied. While a number of deep questions on quadratic polynomials remain, interest expanded to specific (usually complex one- or two- dimensional) families of holomorphic maps, such as quadratic rational maps, cubic polynomials, or other families in which critical orbit relations reduced the space to simple families of maps: for instance, families of polynomials of degrees $d \geq 2$ with a single critical point of higher multiplicity. Only in recent years has the progress achieved so far allowed people to shift interest towards higher-dimensional families of iterated maps, such as general polynomials of degree $d \geq 2$. The study of iterated rational maps, as opposed to polynomials, seems much more difficult, mainly because of lack of a good partition to obtain a good encoding for symbolic dynamics: the superattracting fixed point at ∞, and the dynamic rays emanating from it, are important ingredients for deep studies of polynomials that are not available for general rational maps. A notable exception are rational maps that arise from Newton maps of polynomials: for these, it seems that good partitions for symbolic dynamics are indeed possible.

Transcendental iteration theory has been much less visible for a long time, even though its study goes back to Fatou (we will even treat a question of Euler in Section 4), and a solid body of knowledge has been developed by Baker and coauthors, and later also by Eremenko and Lyubich and by Devaney and coauthors, for more than four decades. Complex dynamics is known for employing methods from many different fields of mathematics, including geometry, complex analysis, algebra and even number theory. Transcendental dynamics unites two different general directions of research: there is a substantial body of knowledge coming from value distribution theory that often yields very general results on large classes of iterated transcendental functions; among the key contributors to this direction of research are Baker, Bergweiler, Eremenko, Rippon, and Stallard. The other direction of research sees transcendental functions as limits of rational functions and employs methods adapted from polynomial or rational iteration; here one usually obtains results on more specific maps or families of maps, most often the prototypical families of exponential or cosine maps; this direction of research was initiated by Devaney and coauthors. Others, like Lyubich and Rempe, have worked from both points of view.

In recent years, transcendental iteration theory has substantially gained interest. There have recently been international conferences specifically on transcendental iteration theory, and at more general conferences transcendental dynamics is obtaining more visibility.

In this survey article, we try to introduce the reader several aspects of transcendental dynamics. It is based on lecture notes of the CIME summer school held in Cosenza/Italy in summer 2008, but substantially expanded. The topic of that course, "dynamics of entire functions", also became the title of this article. We thus focus almost entirely on entire functions: their dynamical theory is much simpler than the theory of general meromorphic transcendental functions, much in the same way that polynomial iteration theory is much simpler (and more successful) than rational iteration theory. However, we believe that some of the successes of the polynomial theory still await to be carried over to the world of entire functions, and that some of the key tools (such as dynamic rays) are currently being developed.

In this survey article, we try to relate the two points of view on entire functions: that on large classes of entire functions and that on specific prototypical families of entire maps: the simplest families of maps are the exponential family $z \mapsto \lambda e^z$ with a single asymptotic value, and the cosine family $z \mapsto ae^z + be^{-z}$ with two critical values. We try to cover several of the key topics in the theory of entire dynamics. This article is written for readers with a solid background in one-dimensional complex analysis, and a nodding acquaintance of complex dynamics of polynomials. Much more than what we need is provided in Milnor's now-classical book [Mi06].

In Section 1, we introduce the basic concepts of complex dynamics, including the Fatou and Julia sets, and more specifically the basic concepts of transcendental dynamics such as singular and asymptotic values. We introduce the important set $I(f)$ of escaping points, review some basic local fixed point theory, and describe the important Zalcman lemma with applications.

Section 2 discusses the possibilities for the Fatou set of entire functions and especially highlight those features that are not known from the rational or polynomial theory: Baker domains, wandering domains, and "Baker wandering domains".

The space of general entire functions is a huge space, and many results require the restriction to smaller families of maps: sometimes because tools are lacking to prove results in greater generality, but sometimes also because the space of entire functions is so big that many different dynamical properties are possible, and satisfactorily strong results are possible only when restricting to maps with specific properties. Section 3 introduces important classes of entire maps that are often useful, especially the Eremenko-Lyubich class B of entire functions of bounded type and the class S of functions of finite type.

Section 4 is an overview on results on the set $I(f)$ of escaping points: just as for polynomials, these points often have useful properties that are comparatively easy to investigate, and thus make it possible to establish interesting properties of the Julia set (and sometimes the Fatou set).

Section 5 discusses a number of properties on the Hausdorff dimension of Julia set; this is a rich and active area with a number of interesting and sometimes surprising results.

Section 6 is a brief introduction to parameter spaces of entire functions. We briefly describe a general result on natural parameter spaces of entire functions, and then discuss exponential parameter space as the best-studied parameter space and prototypical parameter space of entire functions, in analogy to the Mandelbrot set for quadratic polynomials.

Section 7 is not directly concerned with the dynamics of entire functions, but with Newton methods of entire functions: these are special meromorphic functions; we hope that, just as in the rational case, these will be meromorphic functions that can be investigated relatively easily; we describe a number of known results on them.

In Section 8, we list a small number of questions that remain open in the field: some of them have remained open for a long time, while others are new. A research area remains lively as long as it still has open questions.

In a brief appendix, we state a few important theorems from complex analysis that we use throughout.

The research field of complex dynamics is large and active, and many people are working on it from many different points of view. We have selected some topics that we find particularly interesting, and acknowledge that there are a number of other active and interesting topics that deserve no less attention. We mention in particular results on measure theory, including the thermodynamic formalism (see Urbański [Ur03] for a recent survey). Further important omitted areas that we should mention are Siegel disks and their boundaries (see e.g., Rempe [Re04]), questions of linearizations and small cycles (see e.g. Geyer [Ge01]), the construction of entire maps with specific geometric or dynamic properties (here various results of Bergweiler and Eremenko should be mentioned), the relation of transcendental dynamics to Nevanlinna theory, and Thurston theory for transcendental maps (see Selinger [Se09]).

We tried to give many references to the literature, but are acutely aware that the literature is vast, and we apologize to those whose work we failed to mention. Many further references can be found in the 1993 survey article of Bergweiler [Be93] on the dynamics of meromorphic functions.

The illustration on the first page shows the Julia set of a hyperbolic exponential map with an attracting periodic point of period 26. The Fatou set is in white. We thank Lasse Rempe for having provided this picture.

ACKNOWLEDGEMENTS. I would like to thank my friends and colleagues for many interesting and helpful discussions we have had on the field, and in particular on drafts of this manuscript. I would like to especially mention Walter Bergweiler,Jan Cannizzo, Yauhen Mikulich, Lasse Rempe, Phil Rippon, Nikita Selinger, and Gwyneth Stallard. And of course I would like to thank the CIME foundation for having made possible the summer school in Cosenza, and Graziano Gentili and Giorgio Patrizio for having made this such a memorable event!

1 Fatou and Julia Set of Entire Functions

Throughout this text, f will always denote a transcendental entire function $f\colon \mathbb{C} \to \mathbb{C}$. The dynamics is to a large extent determined by the singular values, so we start with their definition.

Definition 1.1 (Singular Value).
A *critical value* is a point $w = f(z)$ with $f'(z) = 0$; the point z is a critical point. An *asymptotic value* is a point $w \in \mathbb{C}$ such that there exists a curve $\gamma\colon [0,\infty) \to \mathbb{C}$ so that $\gamma(t) \to \infty$ and $f(\gamma(t)) \to w$ as $t \to \infty$. The set of *singular values* of f is the closed set

$$S(f) := \overline{\{\text{critical and asymptotic values}\}} \,.$$

This definition is not completely standard: some authors do not take the closure in this definition.

A *postsingular point* is a point on the orbit of a singular value.

In any dynamical system, it is useful to decompose the phase space (in this case, the dynamical plane $\mathbb{C}$) into invariant subsets. In our case, we will mainly consider the *Julia set* $J(f)$, the *Fatou set* $F(f)$, and the *escaping set* $I(f)$, as well as certain subsets thereof.

Definition 1.2 (Fatou and Julia Sets).
The *Fatou set* $F(f)$ is the set of all $z \in \mathbb{C}$ that have a neighborhood U on which the family of iterates $f^{\circ n}$ forms a normal family (in the sense of Montel; see Definition A.1). The *Julia set* $J(f)$ is the complement of the Fatou set: $J(f) := \mathbb{C} \setminus F(f)$. A connected component of the Fatou set is called a *Fatou component*.

Definition 1.3 (The Escaping Set).
The set $I(f)$ is the set of points $z \in \mathbb{C}$ with $f^{\circ n}(z) \to \infty$.

Clearly, $F(f)$ is open and $J(f)$ is closed, while in general, $I(f)$ is neither open nor closed. All three sets $J(f)$, $F(f)$, and $I(f)$ are forward invariant, i.e., $f(F(f)) \subset F(f)$ etc.; this is true by definition. Note that equality may fail when f has omitted values: for instance for $z \mapsto 0.1\,e^z$, the Fatou set contains a neighborhood of the origin, but 0 is an omitted value. It is easy to see that for each $n \geq 1$, $F(f^{\circ n}) = F(f)$ and hence $J(f^{\circ n}) = J(f)$, and of course $I(f^{\circ n}) = I(f)$.

Theorem 1.4 (The Julia Set).
The Julia set is non-empty and unbounded and has no isolated points.

The Fatou set may or may not be empty. An example of an entire function with empty Fatou set is $z \mapsto e^z$ [Mis81]. More generally, the Fatou set is empty for every entire function of finite type (see Definition 3.1) for which all singular values are either preperiodic or escape to ∞ (Corollary 3.14). We will describe the Fatou set, and give examples of different types of Fatou components, in Section 2. The Fatou set is non-empty for instance for any entire function with an attracting periodic point.

The fact that $J(f)$ is non-empty was established by Fatou in 1926 [Fa26]. We give here a simple proof due to Bargmann [Ba99]. We start with a preparatory lemma.

Lemma 1.5 (Existence of Periodic Points).
Every entire function, other than a polynomial of degree 0 *or* 1*, has at least two periodic points of period* 1 *or* 2.

Remark 1.6. This is a rather weak result with a simple proof; a much stronger result is given, without proof, in Theorem 1.21.

Proof. Consider an entire function f and define a meromorphic function via $g(z) := (f \circ f(z) - z)/(f(z) - z)$.

Suppose that g is constant, say $g(z) = c$ for all $z \in \mathbb{C}$, hence $f \circ f(z) - z = c(f(z) - z)$. If $c = 0$, then $f \circ f = \text{id}$, so f is injective and thus a polynomial of degree 1. If $c = 1$, then $f \circ f = f$, so for each $z \in \mathbb{C}$, the value $f(z)$ is a fixed point of f: this implies that either $f = \text{id}$ and every $z \in \mathbb{C}$ is a fixed point, or fixed points of f are discrete and thus f is constant. If $c \notin \{0,1\}$, then differentiation yields $(f' \circ f)f' - 1 = cf' - c$ or $f'(f' \circ f - c) = 1 - c$. Since $c \neq 1$, it follows that f' omits the value 0 and $f' \circ f$ omits the value $c \neq 0$, so f' can assume the value c only at the omitted values of f. By Picard's Theorem A.4, it follows that f' is constant and thus f is a polynomial of degree at most 1.

In our case, f is not a polynomial of degree 0 or 1 by hypothesis, so g is a non-constant meromorphic function. Suppose $p \in \mathbb{C}$ is such that $g(p) \in \{0,1,\infty\}$. If $g(p) = \infty$, then $f(p) = p$; if $g(p) = 0$, then $f(f(p)) = p$; and if $g(p) = 1$, then $f(f(p)) = f(p)$.

If g is transcendental, then by Picard's theorem there are infinitely many $p \in \mathbb{C}$ with $g(p) \in \{0,1,\infty\}$. If g is a non-constant rational map (which in fact never happens), then there are $p_0, p_1, p_\infty \in \overline{\mathbb{C}}$ with $g(p_i) = i$, and at least two of them are in $\mathbb{C}$. □

As an example, the map $f(z) = e^z + z$ has no fixed points; in this case $g(z) = e^{e^z} + 1$ has no $p \in \mathbb{C}$ with $g(p) \in \{1,\infty\}$, but of course infinitely many p with $g(p) = 0$ (corresponding to periodic points of period 2).

Proof of Theorem 1.4. Let f be an entire function other than a polynomial of degree 0 or 1. By Lemma 1.5, f has (at least) two periodic points of period 1 or 2; replacing f by $f \circ f$ if necessary, we may suppose that f has two fixed points, say at $p, p' \in \mathbb{C}$ (note that $J(f) = J(f \circ f)$). If $|f'(p)| > 1$, then $p \in J(f)$. In order to show that $J(f) \neq \emptyset$, we may assume that $p \in F(f)$ and in particular that $|f'(p)| \leq 1$. Let W be the Fatou component containing p.

If $|f'(p)| = 1$, then any limit function of the family of iterates $\{f^{\circ n}|_W\}$ is non-constant. So let $f^{\circ n_j}$ be a subsequence that converges to a non-constant limit function; then $f^{\circ(n_{j+1}-n_j)}$ converges to the identity on W, and this implies that $f|W$ is injective. If $W = \mathbb{C}$, then this is a contradiction to the choice of f, hence $F(f) \neq \mathbb{C}$.

The final case we have to consider is $|f'(p)| < 1$; for convenience, we may suppose that $p = 0$. If $F(f) = \mathbb{C}$, then $f^{\circ n} \to 0$ uniformly on compacts in $\mathbb{C}$. We will show that f is a polynomial of degree at most 1. Let D be an open disk centered at p such that $f(D) \subset D$. For $N \in \mathbb{N}$, let $D_n := f^{\circ(-n)}(D)$; since D is simply connected, it follows that also D_n is simply connected. Let r_n be maximal so that the round circle $\partial D_{r_n}(0) \subset \overline{D_n}$; this implies that $D_{r_n} \subset D_n$. We have $D_n \subset D_{n+1}$ and $\bigcup_n D_n = \mathbb{C}$, hence $r_{n+1} \geq r_n$ and $r_n \to \infty$.

As usual, for $r > 0$, define $M(r;f) := \max\{|f(z)|\colon |z| = r\}$. For each $n \in \mathbb{N}$, define maps $h_n\colon \mathbb{D} \to \mathbb{C}$ via $h_n(z) = M(r_n/2;f)^{-1} f(r_n z)$. All h_n satisfy $h_n(0) = 0$ and $M(1/2;h_n) = 1$. Let

$$c_n := \sup\{r > 0\colon \partial D_r(0) \subset h_n(\mathbb{D})\}.$$

If some c_n satisfies $c_n < 1$, then there are points $a_n, b_n \in \mathbb{C}$ with $|a_n| = 1$, $|b_n| = 2$ so that $h_n(\mathbb{D}) \cap \{a_n, b_n\} = \emptyset$. If this happens for infinitely many n, then we can extract a subsequence with $c_n < 1$, and it follows easily from Montel's theorem that the h_n form a normal family. After extracting another subsequence, we may suppose that the h_n converge to a holomorphic limit function $h\colon \mathbb{D} \to \mathbb{C}$ that inherits from the h_n the properties that $h(0) = 0$ and $M(1/2;h) = 1$. Thus h is non-constant and there is an $r > 0$ with $D_r(0) \subset h(\mathbb{D})$. But then $c_n > r/2$ for all large n. This implies that no subsequence of the c_n tends to 0, and hence that $\inf\{c_n\} > 0$.

There is thus a $c > 0$, and there are $c'_n > c$ with $\partial D_{c'_n}(0) \subset h_n(\mathbb{D})$ for all large n. This implies

$$\partial D_{c'_n M(r_n/2;f)}(0) \subset f(D_{r_n}(0)) \subset f(D_n) \subset D_n$$

and thus, by the definition of r_n,

$$cM(r_n/2;f) < r_n .$$

Therefore, $M(r_n/2;f)/r_n$ is bounded.

Define an entire function g via $g(z) := f(z)/z - f'(0)$ (the isolated singularity at $z = 0$ is removable because $f(0) = 0$, and thus $g(0) = 0$). It has the property that $M(r_n/2;g)$ is bounded; since $r_n \to \infty$, this implies that g is bounded on $\mathbb{C}$ and hence constant. But $g(0) = 0$, hence $f(z) = zf'(0)$ for all z, so f is a polynomial of degree at most 1 as claimed.

We have now shown that, if f is not a polynomial of degree 0 or 1, then $J := J(f) \neq \emptyset$. If $|J| > 1$, then it follows from Picard's Theorem A.4 that J is infinite and unbounded: as soon as J contains at least two points a, a', every neighborhood of ∞ contains infinitely many preimages of a or a'.

We finally have to consider the case that $|J| = 1$, say $J = \{a\}$. In this case, $f(a) = a$, and f must have another fixed point, say p, and $W := \mathbb{C} \setminus \{a\}$ equals the Fatou set. We must have $|f'(p)| \leq 1$, and $|f'(p)| = 1$ would lead to the same contradiction as above. This implies that $|f'(p)| < 1$. But any loop in W starting and ending at p would converge uniformly to p, and by the maximum modulus principle this would imply that W was simply connected, a contradiction. This shows that $|J| > 1$ in all cases, hence that J is always an infinite set (we will show below in Theorem 1.7 that J is always uncountable). □

For an entire function f, an *exceptional point* is a point $z \in \mathbb{C}$ with finite backwards orbit. There can be at most one exceptional point: in fact, any finite set of exceptional points must have cardinality at most 1 (or the complement to the union of their backwards orbits would be forward invariant and hence contained in the Fatou set, by Montel's Theorem A.2, so the Julia set would be finite). If an entire function f has an exceptional point p, it is either an omitted value or a fixed point (such as the point 0 for e^z or ze^z); in such cases, f restricts to a holomorphic self-map of the infinite cylinder $\mathbb{C} \setminus \{p\}$.

Theorem 1.7 (Topological Properties of the Julia Set).
For every entire function f (other than polynomials of degree 0 or 1), the Julia set is the smallest closed backward invariant set with at least 2 points. The Julia set is contained in the backwards orbit of any non-exceptional point in $\mathbb{C}$, it has no isolated points, and it is locally uncountable.

Proof. Suppose $z \in \mathbb{C}$ has a neighborhood U so that $f^{\circ n}(U)$ avoids a. Then it also avoids all points in $f^{-1}(a)$. Unless a is an exceptional point, Montel's Theorem A.2 implies that z is in the Fatou set. The backwards orbit of any non-exceptional point in $\mathbb{C}$ thus accumulates at each point in the Julia set. Any closed backwards invariant set with at least two elements thus contains $J(f)$. The Julia set itself is closed and backward invariant and contains at least two points by Theorem 1.4, so it is the smallest such set.

Since any point in $J(f)$ is the limit point of backwards orbits of points in $J(f)$, it follows that no point in $J(f)$ is isolated. Any closed set without isolated points is locally uncountable. □

Theorem 1.8 (Julia Set and Escaping Set).
For every entire function f (other than polynomials of degree 0 or 1), the set $I(f)$ is infinite, and $J(f) = \partial I(f)$.

Sketch of proof. The hard part consists in showing the existence of an escaping point; this implies that $I(f)$ is infinite. Using Montel's Theorem A.2, this implies that the Fatou set contains every open subset of $\mathbb{C} \setminus I(f)$ (if any). We have $|\mathbb{C} \setminus I(f)| \geq 2$ (for instance, because there exist periodic points; see Lemma 1.5), so for the same reason, any open subset of $I(f)$ is also part of the Fatou set. This implies $J(f) \subset \partial I(f)$. Conversely, if $z \in I(f) \cap F(f)$, then any limit function of the iterates near z must have the value ∞; but locally uniform limits of entire functions are either entire or constant equal to ∞, and $z \in I(f) \setminus \partial I(f)$.

The fact that every transcendental entire function has escaping points was shown by Eremenko [Er89] using Wiman-Valiron-theory (a different proof, based on a theorem of Bohr, is due to Domínguez [Do98]). While the details are quite technical, the idea of Eremenko's proof can be described as follows (here we follow an exposition of Bergweiler). Write $f(z) = \sum_{n\geq 0} a_n z^n$. For a radius $r > 1$, define the *central index* $\nu = \nu(r, f)$ so that $|a_\nu| r^\nu = \max_{n\geq 0}\{|a_n| r^n\}$. Choose z_r so that $|f(z_r)|$ is maximal among all z with $|z| = r$. Then Wiman and Valiron showed that $f(z) \asymp (z/z_r)^\nu f(z_r)$ for z close to z_r, where "$\asymp$" means "the quotient is bounded" (more precisely, choosing some $\tau > 1/2$, this holds if $|z - z_r| < r/\nu^\tau$, and only if r is outside an exceptional set F that is small enough so that $\int_F dt/t < \infty$). We will assume that $\tau \in (1/2, 1)$.

We write $z = z_r e^w$ and obtain $f(z_r e^w) \asymp e^{w\nu} f(z_r)$ if $|w| < 1/\nu^\tau$ and $r \notin F$, in particular if $|\text{Re } w| < \pi/\nu$, $|\text{Im } w| < \pi/\nu$ (if ν is sufficiently large, then $\pi/\nu < 1/\nu^\tau$). The corresponding square around z_r maps to a large annulus around $f(z_r)$ at absolute value much greater than that of z_r. In the image, the argument can be repeated, and this yields an escaping orbit (in each step, we need to exclude values of r from the exceptional set F, and estimate that enough values of r remain). □

This result has been generalized, by a different but related method, by Bergweiler, Rippon, Stallard [BRS08], to show the existence of points that escape through a specific sequence "tracts" of f.

Definition 1.9 (Local Fixed Point Theory).
Consider a local holomorphic function $g \colon U \to \mathbb{C}$ with $U \subset \mathbb{C}$ open, and a fixed point $p \in U$ with derivative $\mu := g'(p)$. Then μ is called the *multiplier* of the fixed point p, and p is called

attracting if $|\mu| < 1$ (and *superattracting* if $\mu = 0$);
repelling if $|\mu| > 1$;

indifferent if $|\mu| = 1$; in particular
rationally indifferent (or parabolic) if μ is a root of unity;
irrationally indifferent if $|\mu|$ is indifferent but not a root of unity.

Theorem 1.10 (Local Fixed Point Theory).
Local holomorphic maps have the following normal forms near fixed points:

- *in a neighborhood of a superattracting fixed point, the map is conformally conjugate to $z \mapsto z^d$ near 0, for a unique $d \geq 2$;*
- *in a neighborhood of an attracting or repelling fixed point with multiplier μ, the map is conformally conjugate to the linear map $z \mapsto \mu z$ near 0 ("attracting and repelling fixed points are locally linearizable");*
- *local normal forms in a neighborhood of parabolic fixed points are complicated (see Abate [Ab]; but within each attracting petal, the map is conformally conjugate to $z \mapsto z+1$ in a right half plane;*
- *an irrationally indifferent fixed point p may or may not be linearizable.*

Remark 1.11. This is a local result; for a proof, see [Mi06]. If an irrationally indifferent fixed point is linearizable, a maximal linearizable neighborhood is called a *Siegel disk*; this is a Fatou component. There is a sufficient condition, due to Yoccoz, that assures linearizability of a local holomorphic map (and in particular of an entire function) with an irrationally indifferent fixed point, depending only on the multiplier [Mi06]. Non-linearizable irrationally indifferent fixed points are called *Cremer points*; they are in the Julia set and not associated to any type of Fatou component. (As a result, if the Julia set equals $\mathbb{C}$, then all fixed points are repelling or Cremer.)

Remark 1.12. The same classification applies of course to periodic points: these are fixed points of appropriate iterates.

In Definition 1.1 we had defined a singular value to be a critical or asymptotic value, or a limit point thereof. Now we discuss singular values somewhat more closely, following Iversen [Iv14]; see also [BE95, Ne53]. Choose a point $a \in \overline{\mathbb{C}}$. For each $r > 0$, let U_r be a component of $f^{-1}(D_\chi(a,r))$ (where $D_\chi(a,r)$ is the r-neighborhood of a with respect to the spherical metric), chosen so that $U_{r'} \subset U_r$ for $r' < r$. Then either $\bigcap_r U_r = \{z\}$ for a unique $z \in \mathbb{C}$, or $\bigcap_r U_r = \emptyset$ (the set $\bigcap_r U_r$ is connected; if it contains more than one point, then f is constant). This collection $\{U_r\}$ is called a *tract* for f. If $\bigcap_r U_r = \{z\}$, then $f(z) = a$, and a is a *regular value* for this tract if $f'(z) \neq 0$, and a is a *critical value* if $f'(z) = 0$. If $\bigcap_r U_r = \emptyset$, then a is an *asymptotic value* for this tract, and there exists a curve γ: $[0,\infty) \to \mathbb{C}$ with $\gamma(t) \to \infty$ through U_r and $f(\gamma(t)) \to a$ (as in Definition 1.1); such a path γ is called an *asymptotic path*. Of course, for the same function f, the same point $a \in \mathbb{C}$ can be a regular or singular value of different types for different tracts.

Critical values are called *algebraic singularities*, and asymptotic values are called *transcendental singularities* (for a particular choice of the tract). Any tract with $\bigcap U_r = \emptyset$ is called an *asymptotic tract*.

An asymptotic value is a *direct singularity* for a tract $\{U_r\}$ if there is an $r > 0$ so that $f(U_r) \not\ni a$, and an *indirect singularity* otherwise. A direct singularity is a *logarithmic singularity* if there is an U_r so that $f\colon U_r \to D_\chi(a,r)\setminus\{a\}$ is a universal covering. In particular, if f is a transcendental entire function, then ∞ is always a direct asymptotic value (see Theorem 1.14 below). For entire functions of bounded type (see Section 3), ∞ is always a logarithmic singularity.

Recall from Definition 1.1 that we defined singular value to be a critical value, an asymptotic value, or a limit point thereof.

Theorem 1.13 (Singular Values).
Any $a \in \mathbb{C}$ that is not a singular value has a neighborhood U so that $f\colon f^{-1}(U) \to U$ is an unbranched covering (i.e., a is a regular value for all tracts).

Proof. Choose a neighborhood U of a, small enough so that it is disjoint from $S(f)$; it thus contains no critical or asymptotic value. If U has an unbounded preimage component, then it is not hard to show that U contains an asymptotic value (successively subdivide U so as to obtain a nested sequence of open sets with diameters tending to 0 but with unbounded preimages). Therefore, a has a simply connected neighborhood for which all preimage components are bounded. If V is such a preimage component, then $f\colon V \to U$ is a branched covering, and if U contains no critical value, then $f\colon V \to U$ is a conformal isomorphism. The claim follows. □

The set of direct asymptotic values is always countable [He57] (but the number of associated asymptotic tracts need not be). There are entire functions for which every $a \in \mathbb{C}$ is an asymptotic value [Gr18b]. On the other hand, according to the Gross Star Theorem [Gr18a], every entire function f has the property that for every $a \in \mathbb{C}$ and for every $b \in \mathbb{C}$ with $f(b) = a$, and for almost every direction, the ray at a in this direction can be lifted under f to a curve starting at b.

The following theorem of Iversen is important.

Theorem 1.14 (Omitted Values are Asymptotic Values).
If some $a \in \overline{\mathbb{C}}$ is assumed only finitely often by some transcendental entire function f, then a is a direct asymptotic value of f. In particular, for every entire function, ∞ is always a direct asymptotic value.

Proof. For any $r > 0$, any bounded component of $f^{-1}(D_\chi(r,a))$ contains a point z with $f(z) = a$. Since by Picard's Theorem A.4, at most one point in $\mathbb{C}$ is assumed only finitely often, it follows that for every $r > 0$, the set $f^{-1}(D_\chi(r,a))$ cannot consist of bounded components only. Therefore, for each $r > 0$, there is at least one unbounded component, and thus at least one asymptotic tract over the asymptotic value a; for such a tract, a is a direct singularity. Since the point ∞ is omitted for each entire function, it is a direct asymptotic value. □

The following definition is of great importance in function theory:

Definition 1.15 (Order of Growth).
The *order* of an entire function f is

$$\operatorname{ord} f := \limsup_{z\to\infty} \frac{\log\log|f|}{\log|z|}\,.$$

(Sometimes, one also uses the lower order of f; for this definition, the lim sup is replaced by the lim inf.)

Remark 1.16. The prototypical example of a function of finite order $d \in \mathbb{N}$ is $z \mapsto \exp(z^d)$. Unlike the properties of finite or bounded type, the property of finite order is *not* preserved under compositions, and was thus thought to be of limited use in dynamics; however, the order of growth has recently found important applications in dynamics, for instance when dynamical properties are controlled using geometric function theory (for instance, in Theorem 4.11 below).

The finite order condition also comes into play in the discussion of direct and indirect asymptotic values.

Theorem 1.17 (Denjoy-Carleman-Ahlfors).
For any function f of finite order, the number of asymptotic values, and even the number of different asymptotic tracts, is at most $\max\{1, 2\operatorname{ord} f\}$ *and hence finite. More precisely, if f has m direct and n indirect asymptotic tracts, then* $2m+n \le \max\{1, 2\operatorname{ord} f\}$.

In fact, between any two different asymptotic tracts of finite asymptotic values, there must be an asymptotic tract of the asymptotic value ∞; and if the latter is very narrow, then f must grow very quickly along this tract; the extra count for direct asymptotic tracts comes in because these, too, need to have certain width (postcomposing f with a Möbius transform preserves the order, and can turn direct tracts over finite asymptotic values to tracts over ∞). Compare [Ne53, Sec. 269].

Theorem 1.18 (Indirect Asymptotic Values and Critical Values).
For any function of finite order, every indirect asymptotic value is a limit point of critical values.

For a proof, see [BE95].

The following result from [Z75] is often useful in complex dynamics, especially transcendental dynamics.

Lemma 1.19 (The Zalcman Lemma).
Let $U \subset \mathbb{C}$ be any domain and let $\mathscr{F}$ be a non-normal family of holomorphic functions from U to $\mathbb{C}$. Then there exist a non-constant entire function $g\colon \mathbb{C} \to \mathbb{C}$ and sequences $f_k \in \mathscr{F}$, $z_k \in U$ and $\rho_k > 0$ with $\rho_k \to 0$ so that $z_k \to z_\infty \in U$ and

$$g_k(\zeta) := f_k(z_k + \rho_k \zeta) \longrightarrow g(\zeta)$$

uniformly on compacts in $\mathbb{C}$. If $a \in U$ is such that the restriction of $\mathscr{F}$ to any neighborhood of a is not normal, then we may require that $z_n \to a$.

Proof. After rescaling, we may suppose that $\overline{\mathbb{D}} \subset U$ and that the point $a = 0$ has the property that the restriction of $\mathscr{F}$ to any neighborhood of 0 is not normal. Let $(f_k) \subset \mathscr{F}$ be any sequence that does not form a normal family when restricted to any neighborhood of a.

According to Marty's criterion (Theorem A.3), a family of maps (f_k) is normal if and only if its spherical derivatives $f^\sharp(z) = |f'(z)|/(1+|f(z)|^2)$ are locally bounded. Since this is not the case, one can extract a subsequence of the f_k and then a sequence $\zeta_k \to 0$ with $f_k^\sharp(\zeta_k) \to \infty$. Choose asequence $(r_k) \subset \mathbb{R}^+$ so that $r_k \to 0$ but $r_k f_k^\sharp(\zeta_k) \to \infty$; we may assume that $|\zeta_k| + 2r_k < 1$ for all k.

Find a sequence $z_k \in \mathbb{C}$ with $|z_k - \zeta_k| < r_k$ so that

$$M_k := f_k^\sharp(z_k)(1 - |z_k - \zeta_k|/r_k) = \sup_{|z-\zeta_k|<r_k} f_k^\sharp(z)(1 - |z - \zeta_k|/r_k)\,.$$

This implies $z_k \to 0$ and $f_k^\sharp(z_k) \geq M_k \geq f_k^\sharp(\zeta_k)$. Extract a subsequence so that $f_k(z_k)$ converges to a limit $b \in \overline{\mathbb{C}}$.

Define $\rho_k := 1/f_k^\sharp(z_k) = (1 - |z_k - \zeta_k|/r_k)/M_k$; then $\rho_k \to 0$. Define functions $g_k \colon \{z \in \mathbb{C} \colon |z| < r_k/\rho_k\} \to \mathbb{C}$ via $g_k(z) := f_k(z_k + \rho_k z)$ (indeed, if $|z| < r_k/\rho_k$, then $|z_k + \rho_k z| < |z_k| + r_k < |\zeta_k| + 2r_k < 1$). Note that $r_k/\rho_k = r_k f_k^\sharp(z_k) \geq r_k f_k^\sharp(\zeta_k) \to \infty$, so every compact subset of $\mathbb{C}$ is contained in the domain of definition of most g_k. We have $g_k^\sharp(0) = \rho_k f_k^\sharp(z_k) = 1$.

The g_k satisfy, for $|z| < r_k/\rho_k$,

$$\begin{aligned} g_k^\sharp(z) &= \rho_k f_k^\sharp(z_k + \rho_k z) \leq \rho_k \frac{M_k}{1 - |z_k + \rho_k z - \zeta_k|/r_k} \\ &\leq \frac{1 - |z_k - \zeta_k|/r_k}{1 - |z_k - \zeta_k|/r_k - |\rho_k z|/r_k} = \frac{1}{1 - \frac{|\rho_k z|/r_k}{1-|z_k-\zeta_k|/r_k}} \\ &= \frac{1}{1 - \frac{|z|}{r_k M_k}}\,. \end{aligned}$$

Since $r_k M_k \geq r_k f_k^\sharp(\zeta_k) \to \infty$, this implies that the $g_k^\sharp$ are uniformly bounded on compact sets, so they converge subsequentially. Since $g_k(0) = f_k(z_k) \to b$, it follows that the g_k converge locally uniformly to a holomorphic limit function with $g^\sharp(0) = 1$, so g is non-constant. But a locally uniform limit of entire functions is entire, hence $b \in \mathbb{C}$. □

Theorem 1.20 (Julia and Fatou Sets).
The Julia set equals the closure of the set of repelling periodic points.

Proof. We give a short proof due to Berteloot and Duval [BD00]. It is clear that the set of repelling periodic points is contained in the Julia set, and thus also their closure.

For the converse, let P_0 be the countable set of points in the forward orbits of the critical points (we do not take the closure here). Let P equal P_0 union the set of exceptional points. Let U be any open set in $\mathbb{C}$ that intersects $J(f)$. We will show that U contains a repelling periodic point of f. Since $J(f)$ is locally uncountable (Theorem 1.4), we may choose a point $a \in (J(f) \cap U) \setminus P$.

Apply the Zalcmann Lemma 1.19 to the family of functions $f^{\circ n}$ based at the point $a \in J(f)$: there exist a sequence $\rho_k \to 0$, a sequence $z_k \to a$, a non-constant entire function $g \colon \mathbb{C} \to \mathbb{C}$, and a subsequence $f^{\circ n_k}$ so that $g_k(\zeta) := f^{\circ n_k}(z_k + \rho_k \zeta) \to g(\zeta)$ uniformly on compacts.

We start with the following claim: *if there exists a disk $D \subset \mathbb{C}$ so that $g\colon D \to g(D)$ is univalent and $a \in g(D) \subset U$, then f has a repelling periodic point in U.*

To see the claim, define a sequence of contractions $r_k(z) := z_k + \rho_k z$. Then $g_k = f^{\circ n_k} \circ r_k$. Let D' be a bounded open disk with $\overline{D'} \subset D$ and $a \in g(D')$. For k sufficiently large, g_k is univalent on D' and $a \in g_k(D') \subset U$, and also $r_k(D') \subset g_k(D') = f^{\circ n_k} \circ r_k(D') \subset U$ (the first assertion follows from the fact that the r_k are contractions towards a). This implies that $(f^{\circ n_k})^{-1}(r_k(D')) \subset r_k(D')$, so $(f^{\circ n_k})^{-1}$ has an attracting fixed point in $r_k(D') \subset U$, by the Schwarz Lemma. This proves the claim.

Our next assertion is that *there always exists a disk D as in the claim.* Let $V \subset \mathbb{C}$ be the set of points z for which there is a disk D so that $g\colon D \to g(D)$ is univalent and $g(D)$ is a neighborhood of z. Every point in $\mathbb{C} \setminus V$ is either an omitted value or a critical value of g. Since g has at most one omitted value by the Picard theorem and at most countably many critical values, it follows that $\mathbb{C} \setminus V$ is at most countable. Clearly, V is open, and it intersects $J(f)$. We need to show that $a \in V$.

By assumption, the backward orbit of a is infinite. Since V is open and intersects $J(f)$, there is a $b \in V \cap U$ and an $n \in \mathbb{N}$ with $f^{\circ n}(b) = a$. Then b has a neighborhood W for which $f^{\circ n}(W)$ is univalent and contains a, so we indeed have $a \in V$, thus proving our assertion.

We have shown that each open set U intersecting $J(f)$ contains a repelling periodic point. Since U is arbitrary, this proves the theorem. □

Theorem 1.21 (Periodic Points of Almost All Periods).
Every entire function has repelling periodic points of all periods, except possibly period 1.

An example of an entire function without periodic points of period 1 is $z \mapsto e^z + z$. Theorem 1.21 was shown by Bergweiler in 1991 [Be91] (even for meromorphic functions), answering a question of Baker from 1960 [Ba60]. A simple proof of the fact that every entire function has periodic points of all periods except possibly 1 can be found in [BB01]. (Already in 1948, Rosenbloom [Ro48] had shown that for every entire function f and every $n \geq 1$, $f^{\circ n}$ has infinitely many fixed points; these are periodic points of f of period n or dividing n.)

Remark 1.22. Meromorphic functions with finitely many poles are often viewed as being similar as entire functions: near ∞, both have no poles. However, there are significant differences: for instance, periodic Fatou components of entire functions are simply connected (Theorem 2.5), but this is not true for meromorphic functions.

2 The Fatou Set of Entire Functions

In this section, we describe the possible types of components of the Fatou set of an entire function. These types are similar as those for polynomials, but there are two extra types: Baker domains (domains at infinity) and wandering domains. We give examples for both.

Theorem 2.1 (Classification of Fatou Components).
Any Fatou component has exactly one of the following types:

- *a periodic component in which the dynamics converges to an attracting or superattracting cycle;*
- *a periodic component in which the dynamics converges to a parabolic cycle;*
- *a periodic component in which the dynamics is conformally conjugate to an irrational rotation on a disk (a Siegel disk);*
- *a periodic component in which the dynamics converges to ∞ (a Baker domain);*
- *a preperiodic component that eventually maps to a periodic component in one of the types above;*
- *a non-periodic component for which all forward iterates are disjoint (a wandering domain).*

Remark 2.2. The difference to the polynomial case is the possibility of Baker domains (these are similar to parabolic domains in the sense that the dynamics converges locally uniformly to a boundary point, except that the parabolic boundary point is replaced by the essential singularity at ∞) and of wandering domains. Rational and meromorphic maps may also have Arnol'd-Herman rings, but these do not exist for entire maps (by the maximum principle).

The proof of the classification of periodic Fatou components [BKL91, EL89] is similar as for rational maps (see Milnor [Mi06]). A difference occurs for Fatou components in which some and thus all orbits converge to the boundary: if f is defined in a neighborhood of a limiting boundary point, then this boundary point must be parabolic; if not, then this boundary point is the essential singularity at ∞ and we have a Baker domain. The only other difference is the possibility that a Fatou component may not be eventually periodic (i.e., a wandering domain): for rational maps, this is ruled out by Sullivan's theorem, but wandering domains do occur for transcendental maps (see Examples 2.10 and 2.11 below). However, there are some families of transcendental entire functions without wandering domains: for instance, see Theorem 3.4 and [EL89, Be93, BHKMT93]. In particular, [BHKMT93] discusses the non-existence of wandering domains in certain cases without using quasiconformal methods: they show that all limit functions of wandering domains must be limit points of the set of singular orbits.

Theorem 2.3 (Singular Values and Fatou Components).
Every cycle of attracting and parabolic Fatou components contains a singular value; and every boundary point of every Siegel disks is a limit point of postsingular points.

The proof of this result is the same as in the rational case [Mi06].

Remark 2.4. The connectivity of a Fatou component U is the number of connected components of $\overline{\mathbb{C}} \setminus U$ (it may be ∞). It is an easy consequence of the Riemann-Hurwitz formula that the connectivity of $f(U)$ is at most that of U. Therefore, every wandering domain has eventually constant connectivity (finite or infinite). Since f has no poles, the image of a multiply connected Fatou component is multiply connected.

If the connectivity of U is finite at least 3 and does not decrease, then by the Riemann-Hurwitz formula U maps homeomorphically onto its image, and the same is then true for the bounded complementary domains of U; this is a contradiction. Therefore, every multiply connected Fatou component is either infinitely connected or eventually doubly connected.

Theorem 2.5 (Multiply Connected Fatou Component).
Every multiply connected Fatou component U of an entire transcendental function has the following properties:

- *U is a wandering domain;*
- *U is bounded;*
- *U has eventual connectivity 2 or ∞;*
- *the iterates converge to ∞ uniformly;*
- *each compact $K \subset \mathbb{C}$ is separated from ∞ by all but finitely many of the iterated images of U;*
- *the map from each component on the orbit of U to its image has a finite mapping degree, and this degree tends to ∞ under iteration;*
- *the $f^{\circ n}(U)$ contain annuli with moduli tending to ∞.*

Remark 2.6. A Fatou component satisfying the properties of Theorem 2.5 is sometimes called a *Baker wandering domain* (caution: this notion must not be confused with that of a Baker domain: a Baker domain is a periodic Fatou component in which the iterates tend to ∞).

Simple connectivity of periodic Fatou components was observed by Baker in [Ba84] (see also Töpfer [Tö39]); the description of multiply connected wandering domains is from [Ba76]. See also [Ri08, Zh06].

Sketch of proof. Let U be a multiply connected Fatou component and let $\gamma_0 \subset U$ be a simple closed curve that is non-contractible. Let U_0 be the bounded domain surrounded by γ_0. Inductively, let U_{n+1} be the largest open domain with $\partial U_{n+1} \subset f(\gamma_n)$ and let $\gamma_{n+1} := \partial U_{n+1}$ (if V_{n+1} is the unique unbounded component of $\mathbb{C} \setminus f(\gamma_n)$, then $U_{n+1} := \mathbb{C} \setminus \overline{V_{n+1}}$). By the maximum principle, it follows that $f(U_n) \subset U_{n+1}$.

The set U_0 intersects the Julia set and hence contains periodic points (Theorem 1.20). Replacing f by an iterate, we may suppose that U_0 contains a fixed point p. Therefore, all γ_n separate p from ∞. Denote the hyperbolic length of γ_n within U_n by $\ell(\gamma_n)$. Then clearly $\ell(\gamma_{n+1}) \leq \ell(\gamma_n)$ for all n.

If U contains a fixed point, then all γ_n have uniformly bounded hyperbolic distances in U from this fixed point, and hence the $f^{\circ n}$ are uniformly bounded on U_0, so γ_0 was contractible in U contrary to the hypothesis. A similar argument applies if U contains any point with bounded orbit, in particular if U contains a periodic point or if U is a parabolic domain, so we are left with the cases of Baker domains and wandering domains.

Now suppose that U is a periodic Baker domain, say of period 1; then $\gamma_n \to \infty$ uniformly. Moreover, the winding number of $f(\gamma_n)$ around p tends to ∞. This implies that each $f(\gamma_n)$ contains a simple closed subcurve, say $\hat{\gamma}_{n+1}$, that surrounds p and with hyperbolic length within U tending to 0. By the Collaring Theorem [DH93], it follows that U contains round annuli, say A_n, that contain $\hat{\gamma}_n$ and with moduli $\operatorname{mod} A_n \asymp 1/\ell(\hat{\gamma}_n)$. Since the Julia set is unbounded, there must be infinitely many n for which A_n and A_{n+1} are non-homotopic in U and so that there is an $a_n \in J(f)$ that is surrounded by A_{n+1} but not by A_n. As the moduli of A_n and A_{n+1} become large, the hyperbolic distances in U between points in $\hat{\gamma}_n$ and in $\hat{\gamma}_{n+1}$ must be unbounded (these distances exceed the distances in $\mathbb{C} \setminus \{0, a_n\}$, which after rescaling is equivalent to $\mathbb{C} \setminus \{0, 1\}$, and the annuli A_n, A_{n+1} rescale to essential annuli in $\mathbb{C} \setminus \{0\}$ with large moduli so that exactly one of them surrounds 1). However, the lengths of $\hat{\gamma}_n$ tend to zero, while the distances between $\hat{\gamma}_n$ and $\hat{\gamma}_{n+1}$ are bounded with respect to the hyperbolic metric in U. This contradiction shows that periodic Baker domains are simply connected.

Finally, we discuss the case that U is a wandering domain. In this case, all γ_n must be disjoint, and no γ_n can be contained in γ_m for $m < n$ (or the entire domain U_n would converge to an attracting periodic point, so γ_n would be contractible). Since $f(U_n) \subset U_{n+1}$, it follows that γ_{n+1} separates γ_n from ∞. The argument now continues as above. □

Corollary 2.7 (Asymptotic Values and Multiple Connectivity).
If an entire function has a finite asymptotic value, then all Fatou components are simply connected.

Example 2.8 (A Baker Domain).
The map $f(z) = z + 1 + e^{-z}$ has a Baker domain containing the right half plane.

Clearly, each $z \in \mathbb{C}$ with $\operatorname{Re} z > 0$ has $\operatorname{Re} f(z) - \operatorname{Re} z > 0$, and this quantity is bounded below uniformly in each half plane $\operatorname{Re} z > \delta > 0$. The Fatou set $F(f)$ is connected and contains the right half plane $\mathbb{H} := \{z \in \mathbb{C} \colon \operatorname{Re} z > 0\}$: we have $F(f) = \bigcup_{n \geq 0} f^{-n}(\mathbb{H})$, and the Julia set is an uncountable collection of curves (see the discussion of Cantor Bouquets in Section 4). This example is due to Fatou [Fa26].

Many more results on Baker domains can be found in the recent survey article [Ri08] by Rippon.

Example 2.9 (A Wandering Domain).
The map $f(z) = z - 1 + e^{-z} + 2\pi i$ has a wandering domain.

This example is due to Herman from around 1985. The map $N \colon z \mapsto z - 1 + e^{-z}$ is the Newton map for the function $z \mapsto e^z - 1$. The basin of the root $z = 0$ is clearly invariant under N, and the basins of the roots $z = 2\pi i n$ for $n \in \mathbb{Z}$ are translates of the basin of $z = 0$ by $2\pi i n$. We have $N(z + 2\pi i) = N(z) + 2\pi i$, so the Fatou and Julia sets are $2\pi i$-invariant. Now $f(z) = N(z) + 2\pi i$, and f preserves $F(N)$ and $J(N)$, so $F(f) \supset F(N)$ by Montel's Theorem A.2. Suppose there was a point $z \in F(f) \cap J(N)$. Let $U_0 \subset F(f)$ be a neighborhood of z. Arbitrarily close to z, there are points converging under N to basins of different roots (Theorem 1.7). This would imply that $F(f)$ was connected, but this leads to the following contradiction: initial points in U_0 at arbitrarily short hyperbolic distance within $F(f)$ would have orbits at least 2π apart, in contradiction to the fact that hyperbolic distances never increase. Therefore, $F(f) = F(N)$. Therefore, all basins of roots for the Newton map N move a distance $2\pi i$ in each iteration of f, so they turn into wandering domains for f.

Example 2.10 (A Simply Connected Wandering Domain).
The map $f(z) = z + \sin z + 2\pi$ has a bounded simply connected wandering domain.

This example is due to [Ba84] (more generally, he considers functions of the type $z \mapsto z + h(z)$, where h is a periodic function). In order to describe this example, first observe that for $g(z) = z + \sin z$, all points $n\pi$ (for odd integers n) are superattracting fixed points, so their immediate basins are disjoint. Moreover, $g(z+2\pi) = g(z) + 2\pi$, so the Julia set is 2π-periodic, and as above $J(f) = J(g)$. Therefore, f maps the immediate basin of $n\pi$ to the immediate basin of $(n+2)\pi$ for odd n, and this is a wandering domain for f. All critical points of g are superattracting fixed points, so it follows easily that these basins are simply connected. It is not hard to see that these basins of g are bounded (the imaginary axis is preserved under g and all points other than zero converge to ∞, and the same is true for 2π-translates, so every Fatou component has bounded real parts, and points with large imaginary parts have images with much larger absolute values).

Example 2.11 (A Baker Wandering Domain).
For appropriate values of $c > 0$ and $r_n \to \infty$, the map

$$f(z) = cz^2 \prod_{n \geq 1} (1 + z/r_n)$$

has a multiply connected wandering domain.

This is the original example from Baker [Ba63, Ba76]. The idea is the following: the radii r_n grow to ∞ very fast. There are annuli A_n with large moduli containing points z with $r_n \ll |z| \ll r_{n+1}$. On them, the factors $(1 + z/r_m)$ with $m > n$ are very close to 1 so that even their infinite product can be ignored, while the finitely many bounded factors essentially equal z/r_n, so on the annuli A_n the map $f(z)$ takes the form $c_n z^{n+2}$. The factors are arranged so that f sends A_n into A_{n+1} with degree $n+2$. Therefore all points in A_n escape to ∞. The point $z = 0$ is a superattracting fixed point. The fact that the A_n are contained in a (Baker) wandering domain, rather than a (periodic) Baker domain (basin at ∞) follows from simple connectivity of Baker domains (Theorem 2.5).

Remark 2.12. Baker wandering domains may have any connectivity, finite or infinite [KS08] (the eventual connectivity will be 2 or ∞). These may coexist with simply connected wandering domains [Be]. These examples use a modified construction from Example 2.11: essentially, there is a sequence of radii $0 < \cdots < r_n < R_n < r_{n+1} < \dots$ and concentric annuli $A_n := \{z \in \mathbb{C} \colon r_n < |z| < R_n\}$, $B_n := \{z \in \mathbb{C} \colon R_n < |z| < r_{n+1}\}$. On B_n, we define the map $g(z) = z^{n+1}(1 + z/\sqrt{R_n r_{n+1}})$, which is very close to z^{n+1} near the inner boundary, and to z^{n+2} near the outer. On the annuli

A_n, we use a C^∞ map that is close to a holomorphic map (quasiregular with small dilatation), and arrange things again so that $f(A_n) \subset A_{n+1}$ so that g sends A_n to its image as a branched cover of degree $n+2$. Using the Measurable Riemann Mapping Theorem, it follows that this C^∞-map g is quasiconformally conjugate to an entire holomorphic map f with similar properties, in particular with a wandering annulus. The construction using quasiconformal surgery gives greater flexibility and allows one to determine the dynamical properties of these maps (for instance, in Baker's example it was not known whether the Baker wandering domain was eventually doubly or infinitely connected).

Kisaka and Shishikura show that every point in the Fatou component of A_n lands in A_N for $N > n$ large enough, and use this to conclude that the Fatou component of A_n is doubly connected. Modifying the construction for small values of n, they obtain Baker wandering domains with any finite, or with infinite connectivity (as well as with many further properties; see [KS08, Be]).

There may even be infinitely many disjoint orbits of wandering domains [Ba84, RS99a].

3 Entire Functions of Finite Type and of Bounded Type

The set of all entire functions is extremely large and accommodates a huge amount of dynamical variety, so in order to have good control and to obtain strong results it is often useful to restrict to smaller classes of functions.

Definition 3.1 (Special Classes of Entire Functions).
The class S ("Speiser class") consists of those entire functions that have only finitely many singular values. The class B ("the Eremenko-Lyubich class of functions of bounded type") consists of those entire functions for which all singular values are contained in a bounded set in $\mathbb{C}$.

Remark 3.2. For an entire function of class B, all transcendental singularities over ∞ are logarithmic.

Remark 3.3. An entire function f has finitely many critical *points* and finitely many asymptotic *tracts* if and only if f has the form $f(z) = c + \int P(z')\exp(Q(z'))\,dz'$ for polynomials P, Q.

Theorem 3.4 (Fatou Components for Class S and Class B).
A function of class S does not have Baker domains or wandering domains. A function of class B does not have Baker domains, and it does not have wandering domains in which the dynamics converges to ∞.

Remark 3.5. This result is due to Eremenko and Lyubich [EL84b, EL92]; for wandering domains in class S, see also Goldberg and Keen [GK86]. The proof is outlined below.

Theorem 3.6 (The Fatou-Shishikura-Inequality).
A function of class S has no more non-repelling periodic cycles than singular grand orbits (and in particular, no more than singular values).

Remark 3.7. A singular grand orbit is any set of singular values for which their orbits intersect pairwise. Theorem 3.6 is due to Eremenko and Lyubich [EL89]; compare also Shishikura's proof of the Fatou-Shishikura-inequality [Sh87]. A different proof is due to Epstein [Ep].

Definition 3.8 (Logarithmic Tract).
For an entire function $f \in B$, a *logarithmic tract* T is a connected component of $f^{-1}(\mathbb{C} \setminus D_R)$, where $D_R \subset \mathbb{C}$ is the closed disk around 0 of radius $R > 0$, and R is chosen so that D_R contains all singular values of f as well as $f(0)$.

Remark 3.9. By construction, the restriction $f\colon T \to \mathbb{C} \setminus D_R$ is a covering. For transcendental entire functions, it is a universal covering: otherwise, the mapping degree from T to $\mathbb{C} \setminus D_R$ would be finite and ∂T would be a circle, and by the maximum modulus principle this would imply that T contained a neighborhood of ∞, so f would be a polynomial.

For a transcendental entire function, the number of logarithmic tracts may be finite or infinite. A function of finite order has only finitely many tracts by Denjoy-Carleman-Ahlfors-theorem (Theorem 1.17). The condition $|f(0)| < R$ is required for the construction of logarithmic coordinates as described below.

Definition 3.10 (Logarithmic Coordinates).
For a function $f \in B$, a corresponding function in logarithmic coordinates is defined as follows: let T_j be the logarithmic tracts of f and let $T'_{j,k}$ be the components of $\log(T_j)$ (so that $\exp(T'_{j,k}) = T_j$ and $T'_{j,k+1} = T'_{j,k} + 2\pi i$). Then a function F corresponding to f in logarithmic coordinates is any holomorphic function

$$F\colon \bigcup_{j,k} T'_{j,k} \to H_{\log R} := \{z \in \mathbb{C}\colon \operatorname{Re} z > \log R\}$$

satisfying $\exp \circ F = f \circ \exp$. Any component $T'_{j,k}$ is called a *tract* of F.

Remark 3.11. Since $\exp\colon H_{\log R} \to \mathbb{C} \setminus D_R$ and $f\colon T_j \to \mathbb{C} \setminus D_R$ are universal covers, while $\exp\colon T'_{j,k} \to T_j$ is a conformal isomorphism, it follows that each restriction $F\colon T'_{j,k} \to H_{\log R}$ is also a conformal isomorphism. Note that F is determined by f only up to additive integer multiples of $2\pi i$, separately for each tract (in general, F is not defined on any right half plane, or on any connected set containing all the tracts).

Entire functions of bounded type are defined so that logarithmic coordinates exist outside any disk D_R that is large enough so as to contain all singular values and the point $f(0)$ (the latter condition is needed in order to assure that $T_j \not\ni 0$, so logarithms can be taken on all tracts). Logarithmic coordinates were introduced to transcendental dynamics by Eremenko and Lyubich [EL84a, EL84b, EL92].

The following fundamental estimate of Eremenko and Lyubich [EL92, Lemma 1] is very useful. Write $\mathbb{H} := \{z \in \mathbb{C} \colon \operatorname{Re} z > 0\}$ for the right half plane.

Lemma 3.12 (Expansion on Logarithmic Tracts).
If T' is a logarithmic tract and $F\colon T' \to \mathbb{H}$ is a conformal isomorphism, then

$$|F'(z)| \geq \frac{1}{4\pi} \operatorname{Re} F(z) \,. \tag{1}$$

Remark 3.13. All we are using about T' is that it is simply connected and disjoint from $T' + 2\pi i k$ for $k \in \mathbb{Z} \setminus \{0\}$.

Proof. Since $F\colon T' \to \mathbb{H}$ is a conformal isomorphism, it has a conformal inverse $G\colon \mathbb{H} \to T'$. Let D be the open disk around $F(z)$ with radius $R = \operatorname{Re} F(z)$. Then by the Koebe 1/4-theorem, $G(D)$ contains a disk around $G(F(z)) = z$ of radius $R|G'(F(z))|/4 = \operatorname{Re} F(z)/4|F'(z)|$; but by periodicity of the tracts in the vertical direction, the image radius must be at most π, and (1) follows. □

Sketch of proof of Theorem 3.4. For a function $F\colon \bigcup T'_{j,k} \to \mathbb{H}$ in logarithmic coordinates, we claim that the set of points z with $\operatorname{Re} F^{\circ n}(z) \to \infty$ has no interior (here we assume for simplicity that, possibly by translating coordinates, $F(T'_{j,k}) = \mathbb{H}$ for all tracts $T'_{j,k}$). The key to this is Lemma 3.12. Suppose that $F\colon \bigcup T'_{j,k} \to \mathbb{H}$ has an open set in the escaping set, say the round disk $D_\varepsilon(z)$ around some point z with $\operatorname{Re} F^{\circ n}(z) \to \infty$. Then (1) implies $|F'(F^{\circ n}(z))| \to \infty$, hence $|(F^{\circ n})'(z)| \to \infty$. Using the Koebe 1/4 theorem it follows that $I(F)$ contains disks of radii $\varepsilon|(F^{\circ n})'(z)|/4$. Since these radii must be bounded by π, it follows that $\varepsilon = 0$. This proves the claim, and this implies that for functions $f \in B$, the escaping set $I(f)$ has empty interior. This proves Theorem 3.4, except for the fact that functions $F \in S$ do not have wandering domains. We do not describe this proof here. It uses the essential ideas of Sullivan's proof in the rational case (see [Mi06]), in particular quasiconformal deformations, and depends on the fact that small perturbations of functions in class S live in a finite-dimensional space. □

Corollary 3.14 (Entire Functions with Empty Fatou Set).
If an entire function f in class B has the property that all its singular values are strictly preperiodic or escape to ∞, then f has empty Fatou set.

Indeed, any Fatou component of f would require an orbit of a singular value that converges to an attracting or parabolic periodic orbit, or accumulates on the boundary of a Siegel disk, by Theorems 2.3 and 3.4.

Functions $f \in B$ have a most useful built-in partition with respect to which symbolic dynamics can be defined for those orbits that stay sufficiently far away from the origin.

Definition 3.15 (Symbolic Dynamics and External Address).
Consider a function $f \in B$ and let $R > 0$ be large enough so that $D_R(0)$ contains all singular values as well as $f(0)$, and let $F\colon T'_{j,k} \to H_{\log R}$ be a function in logarithmic coordinates corresponding to f. Here k ranges through the integers so that $T'_{j,k+1} = T'_{j,k} + 2\pi i$, and j ranges through some (finite or countable) index set J, say.

The *external address* of a point $z \in \mathbb{C}$ so that $F^{\circ n}(z) \in \bigcup_{j,k} T'_{j,k}$ for all $n \geq 0$ is the sequence $(j,k)_n$ so that $f^{\circ n}(z) \in T'_{(j,k)_n}$ for all n.

Our "external address" is an "itinerary" with respect to the partition given by the $T'_{j,k}$; however, in order to achieve notational consistency, we prefer to use the term "itinerary" in a different context: the analogy are dynamic rays of polynomials, our external addresses correspond to external angles (or their expansion in base d, the degree of the polynomials); in many cases, one can define itineraries of dynamic rays, so that different dynamic rays have the same itinerary if and only these two rays land at a common point. See the discussion in [SZ03b, Section 4.5] or [BS].

If $\underline{s} = (j,k)_n$ is an external address, define

$$J_{\underline{s}} := \{z \in \mathbb{C} \colon z \text{ has external address } \underline{s}\} .$$

All sets $J_{\underline{s}}$ are obviously disjoint. Each set $J_{\underline{s}}$ is a closed set, either empty or unbounded: $J_{\underline{s}} \cup \{\infty\}$ is the nested intersection of the non-empty compact sets $\{z \in \mathbb{C} \colon f^{\circ n}(z) \in T'_{(j,k)_n} \text{ for } n = 0,1,\ldots,N \text{ and } N \in \mathbb{N}\}$, so $J_{\underline{s}} \cup \{\infty\}$ is compact and non-empty.

An important special case that simplifies the situation occurs if all $T'_{j,k} \subset H_{\log R}$; this occurs if the original entire function f has an attracting fixed point that attracts all singular values of f. In this case, all $J_{\underline{s}} \cup \{\infty\}$ are nested intersections of non-empty compact and connected sets, hence compact and connected. If the tract boundaries are sufficiently well-behaved (for instances, so that they are eventually parametrized by their real parts, or more generally that their boundaries do not "wiggle" too much), then it can be shown that each non-empty $J_{\underline{s}}$ is a curve that connects some point $z \in \mathbb{C}$ to ∞. These conditions are satisfied in particular when f has finite order in which all singular values are attracted by the same fixed point (see Barański [Bar07]). A prototypical situation in which this had been considered relatively early is the case of exponential maps with attracting fixed points; consider [DK84]. More general results showing that the $J_{\underline{s}}$ are curves, and also providing counterexamples in other cases, are established in [RRRS]. In many cases, all but possibly one point in $J_{\underline{s}}$ are escaping points; a frequent concept that occurs in this case is that of a Cantor bouquet. We will discuss this situation in more detail in Section 4. However, there are entire functions of bounded type, even hyperbolic ones, for which the Julia set does not contain curves. An example of this was constructed in [RRRS].

Theorem 3.16 (Non-Existence of Curves).
There are hyperbolic entire functions of bounded type for which every path component of $J(f)$ is bounded (or even a point).

Here, we give a few ideas of the arguments in Theorems 3.16. The work takes place in logarithmic coordinates and consists of two steps: first we construct a domain $T \subset \mathbb{H}$ with a conformal isomorphism $G \colon T \to \mathbb{H}$ sending the boundary point $\infty \in \partial T$ to the boundary point $\infty \in \partial\mathbb{H}$ and so that all path components of

$$J(G) := \{z \in T \colon G \text{ can be iterated infinitely often starting at } z\}$$

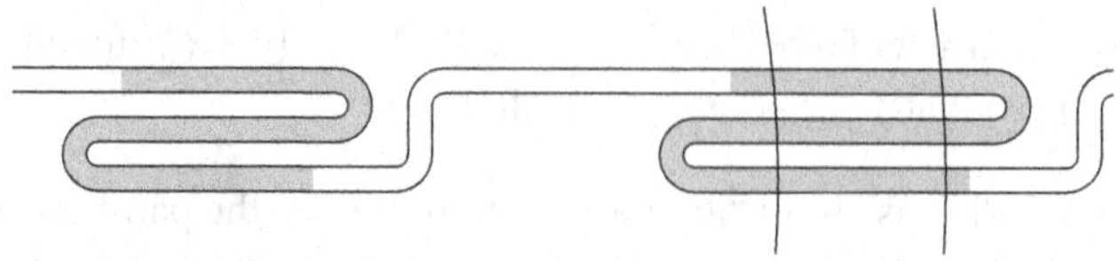

Fig. 1 The wiggles in the proof of Theorem 3.16. Shown is (part of) the tract T with two wiggles (the tract extends to the right towards infinity and has infinitely many wiggles). The shaded domain in the left wiggle maps to a large half annulus that is indicated by two circular arcs, and this image annulus must intersect the second wiggle so that each curve through the second wiggle intersects the image annulus three times

are bounded. Then an approximation method is used to show that there exists an entire function of bounded type f that yields a function F in logarithmic coordinates that is close enough to G so as to have similar properties.

For the first step, the domain T is a long sequence of wiggles as indicated in Figure 1: there are two real parts $x_1 < y_1$ so that T transverses the interval of real parts in $[x_1, y_1]$ three times; this part of T is called the first wiggle. The second wiggle of T is such that $G^{-1}(T)$ sends the second wiggle into the first one so that $G^{-1}(T)$ has to transverse the interval $[x_1, y_1]$ nine times. The third wiggle has the property that its G-preimage in T "wiggles" three times through the second wiggle, and the second preimage wiggles nine times through the first. Iterating this argument, it follows that any curve in $J(G) = \bigcap_{n \geq 0}(G^{-1})^{\circ n}(T)$ must connect real parts x_1 and y_1 infinitely often in alternating order, and this is impossible. Therefore, points in $J(G)$ to the left of the first wiggle have bounded path components. Similarly, it follows that no path component within $J(G)$ can traverse any of the infinitely many wiggles in T further to the right.

The second step in the construction consists of finding an entire function f that has a map F in logarithmic coordinates close to G. This is a classical approximation procedure (see for instance [GE79]). A description of this method can be also found in [RRRS].

The "size" of the wiggles must be large, in the sense that the ratio x_2/x_1 between the ends of the wiggles must tend to ∞. To see this, we decompose $G = G_2 \circ G_1$, where $S = \{z \in \mathbb{C}\colon \operatorname{Re} z > 0, |\operatorname{Im} z| < \pi/2\}$ is a one-sided infinite strip and $G_1\colon T \to S$, $G_2\colon S \to \mathbb{H}$ are conformal isomorphisms fixing the boundary points ∞. Then G_2 is essentially the exponential map, except near the left end, and G_1 can be thought of as an affine map along the long straight pieces of T, and just "bending" the tract T straight into S near the turns.

If a wiggle satisfies $y_n/x_n < \kappa$, then the log of this wiggle (i.e., its preimage under G_2) has length at most $\log \kappa$, and the same is true for the G-preimage. After the next iterated preimage, a wiggle of bounded length at large real part becomes small, and we do not get bounded path components. Therefore, $y_n/x_n \to \infty$ (even extremely fast) is necessary for counterexamples. This implies fast growth of G: if y_n/x_n is large, that means that the tract T bends back far to the left (it has big "wiggles"), and therefore T contains points that are (conformally) far out in T (so that $\operatorname{Re} F(z)$ is large), but $\operatorname{Re} z$ is small.

This condition is related to finite order as follows: if an entire function f has finite order as in Definition 1.15, then the corresponding function in logarithmic coordinates satisfies

$$\lim_{\operatorname{Re} z\to+\infty} \sup \log \operatorname{Re} F(z)/\operatorname{Re} z < \infty \,. \tag{2}$$

We thus use (2) as a definition for a function in logarithmic coordinates to have finite order. Geometric function theory thus implies that large wiggles entail fast growth of F, and this is restricted if F or equivalently f have finite order.

4 The Escaping Set

A significant amount of work on the dynamics of an entire function f has been, and still is, concerned with the set $I(f)$ of escaping points. There are several reasons why the escaping set is interesting:

- $I(f)$ is a non-empty invariant set (and unlike the Julia set, it is always a proper subset of $\mathbb{C}$).
- The set $I(f)$ often has useful structure, such as the union of curves to ∞ ("dynamic rays" and "Cantor Bouquets").
- In many cases, several curves in $I(f)$ connect the same point $z \in \mathbb{C}$ to infinity, even if $z \notin I(f)$ (in other words, several dynamic rays land at a common point). This gives additional structure to the dynamic plane, such as in the famous "puzzle theory" of polynomials [Sch07b, RSch08, Ben].
- $I(f)$ is a subset of the dynamical plane that is often easy to control; for functions of bounded type, it is a subset of the Julia set and thus makes it possible for instance to give lower bounds on the Hausdorff dimension of the Julia set (see Section 5).
- For polynomials, especially with non-escaping critical orbits, $I(f)$ has a very simple structure, and $J(f) = \partial I(f)$; so many descriptions of $J(f)$ for polynomials are based on $I(f)$, and analogous approaches are desired also for entire functions.

While many discussions in the earlier sections hold more generally for iterated meromorphic functions, the discussion of the escaping set (and of functions of bounded type) make sense only for entire functions, in the same sense the polynomial dynamics is much easier to describe than the dynamics of rational mappings.

The fundamental results about the escaping set are due to Eremenko [Er89] (see also Domínguez [Do98]). He established in particular the following.

Theorem 4.1 (The Escaping Set).
Every transcendental entire function f has the following properties:

- $I(f) \cap J(f) \neq \emptyset$;
- $J(f) = \partial I(f)$;
- *all components of* $\overline{I(f)}$ *are unbounded.*

Examples of Fatou components in $I(f)$ have been given in Examples 2.8 (a Baker domain), 2.9, and 2.10 (wandering domains).

We would like to review especially Fatou's Example 2.9 with $f(z) = z + 1 + e^{-z}$. There is a single Fatou component, it contains the right half plane, and in it the orbits converge quite slowly to ∞. However, for integers k, points on $i(2k+1)\pi + \mathbb{R}^-$ converge very quickly to ∞ (at the speed of iterated exponentials); these are countably many curves in $I(f)$ (dynamic rays), and the backwards orbits of these curves form many more curves in $I(f)$. In fact, $I(f)$ contains uncountably many curves to ∞ (a Cantor bouquet).

Eremenko [Er89] raised some fundamental questions on $I(f)$ that inspired further research.

Question 4.2 (Eremenko's Questions on $I(f)$).

Weak version Is every component of $I(f)$ unbounded?
Strong version Can every $z \in I(f)$ be connected to ∞ within $I(f)$?

These questions have inspired a substantial amount of work, and are often referred to as *Eremenko's conjecture* (in its weak and strong form). There are various partial results on them, positive and negative. Rippon and Stallard [RS05a, RS05b] showed the following.

Theorem 4.3 (Baker Wandering Domains and $I(f)$).
Let f be an entire function. Then $I(f)$ always has an unbounded component. If f has a Baker wandering domain, then $I(f)$ is connected.

These results are based on a study of the set $A(f)$ as described below. Rempe [Re07] established sufficient conditions for the weak version of Eremenko's question, as follows.

Theorem 4.4 (Unbounded Components of $I(f)$).
If f is an entire function of bounded type for which all singular orbits are bounded, then every component of $I(f)$ is unbounded.

Substantial attention has been given to the speed of escape for points in $I(f)$. A classical lemma, due to Baker [Ba81], is the following.

Lemma 4.5 (Homogeneous Speed of Escape in Fatou Set).
If $z, w \in I(f)$ are two escaping points from the same Fatou component of an entire function f, then $\log|f^{\circ n}(z)|/\log|f^{\circ n}(w)|$ is bounded as $n \to \infty$. If z is in a (periodic) Baker domain, then $\log|f^{\circ n}(z)| = O(n)$.

This result follows from the observation that the hyperbolic distance between w and z in the hyperbolic metric of $F(f)$ cannot increase, and that this distance is essentially bounded below by the hyperbolic distance of $\mathbb{C} \setminus \mathbb{R}_0^-$.

Unlike other types of periodic Fatou components, Baker domains need not contain singular values. However, if a Baker domain exists, the set of singular values must be unbounded by Theorem 3.4; there is a stronger result by Bargmann [Ba01] and Rippon and Stallard [RS99b] saying that, for every entire function with a Baker domain, there exists a constant $C > 0$ so that every annulus $\{z \in \mathbb{C}\colon r/C < |z| < rC\}$ (for r sufficiently large) intersects the set of singular values of f. More results on Baker domains, including classification results, as well as examples of entire functions with infinitely many Baker domains, can be found in Rippon's survey [Ri08].

There are a number of further interesting subsets of $I(f)$ defined in terms of their speed of escape.

Definition 4.6 (The Sets $A(f)$, $L(f)$ and $Z(f)$).
For an entire function f, the set $A(f)$ is defined as follows: given a large radius $R = R_0 > 0$, define recursively $R_{n+1} := M(R_n, f)$ and set

$$A_R(f) := \{z \in \mathbb{C}\colon |f^{\circ n}(z)| \geq R_n \text{ for all } n \geq 0\},$$

$$A(f) := \bigcup_{n\geq 0} f^{-n}(A_R(f)).$$

The set $Z(f)$ is defined as

$$Z(f) := \left\{z \in \mathbb{C}\colon \frac{\log\log|f^{\circ n}(z)|}{n} \to \infty \text{ as } n \to \infty\right\}.$$

The set $L(f)$ is defined as

$$L(f) := \left\{z \in I(f)\colon \limsup \frac{\log|f^{\circ n}(z)|}{n} < \infty\right\}.$$

The set $A(f)$ describes essentially those points that escape to ∞ as fast as possible; it has been introduced by Bergweiler and Hinkkanen [BH99]. It is quite easy to see that it does not depend on R provided R is large enough so that $J(f) \cap D_R(0) \neq \emptyset$. The set $Z(f)$ describes those points that "zip" to ∞ much faster than what is possible for polynomials (for the latter, $\log\log|f^{\circ n}| = O(n)$). The set $L(f)$ describes slow escape (the letter L stands for s*L*ow or the German *L*angsam). These sets are among those introduced by Rippon and Stallard; they are well on their way towards exhausting the alphabet with interesting sets of different speeds of escape. Unfortunately, the speed of escape is not compatible with the alphabetic order of their letters. The following is a subset of what is known on these sets.

Theorem 4.7 (Speed of Escape).
For every transcendental entire function f, the sets $A(f)$, $L(f)$ and $Z(f)$ satisfy the following:

1. *$J(f) \cap A(f) \neq \emptyset$;*
2. *$A(f) \subset Z(f)$, and all three sets $A(f)$, $L(f)$, and $Z(f)$ are completely invariant;*
3. *$J(f) = \partial A(f) = \partial L(f) = \partial Z(f)$;*
4. *every Fatou component U has either $U \cap Z(f) = \emptyset$ or $U \subset Z(f)$; moreover, it has either $U \cap A(f) = \emptyset$ or $\overline{U} \subset A(f)$, and also either $U \cap L(f) = \emptyset$ or $U \subset L(f)$.*
5. *If f has no wandering domains, then $J(f) = \overline{A(f)} = \overline{Z(f)}$;*
6. *every (periodic) Baker domain lies in $L(f)$;*
7. *if U is a Baker wandering domain, then $\overline{U} \subset A(f)$;*

8. *if f has a Baker wandering domain, then $A(f)$ and $I(f)$ are connected, and each component of $A(f) \cap J(f)$ is bounded; otherwise, each component of $A(f) \cap J(f)$ is unbounded;*
9. *every component of $A(f)$ is unbounded;*
10. *wandering domains may escape slow enough for $L(f)$ and fast enough for $A(f)$.*

Item (1) comes out of Eremenko's proof that $I(f)$ is non-empty (compare Theorem 1.8). In (2), invariance is built into the definition, and the inclusion comes from [BH99]. For Statement (3), note that $\partial A(f)$, $\partial L(f)$, and $\partial Z(f)$ are all closed invariant sets, so they contain $J(f)$ by Theorem 1.7, and the claim reduces to the fact that any Fatou component that intersects $A(f)$, $L(f)$, or $Z(f)$ is contained in these sets. This follows from Lemma 4.5 and also implies (4) (see [RS00]). Statement (5) is from [RS00, BH99]. Statement (6) is due to Baker (see Lemma 4.5). Statements (7) and (9) are from [RS05a], while (8) is unpublished recent work by Rippon and Stallard. Wandering domains in $L(f)$ were given in Examples 2.9 and 2.10, while Baker wandering domains are in $A(f)$; this is (10).

Recent work by Rippon and Stallard is on "spider's webs": these are subsets of $I(f)$ for which every complementary component is connected (such as the orbit of a Baker wandering domain, together with certain parts of $I(f)$ connecting the various wandering domains). Rippon and Stallard propose the idea that this feature may be no less common and prototypical for the dynamics of entire functions as Cantor Bouquets are nowadays often considered to be.

Theorem 4.8 (Slow Escape Possible).
For every real sequence $K_n \to \infty$, there is a $z \in I(f) \cap J(f)$ so that $|f^{\circ n}(z)| < K_n$ for all sufficiently large n.

This result is also due to Rippon and Stallard [RS].

It had been observed by Fatou that $I(f)$ often contains curves to ∞. This was shown for exponential maps in [DGH], and for more general entire functions having logarithmic tracts satisfying certain geometric conditions in [DT86]. This leads to the following.

Definition 4.9 (Dynamic Ray).
A *dynamic ray tail* is an injective curve $\gamma\colon (\tau,\infty) \to I(f)$ so that

- $f^{\circ n}(\gamma(t)) \to \infty$ as $t \to \infty$ for every $n \geq 0$, and
- $f^{\circ n}(\gamma(t)) \to \infty$ as $n \to \infty$ uniformly in t.

A *dynamic ray* is a maximal injective curve $\gamma\colon (0,\infty) \to I(f)$ so that for every $\tau > 0$, the restriction $\gamma|_{(\tau,\infty)}$ is a dynamic ray tail. An *endpoint* of a dynamic ray γ is a point a with $a = \lim_{t \searrow \tau} \gamma(t)$.

Remark 4.10. Dynamic rays are sometimes called "hairs"; we prefer the term "ray" in order to stress the similarity to the polynomial case.

Eremenko asked whether every point in $I(f)$ can be connected to ∞ by a curve in $I(f)$ (compare Conjecture 4.2). This conjecture has been confirmed for exponential maps with attracting fixed points by Devaney and Krych [DK84] (in this case, the Julia set is an uncountable collection of curves), in [SZ03a] for arbitrary exponential maps (including the case when $J(f) = \mathbb{C}$), and in [RoS08] for arbitrary cosine maps $z \mapsto ae^z + be^{-z}$. For functions of bounded type and finite order, this was shown by Barański [Bar07] under the additional condition that there is an attracting fixed point that attracts all singular values, and by Rottenfußer, Rückert, Rempe and Schleicher [RRRS] in general. We state a particular case of the result here (the precise statement in [RRRS] is phrased in geometric terms on the tracts).

Theorem 4.11 (Dynamic Rays, Bounded Type and Finite Order).
For entire functions of bounded type and finite order, or compositions thereof, $I(f)$ consists entirely of rays, possibly with endpoints (the strong version of Eremenko's conjecture holds).

However, there are counterexamples to this question of Eremenko, even for entire functions of bounded type; the counterexamples of course have infinite order, but they can be quite close to having finite order [RRRS] (these had been mentioned in Theorem 3.16).

In many cases, points on dynamic rays are in $A(f)$. This is recent unpublished work by Rippon and Stallard. This is the case if f is a composition of functions of finite order and bounded type, and more generally for all functions in [RRRS] that satisfy an additional "bounded gulfs" condition.

Certain Julia sets, especially of hyperbolic entire functions, are described in terms of *Cantor Bouquets*. Roughly speaking, a Cantor bouquet X is an uncountable union of disjoint curves in the Julia set (dynamic rays with endpoints) so that X has no interior point and each point in X has a neighborhood that intersects X in a set that contains a homeomorphic copy of a Cantor set cross an interval (but every neighborhood of every point in the Cantor bouquet also contains endpoints of curves, so locally X is *not* homeomorphic to a Cantor set cross an interval). Prototypical examples of Cantor bouquets are Julia sets of exponential maps with attracting fixed points [DK84], but there is a substantial body of work identifying Cantor bouquets in many more Julia sets of entire functions (see, for instance, [DT86] or recent work by Rempe and coauthors). Aarts and Oversteegen [AO93] have given an abstract definition of Cantor bouquets, and they have shown some universality properties.

We conclude this section with an amusing result by Mayer [Ma90].

Definition 4.12 (Explosion Set).
A set $X \subset \mathbb{C}$ is called an *explosion set* if $X \cup \{\infty\}$ is connected in $\overline{\mathbb{C}}$, but X is totally disconnected.

Theorem 4.13 (Explosion Set).
For every exponential map $z \mapsto \lambda e^z$ with an attracting fixed point, the landing points of dynamic rays form an explosion set in $\mathbb{C}$.

It seems likely that an analogous result should hold in much greater generality, possibly for bounded type entire functions of finite order, at least when there is an attracting fixed point attracting all singular values.

5 Hausdorff Dimension

In this section, we describe a number of results on the Hausdorff dimension of Julia sets of entire functions. We will start with a few results on specific prototypical families and conclude the section with a collection of results on more general maps. The first two results are due to McMullen [Mc87].

Theorem 5.1 (Hausdorff Dimension 2).
For every map $z \mapsto \lambda e^z$, the Julia set (and even the escaping set) has Hausdorff dimension 2.

Theorem 5.2 (Julia Set of Positive Measure).
For every map $z \mapsto ae^z + be^{-z}$ with $ab \neq 0$, the Julia set (and even the escaping set) has positive 2-dimensional Lebesgue measure.

In this section, we will only sketch some of the proofs (see also the review article [Sch07a]); they use parabolas of the following type (for $\xi > 0$ and $p > 1$):

$$R_{p,\xi} := \{x + iy \in \mathbb{C}\colon |x| > \xi \text{ and } |y| < |x|^p\}.$$

For large p, the set $R_{p,\xi}$ is the complement to a narrow parabola around the imaginary axis.

The proof of Theorem 5.2 goes roughly as follows: we consider a "standard square" $Q \subset R_{p,\xi}$ with sides parallel to the coordinate axes and of side length 2π, where ξ is large depending on a and b. On $R_{p,\xi}$, the map $E(z) = ae^z + be^{-z}$ "essentially" equals ae^z if $\operatorname{Re} z > \xi$, and it equals be^{-z} if $\operatorname{Re} z < -\xi$. Thus if the real parts of Q are in $[x, x+2\pi]$, then the image of Q under E almost equals an annulus between radii e^x and $e^{2\pi}e^x$ (up to factors of $|a|$ or $|b|$), so the fraction (in terms of Lebesgue area) of $E(Q)$ that falls outside of $R_{p,\xi}$ is on the order of $1/\sqrt[p]{e^x} = e^{-x/p}$; the remaining part has real parts of absolute value at least $e^{x/p}$. The image annulus is covered by a large number of standard squares at much bigger real parts, and the argument can be repeated. The idea is that in every step, the relative fraction of area lost into $\mathbb{C} \setminus R_{p,\xi}$ is so small that the total relative loss, in the product over infinitely many steps, is less than one. In other words, a positive fraction of the area within each standard square is in $I(f)$, so $I(f)$ has positive (in fact, infinite) 2-dimensional Lebesgue measure.

The proof of Theorem 5.1 is similar. We consider the truncated quadrant $S_\xi := \{z \in \mathbb{C}\colon |\operatorname{Im} z| < \operatorname{Re} z, \operatorname{Re} z > 0\}$, and the set of orbits that escape within S_ξ: that is, $I_\xi := \{z \in I(f)\colon \text{the orbit of } z \text{ is in } S_\xi\}$. In each step, three quarters of orbits fall outside of S_ξ, so I_ξ has measure zero. We want to estimate its Hausdorff dimension. In this case, a standard square is a square in S_ξ of side length $\pi/2$, again with sides parallel to the coordinate axes, and adjusted vertically so that its images under $E_\lambda(z) = \lambda e^z$ (which are always one quarter of an annulus) are contained again in S_ξ. If Q is a standard square, and $E_\lambda(Q)$ has real parts greater than N, then $E_\lambda(Q)$ intersects approximately N horizontal lines of standard squares, each containing N standard squares (up to bounded factors; of course, the number of standard squares in each line is different). Transporting the standard squares in $E_\lambda(Q)$ back into Q, we replace each standard square by approximately N^2 smaller squares of side length $1/N$. Again, this process is iterated (which increases N and thus makes the bounded factors irrelevant), and this establishes Theorem 5.1.

The next statement combines results of Karpińska, Schleicher, and Barański [Ka99, SZ03a, Bar07]

Theorem 5.3 (Dynamic Rays of Dimension One).
For every exponential or cosine map, the set of dynamic rays has Hausdorff dimension 1. *The same holds for significantly larger classes of maps, including those of bounded type and finite order which have an attracting fixed point that contains all singular values in its immediate basin.*

The leads to the following interesting result discovered by Karpínska: for exponential maps with attracting fixed points, the Julia set is an uncountable union of components, each of which is a dynamic ray with a single endpoint. The dynamic rays have Hausdorff dimension 1 by Theorem 5.3, while the entire Julia set has Hausdorff dimension 2 by Theorem 5.1: therefore, all the dimension sits in the endpoints of the dynamic rays. The situation is even more extreme for cosine maps, where the Julia set has positive Lebesgue measure (Theorem 5.2). For postcritically finite cosine maps, the endpoints of rays are everything but the 1-dimensional set of rays, and we have the following [Sch07b]:

Theorem 5.4 (The Dimension Paradox for Cosine Maps).
Every postcritically finite cosine map $C(z) = ae^z + be^{-z}$ *has the following properties:*

- *the Julia set equals* $\mathbb{C}$;
- *every dynamic ray lands at a unique point in* $\mathbb{C}$;
- *every point in* $\mathbb{C}$ *is either on a dynamic ray, or it is the landing point of one, two, or four dynamic rays;*
- *the set of dynamic rays has Hausdorff dimension* 1;
- *the landing points of these rays are the complement of the one-dimensional set of rays.*

The motor for all the results that the Hausdorff dimension of dynamic rays is 1 is the following lemma of Karpińska [Ka99].

Lemma 5.5 (The Parabola Lemma).
For every map $z \mapsto \lambda e^z$ *or* $z \mapsto ae^z + be^{-z}$*, the set of escaping parameters that escape within the parabola* $P_p = \{z \in \mathbb{C} \colon |\mathrm{Im}\, z| < |\mathrm{Re}\, z|^{1/q}\}$ *has Hausdorff dimension at most* $1 + 1/q$.

The idea of this lemma is that points within this parabola are forced to have fast growing real parts, and standard squares as above are replaced by $N^{1+1/q}$ smaller standard squares (similarly as in the proof of Theorem 5.1, except that the quadrant $\tilde{P}_\xi$ is replaced by the parabola P_q that occupies ever smaller fractions of large centered annuli, as the radii increase). For details, see [Ka99, Sch07a]. The topological claims in Theorem 5.4 use standard contraction arguments using the hyperbolic metric in the finite postsingular set; see [Sch07b].

The results cited so far show that escaping points of exponential maps have the property that all the points on the dynamic rays have dimension 1, while the endpoints have dimension 2. A more precise study, investigating how the endpoints can be subdivided into sets of dimensions between 1 and 2, was done by Karpińska and Urbański [KU06].

We already mentioned in Section 4 that Rippon and Stallard have new results showing that in much greater generality, points on dynamic rays escape to ∞ as fast as possible; this should imply that in many cases, the dimension of dynamic rays equals 1. Barański [Bar08] has shown this for bounded type and finite order maps for which a single attracting fixed point absorbs all singular orbits, and results by Rempe [Re] allow to lift the condition on the attracting fixed point.

The following result on the area of the Fatou set was a conjecture of Milnor and has been shown by Schubert [Schu08]. The proof is in a similar spirit as some of the previous arguments.

Theorem 5.6 (Unbounded Fatou Component of Finite Area).
For the map $z \mapsto \sin z$, the Fatou set is contained in basins of parabolic fixed points. The Fatou set intersects every vertical strip of width 2π in an unbounded set of finite planar Lebesgue measure.

Next we state a useful result of Bock [Bo96].

Theorem 5.7 (Typical Orbits of Entire Functions).
Let f be a non-constant entire function, let $S(f)$ be its set of singular values, and let $P(f) := \overline{\bigcup_{n\geq 0} f^{\circ n}(S_f)}$ be the postsingular set. Then at least one of the following is true:

1. *almost every orbit converges to $P(f) \cup \{\infty\}$;*
2. *$J(f) = \mathbb{C}$ and almost every $z \in \mathbb{C}$ is recurrent.*

This result allows to describe more crisply the behavior of typical orbits. A typical consequence is the following: if an entire function has finitely many singular values, and all of them eventually land on repelling cycles, then if $I(f)$ has positive measure it must have full measure in $\mathbb{C}$ (i.e., $\mathbb{C} \setminus I(f)$ must have measure zero). This applies for instance to postcritically finite maps in the cosine family.

Now we state a number of results on more general entire functions. We start by a result of Stallard [St96].

Theorem 5.8 (Hausdorff Dimension Greater than 1).
For every entire map of bounded type, the Hausdorff dimension of $J(f)$ strictly exceeds 1.

It is still an open question whether there exists an entire function the Julia set of which has Hausdorff dimension 1. A recent result by Bergweiler, Karpińska, and Stallard is the following [BKS].

Theorem 5.9 (Hausdorff Dimension 2).
For every entire map of bounded type and finite order, the Julia set has Hausdorff dimension 2. More generally, if f has bounded type and for every $\varepsilon > 0$ there is a $r_\varepsilon > 0$ so that

$$\log\log|f(z)| \le (\log|z|)^{q+\varepsilon}$$

for $|z| > r_\varepsilon$, then the Julia set has Hausdorff dimension at least $1 + 1/q$.

(Note that functions of finite order satisfy the condition for $q = 1$, so the second statement generalizes the first.)

The following result is due to Stallard [St97].

Theorem 5.10 (Explicit Values of Hausdorff Dimension).
For every $p \in (1,2]$, there is an explicit example of an entire function for which the Julia set has Hausdorff dimension p.

Finally we would like to mention the recent survey article on Hausdorff dimension of entire functions by Stallard [St08].

6 Parameter Spaces

In addition to the study of individual complex dynamical systems, a substantial amount of attention is given to spaces (or families) of maps. This work comes in at least two flavors: in most of the early work, specific usually complex one-dimensional families of maps are considered; in transcendental dynamics, this is most often the family of exponential maps (parametrized as $z \mapsto \lambda e^z$ with $\lambda \in \mathbb{D}^*$ or $z \mapsto e^z + c$ with $c \in \mathbb{C}$). Pioneering work in this direction was by Baker and Rippon [BR84], Devaney, Goldberg, and Hubbard [DGH], and by Eremenko and Lyubich [EL92]. Another flavor is to consider larger "natural" parameter spaces. In rational dynamics, such natural parameter spaces are finite-dimensional and easy to describe explicitly (such as the family of polynomials or rational maps of given degree $d \ge 2$). In transcendental dynamics, reasonable notions of natural parameter spaces are less obvious. Early work in this direction was done for instance by Eremenko and Lyubich [EL92].

We start with a few remarks on general parameter spaces of entire functions and especially a recent theorem by Rempe. We then discuss the family of exponential maps as the space of prototypical entire functions and compare it with the Mandelbrot set as the space of prototypical polynomials. We conclude this section with a question of Euler.

The following definition is often seen as the natural parameter space of transcendental entire functions of bounded type.

Definition 6.1 (Quasiconformally Equivalent Entire Functions).
Two functions f, g of bounded type are called *quasiconformally equivalent near ∞* if there are quasiconformal homeomorphisms $\varphi, \psi \colon \mathbb{C} \to \mathbb{C}$ such that $\varphi \circ f = g \circ \psi$ near ∞.

One way to interpret this definition is as follows. We write $g = (\varphi \circ f \circ \varphi^{-1}) \circ (\varphi \circ \psi^{-1})$, so g is a quasiconformally conjugate function to f, postcomposed with another quasiconformal homeomorphism. In analogy, every quadratic polynomial is conjugate to one of the form $z^2 + c$, so any two quadratic polynomial differ from each other by conjugation, postcomposed with an automorphism of $\mathbb{C}$ (here, there are few enough marked points so that for the postcomposition, it suffices to use complex automorphisms).

Theorem 6.2. *Let f, g be two entire functions of bounded type that are quasiconformally equivalent near ∞. Then there exist $R > 0$ and a quasiconformal homeomorphism $\vartheta\colon \mathbb{C} \to \mathbb{C}$ so that $\vartheta \circ f = g \circ \vartheta$ on*

$$A_R := \{z \in \mathbb{C} \colon |f^{\circ n}(z)| > R \textit{ for all } n \geq 1\} .$$

Furthermore ϑ has zero dilatation on $\{z \in A_R \colon |f^{\circ n}(z)| \to \infty\}$.

In particular, quasiconformally equivalent entire functions of bounded type are quasiconformally conjugate on their escaping sets. This result may be viewed as an analog to Schröder's theorem (that any two polynomials of equal degree are conformally conjugate in a neighborhood of ∞); it is due to Rempe [Re], together with the following corollary.

Corollary 6.3 (No Invariant Line Fields).
Entire functions of bounded type do not support measurable invariant line fields on their sets of escaping points.

A lot of work has been done on the space of the simplest entire functions: that is the space of exponential functions. It plays an important role as a prototypical example, in a similar way as quadratic polynomials play an important prototypical role for polynomials. By now, there is a good body of results on the parameter space of exponential functions, in analogy to the Mandelbrot set. Some of the results go as follows:

Theorem 6.4 (Exponential Parameter Space).
The parameter space of exponential maps $z \mapsto \lambda e^z$ has the following properties:

1. *there is a unique hyperbolic component W of period 1; it is conformally parametrized by a conformal isomorphism $\mu\colon \mathbb{D}^* \to W$, $\mu \mapsto \mu \exp(-\mu)$, so that the map E_λ with $\lambda = \mu \exp(-\mu)$ has an attracting fixed point with multiplier μ;*
2. *for every period $n \geq 2$, there are countably many hyperbolic components of period n; on each component, the multiplier map $\mu\colon W \to \mathbb{D}^*$ is a universal covering;*
3. *for every hyperbolic component W of period ≥ 2, there is a preferred conformal isomorphism $\Phi\colon W \to \mathbb{H}^-$ with $\mu = \exp \circ \Phi$ (where $\mathbb{H}^-$ is the left half plane);*
4. *there is an explicit canonical classification of hyperbolic components and hyperbolic parameters;*

5. *the preferred conformal isomorphism* $\Phi: W \to \mathbb{H}^-$ *extends as a homeomorphism* $\Phi: \overline{W} \to \overline{\mathbb{H}}^-$; *in particular, every hyperbolic component has connected boundary;*
6. *there is an explicit canonical classification of boundary points of hyperbolic components and of exponential maps with indifferent orbits;*
7. *escaping parameters (those for which the singular value escapes to* ∞*) are organized in the form of* parameter rays, *together with landing points of certain parameter rays; this yields an explicit classification of all escaping parameters;*
8. *the Hausdorff dimension of the parameter rays is* 1, *while the Hausdorff dimension of all escaping parameters (parameter rays and some of their landing points) is* 2;
9. *exponential parameter space fails to be locally connected at any point on a parameter ray;*
10. *there is an explicit classification of all parameters for which the singular orbit is finite (i.e., strictly preperiodic);*
11. *the exponential bifurcation locus is connected.*

We will comment on these results, give references, and relate them to results on the Mandelbrot set.

(1), (2), (3), (4) The existence of hyperbolic components of all periods has been shown in several early papers on exponential parameter space [DGH, BR84, EL84b, EL92]; the fact that the multiplier map is a conformal isomorphism onto $\mathbb{D}^*$ for period 1 and a universal cover over $\mathbb{D}^*$ for period 2 and greater is shown there as well. This shows the existence of a conformal isomorphism $\Phi: W \to \mathbb{H}^-$ with $\mu = \exp \circ \Phi$; it is made unique in [Sch00]. Hyperbolic components are given combinatorial labels in [DFJ02], and they are completely classified in [Sch00]; the latter result leads to a complete classification of exponential maps with attracting periodic orbits. The fact that the component of period 1 is special is related to the parametrization $z \mapsto \lambda e^z$: for instance, for the parametrization $z \mapsto e^z + c$, there is still a single hyperbolic component of period 1, but its multiplier map is a universal cover over $\mathbb{D}^*$. All these results are in close analogy to the Mandelbrot set and its degree d cousins, the Multibrot sets: these have finitely many hyperbolic components of each period (with an explicit classification), and the multiplier maps on these have degree $d-1$.

(5), (6) The multiplier map of a hyperbolic component clearly extends continuously to the boundary; it was conjectured in the 1980's (by Eremenko and Lyubich [EL84b] and by Baker and Rippon [BR84]) that every component has connected boundary, or equivalently that Φ extends as a homeomorphism $\Phi: \overline{W} \to \overline{\mathbb{H}}^-$. This conjecture has been confirmed in [Sch04a, RSch09], and this leads to a classification of exponential maps with indifferent orbits. The problem is the following: an *internal parameter ray* of a hyperbolic component is the Φ^{-1}-image of a horizontal line $\gamma_y := \{x+iy \in \mathbb{H}^- : x \in (-\infty, 0)\} \subset \mathbb{H}^-$. It is easy to see that this parameter ray *lands* at some $\lambda_y \in \overline{\mathbb{C}}$, i.e., the limit as $x \to 0$ of $\Phi^{-1}(\gamma_y(x))$ exists. The difficulty is to exclude that the limit $\lambda_y = \infty$: if $\lambda_y \in \mathbb{C}$, then the exponential map E_{λ_y} has an indifferent orbit with multiplier e^{iy}; if not, then the corresponding parameter would be missing from the classification.

(7) Parameter rays in the space of exponential maps are constructed and classified in [FS09]. There are additional parameters for which the singular value escapes; these are landing points of certain parameter rays and are classified in [FRS08].

(8) It is shown in [BBS08] that parameter rays in exponential parameter space have Hausdorff dimension 1, and in [Qiu94] that all escaping parameters have Hausdorff dimension 2. By [FRS08], every escaping parameter is either on a parameter ray or a landing point of one of them.

(9) Contrary to one of the principal conjectures on the Mandelbrot set, and many combinatorial similarities between the parameter spaces of exponential maps and quadratic polynomials, the exponential bifurcation locus is not locally connected at any point on any parameter ray: in fact, any parameter ray is approximated by other parameter rays on both sides, and between any pair of parameter rays there are are hyperbolic components. This destroys local connectivity of the exponential bifurcation locus [RSch08].

(10) For rational maps, there is a fundamental theorem of Thurston [DH93] that characterizes rational maps with finite critical orbit and that is at the basis of most classification theorems in rational dynamics. Unfortunately, there is no analogous result for transcendental maps. Currently, the only extension of Thurston's theorem to the case of transcendental maps is [HSS09] on exponential maps with finite singular orbits (see also [Se09] for work in progress in this direction); the resulting classification of postsingularly finite exponential maps is in [LSV08]. This classification had been expected for a long time [DGH].

(11) The fundamental study of the Mandelbrot starts with Douady and Hubbard's result about connectivity of the Mandelbrot set, or equivalently of its boundary, which is the bifurcation locus in the space of quadratic polynomials. The corresponding result about exponential maps is that the exponential bifurcation locus is connected. Unlike in the polynomial case, where connectivity of the bifurcation locus of quadratic polynomials is the starting point for much of the theory of the Mandelbrot set, this result comes at the end of a detailed study of exponential parameter space. It was shown in [RSch09].

Just like for the whole field of entire dynamics, it is impossible to do justice to the large body of knowledge that has been established from many different points of view even on exponential parameter space. Among further existing work, we would like to mention [GKS04, UZ07, Ye94]. There is also a significant amount of work on other one-dimensional parameter spaces of explicit entire functions; we only mention the work by Fagella, partly with coauthors, on the families $z \mapsto \lambda z e^z$ [Fa99] and on $z \mapsto \lambda z^m e^z$ (with $m \geq 2$) [FaGa07]. Even though these are not entire functions, we would also like to mention work on the family of tangent maps by [KK97].

Conjecture 6.5 (Exponential Parameter Space).

- Hyperbolicity is dense in the space of exponential maps.
- Fibers in exponential parameter space are trivial.

Remark 6.6. Fibers in exponential parameter space are defined in analogy as for the Mandelbrot set (see [RSch08] for the exponential case and [Sch04b] for the Mandelbrot case). The second conjecture says that all non-hyperbolic exponential maps are combinatorially rigid (their landing patterns of periodic dynamic rays differ); it is the analog to the famous conjecture that the Mandelbrot set is locally connected. The second conjecture implies the first [RSch08].

We conclude this section with a question of Euler [E1777] that was raised in 1777, and its generalization to complex numbers.

Question 6.7 (Euler Question: Iterated Exponentiation).
For which real $a > 0$ does the sequence a, a^a, a^{a^a}, $a^{a^{a^a}}$, ..., have a limit?

This question can be rephrased, by setting $\lambda := \log a$, as follows: for which $\lambda \in \mathbb{R}$ does the sequence $x_0 := 0$, $x_{n+1} = \lambda e^{x_n}$, have a limit? In this form, it makes sense for $\lambda \in \mathbb{C}^*$. The answer is surprisingly complicated. Set $E_\lambda(z) := \lambda e^z$. To begin with, we note that if the sequence x_n converges to some $b \in \mathbb{C}$ but is not eventually constant, then b must be an attracting or parabolic fixed point (clearly, no repelling fixed point and no center of a Siegel disk can be a limit point of an orbit unless the latter is eventually constant; and it is not hard to show that a Cremer point cannot be the limit of the unique singular orbit).

The map E_λ has an attracting fixed point if and only if $\lambda = \mu e^{-\mu}$ with $\mu \in \mathbb{D}^*$ (since 0 is the only singular value of E_λ, it follows that in these cases, the orbit of 0 must converge to the attracting fixed point, by Theorem 2.3). Similarly, E_λ has a rationally indifferent fixed point if and only if $\lambda = \mu e^{-\mu}$ and μ is a root of unity. This takes care of all cases where the orbit of 0 converges to a finite limit point in $\mathbb{C}$ without eventually being constant. (The analogous question for periodic limit points leads to the classification of hyperbolic components and their boundaries in exponential parameter space, and thus items (4) and (6) in Theorem 6.4.)

The description of parameters λ in which the orbit of 0 is eventually fixed (or eventually periodic) involves a classification of postsingularly finite exponential maps, and this is settled by item (10) in Theorem 6.4. Finally, the case of parameters λ in which the orbit of 0 converges to ∞ is item (7) in that theorem.

7 Newton Maps of Entire Functions

If f is an entire holomorphic function, then its associated *Newton map* $N_f := \mathrm{id} - f/f'$ is a meromorphic function that naturally "wants to be" iterated. While the iteration of general meromorphic functions falls outside of the scope of this manuscript, there are a number of results specifically on Newton maps of entire functions. In rational iteration theory, polynomials are the easiest maps to work on, and their Newton maps have useful properties that make them easier to investigate than general rational maps. Since we believe that the situation should be similar for in the transcendental world, this section is included.

Definition 7.1 (Basin and Immediate Basin).
For a root α of f, we define its *basin* as $U_\alpha := \{z \in \mathbb{C} \colon N_f^{\circ n}(z) \to \alpha\}$ as $n \to \infty$. The *immediate basin* is the connected component of U_α containing α.

Theorem 7.2 (Immediate Basins Simply Connected).
Every root of f has simply connected immediate basin.

This was shown in [MS06]. It is an open question whether every Fatou component of N_f is simply connected. After work by Taixes, the last open case is whether Baker domains are always simply connected (compare [FJT08]).

Theorem 7.3 (Wandering Newton Domains Simply Connected).
If a Newton map has a wandering domain, then it is simply connected.

This follows from work of Bergweiler and Terglane [BT96]: in analogy with classical work of Shishikura [Sh09], they prove that a multiply connected wandering domain of a transcendental meromorphic map g (such as a Newton map) would require that g has a weakly repelling fixed point; but this is not the case for Newton maps.

Definition 7.4 (Virtual Immediate Basin).
A *virtual immediate basin* is a maximal subset of $\mathbb{C}$ (with respect to inclusion) among all connected open subsets of $\mathbb{C}$ in which the dynamics converges to ∞ locally uniformly and which have an absorbing set. (An *absorbing set* in a domain V is a subset A such that $N_f(\overline{A}) \subset A$ and every compact $K \subset V$ has a $n \geq 0$ so that $N_f^{\circ n}(K) \subset A$.)

Theorem 7.5 (Virtual Immediate Basins Simply Connected).
Every virtual immediate basin is simply connected.

This was also shown in [MS06]. Every virtual immediate basin is contained in a Baker domain; it is an open question whether this basin equals a Baker domain (this is true for simply connected Baker domains). The principal difficulty is the question whether every Baker domain has an absorbing set as in Definition 7.4; this would also imply that every Fatou component of a Newton map is simply connected.

Theorem 7.6 (Two Accesses Enclose Basin).
Let f be an entire function (polynomial or transcendental) and let U_α be the immediate basin of α for N_α. Let $\Gamma_1, \Gamma_2 \subset U_\alpha$ represent two curves representing different invariant accesses to ∞, and let W be an unbounded component of $\mathbb{C} \setminus (\Gamma_1 \cup \Gamma_2)$. Then W contains the immediate basin of a root of f or a virtual immediate basin, provided the following finiteness condition holds: $N_f^{-1}(z) \cap W$ is finite for all $z \in \mathbb{C}$.

Remark 7.7. In the case of a polynomial f, the finiteness condition always holds, and there is no virtual immediate basin. The result thus says that any two accesses of any immediate basins enclose another immediate basin. Theorem 7.6 is due to Rückert and Schleicher [RüS07].

For polynomials, it is known that any two invariant accesses to ∞ within the basin of any root always enclose some other root. The analogous statement is false for entire functions: if $f(z) = ze^{e^z}$, then the Newton map is $N_f(z) = z(1 - 1/(1+e^z))$. The map f has a single root with infinitely many invariant accesses, but these cannot surround any further root: they indeed surround virtual basins. This example is due to S. Mayer [May] (see also [MS06]).

Virtual immediate basins may be thought of as basins of a root at ∞. The prototypical case is $f(z) = \exp(z)$, $N_f(z) = z - 1$: there is no root, all points converge to $-\infty$ under N_f, and indeed f converges to 0 along these orbits. Douady thus asked whether there was a relation between asymptotic values 0 of f and virtual immediate basins. This is indeed often the case. The following two results are due to Buff and Rückert [BR06].

Theorem 7.8 (Logarithmic Singularity Implies Virtual Immediate Basin). *If f has an asymptotic value* 0 *which is a logarithmic singularity, then N_f has a virtual immediate basin.*

There is a partial converse as follows.

Theorem 7.9 (Virtual Immediate Basin and Asymptotic Value). *Let V be a virtual immediate basin with absorbing set A. If the quotient A/N_f has sufficiently large modulus, then f has an asymptotic value* 0.

Remark 7.10. There are indeed counterexamples when A/N_f has small modulus [BDL07].

Theorem 7.11 (Rational Newton Map). *The Newton map N_f of f is rational if and only if $f = pe^q$ for polynomials p and q. In this case, ∞ is a parabolic fixed point with multiplier* 1 *and multiplicity* $\deg(q) + 1$.

This is probably a folklore result; for a proof, which is not difficult, see [RüS07]. Another result that is folklore (at least in the polynomial case) is the following (see again [RüS07]).

Theorem 7.12 (Characterization of Newton Maps). *A meromorphic function g is the Newton map of an entire function if and only if all fixed points ξ of g satisfy $g'(\xi) = (m-1)/m$ for a positive integer m. Two entire functions f_1 and f_2 have identical Newton maps if and only if f_1/f_2 is constant.*

The next result is due to Haruta [Ha99].

Theorem 7.13 (Area of Immediate Basins). *For $f = pe^q$ with polynomials p and q and* $\deg q \geq 3$, *every immediate basin has finite Lebesgue area in the plane.*

We conclude with the following conjecture from [Sch08] on root finding of the Riemann ζ function by Newton's method. Let ξ be the entire function the roots of which are the non-trivial roots of ζ (so that $\xi(s) = \xi(1-s)$).

Conjecture 7.14 (The Riemann ζ function).
There are constants $c,s > 0$ with the following property. If there is a root α of ξ whose immediate basin does not contain one of the points $c_n^{\pm} := \pm 2 + cn/\log|n|$, then the Riemann hypothesis is false and there is a root α' off the critical line with $|\alpha' - \alpha| < s$ and the immediate basin of α' contains a point $c_n^{\pm}$.

Remark 7.15. The preceding conjecture says that the points c_n form an efficient set of starting points for finding all roots of f, so that the first N starting points find at least $c''N$ distinct roots of ξ; and if they do not find all roots, then close to a missed root these starting points find a root that violates the Riemann hypothesis.

More results on the dynamics of Newton maps for entire functions can be found in the survey [Rü08].

8 A Few Open Questions

There has been an enormous amount of work on the dynamics of transcendental entire (and meromorphic) functions, *but* many open questions remain (or rather, *and therefore*, there are many more open questions that await their answers). We just mention a couple of them in order to show that this field continues to be active.

Question 8.1. Is the escaping set $I(f)$ connected for every entire function f?

Question 8.2. Is there an entire function with a wandering domain with bounded orbit?

Eremenko and Lyubich [EL87] constructed an example of a wandering domain on which the iterates do not tend to ∞.

Question 8.3. Is there an entire function of bounded type that has a wandering domain?

(By Theorem 3.4, this entire function could not be of finite type, and the wandering domain could not consist of escaping orbits.)

Question 8.4. If an entire function has order less than $1/2$, does this imply that the Fatou set has no unbounded Fatou components?

This is a conjecture by Baker, and it has inspired a lot of work and partial results, such as by Baker, Stallard, Anderson-Hinkkanen, and others. It is still open even for functions of order 0. Another open question is whether $A(f)$, the set of points that escape to infinity as fast as possible, can have unbounded Fatou components.

Question 8.5. Is there an entire function whose Julia set has Hausdorff dimension 1?

As discussed in Section 5, the Hausdorff dimension of the Julia set always has dimension at least 1, and there are examples of entire functions for which the dimension of the Julia set is arbitrarily close to 1. For functions of bounded type, the Julia set always has Hausdorff dimension strictly greater than 1.

Question 8.6. Does every Newton map of an entire function always have connected Julia set?

This is true for polynomials, and there are positive partial results in the transcendental case. The main difficulty seems to be with Baker domains. (See Section 7.)

We have not discussed irrationally indifferent periodic points of entire functions. The following is an open question even for polynomials of arbitrary degree, but there are complete answers for certain spaces of maps (such as quadratic polynomials, by Yoccoz' linearization condition[Y95], and maps $z \mapsto \lambda z e^z$, by Geyer [Ge01]).

Question 8.7. Give, for various classes of entire functions, a necessary and sufficient condition for an irrationally indifferent periodic point to have a Siegel disk.

We now state a couple of questions that are more specific for the family of exponential maps; but many of these questions can also be asked for larger classes of entire functions: the greater question is "investigate, for various known properties of exponential maps, under which conditions these hold for larger classes of entire functions".

Question 8.8. Suppose for an exponential map $z \mapsto \lambda \exp(z)$, the singular value does not converge to ∞. Is every repelling periodic point the landing point of a periodic dynamic ray? Is there a generalization to larger classes of maps (of bounded type and finite order)?

Question 8.9. Suppose for an exponential map $z \mapsto \lambda e^z$, the Julia set equals $\mathbb{C}$. Does this imply that there is a dynamic ray that does not land? Is there an analogous result for entire functions without critical values? Or for those that have finite asymptotic values?

Question 8.10. Characterize the parameters $\lambda \in \mathbb{C}^*$ for which the set of non-escaping points in $\mathbb{C}$ of $z \mapsto \lambda e^z$ is connected; characterize also those parameters for which the set of non-escaping points union $\{\infty\}$ is connected.

Question 8.11. Show that hyperbolicity is dense in the space of exponential maps.

The analogous question remains open for *any* non-trivial space of holomorphic mappings, even for the prototypical case of quadratic polynomials.

We conclude this section by a question on a very specific map that would have amusing consequences.

Question 8.12. Show that the following map does not have a wandering domain (or at least not one that intersects the real axis):

$$f(z) = z/2 + (1 - \cos \pi z)(z + 1/2)/2 + ((1/2 - \cos \pi z) \sin \pi z)/\pi$$

Remark 8.13. The relevance of the last question is the following: it is known that $\mathbb{Z}$ is in the Fatou set for this function f. Moreover, on $\mathbb{Z}$, the function f coincides with the well-known $3n+1$ problem: $f(n) = n/2$ if n even and $f(n) = (3n+1)/2$ if n is odd. Therefore, solving this question would prove that the $3n+1$ problem does not have an orbit tending to ∞ (it would still be possible for f to have periodic integer orbits other than the cycle $1 \mapsto 2 \mapsto 1$; such as the fixed point -1 or $-5 \mapsto -7 \mapsto -10 \mapsto -5$). See [LSW99] for more details.

A. Background and Notation

In this brief appendix, we collect some of the notation that we are using in this text, and we state some of the standard results from complex analysis that we require.

$\mathbb{D}$	$\{z \in \mathbb{C} \colon \lvert z \rvert < 1\}$ (the complex unit disk)
$\mathbb{D}^*$	$\mathbb{D} \setminus \{0\}$ (the punctured unit disk)
$\overline{\mathbb{C}}$	$\mathbb{C} \cup \{\infty\}$ (the Riemann sphere with marked point at ∞)
$\mathbb{H}$	$\{z \in \mathbb{C} \colon \operatorname{Re} z > 0\}$ (the right half plane)
$D_r(a)$	$\{z \in \mathbb{C} \colon \lvert z - a \rvert < r\}$ (a Euclidean disk in $\mathbb{C}$)
$\chi(a,b)$	the spherical metric between any two points on $\overline{\mathbb{C}}$
$D_\chi(a,r)$	$\{z \in \overline{\mathbb{C}} \colon \chi(a,z) < r\}$ (a disk in $\overline{\mathbb{C}}$ for the spherical metric)
$f^\sharp$	the spherical derivative: $f^\sharp(z) := \lvert f'(z) \rvert / (1 + \lvert f(z) \rvert^2)$
$M(r,f)$	$\max\{\lvert f(z) \rvert \colon \lvert z \rvert = r\}$

Definition A.1 (Normal Family).
Let $U \subset \mathbb{C}$ be a domain. A family (f_k) of holomorphic maps $U \to \overline{\mathbb{C}}$ is called a *normal family* if every sequence of the (f_k) contains a subsequence that converges locally uniformly to a holomorphic limit function $f \colon U \to \overline{\mathbb{C}}$.

Being a normal family is a local property: if every $z \in U$ has a neighborhood on which a family of maps is normal, then the family is normal on all of U (this follows by a standard diagonal argument).

Theorem A.2 (Montel's Theorem).
Any family $(f_k) \colon U \to \overline{\mathbb{C}} \setminus \{a,b,c\}$ of holomorphic functions from a domain $U \subset \overline{\mathbb{C}}$ to the sphere with three distinct punctures a,b,c is normal.

Theorem A.3 (Marty's Criterion).
A family $(f_k) \colon U \to \overline{\mathbb{C}}$ of holomorphic maps is normal if and only if the spherical derivatives are locally bounded.

Theorem A.4 (Picard's Theorem).
If $f \colon \mathbb{D}^* \to \overline{\mathbb{C}}$ has an essential singularity at 0, then f assumes every $z \in \overline{\mathbb{C}}$, with at most two exceptions, infinitely often in every neighborhood of 0.

General references to the results in this appendix include [A98, H64, S93] as well as [Mi06, Be98].

References

[AO93] Aarts, J.M., Oversteegen, L.G.: The geometry of Julia sets. Trans. Am. Math. Soc. **338**(2), 897–918 (1993)

[Ab] Abate, M.: Discrete holomorphic local dynamical systems, course notes, this volume

[A98] Ahlfors, L.: Complex analysis, 3rd edn. International Series in Pure and Applied Mathematics. McGraw-Hill Book Co., New York (1978)

[BBS08] Bailesteanu, M., Balan, V., Schleicher, D.: Hausdorff dimension of exponential parameter rays and their endpoints. Nonlinearity **21**(1), 113–120 (2008)

[Ba60] Baker, I.N.: The existence of fixpoints of entire functions. Math. Zeitschr. **73**, 280–284 (1960)

[Ba63] Baker, I.N.: Multiply connected domains of normality in iteration theory. Math. Z. **81**, 206–214 (1963)

[Ba76] Baker, I.N.: An entire function which has wandering domains. J. Austral. Math. Soc. Ser. A **22**(2), 173–176 (1976)

[Ba81] Baker, I.N.: The iteration of polynomials and transcendental entire functions. J. Austral. Math. Soc. Ser. A **30**, 483–495 (1980/81)

[Ba84] Baker, I.N.: Wandering domains in the iteration of entire functions. Proc. Lond. Math. Soc. (3) **49**(3), 563–576 (1984)

[BKL91] Baker, I.N., Kotus, J., Lü, Y.N.: Iterates of meromorphic functions. III. Preperiodic domains. Ergod. Theor. Dyn. Syst. **11**(4), 603–618 (1991)

[BR84] Baker, I.N., Rippon, P.: Iteration of exponential functions. Ann. Acad. Sci. Fenn. Ser. A I Math. **9**, 49–77 (1984)

[Bar07] Barański, K.: Trees and hairs for some hyperbolic entire maps of finite order. Math. Z. **257**(1), 33–59 (2007)

[Bar08] Barański, K.: Hausdorff dimension of hairs and ends for entire maps of finite order. Math. Proc. Cambridge Philos. Soc. **145**(3), 719–737 (2008)

[Ben] Benini, A.: Combinatorial rigidity for Misiurewicz parameters. Manuscript, in preparation.

[Ba99] Bargmann, D.: Simple proofs of some fundamental properties of the Julia set. Ergod. Theor. Dyn. Syst. **19**(3), 553–558 (1999)

[Ba01] Bargmann, D.: Normal families of covering maps. J. Analyse. Math. **85**, 291–306 (2001)

[BB01] Bargmann, D., Bergweiler, W.: Periodic points and normal families. Proc. Am. Math. Soc. **129**(10), 2881–2888 (2001)

[BD00] Berteloot, F., Duval, J.: Une démonstration directe de la densité des cycles répulsifs dans l'ensemble de Julia. Progr. Math. **188**, Birkhäuser, Basel, 221–222 (2000)

[Be91] Bergweiler, W.: Periodic points of entire functions: proof of a conjecture of Baker. Complex Variables Theory Appl. **17**, 57–72 (1991)

[Be93] Bergweiler, W.: Iteration of meromorphic functions. Bull. Am. Math. Soc. (N.S.) **29**(2), 151–188 (1993)

[Be] Bergweiler, W.: An entire function with simply and multiply connected wandering domains, to appear in Pure Appl. Math. Quarterly

[Be98] Bergweiler, W.: A new proof of the Ahlfors five islands theorem. J. Anal. Math. **76**, 337–347 (1998)

[BDL07] Bergweiler, W., Drasin, D., Langley, J.K.: Baker domains for Newton's method. Ann. Inst. Fourier (Grenoble) **57**(3), 803–814 (2007)

[BE95] Bergweiler, W., Eremenko, A.: On the singularities of the inverse to a meromorphic function of finite order. Rev. Mat. Iberoamericana **11**(2), 355–373 (1995)

[BHKMT93] Bergweiler, W., Haruta, M., Kriete, H., Meier, H., Terglane, N.: On the limit functions of iterates in wandering domains. Ann. Acad. Sci. Fenn. Ser. A I Math. **18**(2), 369–375 (1993)

[BH99] Bergweiler, W., Hinkkanen, A.: On semiconjugation of entire functions. Math. Proc. Cambridge Philos. Soc. **126**(3), 565–574 (1999)

[BKS] Bergweiler, W., Karpińska, B., Stallard, G.: The growth rate of an entire function and the Hausdorff dimension of its Julia set. J. Lond. Math. Soc., doi:10.1112/jlms/jdp042

[BT96] Bergweiler, W., Terglane, N.: Weakly repelling fixpoints and the connectivity of wandering domains. Trans. Am. Math. Soc. **348**(1), 1–12 (1996)

[BRS08] Bergweiler, W., Rippon, P., Stallard, G.: Dynamics of meromorphic functions with direct or logarithmic singularities. Proc. Lond. Math. Soc. **97**, 368-400 (2008)

[Bo96] Bock, H.: On the dynamics of entire functions on the Julia set. Results Math. **30**(1–2), 16–20 (1996)

[BS] Bruin, H., Kaffl, A., Schleicher, D.: Symbolic Dynamcis of Quadratic Polynomials. Monograph, in preparation. Earlier version: Mittag-Leffler Preprint **7** (2001/02)

[BR06] Buff, X., Rückert, J.: Virtual immediate basins of Newton maps and asymptotic values. Int. Math. Res. Not., Art. ID 65498, 18 (2006)

[DGH] Devaney, R., Goldberg, L., Hubbard, J.: A dynamical approximation to the exponential map of polynomials. Preprint, MSRI (1986). See [BDGHHR99].

[BDGHHR99] Bodelon, C., Devaney, R., Goldberg, L., Hayes, M., Hubbard, J., Roberts, G.: Hairs for the complex exponential family. Internat. J. Bif. Chaos **9**(8), 1517–1534 (1999)

[DK84] Devaney, R., Krych, M.: Dynamics of $\exp(z)$. Ergod. Theor. Dyn. Syst. **4**(1), 35–52 (1984)

[DT86] Devaney, R., Tangerman, F.: Dynamics of entire functions near the essential singularity. Ergod. Theor. Dyn. Syst. **6**(4), 489–503 (1986)

[DFJ02] Devaney, R., Fagella, N., Jarque, X.: Hyperbolic components of the complex exponential family. Fund. Math. **174**(3), 193–215 (2002)

[Do98] Domínguez, P.: Dynamics of transcendental meromorphic functions. Annales Academiæ Scientiarum Fennicæ, Mathematica **23**, 225–250 (1998)

[DH93] Douady, A., Hubbard, J.: A proof of Thurston's topological characterization of rational functions. Acta Math. **171**, 263–297 (1993)

[Ep] Epstein, A.: Infinitesimal thurston rigidity and the fatou-shishikura inequality. Stony Brook IMS preprint **1** (1999)

[Er78] Eremenko, A.: The set of asymptotic values of a finite order meromorphic function (Russian). Mat. Zametki **24**(6), 779–783, 893 (1978). (English translation: Math. Notes **24**(5–6), 914–916 (1978))

[Er89] Eremenko, A.: On the iteration of entire functions. Banach Center Publ. **23**, PWN, Warsaw, 339–345 (1989)

[EL84a] Eremenko, A., Lyubich, M.: Iterates of entire functions. Soviet Math. Dokl. **30**(3), 592–594 (1984)

[EL84b] Eremenko, A., Lyubich, M.: Iterates of entire functions (Russian). Preprint, Institute for Low Temperature, Kharkov **6** (1984)

[EL87] Eremenko, A., Lyubich, M.: Examples of entire functions with pathological dynamics. J. Lond. Math. Soc. **36**, 458–468 (1987)

[EL89] Eremenko, A., Lyubich, M.: The dynamics of analytic transformations (Russian). Algebra i Analiz **1**(3), 1–70 (1989); translation in Leningrad Math. J. **1**(3), 563–634 (1990)

[EL92] Eremenko, A., Lyubich, M.: Dynamical properties of some classes of entire functions. Annales Inst. Fourier **42**(4), 989–1020 (1992)

[E1777] Euler, L.: De formulis exponentialibus replicatis. Acta Acad. Petropolitanae, **1**, 38–60 (1777)

[Fa99] Fagella, N.: Dynamics of the complex standard family. J. Math. Anal. Appl. **229**(1), 1–31 (1999)

[FaGa07] Fagella, N., Garijo, A.: The parameter planes of $\lambda z^m \exp(z)$ for $m \geq 2$. Comm. Math. Phys. **273**(3), 755–783 (2007)

[FJT08] Fagella, N., Jarque, X., Taixés, J.: On connectivity of Julia sets of transcendental meromorphic maps and weakly repelling fixed points I. Proc. Lond. Math. Soc. **97**(3), 599–622 (2008)

[Fa26] Fatou, P.: Sur l'itération des fonctions transcendantes entières. Acta Math. **47**, 337–370 (1926)

[FS09] Förster, M., Schleicher, D.: Parameter rays in the space of exponential maps. Ergod. Theor. Dyn. Syst. **29**, 515–544 (2009)

[FRS08] Förster, M., Rempe, L., Schleicher, D.: Classification of escaping exponential maps. Proc. Am. Math. Soc. **136**, 651–663 (2008)

[Ge01] Geyer, L.: Siegel discs, Herman rings and the Arnold family. Trans. Am. Math. Soc. **353**(9), 3661–3683 (2001)

[GE79] Gol'dberg, A., Eremenko, A.: Asymptotic curves of entire functions of finite order. Mat. Sb. (N.S.) **109(151)**(4), 555–581, 647 (1979)

[GK86] Goldberg, L., Keen, L.: A finiteness theorem for a dynamical class of entire functions. Ergod. Theor. Dyn. Syst. **6**, 183–192 (1986)

[GKS04] Graczyk, J., Kotus, J., Świátek, G.: Non-recurrent meromorphic functions. Fund. Math. **182**(3), 269–281 (2004)

[Gr18a] Gross, W.: Über die Singularitäten analytischer Funktionen. Monatsh. Math. Phys. **29**(1), 3–47 (1918)

[Gr18b] Wilhelm, G.: Eine ganze Funktion, für die jede komplexe Zahl Konvergenzwert ist (German). Math. Ann. **79**(1–2), 201–208 (1918)

[Ha99] Haruta, M.: Newton's method on the complex exponential function. Trans. Am. Math. Soc. **351**(6), 2499–2513 (1999)

[H64] Hayman, W.: Meromorphic Functions. Oxford Mathematical Monographs, Clarendon Press, Oxford (1964)

[He57] Heins, M.: Asymptotic spots of entire and meromorphic functions. Ann. Math. (2) **66**, 430–439 (1957)

[HSS09] Hubbard, J., Schleicher, D., Shishikura, M.: Exponential Thurston maps and limits of quadratic differentials. J. Am. Math. Soc. **22**, 77–117 (2009)

[Iv14] Iversen, F.: Recherches sur les fonctions inverses des fonctions méromorphes. Thèse, Helsingfors, 1914 (see [Ne53]).

[Ka99] Karpińska, B.: Hausdorff dimension of the hairs without endpoints for $\lambda\exp(z)$. C. R. Acad. Sci. Paris Séer. I Math. **328**, 1039–1044 (1999)

[KU06] Karpińska, B., Urbański, M.: How points escape to infinity under exponential maps. J. Lond. Math. Soc. (2) **73**(1), 141–156 (2006)

[KK97] Keen, L., Kotus, J.: Dynamics of the family $\lambda \tan z$. Conform. Geom. Dyn. **1**, 28–57 (1997)

[KS08] Kisaka, M., Shishikura, M.: On multiply connected wandering domains of entire functions. In: Phil Rippon, Gwyneth Stallard (eds.) Transcendental Dynamics and Complex Analysis, London Math. Soc. Lecture Note Ser. 348, Cambridge University Press, Cambridge (2008)

[LSV08] Laubner, B., Schleicher, D., Vicol, V.: A combinatorial classification of postsingularly preperiodic complex exponential maps. Discr. Cont. Dyn. Syst. **22**(3), 663–682 (2008)

[LSW99] Letherman, S., Schleicher, D., Wood, R.: On the $3n+1$-problem and holomorphic dynamics. Exp. Math. **8**(3), 241–251 (1999)

[Ma90] Mayer, J.: An explosion point for the set of endpoints of the Julia set of $\lambda\exp(z)$. Ergod. Theor. Dyn. Syst. **10**(1), 177–183 (1990)

[May] Mayer, S.: Newton's Method for Entire Functions. Diplomarbeit, Technische Universität München (2002)

[MS06] Mayer, S., Schleicher, D.: Immediate and virtual basins of Newton's method for entire functions. Ann. Inst. Fourier (Grenoble) **56**(2), 325–336 (2006)

[Mc87] McMullen, C.: Area and Hausdorff dimension of Julia sets of entire functions. Trans. Am. Math. Soc. **300**(1), 329–342 (1987)

[Mi06] Milnor, J.: Dynamics in one complex variable, 3rd ed. Annals of Mathematics Studies **160**, Princeton University Press, Princeton, NJ (2006)

[Mis81] Misiurewicz, M.: On iterates of e^z. Ergod. Theor. Dyn. Syst. **1**(1), 103–106 (1981)

[Ne53] Nevanlinna, R.: Eindeutige Analytische Funktionen, Zweite Auflage. Springer Verlag, Göttingen (1953)

[Qiu94] Qiu, W.: Hausdorff dimension of the M-set of $\lambda\exp(z)$. Acta Math. Sinica (N.S.) **10**(4), 362–368 (1994)

[Re04] Rempe, L.: On a question of Herman, Baker and Rippon concerning Siegel disks. Bull. Lond. Math. Soc. **36**(4), 516–518 (2004)

[Re07] Rempe, L.: On a question of Eremenko concerning escaping components of entire functions. Bull. Lond. Math. Soc. **39**, 661–666 (2007)

[Re] Rempe, L.: Rigidity of escaping dynamics for transcendental entire functions. Acta Math. **203** 235–267 (2009)

[RSch08] Rempe, L., Schleicher, D.: Bifurcation loci of exponential maps and quadratic polynomials: local connectivity, triviality of fibers, and density of hyperbolicity. In: M. Lyubich, M. Yampolsky (eds.) Holomorphic Dynamics and Renormalization: A Volume in Honour of John Milnor's 75th birthday. Fields Institute Communications **53**, 177–196 (2008)

[RSch09] Rempe, L., Schleicher, D.: Bifurcations in the space of exponential maps. Invent. Math. **175**(1), 103–135 (2009)

[Ri08] Rippon, P.: Baker domains. In: Phil Rippon, Gwyneth Stallard (eds.) Transcendental Dynamics and Complex Analysis. Lond. Math. Soc. Lecture Note Ser. 348, Cambridge University Press, Cambridge, 371–395 (2008)

[RS99a] Rippon, P.J., Stallard, G.M.: Families of Baker domains: I. Nonlinearity **12**, 1005–1012 (1999a)

[RS99b] Rippon, P.J., Stallard, G.M.: Families of Baker domains II, Conform. Geom. Dyn., **3**, 67–78 (1999b)

[RS00] Rippon, P.J., Stallard, G.M.: On sets where iterates of a meromorphic function zip to infinity. Bull. Lond. Math. Soc. **32**, 528–536 (2000)

[RS05a] Rippon, P.J., Stallard, G.M.: On questions of Fatou and Eremenko. Proc. Am. Math. Soc. **133**, 1119–1126 (2005a)

[RS05b] Rippon, P.J., Stallard, G.M.: Escaping points of meromorphic functions with a finite number of poles. J. d'Analyse Math. **96**, 225–245 (2005b)

[RS] Phil J. Rippon and Gwyneth M. Stallard, *Slow escaping points of meromorphic functions*, Transactions of the Amer. Math. Soc., to appear, ArXiv:0812.2410v1

[Ro48] Rosenbloom, P.C.: L'itération des fonctions entières. C. R. Acad. Sci. Paris Sér. I Math. **227**, 382–383 (1948)

[RoS08] Rottenfußer, G., Schleicher, D.: Escaping points of the cosine family. In: Phil Rippon, Gwyneth Stallard (eds.) Transcendental Dynamics and Complex Analysis. Lond. Math. Soc. Lecture Note Ser. **348**. Cambridge University Press, Cambridge, 396–424 (2008)

[RRRS] Rottenfußer, G., Rückert, J., Rempe, L., Schleicher, D.: Dynamic rays of bounded-type entire functions. Ann. Math. to appear

[Rü08] Rückert, J.: Rational and transcendental Newton maps. In: M. Lyubich, M. Yampolsky (eds.) Holomorphic Dynamics and Renormalization: A Volume in Honour of John Milnor's 75th birthday. Fields Institute Communications **53**, 197–211 (2008)

[RüS07] Rückert, J., Schleicher, D.: On Newton's method for entire functions. J. Lond. Math. Soc. (2) **75**, 659–676 (2007)

[S93] Schiff, J.: Normal Families. Universitext. Springer Verlag, New York (1993)

[Sch00] Schleicher, D.: Attracting dynamics for exponential maps. Ann. Acad. Sci. Fenn. **28**(1), 3–34 (2003)

[Sch04a] Schleicher, D.: Hyperbolic components in exponential parameter space. Comptes Rendus — Mathématiques **339**(3), 223–228 (2004)

[Sch04b] Schleicher, D.: On fibers and local connectivity of Mandelbrot and Multibrot sets. In: M. Lapidus, M. van Frankenhuysen (eds.) Fractal Geometry and Applications: A Jubilee of Benoit Mandelbrot. Proceedings of Symposia in Pure Mathematics **72**, pp. 477–507. American Mathematical Society, Providence, RI (2004)

[Sch07a] Schleicher, D.: Hausdorff dimension, its properties, and its surprises. Am. Math. Monthly **114**(6), 509–528 (2007)

[Sch07b] Schleicher, D. The dynamical fine structure of iterated cosine maps and a dimension paradox. Duke Math. J. **136**(2), 343–356 (2007)

[Sch08] Schleicher, D.: Newton's method as a dynamical system: efficient root finding of polynomials and the Riemann ζ function. In: M. Lyubich, M. Yampolsky (eds.) Holomorphic Dynamics and Renormalization: a Volume in Honour of John Milnor's 75th birthday. Fields Institute Communications **53**, 213–224 (2008)

[SZ03a] Schleicher, D., Zimmer, J.: Escaping points for exponential maps. J. Lond. Math. Soc. (2) **67**, 380–400 (2003)

[SZ03b] Schleicher, D., Zimmer, J.: Periodic points and dynamic rays of exponential maps. Ann. Acad. Sci. Fenn. **28**(2), 327–354 (2003)

[Schu08] Schubert, H.: Area of Fatou sets of trigonometric functions. Proc. Am. Math. Soc. **136**, 1251–1259 (2008)

[Se09] Selinger, N.: On the Boundary Behavior of Thurston's Pullback Map. In: Dierk Schleicher (ed.) Complex Dynamics: Families and Friends, pp. 585–595, Chapter 16. A K Peters, Wellesley, MA (2009)

[Sh87] Shishikura, M.: On the quasiconformal surgery of rational functions. Ann. Sci. Éc. Norm. Sup. 4e Ser. **20**, 1–29 (1987)

[Sh09] Shishikura, M.: The connectivity of the Julia set and fixed points. In: Dierk Schleicher (ed.) Complex Dynamics: Families and Friends, pp. 257–276. Chapter 6. A K Peters, Wellesley, MA (2009)

[St96] Stallard, G.: The Hausdorff dimension of Julia sets of entire functions. II. Math. Proc. Cambridge Philos. Soc. **119**(3), 513–536 (1996)

[St97] Stallard, G.: The Hausdorff dimension of Julia sets of entire functions. III. J. Math. Proc. Camb. Philos. Soc. **122**(2), 223–244 (1997)

[St08] Stallard, G.: Dimensions of Julia sets of transcendental meromorphic functions. In: Transcendental dynamics and complex analysis, Lond. Math. Soc. Lecture Note Ser. **348**, pp. 425–446. Cambridge University Press, Cambridge 2008

[Ur03] Urbański, M.: Measures and dimensions in conformal dynamics. Bull. Am. Math. Soc. (N.S.) **40**(3), 281–321 (2003)

[UZ07] Urbański, M., Zdunik, A.: Instability of exponential Collet-Eckmann maps. Israel J. Math. **161**, 347–371 (2007)

[Tö39] Töpfer, H.: Über die Iteration der ganzen transzendenten Funktionen, insbesondere von sin z und cos z (German), Math. Ann. **117**, 65–84 (1939)

[Ye94] Ye, Z.: Structural instability of exponential functions. Trans. Am. Math. Soc. **344**(1), 379–389 (1994)

[Y95] Yoccoz, J.-C.: Théorème de Siegel, nombres de Bruno et polynômes quadratiques. Petits diviseurs en dimension 1. Astérisque **231**, 3–88 (1995)

[Zh06] Zheng, J.-H.: On multiply-connected Fatou components in iteration of meromorphic functions. J. Math. Anal. Appl. **313**(1), 24–37 (2006)

[Z75] Zalcman, L.: A heuristic principle in complex function theory. Am. Math. Monthly **82**(8), 813–817 (1975)

List of Participants

1. Abate Marco, Italy,
   abate@dm.unipi.it
2. Arosio Leandro, Italy,
   arosio@mat.uniroma1.it
3. Bayraktar Turgay, USA,
   tbayrakt@indiana.edu
4. Bedford Eric, USA,
   bedford@indiana.edu
5. Bisi Cinzia, Italy,
   bisi@mat.unical.it
6. Brunella Marco, France,
   Marco.Brunella@u-bourgogne.fr
7. Cannizzo Jan, Germany,
   j.cannizzo@jacobs-university.de
8. Casavecchia Tiziano, Italy,
   casavecc@mail.dm.unipi.it
9. Corvaja Pietro, Italy,
   corvaja@dimi.uniud.it
10. de Fabritiis Chiara, Italy,
    fabritiis@dipmat.univpm.it
11. Deserti Julie, France,
    deserti@math.jussieu.fr
12. Dini Gilberto, Italy,
    dini@math.unifi.it
13. Francaviglia Stefano, Italy,
    s.francaviglia@sns.it
14. Frosini Chiara, Italy,
    frosini@math.unifi.it
15. Gentili Graziano, Italy,
    gentili@math.unifi.it

16. Guenot Jacques, Italy,
guenot@libero.it
17. Manjarin Monica, Spain,
manjarin@ub.edu
18. Molino Laura, Italy,
laura.molino@unipr.it
19. Nisoli Isaia, Italy,
nisoli@mail.dm.unipi.it
20. Parrini Carla, Italia,
parrini@math.unifi.it
21. Patrizio Giorgio, Italy,
patrizio@math.unifi.it
22. Perelli Alberto, Italia,
perelli@dima.unige.it
23. Radu Remus, USA,
remusradu@math.cornell.edu
24. Raissy Jasmin, Italy,
raissy@mail.dm.unipi.it
25. Rosati Lilia, Italy,
rosati@math.unifi.it
26. Ruggiero Matteo, Italy,
matteo.ruggiero@sns.it
27. Schleicher Dierk, Germany,
dierk@jacobs-university.de
28. Sibony Nassim, France,
nessim.sibony@math.u-psud.fr
29. Stoppato Caterina, Italy,
stoppato@math.unifi.it
30. Taflin Johan, France,
johan.taflin@gmail.com
31. Tanase Raluca, USA,
ralucat@math.cornell.edu
32. Villarini Massimo, Italy,
massimo.villarini@unimore.it
33. Vitali Sara, Italy,
vitali@mat.uniroma2.it
34. Vlacci Fabio, Italy,
vlacci@math.unifi.it

Lecture Notes in Mathematics

For information about earlier volumes
please contact your bookseller or Springer
LNM Online archive: springerlink.com

Vol. 1816: S. Albeverio, W. Schachermayer, M. Talagrand, Lectures on Probability Theory and Statistics. Ecole d'Eté de Probabilités de Saint-Flour XXX-2000. Editor: P. Bernard (2003)
Vol. 1817: E. Koelink, W. Van Assche (Eds.), Orthogonal Polynomials and Special Functions. Leuven 2002 (2003)
Vol. 1818: M. Bildhauer, Convex Variational Problems with Linear, nearly Linear and/or Anisotropic Growth Conditions (2003)
Vol. 1819: D. Masser, Yu. V. Nesterenko, H. P. Schlickewei, W. M. Schmidt, M. Waldschmidt, Diophantine Approximation. Cetraro, Italy 2000. Editors: F. Amoroso, U. Zannier (2003)
Vol. 1820: F. Hiai, H. Kosaki, Means of Hilbert Space Operators (2003)
Vol. 1821: S. Teufel, Adiabatic Perturbation Theory in Quantum Dynamics (2003)
Vol. 1822: S.-N. Chow, R. Conti, R. Johnson, J. Mallet-Paret, R. Nussbaum, Dynamical Systems. Cetraro, Italy 2000. Editors: J. W. Macki, P. Zecca (2003)
Vol. 1823: A. M. Anile, W. Allegretto, C. Ringhofer, Mathematical Problems in Semiconductor Physics. Cetraro, Italy 1998. Editor: A. M. Anile (2003)
Vol. 1824: J. A. Navarro González, J. B. Sancho de Salas, $\mathcal{C}^\infty$ – Differentiable Spaces (2003)
Vol. 1825: J. H. Bramble, A. Cohen, W. Dahmen, Multiscale Problems and Methods in Numerical Simulations, Martina Franca, Italy 2001. Editor: C. Canuto (2003)
Vol. 1826: K. Dohmen, Improved Bonferroni Inequalities via Abstract Tubes. Inequalities and Identities of Inclusion-Exclusion Type. VIII, 113 p, 2003.
Vol. 1827: K. M. Pilgrim, Combinations of Complex Dynamical Systems. IX, 118 p, 2003.
Vol. 1828: D. J. Green, Grbner Bases and the Computation of Group Cohomology. XII, 138 p, 2003.
Vol. 1829: E. Altman, B. Gaujal, A. Hordijk, Discrete-Event Control of Stochastic Networks: Multimodularity and Regularity. XIV, 313 p, 2003.
Vol. 1830: M. I. Gil', Operator Functions and Localization of Spectra. XIV, 256 p, 2003.
Vol. 1831: A. Connes, J. Cuntz, E. Guentner, N. Higson, J. E. Kaminker, Noncommutative Geometry, Martina Franca, Italy 2002. Editors: S. Doplicher, L. Longo (2004)
Vol. 1832: J. Azéma, M. Émery, M. Ledoux, M. Yor (Eds.), Séminaire de Probabilités XXXVII (2003)
Vol. 1833: D.-Q. Jiang, M. Qian, M.-P. Qian, Mathematical Theory of Nonequilibrium Steady States. On the Frontier of Probability and Dynamical Systems. IX, 280 p, 2004.
Vol. 1834: Yo. Yomdin, G. Comte, Tame Geometry with Application in Smooth Analysis. VIII, 186 p, 2004.
Vol. 1835: O.T. Izhboldin, B. Kahn, N.A. Karpenko, A. Vishik, Geometric Methods in the Algebraic Theory of Quadratic Forms. Summer School, Lens, 2000. Editor: J.-P. Tignol (2004)
Vol. 1836: C. Năstăsescu, F. Van Oystaeyen, Methods of Graded Rings. XIII, 304 p, 2004.
Vol. 1837: S. Tavaré, O. Zeitouni, Lectures on Probability Theory and Statistics. Ecole d'Eté de Probabilités de Saint-Flour XXXI-2001. Editor: J. Picard (2004)
Vol. 1838: A.J. Ganesh, N.W. O'Connell, D.J. Wischik, Big Queues. XII, 254 p, 2004.
Vol. 1839: R. Gohm, Noncommutative Stationary Processes. VIII, 170 p, 2004.
Vol. 1840: B. Tsirelson, W. Werner, Lectures on Probability Theory and Statistics. Ecole d'Eté de Probabilités de Saint-Flour XXXII-2002. Editor: J. Picard (2004)
Vol. 1841: W. Reichel, Uniqueness Theorems for Variational Problems by the Method of Transformation Groups (2004)
Vol. 1842: T. Johnsen, A. L. Knutsen, K_3 Projective Models in Scrolls (2004)
Vol. 1843: B. Jefferies, Spectral Properties of Noncommuting Operators (2004)
Vol. 1844: K.F. Siburg, The Principle of Least Action in Geometry and Dynamics (2004)
Vol. 1845: Min Ho Lee, Mixed Automorphic Forms, Torus Bundles, and Jacobi Forms (2004)
Vol. 1846: H. Ammari, H. Kang, Reconstruction of Small Inhomogeneities from Boundary Measurements (2004)
Vol. 1847: T.R. Bielecki, T. Bjrk, M. Jeanblanc, M. Rutkowski, J.A. Scheinkman, W. Xiong, Paris-Princeton Lectures on Mathematical Finance 2003 (2004)
Vol. 1848: M. Abate, J. E. Fornaess, X. Huang, J. P. Rosay, A. Tumanov, Real Methods in Complex and CR Geometry, Martina Franca, Italy 2002. Editors: D. Zaitsev, G. Zampieri (2004)
Vol. 1849: Martin L. Brown, Heegner Modules and Elliptic Curves (2004)
Vol. 1850: V. D. Milman, G. Schechtman (Eds.), Geometric Aspects of Functional Analysis. Israel Seminar 2002-2003 (2004)
Vol. 1851: O. Catoni, Statistical Learning Theory and Stochastic Optimization (2004)
Vol. 1852: A.S. Kechris, B.D. Miller, Topics in Orbit Equivalence (2004)
Vol. 1853: Ch. Favre, M. Jonsson, The Valuative Tree (2004)
Vol. 1854: O. Saeki, Topology of Singular Fibers of Differential Maps (2004)
Vol. 1855: G. Da Prato, P.C. Kunstmann, I. Lasiecka, A. Lunardi, R. Schnaubelt, L. Weis, Functional Analytic Methods for Evolution Equations. Editors: M. Iannelli, R. Nagel, S. Piazzera (2004)
Vol. 1856: K. Back, T.R. Bielecki, C. Hipp, S. Peng, W. Schachermayer, Stochastic Methods in Finance, Bressanone/Brixen, Italy, 2003. Editors: M. Fritelli, W. Runggaldier (2004)
Vol. 1857: M. Émery, M. Ledoux, M. Yor (Eds.), Séminaire de Probabilités XXXVIII (2005)
Vol. 1858: A.S. Cherny, H.-J. Engelbert, Singular Stochastic Differential Equations (2005)
Vol. 1859: E. Letellier, Fourier Transforms of Invariant Functions on Finite Reductive Lie Algebras (2005)

Vol. 1860: A. Borisyuk, G.B. Ermentrout, A. Friedman, D. Terman, Tutorials in Mathematical Biosciences I. Mathematical Neurosciences (2005)

Vol. 1861: G. Benettin, J. Henrard, S. Kuksin, Hamiltonian Dynamics – Theory and Applications, Cetraro, Italy, 1999. Editor: A. Giorgilli (2005)

Vol. 1862: B. Helffer, F. Nier, Hypoelliptic Estimates and Spectral Theory for Fokker-Planck Operators and Witten Laplacians (2005)

Vol. 1863: H. Führ, Abstract Harmonic Analysis of Continuous Wavelet Transforms (2005)

Vol. 1864: K. Efstathiou, Metamorphoses of Hamiltonian Systems with Symmetries (2005)

Vol. 1865: D. Applebaum, B.V. R. Bhat, J. Kustermans, J. M. Lindsay, Quantum Independent Increment Processes I. From Classical Probability to Quantum Stochastic Calculus. Editors: M. Schürmann, U. Franz (2005)

Vol. 1866: O.E. Barndorff-Nielsen, U. Franz, R. Gohm, B. Kümmerer, S. Thorbjønsen, Quantum Independent Increment Processes II. Structure of Quantum Lévy Processes, Classical Probability, and Physics. Editors: M. Schürmann, U. Franz, (2005)

Vol. 1867: J. Sneyd (Ed.), Tutorials in Mathematical Biosciences II. Mathematical Modeling of Calcium Dynamics and Signal Transduction. (2005)

Vol. 1868: J. Jorgenson, S. Lang, $\text{Pos}_n(\text{R})$ and Eisenstein Series. (2005)

Vol. 1869: A. Dembo, T. Funaki, Lectures on Probability Theory and Statistics. Ecole d'Eté de Probabilités de Saint-Flour XXXIII-2003. Editor: J. Picard (2005)

Vol. 1870: V.I. Gurariy, W. Lusky, Geometry of Mntz Spaces and Related Questions. (2005)

Vol. 1871: P. Constantin, G. Gallavotti, A.V. Kazhikhov, Y. Meyer, S. Ukai, Mathematical Foundation of Turbulent Viscous Flows, Martina Franca, Italy, 2003. Editors: M. Cannone, T. Miyakawa (2006)

Vol. 1872: A. Friedman (Ed.), Tutorials in Mathematical Biosciences III. Cell Cycle, Proliferation, and Cancer (2006)

Vol. 1873: R. Mansuy, M. Yor, Random Times and Enlargements of Filtrations in a Brownian Setting (2006)

Vol. 1874: M. Yor, M. Émery (Eds.), In Memoriam Paul-Andr Meyer - Sminaire de Probabilits XXXIX (2006)

Vol. 1875: J. Pitman, Combinatorial Stochastic Processes. Ecole d'Et de Probabilits de Saint-Flour XXXII-2002. Editor: J. Picard (2006)

Vol. 1876: H. Herrlich, Axiom of Choice (2006)

Vol. 1877: J. Steuding, Value Distributions of *L*-Functions (2007)

Vol. 1878: R. Cerf, The Wulff Crystal in Ising and Percolation Models, Ecole d'Et de Probabilités de Saint-Flour XXXIV-2004. Editor: Jean Picard (2006)

Vol. 1879: G. Slade, The Lace Expansion and its Applications, Ecole d'Et de Probabilits de Saint-Flour XXXIV-2004. Editor: Jean Picard (2006)

Vol. 1880: S. Attal, A. Joye, C.-A. Pillet, Open Quantum Systems I, The Hamiltonian Approach (2006)

Vol. 1881: S. Attal, A. Joye, C.-A. Pillet, Open Quantum Systems II, The Markovian Approach (2006)

Vol. 1882: S. Attal, A. Joye, C.-A. Pillet, Open Quantum Systems III, Recent Developments (2006)

Vol. 1883: W. Van Assche, F. Marcellàn (Eds.), Orthogonal Polynomials and Special Functions, Computation and Application (2006)

Vol. 1884: N. Hayashi, E.I. Kaikina, P.I. Naumkin, I.A. Shishmarev, Asymptotics for Dissipative Nonlinear Equations (2006)

Vol. 1885: A. Telcs, The Art of Random Walks (2006)

Vol. 1886: S. Takamura, Splitting Deformations of Degenerations of Complex Curves (2006)

Vol. 1887: K. Habermann, L. Habermann, Introduction to Symplectic Dirac Operators (2006)

Vol. 1888: J. van der Hoeven, Transseries and Real Differential Algebra (2006)

Vol. 1889: G. Osipenko, Dynamical Systems, Graphs, and Algorithms (2006)

Vol. 1890: M. Bunge, J. Funk, Singular Coverings of Toposes (2006)

Vol. 1891: J.B. Friedlander, D.R. Heath-Brown, H. Iwaniec, J. Kaczorowski, Analytic Number Theory, Cetraro, Italy, 2002. Editors: A. Perelli, C. Viola (2006)

Vol. 1892: A. Baddeley, I. Bárány, R. Schneider, W. Weil, Stochastic Geometry, Martina Franca, Italy, 2004. Editor: W. Weil (2007)

Vol. 1893: H. Hanßmann, Local and Semi-Local Bifurcations in Hamiltonian Dynamical Systems, Results and Examples (2007)

Vol. 1894: C.W. Groetsch, Stable Approximate Evaluation of Unbounded Operators (2007)

Vol. 1895: L. Molnár, Selected Preserver Problems on Algebraic Structures of Linear Operators and on Function Spaces (2007)

Vol. 1896: P. Massart, Concentration Inequalities and Model Selection, Ecole d'Été de Probabilités de Saint-Flour XXXIII-2003. Editor: J. Picard (2007)

Vol. 1897: R. Doney, Fluctuation Theory for Lévy Processes, Ecole d'Été de Probabilités de Saint-Flour XXXV-2005. Editor: J. Picard (2007)

Vol. 1898: H.R. Beyer, Beyond Partial Differential Equations, On linear and Quasi-Linear Abstract Hyperbolic Evolution Equations (2007)

Vol. 1899: Séminaire de Probabilités XL. Editors: C. Donati-Martin, M. Émery, A. Rouault, C. Stricker (2007)

Vol. 1900: E. Bolthausen, A. Bovier (Eds.), Spin Glasses (2007)

Vol. 1901: O. Wittenberg, Intersections de deux quadriques et pinceaux de courbes de genre 1, Intersections of Two Quadrics and Pencils of Curves of Genus 1 (2007)

Vol. 1902: A. Isaev, Lectures on the Automorphism Groups of Kobayashi-Hyperbolic Manifolds (2007)

Vol. 1903: G. Kresin, V. Maz'ya, Sharp Real-Part Theorems (2007)

Vol. 1904: P. Giesl, Construction of Global Lyapunov Functions Using Radial Basis Functions (2007)

Vol. 1905: C. Prévôt, M. Röckner, A Concise Course on Stochastic Partial Differential Equations (2007)

Vol. 1906: T. Schuster, The Method of Approximate Inverse: Theory and Applications (2007)

Vol. 1907: M. Rasmussen, Attractivity and Bifurcation for Nonautonomous Dynamical Systems (2007)

Vol. 1908: T.J. Lyons, M. Caruana, T. Lévy, Differential Equations Driven by Rough Paths, Ecole d'Été de Probabilités de Saint-Flour XXXIV-2004 (2007)

Vol. 1909: H. Akiyoshi, M. Sakuma, M. Wada, Y. Yamashita, Punctured Torus Groups and 2-Bridge Knot Groups (I) (2007)

Vol. 1910: V.D. Milman, G. Schechtman (Eds.), Geometric Aspects of Functional Analysis. Israel Seminar 2004-2005 (2007)

Vol. 1911: A. Bressan, D. Serre, M. Williams, K. Zumbrun, Hyperbolic Systems of Balance Laws. Cetraro, Italy 2003. Editor: P. Marcati (2007)

Vol. 1912: V. Berinde, Iterative Approximation of Fixed Points (2007)

Vol. 1913: J.E. Marsden, G. Misiołek, J.-P. Ortega, M. Perlmutter, T.S. Ratiu, Hamiltonian Reduction by Stages (2007)

Vol. 1914: G. Kutyniok, Affine Density in Wavelet Analysis (2007)

Vol. 1915: T. Bıyıkoğlu, J. Leydold, P.F. Stadler, Laplacian Eigenvectors of Graphs. Perron-Frobenius and Faber-Krahn Type Theorems (2007)

Vol. 1916: C. Villani, F. Rezakhanlou, Entropy Methods for the Boltzmann Equation. Editors: F. Golse, S. Olla (2008)

Vol. 1917: I. Veselić, Existence and Regularity Properties of the Integrated Density of States of Random Schrdinger (2008)

Vol. 1918: B. Roberts, R. Schmidt, Local Newforms for GSp(4) (2007)

Vol. 1919: R.A. Carmona, I. Ekeland, A. Kohatsu-Higa, J.-M. Lasry, P.-L. Lions, H. Pham, E. Taflin, Paris-Princeton Lectures on Mathematical Finance 2004. Editors: R.A. Carmona, E. inlar, I. Ekeland, E. Jouini, J.A. Scheinkman, N. Touzi (2007)

Vol. 1920: S.N. Evans, Probability and Real Trees. Ecole d'Été de Probabilités de Saint-Flour XXXV-2005 (2008)

Vol. 1921: J.P. Tian, Evolution Algebras and their Applications (2008)

Vol. 1922: A. Friedman (Ed.), Tutorials in Mathematical BioSciences IV. Evolution and Ecology (2008)

Vol. 1923: J.P.N. Bishwal, Parameter Estimation in Stochastic Differential Equations (2008)

Vol. 1924: M. Wilson, Littlewood-Paley Theory and Exponential-Square Integrability (2008)

Vol. 1925: M. du Sautoy, L. Woodward, Zeta Functions of Groups and Rings (2008)

Vol. 1926: L. Barreira, V. Claudia, Stability of Nonautonomous Differential Equations (2008)

Vol. 1927: L. Ambrosio, L. Caffarelli, M.G. Crandall, L.C. Evans, N. Fusco, Calculus of Variations and Non-Linear Partial Differential Equations. Cetraro, Italy 2005. Editors: B. Dacorogna, P. Marcellini (2008)

Vol. 1928: J. Jonsson, Simplicial Complexes of Graphs (2008)

Vol. 1929: Y. Mishura, Stochastic Calculus for Fractional Brownian Motion and Related Processes (2008)

Vol. 1930: J.M. Urbano, The Method of Intrinsic Scaling. A Systematic Approach to Regularity for Degenerate and Singular PDEs (2008)

Vol. 1931: M. Cowling, E. Frenkel, M. Kashiwara, A. Valette, D.A. Vogan, Jr., N.R. Wallach, Representation Theory and Complex Analysis. Venice, Italy 2004. Editors: E.C. Tarabusi, A. D'Agnolo, M. Picardello (2008)

Vol. 1932: A.A. Agrachev, A.S. Morse, E.D. Sontag, H.J. Sussmann, V.I. Utkin, Nonlinear and Optimal Control Theory. Cetraro, Italy 2004. Editors: P. Nistri, G. Stefani (2008)

Vol. 1933: M. Petkovic, Point Estimation of Root Finding Methods (2008)

Vol. 1934: C. Donati-Martin, M. Émery, A. Rouault, C. Stricker (Eds.), Séminaire de Probabilités XLI (2008)

Vol. 1935: A. Unterberger, Alternative Pseudodifferential Analysis (2008)

Vol. 1936: P. Magal, S. Ruan (Eds.), Structured Population Models in Biology and Epidemiology (2008)

Vol. 1937: G. Capriz, P. Giovine, P.M. Mariano (Eds.), Mathematical Models of Granular Matter (2008)

Vol. 1938: D. Auroux, F. Catanese, M. Manetti, P. Seidel, B. Siebert, I. Smith, G. Tian, Symplectic 4-Manifolds and Algebraic Surfaces. Cetraro, Italy 2003. Editors: F. Catanese, G. Tian (2008)

Vol. 1939: D. Boffi, F. Brezzi, L. Demkowicz, R.G. Durán, R.S. Falk, M. Fortin, Mixed Finite Elements, Compatibility Conditions, and Applications. Cetraro, Italy 2006. Editors: D. Boffi, L. Gastaldi (2008)

Vol. 1940: J. Banasiak, V. Capasso, M.A.J. Chaplain, M. Lachowicz, J. Miękisz, Multiscale Problems in the Life Sciences. From Microscopic to Macroscopic. Będlewo, Poland 2006. Editors: V. Capasso, M. Lachowicz (2008)

Vol. 1941: S.M.J. Haran, Arithmetical Investigations. Representation Theory, Orthogonal Polynomials, and Quantum Interpolations (2008)

Vol. 1942: S. Albeverio, F. Flandoli, Y.G. Sinai, SPDE in Hydrodynamic. Recent Progress and Prospects. Cetraro, Italy 2005. Editors: G. Da Prato, M. Rckner (2008)

Vol. 1943: L.L. Bonilla (Ed.), Inverse Problems and Imaging. Martina Franca, Italy 2002 (2008)

Vol. 1944: A. Di Bartolo, G. Falcone, P. Plaumann, K. Strambach, Algebraic Groups and Lie Groups with Few Factors (2008)

Vol. 1945: F. Brauer, P. van den Driessche, J. Wu (Eds.), Mathematical Epidemiology (2008)

Vol. 1946: G. Allaire, A. Arnold, P. Degond, T.Y. Hou, Quantum Transport. Modelling, Analysis and Asymptotics. Cetraro, Italy 2006. Editors: N.B. Abdallah, G. Frosali (2008)

Vol. 1947: D. Abramovich, M. Mariño, M. Thaddeus, R. Vakil, Enumerative Invariants in Algebraic Geometry and String Theory. Cetraro, Italy 2005. Editors: K. Behrend, M. Manetti (2008)

Vol. 1948: F. Cao, J-L. Lisani, J-M. Morel, P. Mus, F. Sur, A Theory of Shape Identification (2008)

Vol. 1949: H.G. Feichtinger, B. Helffer, M.P. Lamoureux, N. Lerner, J. Toft, Pseudo-Differential Operators. Quantization and Signals. Cetraro, Italy 2006. Editors: L. Rodino, M.W. Wong (2008)

Vol. 1950: M. Bramson, Stability of Queueing Networks, Ecole d'Eté de Probabilits de Saint-Flour XXXVI-2006 (2008)

Vol. 1951: A. Moltó, J. Orihuela, S. Troyanski, M. Valdivia, A Non Linear Transfer Technique for Renorming (2009)

Vol. 1952: R. Mikhailov, I.B.S. Passi, Lower Central and Dimension Series of Groups (2009)

Vol. 1953: K. Arwini, C.T.J. Dodson, Information Geometry (2008)

Vol. 1954: P. Biane, L. Bouten, F. Cipriani, N. Konno, N. Privault, Q. Xu, Quantum Potential Theory. Editors: U. Franz, M. Schuermann (2008)

Vol. 1955: M. Bernot, V. Caselles, J.-M. Morel, Optimal Transportation Networks (2008)

Vol. 1956: C.H. Chu, Matrix Convolution Operators on Groups (2008)

Vol. 1957: A. Guionnet, On Random Matrices: Macroscopic Asymptotics, Ecole d'Eté de Probabilits de Saint-Flour XXXVI-2006 (2009)

Vol. 1958: M.C. Olsson, Compactifying Moduli Spaces for Abelian Varieties (2008)

Vol. 1959: Y. Nakkajima, A. Shiho, Weight Filtrations on Log Crystalline Cohomologies of Families of Open Smooth Varieties (2008)
Vol. 1960: J. Lipman, M. Hashimoto, Foundations of Grothendieck Duality for Diagrams of Schemes (2009)
Vol. 1961: G. Buttazzo, A. Pratelli, S. Solimini, E. Stepanov, Optimal Urban Networks via Mass Transportation (2009)
Vol. 1962: R. Dalang, D. Khoshnevisan, C. Mueller, D. Nualart, Y. Xiao, A Minicourse on Stochastic Partial Differential Equations (2009)
Vol. 1963: W. Siegert, Local Lyapunov Exponents (2009)
Vol. 1964: W. Roth, Operator-valued Measures and Integrals for Cone-valued Functions and Integrals for Cone-valued Functions (2009)
Vol. 1965: C. Chidume, Geometric Properties of Banach Spaces and Nonlinear Iterations (2009)
Vol. 1966: D. Deng, Y. Han, Harmonic Analysis on Spaces of Homogeneous Type (2009)
Vol. 1967: B. Fresse, Modules over Operads and Functors (2009)
Vol. 1968: R. Weissauer, Endoscopy for GSP(4) and the Cohomology of Siegel Modular Threefolds (2009)
Vol. 1969: B. Roynette, M. Yor, Penalising Brownian Paths (2009)
Vol. 1970: M. Biskup, A. Bovier, F. den Hollander, D. Ioffe, F. Martinelli, K. Netočný, F. Toninelli, Methods of Contemporary Mathematical Statistical Physics. Editor: R. Kotecký (2009)
Vol. 1971: L. Saint-Raymond, Hydrodynamic Limits of the Boltzmann Equation (2009)
Vol. 1972: T. Mochizuki, Donaldson Type Invariants for Algebraic Surfaces (2009)
Vol. 1973: M.A. Berger, L.H. Kauffmann, B. Khesin, H.K. Moffatt, R.L. Ricca, De W. Sumners, Lectures on Topological Fluid Mechanics. Cetraro, Italy 2001. Editor: R.L. Ricca (2009)
Vol. 1974: F. den Hollander, Random Polymers: École d'Été de Probabilités de Saint-Flour XXXVII – 2007 (2009)
Vol. 1975: J.C. Rohde, Cyclic Coverings, Calabi-Yau Manifolds and Complex Multiplication (2009)
Vol. 1976: N. Ginoux, The Dirac Spectrum (2009)
Vol. 1977: M.J. Gursky, E. Lanconelli, A. Malchiodi, G. Tarantello, X.-J. Wang, P.C. Yang, Geometric Analysis and PDEs. Cetraro, Italy 2001. Editors: A. Ambrosetti, S.-Y.A. Chang, A. Malchiodi (2009)
Vol. 1978: M. Qian, J.-S. Xie, S. Zhu, Smooth Ergodic Theory for Endomorphisms (2009)
Vol. 1979: C. Donati-Martin, M. Émery, A. Rouault, C. Stricker (Eds.), Śeminaire de Probablitiés XLII (2009)
Vol. 1980: P. Graczyk, A. Stos (Eds.), Potential Analysis of Stable Processes and its Extensions (2009)
Vol. 1981: M. Chlouveraki, Blocks and Families for Cyclotomic Hecke Algebras (2009)
Vol. 1982: N. Privault, Stochastic Analysis in Discrete and Continuous Settings. With Normal Martingales (2009)
Vol. 1983: H. Ammari (Ed.), Mathematical Modeling in Biomedical Imaging I. Electrical and Ultrasound Tomographies, Anomaly Detection, and Brain Imaging (2009)
Vol. 1984: V. Caselles, P. Monasse, Geometric Description of Images as Topographic Maps (2010)
Vol. 1985: T. Linß, Layer-Adapted Meshes for Reaction-Convection-Diffusion Problems (2010)
Vol. 1986: J.-P. Antoine, C. Trapani, Partial Inner Product Spaces. Theory and Applications (2009)
Vol. 1987: J.-P. Brasselet, J. Seade, T. Suwa, Vector Fields on Singular Varieties (2010)
Vol. 1988: M. Broué, Introduction to Complex Reflection Groups and Their Braid Groups (2010)
Vol. 1989: I.M. Bomze, V. Demyanov, Nonlinear Optimization. Cetraro, Italy 2007. Editors: G. di Pillo, F. Schoen (2010)
Vol. 1990: S. Bouc, Biset Functors for Finite Groups (2010)
Vol. 1991: F. Gazzola, H.-C. Grunau, G. Sweers, Polyharmonic Boundary Value Problems (2010)
Vol. 1992: A. Parmeggiani, Spectral Theory of Non-Commutative Harmonic Oscillators: An Introduction (2010)
Vol. 1993: P. Dodos, Banach Spaces and Descriptive Set Theory: Selected Topics (2010)
Vol. 1994: A. Baricz, Generalized Bessel Functions of the First Kind (2010)
Vol. 1995: A.Y. Khapalov, Controllability of Partial Differential Equations Governed by Multiplicative Controls (2010)
Vol. 1996: T. Lorenz, Mutational Analysis. A Joint Framework for Cauchy Problems *In* and *Beyond* Vector Spaces (2010)
Vol. 1997: M. Banagl, Intersection Spaces, Spatial Homology Truncation, and String Theory (2010)
Vol. 1998: M. Abate, E. Bedford, M. Brunella, T.-C. Dinh, D. Schleicher, N. Sibony, Holomorphic Dynamical Systems. Cetraro, Italy 2008. Editors: G. Gentili, J. Guenot, G. Patrizio (2010)
Vol. 1999: H. Schoutens, The Use of Ultraproducts in Commutative Algebra (2010)
Vol. 2000: H. Yserentant, Regularity and Approximability of Electronic Wave Functions (2010)

Recent Reprints and New Editions

Vol. 1702: J. Ma, J. Yong, Forward-Backward Stochastic Differential Equations and their Applications. 1999 – Corr. 3rd printing (2007)
Vol. 830: J.A. Green, Polynomial Representations of GL_n, with an Appendix on Schensted Correspondence and Littelmann Paths by K. Erdmann, J.A. Green and M. Schoker 1980 – 2nd corr. and augmented edition (2007)
Vol. 1693: S. Simons, From Hahn-Banach to Monotonicity (Minimax and Monotonicity 1998) – 2nd exp. edition (2008)
Vol. 470: R.E. Bowen, Equilibrium States and the Ergodic Theory of Anosov Diffeomorphisms. With a preface by D. Ruelle. Edited by J.-R. Chazottes. 1975 – 2nd rev. edition (2008)
Vol. 523: S.A. Albeverio, R.J. Høegh-Krohn, S. Mazzucchi, Mathematical Theory of Feynman Path Integral. 1976 – 2nd corr. and enlarged edition (2008)
Vol. 1764: A. Cannas da Silva, Lectures on Symplectic Geometry 2001 – Corr. 2nd printing (2008)

LECTURE NOTES IN MATHEMATICS

Springer

Edited by J.-M. Morel, F. Takens, B. Teissier, P.K. Maini

Editorial Policy (for Multi-Author Publications: Summer Schools/Intensive Courses)

1. Lecture Notes aim to report new developments in all areas of mathematics and their applications - quickly, informally and at a high level. Mathematical texts analysing new developments in modelling and numerical simulation are welcome. Manuscripts should be reasonably self-contained and rounded off. Thus they may, and often will, present not only results of the author but also related work by other people. They should provide sufficient motivation, examples and applications. There should also be an introduction making the text comprehensible to a wider audience. This clearly distinguishes Lecture Notes from journal articles or technical reports which normally are very concise. Articles intended for a journal but too long to be accepted by most journals, usually do not have this "lecture notes" character.
2. In general SUMMER SCHOOLS and other similar INTENSIVE COURSES are held to present mathematical topics that are close to the frontiers of recent research to an audience at the beginning or intermediate graduate level, who may want to continue with this area of work, for a thesis or later. This makes demands on the didactic aspects of the presentation. Because the subjects of such schools are advanced, there often exists no textbook, and so ideally, the publication resulting from such a school could be a first approximation to such a textbook. Usually several authors are involved in the writing, so it is not always simple to obtain a unified approach to the presentation.

   For prospective publication in LNM, the resulting manuscript should not be just a collection of course notes, each of which has been developed by an individual author with little or no co-ordination with the others, and with little or no common concept. The subject matter should dictate the structure of the book, and the authorship of each part or chapter should take secondary importance. Of course the choice of authors is crucial to the quality of the material at the school and in the book, and the intention here is not to belittle their impact, but simply to say that the book should be planned to be written by these authors jointly, and not just assembled as a result of what these authors happen to submit.

   This represents considerable preparatory work (as it is imperative to ensure that the authors know these criteria before they invest work on a manuscript), and also considerable editing work afterwards, to get the book into final shape. Still it is the form that holds the most promise of a successful book that will be used by its intended audience, rather than yet another volume of proceedings for the library shelf.
3. Manuscripts should be submitted either online at www.editorialmanager.com/lnm/ to Springer's mathematics editorial, or to one of the series editors. Volume editors are expected to arrange for the refereeing, to the usual scientific standards, of the individual contributions. If the resulting reports can be forwarded to us (series editors or Springer) this is very helpful. If no reports are forwarded or if other questions remain unclear in respect of homogeneity etc, the series editors may wish to consult external referees for an overall evaluation of the volume. A final decision to publish can be made only on the basis of the complete manuscript; however a preliminary decision can be based on a pre-final or incomplete manuscript. The strict minimum amount of material that will be considered should include a detailed outline describing the planned contents of each chapter.

   Volume editors and authors should be aware that incomplete or insufficiently close to final manuscripts almost always result in longer evaluation times. They should also be aware that parallel submission of their manuscript to another publisher while under consideration for LNM will in general lead to immediate rejection.

4. Manuscripts should in general be submitted in English. Final manuscripts should contain at least 100 pages of mathematical text and should always include
   - a general table of contents;
   - an informative introduction, with adequate motivation and perhaps some historical remarks: it should be accessible to a reader not intimately familiar with the topic treated;
   - a global subject index: as a rule this is genuinely helpful for the reader.

   Lecture Notes volumes are, as a rule, printed digitally from the authors' files. We strongly recommend that all contributions in a volume be written in the same LaTeX version, preferably LaTeX2e. To ensure best results, authors are asked to use the LaTeX2e style files available from Springer's web-server at

   ftp://ftp.springer.de/pub/tex/latex/svmonot1/ (for monographs) and
   ftp://ftp.springer.de/pub/tex/latex/svmultt1/ (for summer schools/tutorials).

   Additional technical instructions are available on request from: lnm@springer.com.
5. Careful preparation of the manuscripts will help keep production time short besides ensuring satisfactory appearance of the finished book in print and online. After acceptance of the manuscript authors will be asked to prepare the final LaTeX source files and also the corresponding dvi-, pdf- or zipped ps-file. The LaTeX source files are essential for producing the full-text online version of the book. For the existing online volumes of LNM see: http://www.springerlink.com/openurl.asp?genre=journal&issn=0075-8434.

   The actual production of a Lecture Notes volume takes approximately 12 weeks.
6. Volume editors receive a total of 50 free copies of their volume to be shared with the authors, but no royalties. They and the authors are entitled to a discount of 33.3% on the price of Springer books purchased for their personal use, if ordering directly from Springer.
7. Commitment to publish is made by letter of intent rather than by signing a formal contract. Springer-Verlag secures the copyright for each volume. Authors are free to reuse material contained in their LNM volumes in later publications: a brief written (or e-mail) request for formal permission is sufficient.

Addresses:

Professor J.-M. Morel, CMLA,
École Normale Supérieure de Cachan,
61 Avenue du Président Wilson,
94235 Cachan Cedex, France
E-mail: Jean-Michel.Morel@cmla.ens-cachan.fr

Professor F. Takens, Mathematisch Instituut,
Rijksuniversiteit Groningen, Postbus 800,
9700 AV Groningen, The Netherlands
E-mail: F.Takens@rug.nl

Professor B. Teissier,
Institut Mathématique de Jussieu,
UMR 7586 du CNRS,
Équipe "Géométrie et Dynamique",
175 rue du Chevaleret,
75013 Paris, France
E-mail: teissier@math.jussieu.fr

For the "Mathematical Biosciences Subseries" of LNM:

Professor P.K. Maini, Center for Mathematical Biology,
Mathematical Institute, 24-29 St Giles,
Oxford OX1 3LP, UK
E-mail: maini@maths.ox.ac.uk

Springer, Mathematics Editorial I, Tiergartenstr. 17,
69121 Heidelberg, Germany,
Tel.: +49 (6221) 487-8259
Fax: +49 (6221) 4876-8259
E-mail: lnm@springer.com